Climate Change and Microbial Ecology

Current Research and Future Trends (Second Edition)

Editor:

Jürgen Marxsen
Justus Liebig University
iFZ Research Centre for Biosystems, Land Use and Nutrition
Department of Animal Ecology
Heinrich-Buff-Ring 26
35392 Giessen
Germany

Cover Illustration

Photomicrographs of cyanobacteria (the illustration is derived from Chapter 1, Fig. 1.2).

Caister Academic Press www.caister.com

Caister Academic Press, UK
www.caister.com

ISBN (paperback): 978-1-913652-57-9
ISBN (ebook): 978-1-913652-58-6

DOI: https://doi.org/10.21775/9781913652579

Contents

1 Impacts of Climate Change on Cyanobacteria in Aquatic Environments 1
Hans W. Paerl

2 Climate Change Effects on Planktonic Bacterial Communities in the Ocean: From Structure and Function to Long-term and Large-scale Observations 37
Ingrid Brettar, Manfred G. Höfle, Carla Pruzzo and Luigi Vezzulli

3 Climate Change, Microbial Communities and Agriculture in Semiarid and Arid Ecosystems 71
Felipe Bastida, Alfonso Vera, Marta Díaz, Carlos García, Antonio Ruíz-Navarro and José Luis Moreno

4 Responses of Aquatic Protozoans to Climate Change 107
Hartmut Arndt and Mar Monsonís Nomdedeu

5 Terrestrial Fungi and Global Climate Change 133
Irina Sidorova and Elena Voronina

6 Impact of Climate Change on Aquatic Hyphomycetes 193
Verónica Ferreira

7 Aquatic Viruses and Climate Change 215
Rui Zhang, Markus G. Weinbauer and Peter Peduzzi

8 Microbes in Aquatic Biofilms under the Effect of Changing Climate 239
Anna M. Romaní, Stéphanie Boulêtreau, Verónica Díaz Villanueva, Frédéric Garabetian, Jürgen Marxsen, Helge Norf, Elisabeth Pohlon and Markus Weitere

9 Climate Change, Microbes, and Soil Carbon Cycling 267
Timothy H. Keitt, Colin R. Addis, Daniel L. Mitchell, Andria Salas and Christine V. Hawkes

10 Environmental Change and Microbial Contributions to Carbon Cycle Feedbacks 297
Lei Qin, Hojeong Kang, Chris Freeman, Juanita Mora-Gómez and Ming Jiang ....

11 Climate Change and Nitrogen Turnover in Soils and Aquatic Environments 327
Gero Benckiser

12 Changes in Precipitation Patterns: Responses and Strategies from Streambed Sediment and Soil Microbes 391
Giulia Gionchetta, Aline Frossard, Luis Bañeras and Anna Maria Romaní

13 Groundwater Microbial Communities in Times of Climate Change 421
Alice Retter, Clemens Karwautz and Christian Griebler

14 Ecosystem Metabolism in River Networks and Climate Change 451
Vicenç Acuña, Anna Freixa, Rafael Marcé and Xisca Timoner

15 Microbial Communities and Processes under Climate and Land-use Change in the Tropics 483
Stephen A. Wood, Krista McGuire and Jonathan E. Hickman

16 Geoengineering the Climate via Microorganisms: a Peatland Case Study 517
Christian Dunn, Nathalie Fenner, Anil Shirsat and Chris Freeman

Preface to the 1st edition

When I was asked by the publisher some time ago, whether I was willing to edit a book on *Climate Change and Microbial Ecology*, I agreed because I felt that it was timely to have a comprehensive overview on the advancements in this field. I thought that it was appropriate to summarize the current knowledge on how the microorganisms on earth are affected by global climate change and vice versa how they themselves affect the development of global climate change, - by viewing from the perspective of the different groups of microorganisms like bacteria including cyanobacteria, fungi, and protozoa, also viruses, as well as by looking in detail on the different ecosystems on earth like oceans, inland waters with rivers, lakes, and groundwater, and soil. Thus I am happy that a broad range of renowned scientists provided their expertise demonstrating not only the actual status but also the imminent need to increase our knowledge on the role of microbial communities with respect to global climate change.

The reader will observe that the style of the chapters is not always consistent between the different authors. Some of the chapters are short and concentrated, whereas other chapters go into great detail. However, I decided the differences to be maintained in order to allow the authors to present these review papers using their personal preferences. Unfortunately, there are a few gaps in this book, which prohibits the presentation of a complete suite of the major aspects within the general topic of "Climate Change and Microbial Ecology", mainly because the manuscripts from a few authors were not received within an acceptable time frame.

Nevertheless, I am convinced that the list of contributions to this book covers most of the important areas from the book's title *Climate Change and Microbial Ecology: Current Research and Future Trends* and that the volume will be helpful not only for every microbial ecologist from the PhD student to the experienced scientist, but also for every one interested in the field of global climate change.

Finally, I would like to express my thanks to all the authors for their kind cooperation. They did a great job in presenting a timely overview on topics of climate change and microbial ecology in their special fields of expertise. I am also indebted to Dr Thomas Horvath (University of Koblenz-Landau, Landau, Germany) who improved some manuscripts prepared by authors who were not native English speakers, and also to the publisher who was especially patient with the many delays occurring through the preparation of this volume.

Jürgen Marxsen, Giessen, Germany

Preface to the 2nd edition

Climate change is continuing unabated. I think there can be no doubt. Thus I am very happy to be able to present now to the scientific community and everybody who is interested an enlarged and updated book on *Climate Change and Microbial Ecology: Current Research and Future Trends*. The first edition of the book appeared in 2016, and I was surprized that the publisher asked me very soon to prepare a new edition. This meant to me that the interest of readers in the book and its topic could not have been minor, and I enjoyed agreeing. Now, about five years after the book had appeared for the first time, we are able to present this new edition containing now even four more chapters. Many of the authors revised and updated or even expanded their chapters, only few chapters had to be printed unchanged. With the assistance of renowned colleagues who entered the project and prepared new chapters it was possible to close at least some of the gaps from the first edition although not all. So this new edition covers more of the important fields from the book's title and hopefully will be even more helpful for the audience addressed.

Let me again express my thanks to all the authors for their kind cooperation and for their great engagement in preparing new or revised contributions with timely overviews on topics of climate change and microbial ecology in their field of expertise. Also warmest thanks to the publisher who encouraged me to start working on the preparation of the new edition and as I am firmly convinced, will produce this new edition with the same high standard as the first one.

Jürgen Marxsen, Giessen, Germany

Introduction

The structure of microbial communities and their functions play a crucial role for the flow of matter in the Earth's biogeochemical cycles. Effects of microbial communities on the carbon and nitrogen cycles are particularly important for producing climate gases such as CO_2, CH_4, or N_2O, thus enhancing or hampering global climate change. However, the biogeochemical cycles are reversely impacted by global climate change themselves, for example by increasing temperature, increasing CO_2 concentration, or changing soil humidity. Microbes may respond differently to human-caused climate change. They may act as potent amplifiers, but may also dampening human-caused effects on climate. Nevertheless, understanding of microbial ecology in the different ecosystems on Earth, such as soil, oceans, inland waters, is essential for our ability to assess the importance of biogeochemical cycles–climate feedbacks (Bardgett et al., 2008). Unfortunately microbial communities are extremely complex in structure and function and can be affected by climate and other global changes in many ways, which impedes our ability to draw reliable conclusions.

The *IPCC Special Report on the Impacts of Global Warming of 1.5°C above Pre-Industrial Levels* (2018) emphasized that climate change continues unabated. "Emissions of greenhouse gases due to human activities, the root cause of global warming, continue to increase, year after year." (IPCC 2018) This underlines the importance of the main objective of this book, to present in its 2nd edition again a timely overview of advancements in the field of climate change and microbial ecology. Individual chapters cover the various types of ecosystems, the role of different groups of microorganisms, and the complex effects on and interrelationships with the various cycles of matter. A few special chapters cover applied aspects such as land-use and geoengineering.

The volume starts in its initial chapters with contributions on the different microbial groups occurring in the Earth's ecosystems.

Hans W. Paerl (Chapter 1) shows the impact of climate change on the Earth's oldest oxygenic phototrophs, the cyanobacteria. Their long evolutionary history was of special advantage for adapting to recent anthropogenic modifications of their environment, such as eutrophication, water diversion, withdrawal, and salinization. Harmful cyanobacterial blooms were promoted worldwide by eutrophication, but manifestations of climate change, particularly temperature rise, play synergistic roles in promoting blooms of cyanobacteria, as Paerl can demonstrate in his revised and extended chapter.

Ingrid Brettar, Manfred G. Höfle, Carla Pruzzo, and Luigi Vezzulli (Chapter 2) present a paper on climate change effects on bacteria in aquatic environments - with the focus on marine planktonic communities, because very little is known about climate change effects on inland water as well as on sediment communities in both, inland and marine systems. Beside effects of climate change on bacterial growth and community composition, the authors consider effects on bacteria-mediated biogeochemical cycling and potential hazards by increased abundance of pathogenic bacteria.

The new contribution of Felipe Bastida, Alfonso Vera, Marta Díaz, Carlos García, Antonio Ruíz-Navarro, and José Luis Moreno (Chapter 3) is focusing on soil microbial communities in semiarid and arid environments. Climate change strongly impacts the diversity of soil microbial communities and their ecosystem functioning, but agricultural activities may enhance the effects of climate change. The maintenance of productivity often requires the application of nutrients (organic amendments and fertilizers) and alternative water sources for irrigation, which can be accompanied by contamination. The current impacts and perspectives of these practices, together with climate change, on microbial communities in semiarid and arid soils are described in this chapter.

Heterotrophic protozoans play a key role in marine and freshwater microbial food chains. The major effects of increased temperature on free-living heterotrophic protists are summarized by Hartmut Arndt and Mar Monsonís Nomdedeu (Chapter 4, updated and extended), who also illustrate the complexity of temperature impact on the background of complex microbial food web interactions.

Irina Sidorova and Elena Voronina completely revised and substantially enlarged the chapter on terrestrial fungi (Chapter 5). They give a review on different groups of saprotrophic, mycorrhizal, and pathogenic fungi addressing direct and indirect effects of climate change including responses to warming, extreme weather events, elevated carbon dioxide, and nitrogen concentrations. Although a huge variety across taxa and functional guilds occurs, the authors discuss possible mechanisms underlying climate change effects observed or predicted.

Aquatic hyphomycetes are important components of heterotrophic food webs in woodland streams as pioneer colonizers of submerged leaf litter from the surrounding environment. They mineralize litter carbon and nutrients and convert dead organic matter into biomass, establishing the link between basal resources and higher trophic levels. Verónica Ferreira (Chapter 6) addresses direct and indirect effects of climate change on their community composition, growth, reproduction, metabolism, and decomposing activity and discusses the consequences for the functioning of woodland streams.

The current knowledge on potential climate-related consequences for viral assemblages, virus-host interactions, and virus functions, and in turn viral processes contributing to climate change is synthesized by Rui Zhang, Markus G. Weinbauer, and Peter Peduzzi in a revised and extended contribution (Chapter 7). They show that viruses have the potential to significantly influence carbon and nutrient cycles as well as food webs in aquatic ecosystems and that viruses are clearly needed to be incorporated into future ocean and inland water climate models.

Biofilms are complex and dynamic assemblages of microorganisms, important particularly in many aquatic ecosystems. How they are affected through warming and desiccation is reviewed by Anna M. Romaní, Stéphanie Boulêtreau, Verónica Díaz Villanueva, Frédéric Garabetian, Jürgen Marxsen, Helge Norf, Elisabeth Pohlon, and Markus Weitere (Chapter 8). Commonly observed effects of warming on biofilms include changes in the autotrophic and heterotrophic community composition and extracellular polymeric substances. Photosynthesis, respiration, denitrification, and extracellular enzyme activity show different sensitivity to temperature. However desiccation may produce more permanent changes, more on the biofilm microbial structure than on activities.

The second part of the book is focussed on climate change and its effect on microbes in biogeochemical cycles.

The importance of microbial responses for the balance of soil carbon loss and storage under future temperature and precipitation conditions is examined by Timothy H. Keitt, Colin Addis, Daniel Mitchell, Andria Salas, and Christine V. Hawkes (Chapter 9). The authors propose four classes of response mechanisms allowing for a more general understanding of microbial climate responses. They find that moisture has large effects on predictions for soil carbon and microbial pools.

Soil microbes play a central role in the global carbon cycle; they metabolize organic matter, thereby releasing more than 60 Pg C annually. The composition and activity of microbial communities are strongly influenced by environmental conditions. Thus, global climate change may provoke climate-microbial feedbacks to accelerate or alleviate greenhouse gas emission. Lei Qin, Hojeong Kang, Chris Freeman, Juanita Mora-Gómez, and Ming Jiang (Chapter 10; updated and extended) review the effects of elevated CO_2 concentrations, temperature increase, and precipitation changes on soil microbial community composition and metabolism. They further suggest topics important to be addressed for better understanding the implications of microbial feedback to climate change.

The contribution of Benckiser (Chapter 11) is an updated and extended review of the current knowledge about N cycling in terrestrial and aquatic environments, beginning with the introduction by N_2-fixation followed by the uptake of NH_4 into cells and its transformation, the oxidation to NO_3^-, and the energy conserving reduction of NO_3^- to N_2O and N_2. The author discuss how a less N_2O polluted atmosphere could be achieved under the aspect of climate change perspectives via less N overloaded agri- and aquaculture systems and less nitrate polluted groundwaters, particularly in the industrialized world.

Giulia Gionchetta, Aline Frossard, Luis Bañeras, and Anna Maria Romaní (Chapter 12) provide a new chapter, which reviews the responses and strategies of streambed and soil microbes to changes in precipitation patterns. Such communities contribute greatly to global biogeochemical cycles and thus it is crucial to understand their response mechanisms to increasing dryness. The authors regard the responses to different organizational levels, from cells to whole communities and emphasize the importance of habitat heterogeneity in support of microbial resistance and resilience against intensification of dry-wet extreme episodes.

Alice Retter, Clemens Karwautz, and Christian Griebler, in another new contribution (Chapter 13), summarize actual and expected impacts of climate change on subsurface ecosystems, the earth largest reservoir of freshwater. Significant increase in temperature and serious consequences from extreme hydrological events already observed are altering the composition of the specialized microbiome and fauna and endangering important ecosystems functions such as the cycling of carbon and nitrogen in the typically vulnerable groundwater systems. Retter et al. emphasize that understanding the interplay of biotic and abiotic drivers is required to anticipate future effects of climate change to groundwater resources and habitats.

Ecosystem respiration is typically the dominant process in river networks because of the fuelling by organic carbon from the terrestrial environment. Vicenç

Acuña, Anna Freixa, Rafael Marcé, and Xisca Timoner (Chapter 14) outline in their updated and extended contribution that the mineralization of organic carbon within river networks will be highly sensitive to global climate change because of major increases in the extent of non-flowing periods as well as in flood frequency and magnitude. The alterations in flow regime will increase organic carbon export, whereas temperature rise will increase organic carbon mineralization rates.

The final two chapters discuss selected applied aspects of the topic "Climate Change and Microbial Ecology". One contribution regards the combined impacts of land-use and climate change; the other one presents an example of how our knowledge on microbial ecology could be used for climate engineering.

Together with climate change, land-use is among the most important drivers impacting soil microbial communities and several microbially mediated biogeochemical processes. Stephen A. Wood, Krista McGuire, and Jonathan E. Hickman (Chapter 15) found evidence that both climate change and land-use change have strong impacts on the composition and functioning of tropical microbial communities, potentially amplifying the effects of climate change. The authors suggest research priorities which could improve our understanding of microbial responses to climate and land-use change including these drivers' interactions.

Christian Dunn, Nathalie Fenner, Anil Shirsat, and Chris Freeman (Chapter 16) propose that the "enzymic latch" in the breakdown of organic matter, particularly by inhibitory effects of phenolic compounds, could be used for a number of peatland based geoengineering schemes maximizing their abilities to store and capture carbon. Peatlands contain more than twice as much carbon that is contained in the Earth's forests. As with all geoengineering approaches, peatland geoengineering is not a 'magic bullet' to reverse the effects of climate change, but it has numerous advantages over other proposed schemes. The authors illustrate in their revised and updated chapter that peatland geoengineering offers a realistic concept (as "Plan B") for saving the Earth from the outcomes of anthropogenic climate change.

References

Bardgett, R.D., Freeman, C., and Ostle, N.J. (2008). Microbial contributions to climate change through carbon cycle feedbacks. ISME J. *2*, 805–814.

IPCC - Intergovernmental Panel on Climate Change (2018). Global Warming of 1.5°C. An IPCC Special Report on the Impacts of Global Warming of 1.5°C above Pre-Industrial Levels and Related Global Greenhouse Gas Emission Pathways, in the Context of Strengthening the Global Response to the Threat of Climate Change, Sustainable Development, and Efforts to Eradicate Poverty. V. Masson-Delmotte, P. Zhai, H.O. Pörtner, D. Roberts, J. Skea, P.R. Shukla, A. Pirani, W. Moufouma-Okia, C. Péan, R. Pidcock, S. Connors, J.B.R. Matthews, Y. Chen, X. Zhou, M.I. Gomis, E. Lonnoy, T. Maycock, M. Tignor and T. Waterfield, eds. (Geneva, Switzerland: World Meteorological Organization).

Current books of interest

- Alphaherpesviruses 2020
- Legionellosis Diagnosis and Control in the Genomic Era 2020
- Bacterial Viruses: Exploitation for Biocontrol and Therapeutics 2020
- Microbial Biofilms: Current Research and Practical Implications 2020
- Astrobiology: Current, Evolving and Emerging Perspectives 2020
- *Chlamydia* Biology: From Genome to Disease 2020
- Bats and Viruses: Current Research and Future Trends 2020
- SUMOylation and Ubiquitination: Current and Emerging Concepts 2019
- Avian Virology: Current Research and Future Trends 2019
- Microbial Exopolysaccharides: Current Research and Developments 2019
- Polymerase Chain Reaction: Theory and Technology 2019
- Pathogenic Streptococci: From Genomics to Systems Biology and Control 2019
- Insect Molecular Virology: Advances and Emerging Trends 2019
- Methylotrophs and Methylotroph Communities 2019
- Prions: Current Progress in Advanced Research (Second Edition) 2019
- Microbiota: Current Research and Emerging Trends 2019
- Microbial Ecology 2019
- Porcine Viruses: From Pathogenesis to Strategies for Control 2019
- *Lactobacillus* Genomics and Metabolic Engineering 2019
- Cyanobacteria: Signaling and Regulation Systems 2018
- Viruses of Microorganisms 2018
- Protozoan Parasitism: From Omics to Prevention and Control 2018
- Genes, Genetics and Transgenics for Virus Resistance in Plants 2018
- Plant-Microbe Interactions in the Rhizosphere 2018
- DNA Tumour Viruses: Virology, Pathogenesis and Vaccines 2018
- Pathogenic *Escherichia coli*: Evolution, Omics, Detection and Control 2018
- Postgraduate Handbook 2018
- Enteroviruses: Omics, Molecular Biology, and Control 2018
- Molecular Biology of Kinetoplastid Parasites 2018
- Bacterial Evasion of the Host Immune System 2017
- Illustrated Dictionary of Parasitology in the Post-Genomic Era 2017
- Next-generation Sequencing and Bioinformatics for Plant Science 2017
- Brewing Microbiology: Current Research, Omics and Microbial Ecology 2017
- Metagenomics: Current Advances and Emerging Concepts 2017
- The CRISPR/Cas System: Emerging Technology and Application 2017
- *Bacillus*: Cellular and Molecular Biology (Third edition) 2017
- Cyanobacteria: Omics and Manipulation 2017
- Foot-and-Mouth Disease Virus: Current Research and Emerging Trends 2017

www.caister.com

Chapter 1

Impacts of Climate Change on Cyanobacteria in Aquatic Environments

Hans W. Paerl

Institute of Marine Sciences, The University of North Carolina at Chapel Hill, Morehead City, NC, USA

Email: hpaerl@email.unc.edu

DOI: https://doi.org/10.21775/9781913652579.01

Abstract

Cyanobacteria are the Earth's oldest oxygenic phototrophs and they have had major impacts on shaping its biosphere; starting with the formation of an oxic atmosphere. Their long evolutionary history (>2.5 billion years) has enabled them to adapt to geochemical and climatic changes, including numerous cooling and warming periods, volcanism and accompanying atmospheric physical-chemical changes, extreme dry and wet periods, and recent anthropogenic modifications of aquatic environments, including nutrient over-enrichment (eutrophication), chemical pollution, water diversions, withdrawal and salinization. Combined, these modifications have promoted a worldwide proliferation of cyanobacterial blooms that are harmful to ecological and animal (including human) health. In addressing steps needed to stem and reverse this troubling trend, nutrient input reductions are a "bottom line" necessity, regardless of other physical-chemical-biotic control strategies that are applied. Cyanobacteria exhibit optimal growth rates and bloom potentials at relatively high water temperatures; hence global warming plays a key interactive role in their expansion and persistence. Additional manifestations of climatic change, including increased vertical stratification, salinization, and intensification of storms and droughts and their impacts on nutrient delivery and flushing characteristics of affected water bodies, play synergistic roles in promoting bloom frequency, intensity, geographic distribution and duration. Rising temperatures cause shifts in critical nutrient thresholds at which cyanobacterial blooms can develop; thus nutrient reductions for bloom control may need to be more aggressively pursued in response to climatic changes taking place worldwide. Cyanobacterial bloom control must consider both N and P loading reductions formulated within the context of altered thermal and hydrologic regimes associated with climate change.

Introduction

Cyanobacteria are the Earth's oldest known prokaryotic oxygenic phototrophs, having appeared over 2.5 billion years ago (Schopf, 2000). They have witnessed major biogeochemical and climatic changes, including extreme swings in irradiance (visible and UV light) and temperature (ice ages as well as warm periods), rising oxygen levels (in large part due their own photosynthetic activities) and major changes in chemical composition during the evolution of the Earth's atmosphere (Schopf, 2000). They have witnessed periods of high and low nutrient (N, P, minor elements) abundance, and a great deal of variability in climatic conditions, including extremely wet and dry periods, combined with major changes in the Earth's surface temperature and irradiance. In addition, geophysical processes such as volcanism and continental drift have impacted their habitats and have exerted eco-physiological stresses over a wide range of time scales.

As a major phylogenetic group, cyanobacteria have "seen it all" when considering potential physical-chemical impacts and their biotic ramifications on Earth. This is probably a key reason why cyanobacteria exhibit an extremely broad

Figure 1.1 Cyanobacterial blooms, viewed for space and in the field. 1A: NASA SeaWiFS image of a *Trichodesmium* spp. bloom in the tropical Atlantic Ocean (Courtesy NASA). 1B: ASTER-TERRA image of a *Lyngbya* sp. Bloom in Lake Atitlan, Guatamala (Courtesy NASA). 1C: MODIS image of *Microcystis*-dominated blooms in the Western Basin of lake Erie and southern region of Saginaw Bay, Laurentian Great Lakes during the summer of 2009 (Courtesy NASA and NOAA Coastwatch-Great Lakes). 1D: Bloom of the benthic Cyanobacterium *Lyngbya wollei* at Silver Glen Springs, Florida (Photo, Hans Paerl). 1E: View of a *Microcystis*-dominated bloom in Meiliang Bay, Lake Taihu during summer 2009 (Photo, Hans Paerl). 1F: Hans Paerl sampling the Taihu bloom during 2007. 1G: A *Trichodesmium* sp. bloom in South Pacific Ocean waters near Fiji (Photo, courtesy Ryan Paerl). 1H: Mixed *Microcystis*, *Anabaena*, and *Aphanizomenon* sp. bloom in the St. Johns River, Florida, summer 1999 (Photo, courtesy John Burns). 1I: Aircraft view of an *Anabaena* bloom on the St. Johns River (Photo, courtesy Bill Yates/CYPIX). 1J: A benthic *Lyngbya* sp. bloom floating to the surface of Weeki Watchee Springs, Florida during summer 2003 (Photo, Hans Paerl). 1K: A marine *Lyngbya* sp. bloom attached to a reef near Puerto Rico (Photo Courtesy, Valerie Paul). 1L: Mixed *Microcystis* and *Anabaena* bloom at a development near the Indian River Lagoon, Florida (Photo, John Burns).

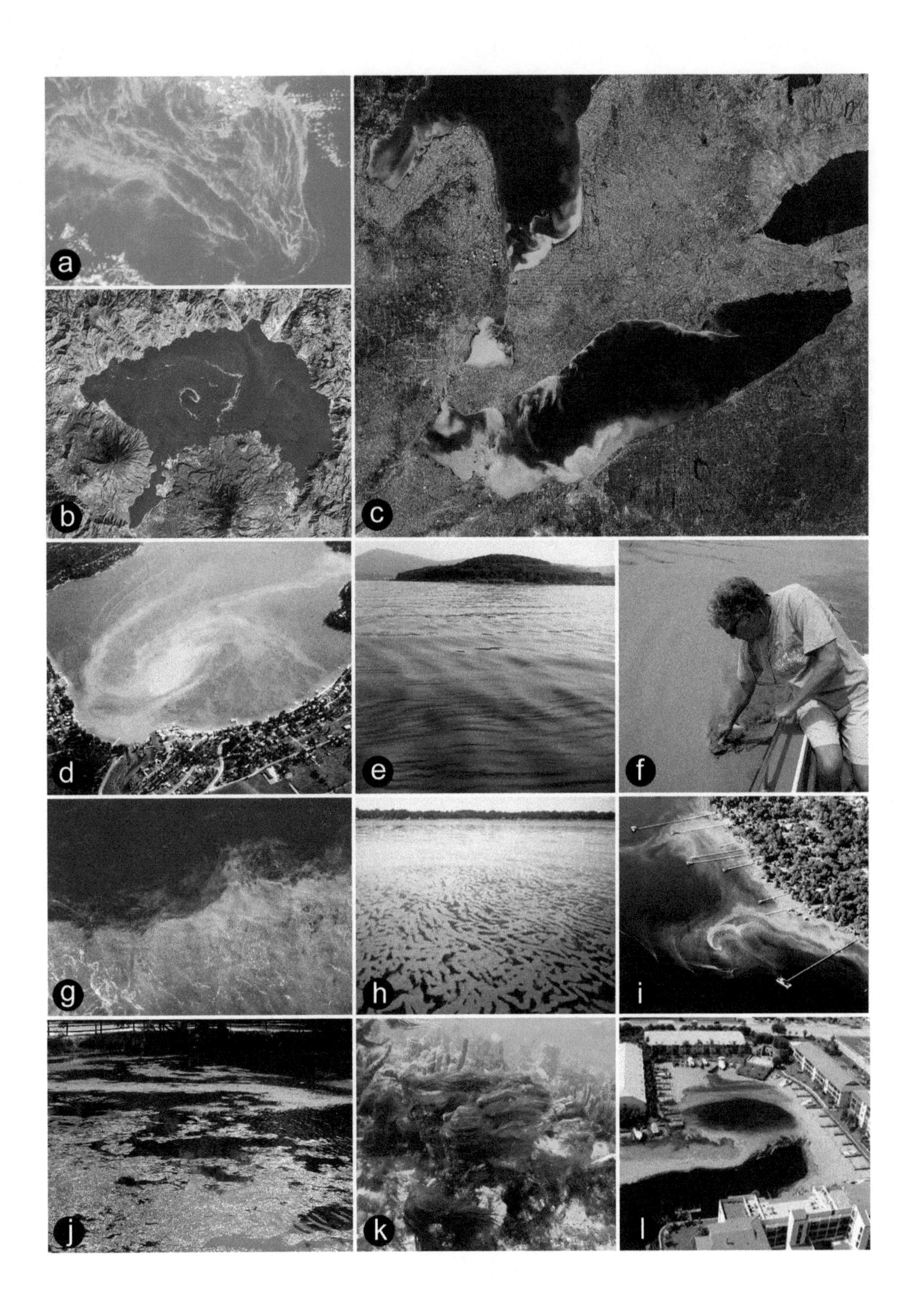
a
b
c
d
e
f
g
h
i
j
k
l

geographic distribution, ranging from polar to tropical regions, from subsurface aquatic to alpine habitats (Potts and Whitton, 2000; Mishra et al., 2019). They can be found in virtually all terrestrial and aquatic habitats, ranging from deserts to tropical rain forests and from the ultraoligotrophic open ocean to hypereutrophic lakes (Potts and Whitton, 2000; Whitton, 2012). Over geological and biological time scales, cyanobacteria reveal a remarkable ability to both counter extreme climatic conditions and thrive under them.

Diverse cyanobacterial taxa exhibit widespread adaptations climatic extremes, including the formation of heat and desiccation-tolerant resting cells, or akinetes, cysts, the presence of photoprotective and desiccation-resistant sheaths and capsules, a wide array of photoprotective (including UV protective) cellular pigments, the ability to glide and (in planktonic environments) use buoyancy regulation to adjust and optimize their position in the water column in response to irradiance and nutrient gradients (Potts and Whitton, 2000; Reynolds, 2006; Huisman et al., 2018; Mishra et al., 2019). They have also developed a wide array of physiological adaptations to periodic nutrient deplete conditions, including the ability to convert or “fix” atmospheric nitrogen (N_2) into biologically-available ammonia (Gallon, 1992), sequester (by chelation) iron (Wilhelm and Trick, 1994), store phosphorus, nitrogen and other essential nutrients (Healy, 1982; Reynolds, 2006), and produce metabolites that enhance their ability to counter potentially adverse conditions in their immediate environment, including photooxidation, and serve yet to be discovered protective and adaptive functions (Paerl and Millie, 1996; Huisman et al., 2005, 2018; Paerl and Otten, 2013 a, b). Cyanobacteria have a diverse suite of mutualistic and symbiotic associations with prokaryotic and eukaryotic microbes, plants and animals, that help ensure their (as well as their partners’) survival in environments too hostile for individual members to survive in (Paerl, 1982; 1986; Hooker et al., 2019).

It should therefore come as no great surprise that a rich “playbook” of ecological strategies aimed at surviving and at times thriving as massive growths or “blooms” under these conditions has enabled cyanobacteria to take advantage of more recent human alterations of aquatic environments; including nutrient over-enrichment (eutrophication), hydrologic alterations due to water withdrawal (for drinking, irrigation, industrial use) from streams, rivers and lakes, dams/reservoirs, artificial waterway, lagoon and marina construction, and alterations of a variety of benthic and planktonic habitats (Fig. 1.1) (Paerl et al., 2016a; 2019a,b).

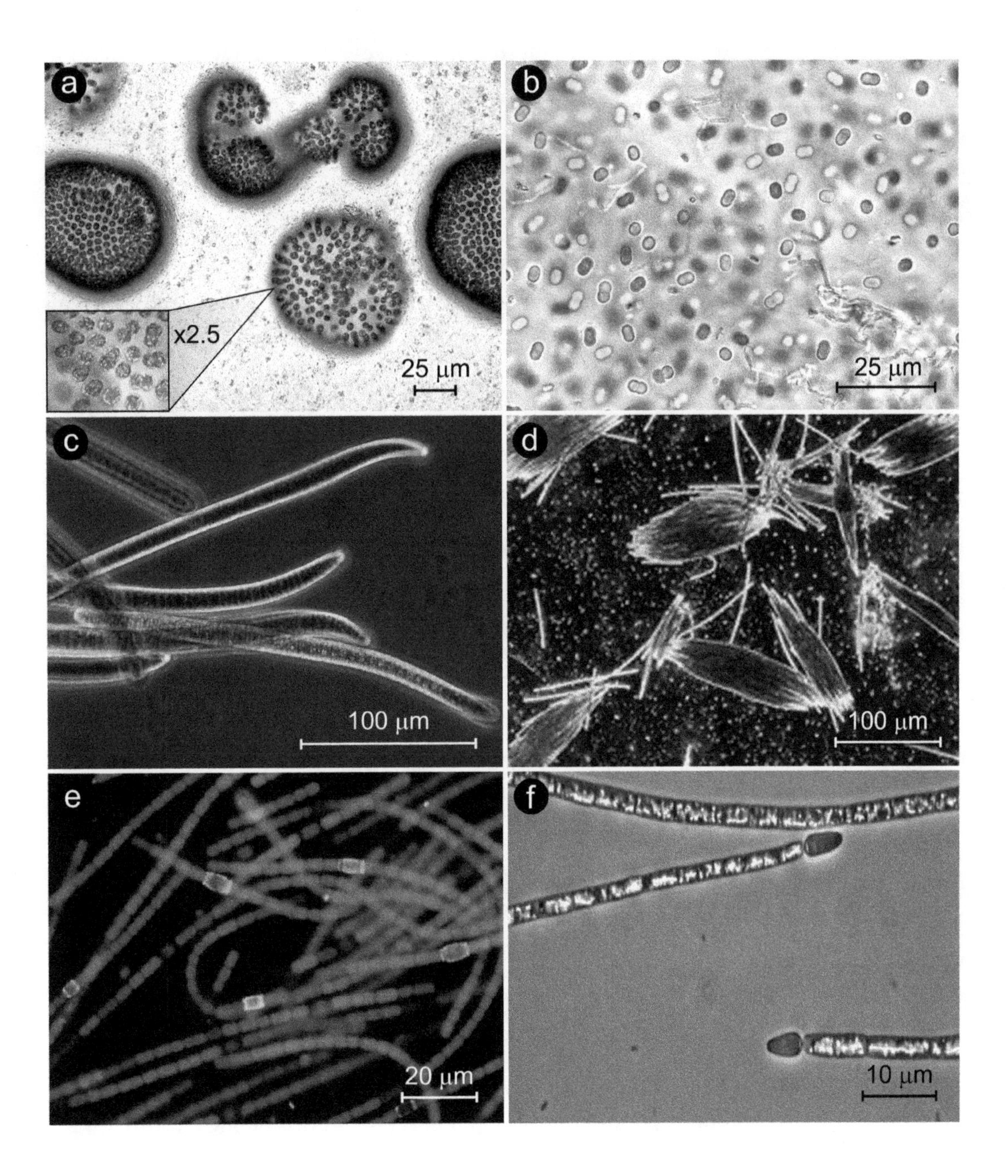

Figure 1.2 Photomicrographs of genera representing the three major cyanobacterial morphological groups. These include representative coccoid (1A, 1B), filamentous non-heterocystous (1C, 1D) and filamentous heterocystous (1E, 1F) CyanoHAB genera. 1A: *Microcystis* spp. (Photo, John Wehr). 1B: *Synechococcus* sp. (Photo, Chris Carter). 1C: *Oscillatoria* sp. (Photo, Hans Paerl). 1D: *Trichodesmium* sp. (Photo, Hans Paerl). 1E: *Anabaena oscillarioides* immunofluorescence micrograph, showing the nitrogen-fixing heterocysts (green) (Photo, Hans Paerl). 1F: *Cylindrospermopsis raciborskii* (Photo, Hans Paerl).

Climatic variability and change have been features of the Earth's atmosphere and biosphere ever since the evolution of life and cyanobacterial diversification (Schopf, 2000). Key features of climate change and their potential impacts on cyanobacterial activity, community structure and function that will be discussed in this chapter include changes in temperature and precipitation (including more extreme rainfall and drought events), and the combination of these factors. I will discuss these impacts and their ramifications for cyanobacterial eco-physiology, habitats and community composition and function, with an emphasis on biogeochemical cycling, water quality and trophodynamics along the freshwater to marine continuum.

The cyanobacterial "players"

In aquatic ecosystems, cyanobacteria exist in 3 major morphologically-distinct groups (Fig. 1.2). These include: 1) Coccoid cells, ranging from solitary (e.g. *Synechococcus*, *Chroococcus*) (< 3 μm diameter), largely non-N_2 fixing coccoid to ovoid cyanobacterial genera, make up an important, and at times dominant (>50%), fraction of freshwater, estuarine and marine phytoplankton biomass (Kuosa, 1991; Sánchez-Baracaldo et al., 2008; Gaulke et al., 2010; Paerl, 2012). Other coccoid forms are aggregated in colonies that are widespread and sometimes dominate as "blooms" (e.g., *Microcystis*) in planktonic and benthic environments over a wide range of trophic states (ultraoligotrophic to hypereutrophic) (Fig. 1.1). Most of these genera do not fix nitrogen, and hence are dependent on combined nitrogen supplies. Some genera (e.g., *Microcystis*) can produce secondary metabolites that are toxic to inhabitants and consumers, ranging from zooplankton to fish to humans (Huisman et al., 2018). 2) Filaments of mostly undifferentiated cells. This group is largely comprised of non-N_2 fixing genera (e.g. *Oscillatoria*, *Planktothrix*); however, some N_2 fixing genera also exists (e.g. *Lyngbya*, *Trichodesmium*), and these genera can at times dominate, as blooms, in benthic and planktonic environments. 3) Filaments with highly differentiated, biochemically-specialized N_2 fixing, cells called heterocysts (Wolk, 1996). Heterocystous cyanobacteria are considered morphologically most advanced because heterocysts appear to be an adaptation to ambient oxygen-rich conditions (Wolk, 1996), which the cyanobacteria brought about during their proliferation on Earth (Schopf, 2000). There are numerous bloom-forming genera in this group (e.g., *Anabaena*, recently renamed *Dolichospermum*, *Aphanizomenon*, *Cylindrospermopsis*, *Nodularia*). In addition to planktonic bloom formers, benthic filamentous genera (*Calothrix*, *Rivularia*, *Scytonema*, *Lyngbya*, *Oscillatoria*) can undergo explosive growths as epiphytes, mats and biofilms (Fig. 1.1). Each group contains species that produce secondary metabolites that can be toxic to a variety of animal consumers, ranging from zooplankton to fish to mammals, including humans (Carmichael, 1997, 1998; Paerl and Otten, 2013; Huisman et al., 2018).

Climate change, cyanobacterial ecology and dominance

While there is a rich literature showing a clear link between nutrient (N, P, trace metals) availability and the composition, distribution and abundance of cyanobacterial taxa in aquatic ecosystems (Vincent, 1987; Potts and Whitton, 2000; Huisman et al., 2005; 2018; Paerl and Fulton, 2006; Burford et al., 2019), climate change plays an additional modulating role. Rising global temperatures, altered precipitation patterns and changes in hydrologic properties (i.e., freshwater discharge or flushing rates) of water bodies strongly influence growth rates, composition and bloom dynamics of cyanobacteria (Jöhnk et al., 2008; Paerl and Huisman, 2008, 2009; Paerl et al. 2011a, b; Paul, 2008; Kosten et al., 2012; Burford et al., 2019; Wells et al., 2019). Warmer temperatures favor surface bloom-forming cyanobacterial genera because as prokaryotes, they tend to show a strong preference for relatively warm conditions, and their maximal growth rates occur at relatively high temperatures; often in excess of 25 °C (Foy et al., 1976; Robarts and Zohary, 1987; Butterwick et al., 2005) (Fig. 1.3). At these elevated temperatures, cyanobacteria can outcompete eukaryotic algae (Weyhenmeyer, 2001; Elliot, 2010). Specifically, as the growth rates of the eukaryotic taxa reach their maxima or decline in response to warming, cyanobacterial growth rates reach their optima (Paerl, 2017; Paerl et al., 2011; 2019a,b) (Fig. 1.3).

Warmer surface waters are also prone to more intense vertical stratification. The strength of vertical stratification depends on the density difference between the relatively warm surface layer and the cold water beneath (Fig. 1.4). In marine systems, salinity gradients also induce stratification. As mean temperatures rise, waters will begin to stratify earlier in the spring, and stratification will persist longer into the fall (Stüken et al., 2006; Peeters et al., 2007; Suikkanen et al., 2007; Wiedner et al., 2007; Wagner and Adrian, 2009). Northern lakes, rivers and estuarine ecosystems have shown warming of surface waters, leading to earlier "ice out" and later "ice on" periods and stronger vertical temperature stratification. This has

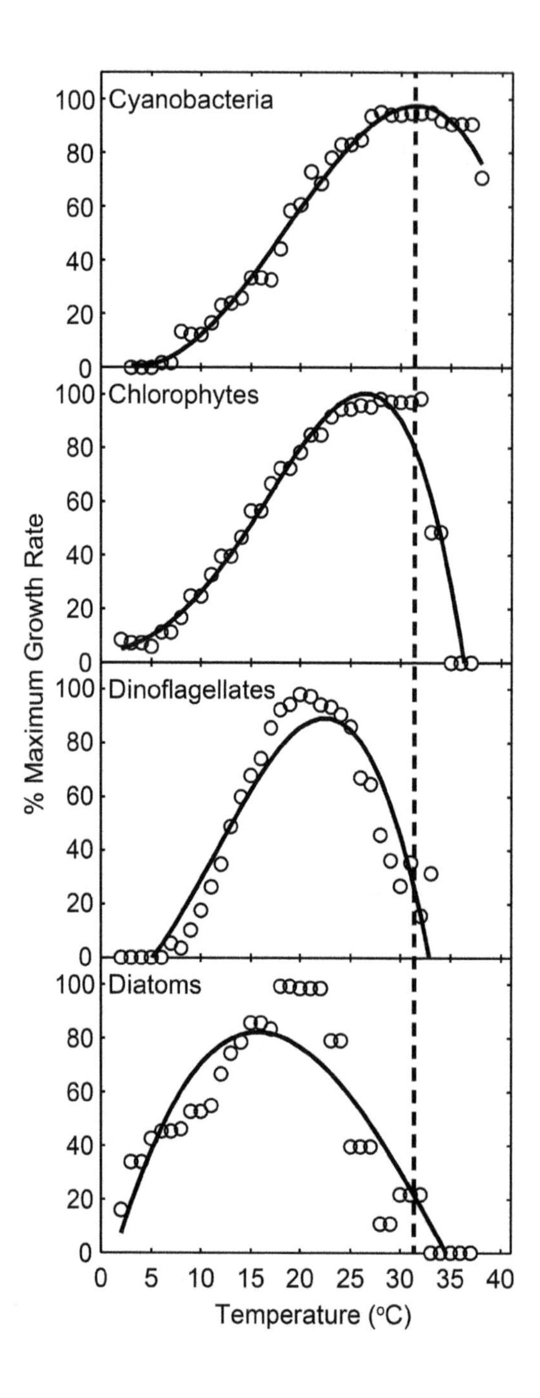

Figure 1.3 Relationships between temperature and specific growth rates of cyanobacterial species and eukaryotic phytoplankton in three different taxonomic groups (chlorophytes, dinoflagellates and diatoms). The dashed line is for comparison of optimal cyanobacterial growth temperature with temperature-growth relationships in other groups. Data points are 5° C running bin averages of percent maximum growth rates from 3-4 species within each group. Fitted lines are third order polynomials and are included to emphasize the shape of the growth versus temperature relationship. Figure adapted from Paerl et al., (2011a). Data sources and percent maximum growth rates of individual species are provided in Paerl et al. (2011a).

extended both the periodicity and range of cyanobacterial species, especially bloom-forming ones. Evidence for this can be obtained from lakes in northern Europe and North America, some of which no longer have ice on them during winter months (Wiedner et al., 2007; Wagner and Adrian, 2009).

Even in the Polar Regions, relatively small increases in warming can have significant impacts on the activities, biogeochemical cycling and trophodynamics associated with cyanobacterial communities. Along the margins of the Antarctic continent, cyanobacteria are found associated with exposed soils, glaciers, ice shelves, frozen lakes and stream beds, where they are often comprise most of the ecosystem biomass (Vincent, 1988). These communities are most commonly found as desiccated, frozen mats comprised of both coccoid and filamentous genera. However, during the Austral summer, there is a brief "window" (usually several weeks to a month) during which surface temperatures are high enough to melt the ice, allowing cyanobacteria and associated microflora access to liquid water. Virtually all the photosynthetic activity (primary production), nutrient cycling and trophic transfer is confined to this ice-free period (Vincent, 1988; Priscu, 1998; Priscu et al., 1998). Global warming will likely have a positive effect on the cyanobacterial communities by providing an extended "window" of liquid water conditions. In addition to increasing habitat availability, polar cyanobacteria are also highly responsive to warmer temperatures (Vincent and Quesada, 2012). Clearly, these effects have ramifications for nutrient and carbon cycling in the aquatic and terrestrial ecosystems experiencing this manifestation of warming.

In the extensive Arctic tundra environments, warming has likewise been documented, to the extent that near-surface permafrost is thawing and the temporal extent to which liquid water persists in these environments is increasing. Cyanobacterial communities are already abundant in surface soils, wetlands, stream and lakes comprising these environments (Vincent, 2000; Zahkia et al., 2008). The opportunities for these communities to increase their productivity, abundance and distributions are clearly increasing, given longer periods of ice-free, liquid water conditions.

In both Arctic and Antarctic environments it is likely that warming habitats are opening up to cyanobacterial and associated microbial expansion. This includes the polar seas, which at present contain undetectable or sparse populations of cyanobacteria, in contrast to most of the world oceans (Vincent and Quesada, 2012).

The overall biogeochemical and trophic importance of this ramification of climate change needs to be addressed. It could have a highly significant impact on C and nutrient cycling as well as greenhouse gas fluxes between the atmosphere and the massive amount of terrestrial, wetland and aquatic surface area characterizing these regions.

Another symptom of climatic changes potentially impacting cyanobacterial communities is increasing variability and more extremeness in precipitation amounts and patterns. Storm events, including tropical cyclones, nor'easters, and summer thunderstorms, are becoming more extreme, in that they contain higher amounts and intensities of rainfall (Webster, 2005; IPCC, 2014; Holland and Webster, 2007; Allan and Soden, 2008; Bender et al., 2010; Wuebbles et al., 2014; Kossin et al., 2017; Paerl et al., 2019). Conversely, droughts are becoming more severe and protracted (Trenberth, 2005; National Academy of Sciences USA, 2016). These events cause large changes in hydrologic variability, i.e., wetter wet periods and drier dry periods. This has led to more episodic "flashy" discharge periods in which large amounts of nutrients and organic matter are captured and transported in runoff events that can lead to rapid and profound nutrient enrichment of receiving waters (Bianchi et al, 2013; Paerl et al., 2018a; Hounshell et al., 2019). If such events are followed by periods of extended drought in which freshwater flow decreases dramatically and residence time of receiving waters increases, conditions favoring cyanobacterial dominance and bloom formation will greatly improve (Paerl et al., 2016a). This will be particularly effective if it is accompanied by warming, since as a phytoplankton group, cyanobacteria have relatively slow growth rates at moderate temperatures (Butterwick et al., 2005) (Fig. 1.3), which would increase in a warmer regime (Paul, 2008; Paerl and Paul, 2011). The combination of episodic loads of nutrients (e.g., spring runoff period), followed by a protracted warm (summer), low discharge period (long residence time) can promote cyanobacterial growth and bloom potentials in geographically-diverse regions (Paerl et al., 2011a; 2016b; Bargu et al., 2019).

Examples of this sequence of events include the Swan River and Estuary (Australia), Hartbeespoortdam (South Africa), the Neuse and Chowan River Estuaries (North Carolina, USA), the Potomac River (Chesapeake Bay, USA), and Lake Taihu (China) (Paerl et al., 2011a, 2016b, 2018b, 2019a). Attempts to regulate discharge of rivers and lakes by dams and sluices may increase residence time, and thus further enhance cyanobacterial bloom proliferation (Mitrovic et al., 2003).

Higher amounts of freshwater runoff can also enhance vertical density stratification (reduced vertical mixing) in waters having appreciable salinity, including estuarine and coastal waters as well as saline lakes and rivers, by allowing relatively light freshwater lenses to establish themselves on top of heavier (denser) saltwater. The resultant enhanced vertical stratification will favor phytoplankton capable of vertical migration to position themselves at physically-chemically optimal depths (Paerl and Huisman, 2009; Huisman et al., 2018). Bloom-forming cyanobacteria are capable of rapidly altering their buoyancy in response to varying light, temperature and nutrient regimes, by periodically forming blooms in surface

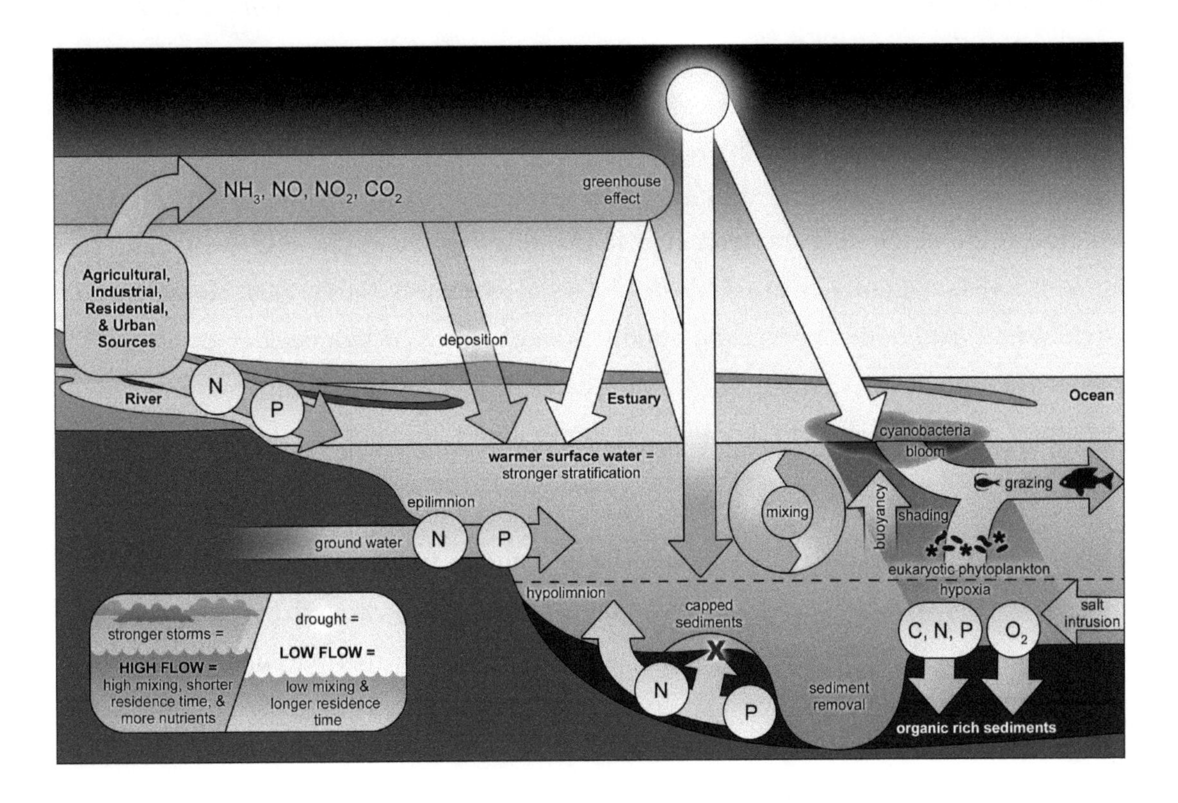

Figure 1.4 Conceptual diagram illustrating the various external and internal environmental and ecological factors controlling growth, accumulation (as blooms) and fate of Cyanobacteria in aquatic ecosystems. Factors can act individually or in combined (synergistic, antagonistic) ways. They include; surface and subsurface as well as atmospheric nutrient inputs, physical controls, including mixing/circulation, freshwater inputs and flushing (i.e., residence time), light, temperature (including greenhouse gas mediated warming), grazing, and numerous within-system feedbacks, such as stratification and organic matter driven hypoxia, nutrient regeneration and light shading by blooms of subsurface phytoplankton populations. Lastly, physical forcing such as wind-driven vertical mixing, can lead to sediment resuspension, which will impact light and nutrient availability.

waters (Walsby et al., 1997). Surface blooms are inhospitable to grazers and eukaryotic taxa that cannot handle the excessive irradiance in these waters. Many bloom taxa have photoprotective pigments, enabling them persist as surface blooms (Paerl et al., 1983), while sub-surface algal taxa will be shaded by surface blooms, leaving them in sub-optimal light conditions (Fig. 1.4).

More extensive summer droughts, rising sea levels, and increased use of freshwater for agricultural irrigation can lead to salinization, and this phenomenon has increased worldwide. Numerous cyanobacterial genera are salt-tolerant, even though they may be most common to freshwater ecosystems (probably because these are often nutrient enriched). These genera include the N_2 fixers *Anabaenopsis*, *Dolichospermum Nodularia*, and some species of *Lyngbya* and *Oscillatoria*, as well as non-N_2 fixing genera, including *Microcystis*, *Oscillatoria*, *Phormidium* and picoplanktonic genera (*Synechococcus*, *Chroococcus*). Some strains of *Microcystis aeruginosa* remain unaffected by salinities up to 10, nearing 30% of that of seawater (Tonk et al., 2007), and in Patos Lagoon, Brazil, it can thrive under "mixohaline" conditions. Some *Dolichospermun* and *Anabaenopsis* species can withstand salinities up to 15 (Montagnolli et al., 2004), while the common Baltic Sea bloom-former *Nodularia spumigena* can tolerate salinities exceeding 20 (Moisander et al., 2002; Mazur-Marsec et al., 2005). These salt-tolerant species are common to brackish systems; presumably spurred on by a combination of nutrient over-enrichment, climatically- or anthropogenically-driven flooding, water withdrawal and salinization.

Examples of cyanobacterial expansion in a climatically-changing world

Over the past several decades, field studies have demonstrated the expansion of cyanobacterial taxa in response to changing climatic (thermal and hydrologic) conditions. It should be noted that in certain cases, such as altered hydrology, the drivers cannot be solely attributed to climate change, but rather reflect the complex interactions of human alteration of hydrology as well as changing rainfall patterns and amounts.

Cylindrospermopsis raciborskii

The filamentous, bloom-forming and toxin-producing diazotroph *Cylindrospermopsis raciborskii* has undergone recent expansion of its geographical range (Fig. 1.5). Its expansion initially gained widescale attention following an outbreak of severe hepatitis-like disease on Palm Island (Australia), the so-called "Palm Island mystery disease" (Carmichael, 2001). This outbreak followed treatment of a local water supply reservoir in Australia, with copper sulfate. Epidemiological studies confirmed the linkage between the "mystery disease" and the newfound presence of *Cylindrospermopsis* (Carmichael, 2001). Lysis of the *Cylindrospermopsis* bloom released the highly stable cyanotoxin cylindrospermopsin into the water supply.

Cylindrospermopsis has traditionally been described as a tropical/subtropical genus (Padisak, 1997). However, *C. raciborskii* was documented in Europe during the 1930s, and showed a progressive colonization from Greece and Hungary towards higher latitudes near the end of the 20th century (Padisak, 1997; Kokociński et al, 2017; Weithoff et al., 2017). It was described in France in 1994, in the Netherlands in 1999, and it is now widespread in lakes in northern Germany (Stuken et al., 2006; Wiedner et al., 2007). It has also been detected in Canada (Hamilton et al., 2005) and is proliferating in South Africa (Van Vuuren and Kriel, 2008). *C. raciborskii* was noted in Florida almost 35 years ago, after which it aggressively proliferated throughout lakes and rivers (Chapman and Schelske, 1997). It is now present

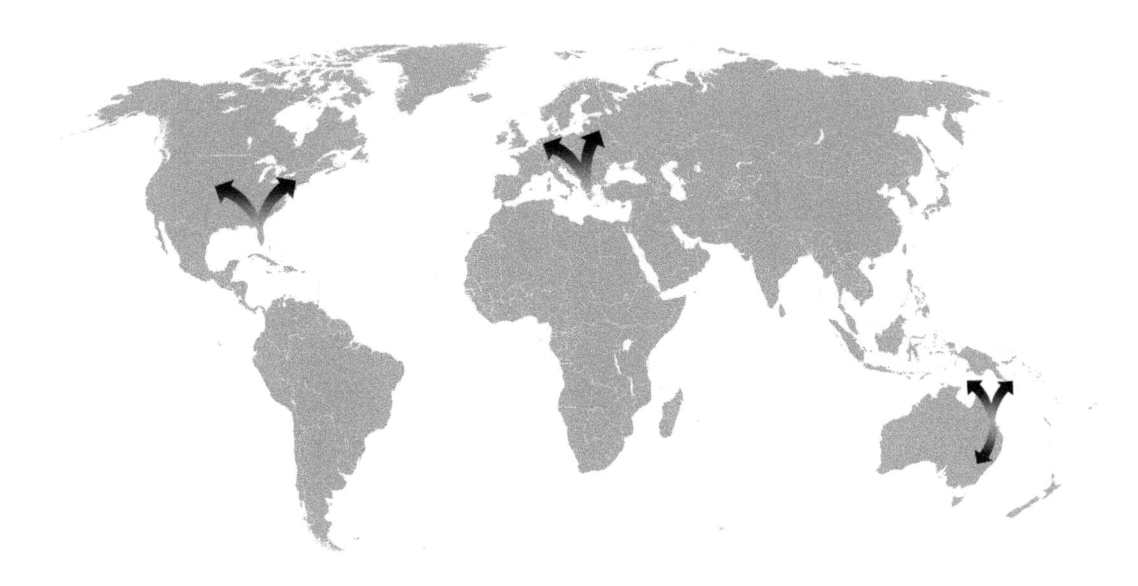

Figure 1.5 Map showing the geographic expansion of the harmful (toxic) cyanobacterial species *Cylindrospermopsis raciborskii*. Expansion is best documented in Europe, North America and Australia, where populations have expanded both in response to altered hydrologic regimes (e.g., droughts and poor flushing in Australian reservoirs) and warming (northerly directions in Europe and North America).

throughout the US in reservoirs, lakes, rivers and even oligohaline estuarine waters experiencing various degrees of eutrophication and loss of water clarity (Paerl and Fulton, 2006; Calandrino and Paerl, 2011). This combination is significant because *C. raciborskii* is adapted to low light conditions (high turbidity) typifying eutrophic waters. It also prefers water temperatures above 20°C, and survives adverse conditions using specialized vegetative resting cells (akinetes). This scenario hints of a link to eutrophication *and* global warming. The activation of akinetes in this and other heterocystous species (e.g., *Aphanizomenon ovalisporum*) is strongly temperature regulated (Cires et al., 2012). Increases in ambient temperatures may therefore play an important role in the geographic dispersal strategy, and potential expansion of this and other akinete-forming genera (*Anabaenopsis*, *Dolichospermum*, *Nodularia*).

Lyngbya spp.

Blooms of the filamentous, non-heterocystous, toxin-producing bloom former *Lyngbya* spp. have become increasingly common and problematic in nutrient-enriched freshwater and marine ecosystems; including those that have experienced human disturbances such as dredging, municipal waste inputs and the discharge of nutrient-laden agricultural runoff (Watkinson et al., 2005; Osborne et al., 2007). *Lyngbya* is a ubiquitous genus, with various species occurring in planktonic and benthic habitats. *L. majuscula* (marine-benthic) and *L. wollei* (freshwater-benthic and planktonic) are opportunistic invaders. Following large climatic and hydrologic perturbations such as tropical cyclones, *L. wollei* has proven to be an aggressive initial colonizer of perturbed systems (Paerl, 2012; Paerl and Fulton, 2006). *Lyngbya* blooms can proliferate as dense, attached or floating mats that shade other primary producers, enabling it to dominate the system by effectively outcompeting them for light (Fig. 1.6). As is the case with *Cylindrospermopsis* and *Microcystis*, this genus benefits from *both* human and climate-induced environmental change.

Managing cyanobacterial bloom dynamics in a climatically-changing world

Cyanobacterial growth and bloom formation thrive on the synergistic interactions between human- and climatically-altered physical-chemical alterations of aquatic ecosystems (Paerl and Otten 2003; Huisman et al., 2018; Burford et al., 2019). This presents a formidable challenge to water quality managers because ecosystem level, physical, chemical and biotic regulatory variables often co-occur and interact

Figure 1.6 Examples of expanding freshwater (left) and coastal marine (right) blooms of benthic *Lyngbya* spp. blooms. Left hand side: a) *Lyngbya* spp. bloom in Silver Spring, Florida. b) close-up photograph of *Lyngbya* aggregates covering freshwater macrophytes. c) Photomicrograph of dominant *Lyngbya* species forming the bloom. Right hand side: d) *Lyngbya* spp. bloom covering a seagrass bed, near Sanibel Island, Florida. e) Close-up photograph of *Lyngbya* aggregates clinging to seagrass leaves. d) Photomicrograph of *Lyngbya* sp. responsible for the bloom.

synergistically and/or antagonistically to control the activities (N_2 fixation, photosynthesis) and growth of harmful (toxic, food web disrupting, hypoxia-generating) bloom-forming cyanobacteria, or CyanoHABs (Paerl, 1988; Paerl and Millie, 1996) (Fig. 1.4). How to best address this challenge?

Nutrient input reductions are the most obvious targets, which can be altered and as such should be a central part of CyanoHAB mitigation strategies in both freshwater and marine environments (Figs. 1.4, 1.7) (Hamilton et al., 2016). We have long been aware that P input reduction is an effective means of reducing cyanobacterial dominance in aquatic, and especially freshwater, ecosystems. However, there are increasing instances where N input reductions are also needed. This is especially the case in eutrophic, CyanoHAB susceptible, lakes, rivers, estuaries and coastal waters which are capable of assimilating more N and increasing their trophic state (Paerl and Scott, 2010; Paerl, 2013; Paerl et al., 2016b; Scott et al., 2019). A key management priority is establishing N and P input thresholds (e.g., Total Maximum Daily Loads; TMDLs), below which CyanoHABs can be controlled in terms of magnitude, temporal and spatial coverage (Paerl, 2013). The ratios of N to P inputs should be considered when developing these thresholds (Smith 1983). Ideal input ratios are those that do not favor CyanoHAB species over more desirable taxa, but there does not appear to be a universal ratio- above or below - which CyanoHABs can be consistently and reliably controlled (Paerl et al., 2019a). For this reason, total nutrient loads *and* concentrations need to be considered in CyanoHAB management (Paerl and Scott, 2010; Scott et al., 2013). For example, it is generally thought that total molar N:P ratios above ~15 discourage CyanoHAB dominance (Smith, 1983; Smith and Schindler, 2009). However, if the nutrient load and internal concentrations of N or P are extremely high (i.e. above saturation levels), a ratio approach for reducing CyanoHABs is not likely to be effective (Paerl and Otten 2013a; Paerl et al., 2019a).

There are many ways to reduce nutrient inputs on a lake or larger ecosystem scale (c.f., Smith and Schindler, 2009; US EPA, 2011; Hamilton et al., 2016). Nutrient inputs have been classified as point source and non-point source. Point sources are often associated with well-defined and identifiable discharge sites; therefore, these nutrient inputs are relatively easy to control. Targeting point sources is often attractive, because they can account for a highly significant share of P and N loading, they are readily identifiable, accessible, and hence from a regulatory

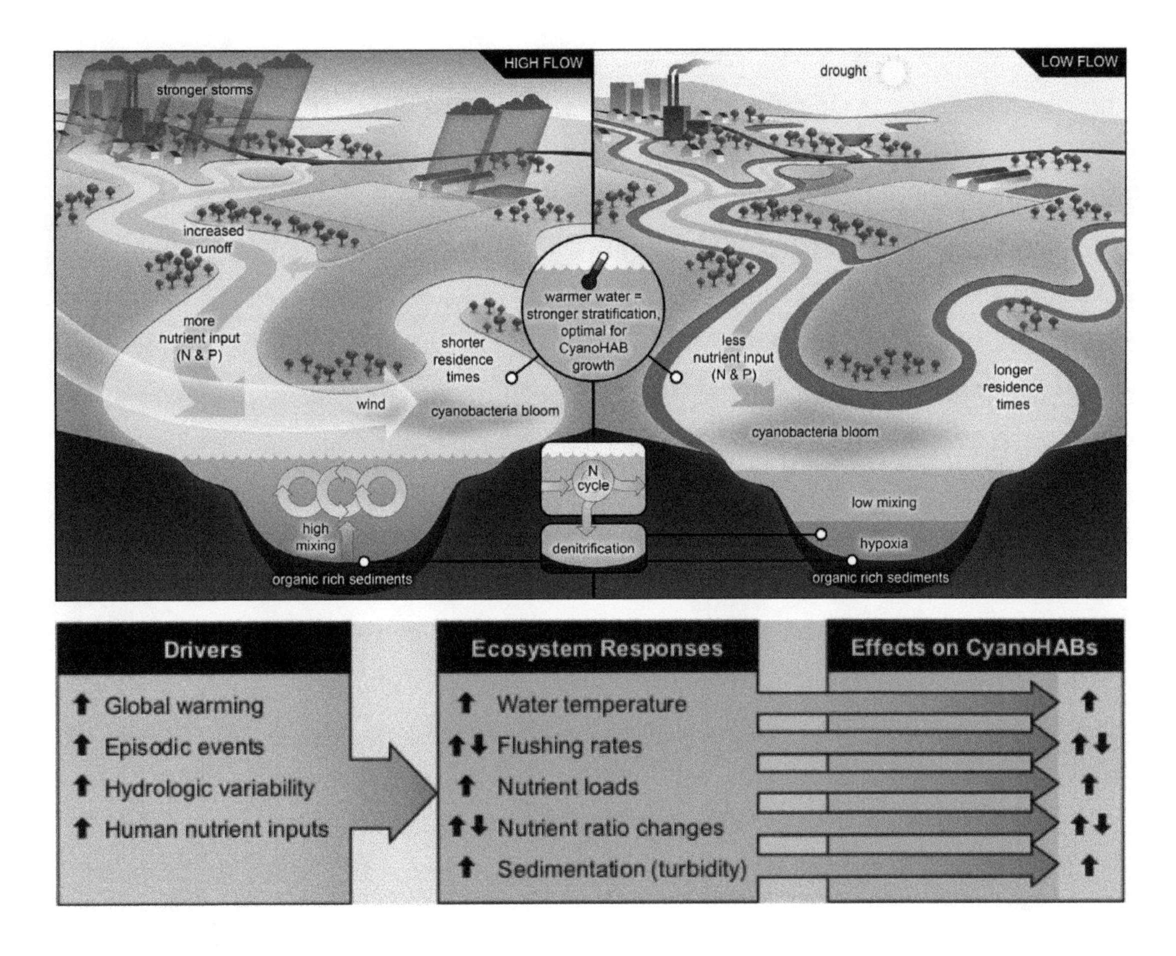

Figure 1.7 Conceptual diagrams, showing (top) the ecosystem-scale roles of physical drivers (freshwater input and flushing/residence times, wind mixing) and nutrient (N and P) input controls on harmful cyanobacterial bloom (CyanoHAB) potentials. The lower diagram summarizes the interactive human and climatic factors affecting CyanoHAB growth responses, aquatic ecosystem responses and positive/negative ecological effects. Figure adapted from Paerl et al., 2016a.

perspective, easiest to control. The major challenge that remains in many watersheds is targeting and controlling nonpoint sources, which in many instances are the largest sources of nutrients (US EPA, 2011; Hamilton et al., 2016; Paerl et al., 2019a,b); hence, their controls are likely to play a critical role in mitigating CyanoHABs in the context of human and climatically-driven environmental changes currently taking place.

Phosphorus management

Phosphorus inputs to aquatic ecosystems are dominated by; 1) non-point source surface runoff, and 2) point sources such as effluents from wastewater treatment plants, industrial and municipal discharges, and 3) subsurface drainage from septic systems and groundwater (Hamilton et al., 2016). Among these, point sources have been the focus of P reductions. In agricultural and urban watersheds, non-point surface and subsurface P inputs are of increasing concern. Increased P fertilizer use, generation and discharge of animal waste, soil disturbance and erosion, conversion of forests and grasslands to row-crop and other intensive farming operations, and the proliferation of septic systems accompanying human population growth are rapidly increasing non-point P loading (Sharpley et al., 2010; Hamilton et al., 2016). In agricultural and urban watersheds, non-point sources can account for at least 50% of annual P loading (Sharpley et al., 2010). Because of the diffuse nature of these loadings, they are more difficult to identify and address from a nutrient management perspective. They are also very susceptible to mobilization due to an increase in episodic rainfall events, including nor'easters and tropical cyclones.

The manner in which P is discharged to P-sensitive waters pays a role in CyanoHAB proliferation and management. Considerations include; 1) total annual (i.e., chronic) P loading, 2) shorter-term seasonal and event-based pulse (i.e., acute) P loadings, 3) particulate vs. dissolved P loading, and 4) inorganic vs. organic P loading. In terms of overall ecosystem P budgets and long-term responses to P loadings (and reductions), annual P inputs are of fundamental importance. When and where P enrichment occurs can determine the difference between bloom-plagued vs. bloom-free conditions. For example, if a large spring P discharge event precedes a summer of dry, stagnant (stratified) conditions in a relatively long residence time water body, the spring P load will be available to support summer bloom development and persistence (Paerl et al., 2016a). Effective exchange and cycling between the water column and bottom sediments can retard P transport and hence retain P (Wetzel 2001) (Fig. 1.4). As a result, acute P inputs due to high flow events and periods may be retained longer than would be estimated based on water flushing time alone (Paerl et al., 2019b; Wurtsbaugh et al., 2019). As such, water bodies exhibit both rapid biological responses to and a "memory" for acute P loads.

Unlike N, which can exist in dissolved gaseous forms, P exists only in dissolved ionic and particulate forms in natural waters. Therefore, the main concern is with dissolved vs. particulate forms of inorganic and organic P. Dissolved inorganic P (DIP) exists as orthophosphate (PO_4^{3-}), which is readily assimilated by all CyanoHAB taxa. Cyanobacteria can accumulate and store assimilated P intracellularly as polyphosphates, which can be available for subsequent use during times of P depletion (Healy, 1982). Dissolved organic P (DOP) can also be a significant fraction of the total dissolved P pool. DOP can be assimilated by bacteria, microalgae and cyanobacteria, although not as rapidly as PO_4^{3-} (Lean, 1973). A large fraction of the assimilated DOP is microbially recycled to DIP, enhancing P availability. The role of particulate P (as inorganic or organic forms) in aquatic production and nutrient cycling dynamics is less well understood. Particulate P (PP) may provide a source of DIP and DOP via desorption and leaching, and it may serve as a "slow release" source of DIP (Hamilton et al., 2016). In this manner a fraction PP can serve as a source of biologically-available P and hence play a role in CyanoHAB control. On the ecosystem-scale, sedimented PP serves as an important source of stored P for subsequent release, especially during hypoxic/anoxic periods. It is therefore essential to include *both* dissolved and particulate P when managing P inputs, especially under more extreme hydrologically-variable conditions predicted with climate change.

Nitrogen management

Nitrogen exists in multiple dissolved, particulate and gaseous forms. Many of these forms are biologically-available and readily exchanged within and between the atmosphere, water column and sediments (Galloway et al., 2004). In addition, biological nitrogen (N_2) fixation and denitrification control the exchange between inert gaseous atmospheric N_2 and biologically-available combined N forms. Combined forms of N include dissolved inorganic N (DIN; including ammonium (NH_4^+), nitrate (NO_3^-) and nitrite (NO_3^-)), dissolved organic N (DON; e.g., amino acids and peptides, urea, organo-nitrates), and particulate organic N (PON; polypeptides, proteins, organic detritus). These forms can be supplied from non-point and point sources. Non-point sources include surface runoff, atmospheric deposition and groundwater, while point sources are dominated by municipal, agricultural and industrial wastewater. In rural and agricultural settings, non-point N inputs tend to dominate (>50% of total N loading), while in urban centers, point sources often dominate (US EPA, 2011). All sources contain diverse organic and inorganic N species in dissolved and particulate forms; representing a mixture of biologically-available DIN, DON and PON.

Nitrogen inputs are dynamic, reflecting land use, population and economic growth and hydrologic conditions (Galloway et al., 2002). The means and routes by which human N sources impact eutrophication are changing (US EPA, 2011; Hamilton et al., 2016; Wurtsbaugh et al., 2019). Among the most rapidly-growing (amount and geographic scale) sources of human N loading are surface runoff, groundwater and atmospheric deposition. Atmospheric N loading and groundwater are often-overlooked, but expanding sources of N input to N-sensitive waters (Paerl et al., 2002; US EPA, 2011). As with P, N input and cycling dynamics are sensitive to patterns and intensities of precipitation, freshwater flow, which control mobilization in the watershed and discharge to N-sensitive waters.

Recent work has shown that nitrogen availability plays an increasingly important role in both freshwater and marine eutrophication, including cyanobacterial bloom dynamics and control (Dodds and Smith, 2016; Paerl et al., 2016b; Paerl et al., 2018a; Scott et al., 2019; Newell et al., 2019; Wurtsbaugh et al., 2019). From both biogeochemical and climatic perspectives, this is an important set of findings, because human watershed N loadings are increasing with expanding agricultural and urban activities, and more extreme storm events, yielding larger amounts and more intense precipitation are leading to an increase in N discharge to N-sensitive waters, ranging from alpine lakes to the coastal ocean (Paerl et al., 2018b). This calls for stricter watershed N management and retention efforts, especially in agricultural and urbanizing regions (Hamilton et al., 2016).

The interactions of physical, biological and nutrient controls of cyanobacteria in a climatically-changing world

Physical factors

Physical factors, including altered turbulence, vertical mixing and hydrologic flushing play key roles in cyanobacterial bloom dynamics (Paerl, 1988; Reynolds, 2006; Paerl et al. 2016a). It is well known that vertical stability (thermal or salinity stratification), and long water replacement (residence or flushing) times favor cyanobacteria over eukaryotic phytoplankton; hence disruption of these conditions can, under certain circumstances and in specific systems modulate cyanobacterial bloom dynamics (Figs. 1.4, 1.7). Vertical mixing devices, bubblers and other means of breaking down destratification have proven effective in controlling outbreaks and persistence of

CyanoHABs in relatively small impoundments such as farm and fish ponds (Visser et al., 1996; Huisman et al., 2005). These devices have limited applicability in large lake, estuarine and coastal waters however, because they cannot exert their forces over very large areas and volumes.

Increasing the flushing rates, and thereby decreasing water residence time (or water age), can be effective in reducing or controlling bloom taxa; mainly because cyanobacteria exhibit relatively slow growth rates, relative to eukaryotes (Mitrovic et al., 2003; Paerl et al., 2016a). Horizontal flushing, by increasing the water flow through lakes or estuaries, can reduce the time for cyanobacterial bloom development (Maier et al., 2004). While these approaches can yield positive results (i.e. suppression of CyanoHAB intensity), hydrologic changes can be quite expensive and restricted to relatively small water bodies.

Water quality managers must ensure that the flushing water is relatively low in nutrient content, so as not to worsen the enrichment problem, especially in large water bodies, which tend to have a long "memory" for nutrient inputs. For example, in hypereutrophic Lake Taihu, China, efforts to reduce *Microcystis*-dominated blooms by flushing this large lake with nearby Yangtze River water, which reduced the lake's overall residence time from ~ 1 Year to ~ 200 days, have not had a significant impact on reducing blooms, largely because Yangtze River water is exceedingly high in dissolved N and P compounds. Furthermore, the inflow pattern of Yangtze River water has altered the circulation regime of Taihu, and entrained or "trapped" blooms in the lake's northern bays, where they are most intense to begin with (Qin et al., 2010). Lastly, few catchments have the luxury of being able to use precious water resources normally reserved for drinking or irrigation water for flushing purposes. This is especially true of regions where freshwater runoff is limited and/or is periodically impacted by droughts (Paerl et al., 2011a, 2016a, 2019a, b).

Non-nutrient chemical controls

Chemical treatments have been used to control cyanobacterial blooms. These include the applications of algaecides, the most common of which is copper sulfate. Copper sulfate is effective, but it can be toxic to a wide variety of plant and animal species and its residue in the sediments is problematic as a legacy pollutant. Hydrogen peroxide (Matthijs et al., 2013) has been shown to be an effective algaecide. It is an attractive alternative to copper sulfate because it is selective for cyanobacteria (vs. eukaryotic algae and higher plants), and poses no serious long-term pollution problem. Both of these treatments are restricted to fairly small impoundments. It must be cautioned however that all algaecides will lyse cells and thereby release endotoxins into the aqueous bulk phase, making them more difficult to remove by drinking water plants and other sources of palatable water supplies (e.g., irrigation, food processing, brewing).

An alternative (to algaecides) chemical approach is to employ precipitation of P, thereby keeping it "locked up" in the sediments (Fig. 1.4). One treatment, called "Phoslock" uses a bentonite clay infused with the rare earth element lanthanum (Robb et al., 2003). The lanthanum ions are electrostatically bound to the bentonite, while also strongly binding to phosphate anions. The bound phosphate then settles out of the water column and the thin layer (~1 mm) of Phoslock on the sediment surface forms a barrier to phosphate diffusing out of the sediments (Robb et al., 2003). Phoslock has been shown to work well in small reservoirs, where it can lead to P-limited conditions that can control CyanoHAB production (Robb et al., 2003). Additionally, the thin Phoslock layer increases the critical erosional velocity of fine-grained surficial sediments, which should reduce the frequency of resuspension events and associated pulse nutrient loading (although, this will largely be effective in relatively small, vertically-stratified deep lakes). Sediment stabilization and reduced phytoplankton biomass may also aid restoration of macrophyte communities in shallow, eutrophic systems where light limitation and low root anchoring capacity of fine-grained, organic-rich sediments often synergistically determine CyanoHAB dominance. Shallow lakes in which wind-driven sediment resuspension may be a common feature may not be good candidates for precipitation techniques like Phoslock.

Altering Sediment Dynamics

Even when external nutrient inputs are reduced, the legacy of eutrophication in sediments can perpetuate high internal nutrient loads and provide a steady inoculum of algal spores or cysts that can continue to fuel CyanoHABs (Petersen, 1982; Cronberg, 1982; Robb et al., 2003; Wurtsbaugh et al., 2019). Therefore, either removing sediments or capping them so that sediment-water column exchange of nutrients and algal cells is restricted has been used to control CyanoHABs (Fig. 1.4).

Sediment removal involves expensive dredging, disturbance of lake bottoms, which can lead to additional nutrient (and potentially toxic substances) release and destruction of benthic flora and fauna (Petersen, 1982). There are examples of successful eradication of CyanoHABs using this approach, e.g., Lake Trummen, Sweden, a small (~ 1 km^2, mean depth 1.6 m) lake that experienced CyanoHAB related water quality degradation in response to domestic sewage and industrial nutrient inputs during the mid-1900's (Cronberg, 1982). Suction dredging the upper half meter of sediments during a 2 year period led to highly significant decreases in nutrient concentrations and CyanoHABs (Cronberg, 1982). The Lake Trummen success can largely be attributed to its small, easily manipulatable size, and the ability to effectively target reductions of external nutrient loads from its small (13 km^2) watershed, following dredging. In other sediment dredging efforts on sections of large lakes, results have been far less successful or not noticeable at all (e.g., Lake Taihu, China) (Qin et al., 2019).

Biological controls

Biological controls include a number of approaches to change the aquatic food web to increase grazing pressure on cyanobacteria or to reduce recycling of nutrients. Biomanipulation approaches can include introducing fish and benthic filter feeders capable of consuming cyanobacteria, or introduction of lytic bacteria and viruses. The most common biomanipulation approaches are intended to increase the abundance of herbivorous zooplankton by removing zooplanktivorous fish or introducing piscivorous fish. Alternatively, removal of benthivorous fish can reduce resuspension of nutrients from the bottom sediments. Questions have been raised about the long-term efficacy of curtailing cyanobacterial blooms by increasing grazing pressure, because this may lead to dominance by ungrazable or toxic strains (McQueen, 1990; Ghadouani et al., 2003). Presently, biomanipulation is viewed as one component of an integrated approach to water quality management in circumstances in which nutrient reductions alone are insufficient to restore water quality (Moss et al., 1996; Scheffer, 1998; Elser, 1999; Jeppesen et al., 2007a, b). Otherwise, nutrient management is the most practical, economically feasible, environmentally-friendly, long-term option.

In an overwhelming number of cases, nutrient input reductions are the most direct, simple, and ecologically/economically feasible cyanobacterial management strategy; this is especially true for ecosystems experiencing effects of climate change, including warming and/or increased hydrologic variability and extremes. Nutrient input reductions that can decrease cyanobacterial competitive abilities, possibly combined with physical controls (in systems that are amenable to those controls) are often the most effective strategies. Nutrient (specifically N) treatment costs can be prohibitive however, in which case, alternative nutrient removal strategies may prove attractive; including construction of wetlands, cultivation and stimulation of macrophytes, stocking of herbivorous (and specifically cyanobacteria consuming) fish and shellfish species (Jeppesen et al., 2007a, b; Havens et al., 2016).

Conclusions

Cyanobacteria are globally distributed and their activities and relative roles in production and nutrient cycling dynamics are controlled by a complex set of environmental variables that are heavily influenced by human and climatic perturbations. Their long evolutionary history has enabled them to structurally and functionally diversify, which in turn has enabled them to adapt to short-term (i.e., diel, seasonal, decadal) and longer term (geological) environmental perturbations and more gradual changes. Because they have experienced major and extreme climatic shifts over these time scales, they are well suited to deal with and take advantage of various climatic changes that we are now experiencing, including warming, altered rainfall patterns and amounts, resultant changes in freshwater runoff, flushing and vertical stratification.

In addition to climatically-driven environmental changes known to influence cyanobacterial growth and dominance, the most significant anthropogenically-influenced factors include; 1) nutrient (especially N and P) enrichment, 2) hydrological changes, including freshwater diversions, the construction of impoundments such as reservoirs, water use for irrigation, drinking, flood control, all of which affect water residence time or flushing rates, 3) biological alterations of aquatic ecosystems, including manipulations of grazers (from zooplankton to fish), and lastly 4) the discharge of xenobiotic compounds e.g., heavy metals, herbicides and pesticides, industrial and domestic chemicals, antibiotics and other synthetic growth regulators, all of which affect phytoplankton community growth and composition.

Effective long term management of CyanoHABs must address the above-mentioned suites of environmental factors, along with knowledge of the ecological and physiological adaptations that certain taxa possess to circumvent controls derived from our knowledge of these factors. Examples include; 1) the ability of N_2 fixing taxa to exploit N-limited conditions, 2) the ability of certain buoyant taxa to counteract mixing and other means of man-induced destratification aimed at minimizing cyanobacterial dominance, 3) specific mutualistic and symbiotic associations that cyanobacteria have with other microorganisms, higher plants and animals, which may provide clues as to the roles toxins and other chemical factors play in shaping biotic community structure and function.

In an overwhelming number of cases, nutrient input reductions are the most direct, simple, and ecologically/economically feasible cyanobacterial management strategy; this is especially true for ecosystems experiencing effects of climate change, including warming and/or increased hydrologic variability and extremes (Fig. 1.7). Nutrient input reductions that can decrease cyanobacterial competitive abilities, possibly combined with physical controls (in systems that are amenable to those controls) are often the most effective strategies. Nutrient (specifically N) treatment costs can be prohibitive however, in which case, alternative nutrient removal strategies may prove attractive; including construction of wetlands, cultivation and stimulation of macrophytes, stocking of herbivorous (and specifically cyanobacteria consuming) fish and shellfish species.

Water quality managers will have to accommodate the hydrological and physical-chemical effects of climatic change in their strategies. Given the competitive advantages (over eukaryotic algae) that cyanobacteria enjoy in a more climatically-extreme period we are now experiencing, efforts aimed at control and management of cyanobacteria will need to be flexible enough to incorporate this extremeness. For example, nutrient input reductions aimed at stemming eutrophication and cyanobacterial bloom potentials will need to be carefully gaged and potentially changed to accommodate higher cyanobacterial growth potentials due to warming and increasing bloom potentials due to stronger vertical stratification and positive nutrient cycling feedbacks.

Lastly, without a comprehensive strategy to reduce greenhouse gas emissions, future warming trends and their impacts on aquatic ecosystems will likely only lead to further expansion and dominance of these ecosystems by cyanobacteria.

Acknowledgements

I appreciate the technical assistance and input of A. Joyner, R. Fulton, N. Hall, T. Otten, B. Peierls, and W. Vincent. Research discussed in this chapter was partially supported by the National Science Foundation (OCE-1840715; CBET 1230543, ENG 1803679 and Dimensions of Biodiversity 1831096), National Institutes of Health Project 1P01ES028939-01, the North Carolina Dept. of Natural Resources and Community Development/UNC Water Resources Research Institute (Neuse River

Estuary Monitoring and Modeling Project, ModMon), and the St. Johns Water Management District, Florida.

References

Allan, R.P., and Soden, B.J. (2008). Atmospheric warming and the amplification of precipitation extremes. Science 321, 1481-1484.

Bargu, S., Justic, D., White, J., Lane, R., Day, J., Paerl, H., and Raynie. R. (2019). Mississippi River diversions and phytoplankton dynamics in deltaic Gulf of Mexico estuaries: A review. Estuarine, Coastal and Shelf Science 221, 39-52.

Bender, M.A., Knutson, T.R., Tuleya, R.E., Sirutis, J.J., Vecchi, G.A., Garner, S.T., and Held, I.M. (2010). Modeled impact of anthropogenic warming on the frequency of intense Atlantic hurricanes. Science 327, 454-458.

Bianchi, T.S., Garcia-Tigreros, F., Yvon-Lewis, S.A., Shields, M., Mills, H.J., Butman, D., and Walker, N. (2013) Enhanced transfer of terrestrially derived carbon to the atmosphere in a flooding event. Geophys. Res. Lett. 40, 116-122.

Burford, M.A., Carey, C.C., Hamilton, D.P, Huisman, J., Paerl, H.W., Wood, S.A., and Wulff, A. (2019). Perspective: Advancing the research agenda for improving understanding of cyanobacteria in a future of global change. Harmful Algae DOI: 10.1016/j.hal.2019.04.004

Butterwick, C., Heaney, S.I., and Talling, J.F. (2005). Diversity in the influence of temperature on the growth rates of freshwater algae, and its ecological relevance. J. Freshwater Biol. 50, 291-300.

Calandrino, E.S., and Paerl, H.W. (2011). Determining the potential for the proliferation of the harmful cyanobacterium *Cylindrospermopsis raciborskii* in Currituck Sound, North Carolina. Harmful Algae 11, 1-9.

Carmichael, W.W. (1997). The cyanotoxins. Adv. Bot. Res. 27, 211-256.

Carmichael, W.W. (1998). Microcystin concentrations in human livers, estimation of human lethal dose-lessons from Caruaru, Brazil. In Proceedings of the 4th International Conference on Toxic Cyanobacteria, Beaufort, NC, September 1999. Paerl, H.W., and Carmichael, W.W., eds.

Carmichael, W.W. (2001). Health effects of toxin producing cyanobacteria: the cyanoHABs. Human Ecological Risk Assessment 7, 1393-1407.

Chapman, A.D., and Schelske, C,L. (1997). Recent appearance of *Cylindrospermopsis* (Cyanobacteria) in five hypereutrophic Florida lakes. J. Phycol. 33, 191-195.

Cirés, S., Wörmer, L., Wiedner, C., and Quesada, A. (2012). Temperature-dependent dispersal strategies of *Aphanizomenon ovalisporum* (Nostocales, Cyanobacteria):Implications for the annual life cycle. Microb. Ecol. DOI: 10.1007/ s00248-012-0109-8

Conley, D.J., Paerl, H.W., Howarth, R.W., Boesch, D.F., Seitzinger, S.P., Havens, K.E., Lancelot, C., and Likens, G.E. (2009). Controlling eutrophication: nitrogen and phosphorus. Science 323, 1014-1015.

Cronberg, G. (1982). Changes in the phytoplankton of Lake Trummen induced by restoration. Hydrobiologia 86, 185-193.

Dodds, W. K., and Smith, V.H. (2016) Nitrogen, phosphorus, and eutrophication in streams. Inland Waters 6, 155−164.

Elliott, J.A. (2010). The seasonal sensitivity of cyanobacteria and other phytoplankton to changes in flushing rate and water temperature. Global Change Biol. 16, 864-876.

Elser, J.J. (1999). The pathway to noxious cyanobacteria blooms in lakes: the food web as the final turn. Freshwater Biol. 42, 537-543.

Foy, R.H., Gibson, C.E., and Smith, R.V. (1976). The influence of daylength, light intensity and temperature on the growth rates of planktonic blue-green algae. Eur. J. Phycol. 11, 151-163.

Gallon, J.R. (1992). Tansley Review No. 44/ Reconciling the incompatible: N_2 fixation and O_2. New Phytol. 122, 571-609.

Galloway, J.N., Cowling, E.B., Seitzinger, S.P., and Sokolow, R.H. (2002). Reactive nitrogen: too much of a good thing. Ambio. 31, 60-66.

Galloway, J.N., Dentener, F.J., Capone, D.G., Boyer, E.W., Howarth, R.W., Seitzinger, S.P., Asner, G.P., Cleveland, C.C., Green, P.A., Holland, E.A., Karl, D.M., Michaels, A.F., Porter, J.H., Townsend, A.R., and Vorosmarty, C.J. (2004). Nitrogen cycles: past, present, and future. Biogeochem. 70, 153-226.

Gaulke, A.K., Wetz, M.S., and Paerl, H.W. (2010). Picoplankton: A major contributor to planktonic biomass and primary production in a eutrophic, river-dominated estuary. Estuar. Coast. Shelf Science 90, 45-54.

Ghadouani, A., Pinel-Alloul, B., and Prepas, E.E. (2003). Effects of experimentally induced cyanobacterial blooms on crustacean zooplankton communities. Freshwater Biol. 48, 363-38.

Hamilton, D.P., Salmaso, N., and Paerl, H.W. (2016). Mitigating harmful cyanobacterial blooms: strategies for control of nitrogen and phosphorus loads. Aquatic Ecology 50, 351–366.

Hamilton, P., Ley M., Dean S., and Pick, F. (2005). The occurrence of the cyanobacterium *Cylindrospermopsis raciborskii* in Constance Lake: an exotic cyanoprokaryote new to Canada. Phycologia 44, 17-25.

Healy, F.P. (1982). Phosphate. In The biology of cyanobacteria, N.G. Carr and B.A. Whitton, eds. (Oxford: Blackwell Scientific Publications), pp. 105124.

Havens, K., Paerl, H., Phlips, E., Zhu, M., Beaver, J., and Srifa, A. (2016). Extreme weather events and climate variability provide a lens to how shallow lakes may respond to climate change. Water 8, DOI: 10.3390/w8060229

Holland, G.J. and Webster, P.J. (2007). Heightened tropical cyclone activity in the North Atlantic: natural variability of climate trend? Philos. T. Roy. Soc. A. DOI: 10.1098/rsta.2007.2083

Hooker, K.V., Li, C., Cai, H., Krumholz, L.R., Hambright, K.D., Paerl, H.W., Steffen, M.M., Wilson, A.E., Burford, M., Grossart, H.-P., Hamilton, P., Jiang, H., Sukenik, A., Latour, D., Meyer, E.I., .Padisák, J., Qin, B., .Zamor, R.M., and Zhu, G. (2019). The global *Microcystis* interactome. Limnol. Oceanogr. DOI: 10.1002/lno.11361

Huisman, J.M., Matthijs, H.C.P., and Visser, P.M. (2005). Harmful cyanobacteria. Springer Aquatic Ecology Series 3 (Dordrecht, The Netherlands: Springer).

Huisman, J., Codd, G., Paerl, H., Ibelings, B., Verspagen, J., and Visser, P. (2018). Cyanobacterial blooms. Nature Rev. Microbiol. DOI: 10.1038/s41579-018-0040-1

Intergovernment Panel on Climate Change (IPCC) (2012). Managing the risks of extreme events and disasters to advance climate change adaptation. A Special Report of Working Groups I and II of the Intergovernmental Panel on Climate Change, C. Field, V. Barros, T.F. Stocker, D. Qin, D.J. Dokken, K.L. Ebi, M.D. Mastrandrea, K.J. Mach, G.K. Plattner, S.K. Allen, M. Tignor, and P.M. Midgley, eds. (Cambridge, UK and New York, NY, USA: Cambridge University Press).

Intergovernment Panel on Climate Change (IPCC) (2014). Synthesis Report. Contribution of Working Groups I, II and III to the Fifth Assessment Report of the Intergovernmental Panel on Climate Change, R.K. Pachauri and L.A. Meyer, eds. (Geneva: IPCC) pp. 151.

Jeppesen, E., Meerhoff, M., Jacobsen, B.A., Hansen, R.S., Søndergaard, M., Jensen, J.P., Lauridsen, T.L., Mazzeo, N., and Branco, C.W.C. (2007a). Restoration of shallow lakes by nutrient control and biomanipulation: the successful strategy varies with lake size and climate. Hydrobiologia 581, 269-285.

Jeppesen, E., Søndergaard, M., Meerhoff, M., Lauridsen, T.L., and Jensen, J.P. (2007b). Shallow lake restoration by nutrient loading reduction: some recent findings and challenges ahead. Hydrobiologia 584, 239-252.

Jöhnk, K.D., Huisman, J., Sharples, J., Sommeijer, B., Visser, P.M., and Stroom, J.M. (2008). Summer heatwaves promote blooms of harmful cyanobacteria. Global Change Biol. 14, 495-512.

Kokociński, M., Gągała, I., Jasser, I., Karosienė, J., Kasperovičienė, J., Kobos, J., Koreivienė, J., Soininen, J., Szczurowska, A., and Woszczyk, M., (2017). Distribution of invasive *Cylindrospermopsis raciborskii* in the East-Central Europe is driven by climatic and local environmental variables. FEMS Microb.Ecol. 93, DOI: 10.1093/femsec/fix035

Kossin, J.P., Hall, T., Knutson, T., Kunkel, K.E.T., Trapp, R.J., Waliser, D.E., and Wehner, M.F. (2017). Extreme storms. In Climate Science Special Report: Fourth National Climate Assessment, Volume I, D.J. Wuebbles, D.W. Fahey, K.A. Hibbard, D.J. Dokken, B.C. Stewart, and T.K. Maycock, eds. U.S. Global Change Research Program, Washington, DC, USA.

Kosten, S., Huszar, V.L.M., Bécares, E., Costa, L.S., van Donk, E., Hansson, L.A., Jeppesen, E., Kruk, C., Lacerot, G., Mazzeo, N., De Meester, L., Moss, B., Lürling, M., Nõges, T., Romo, S., and Scheffer, M. (2012). Warmer climates boost cyanobacterial dominance in shallow lakes. Global Change Biol. 18, 118-126.

Kuosa, H. (1991). Picoplanktonic algae in the northern Baltic Sea: seasonal dynamics and flagellate grazing. Mar. Ecol. Prog. Ser. 73, 269-276.

Lean, D.R.S. (1973). Movement of phosphorus between its biologically-important forms in lakewater. Journal of the Fisheries Resources Board Canada 30, 1525-1536.

Maier, H.R., Kingston, G.B., Clark, T., Frazer, A., and Sanderson, A. (2004). Risk-based approach for assessing the effectiveness of flow management in controlling cyanobacterial blooms in rivers. River Research and Applications 20, 459-471.

Matthijs, H.C.P., Visser, P.M., Reeze, B., Meeuse, J., Slot, P.C., Wjin, G., Talens, R., and Huisman J. (2012). Selective suppression of harmful cyanobacteria in an entire lake with hydrogen peroxide. Water Res. 46, 1460-1472.

Mazur-Marzec, H., Żeglińska, L., and Pliński, M. (2005). The effect of salinity on the growth, toxin production, and morphology of *Nodularia spumigena* isolated from the Gulf of Gdansk, southern Baltic Sea. J. Appl. Phycol. 17, 171-175.

McQueen, D.J. (1990). Manipulating lake community structure: where do we go from here? Freshwater Biol. 23, 613-620.

Mishra, A.T., Tiwari, D.N., and Rai, A.N., eds. (2019). Cyanobacteria: From Basic Science to Applications (Amsterdam, The Netherlands: Elsevier Science Direct).

Mitrovic, S. M., Oliver, R. L., Rees, C., Bowling, L. C., and Buckney, R. T. (2003). Critical velocities for the growth and dominance of *Anabaena circinalis* in some turbid freshwater rivers. Freshw. Biol. 48, 164–174.

Moisander, P.H., McClinton, III E., and Paerl, H.W. (2002). Salinity effects on growth, photosynthetic parameters, and nitrogenase activity in estuarine planktonic cyanobacteria. Microb. Ecol. 43, 432-442.

Montangnolli, W., Zamboni, A., Luvizotto-Santos, R., and Yunes, J.S. (2004). Acute effects of *Microcystis aeruginosa* from the Patos Lagoon estuary, southern Brazil, on the microcrustacean *Kalliapseudes schubartii* (Crustacea: Tanaidacea). Arch. Environ. Contam. Toxicol. 46, 463-469.

Moss, B., Madgwick, J., and Phillips, J.G. (1996). A guide to the restoration of nutrient-enriched shallow lakes. (United Kingdom: W.W. Hawes).

National Academy of Sciences USA (2016). Attribution of Extreme Weather Events in the Context of Climate Change. The National Academies Press, Washington, D.C.

Newell, S.E., Davis, T.W., Johengen, T.H., Gossiaux, D., Burtner, A., Palladino, D., and McCarthy, M.J. (2019). Reduced forms of nitrogen are a driver of non-nitrogen-fixing harmful cyanobacterial blooms and toxicity in Lake Erie. Harmful Algae 81, 86-93.

Osborne, N.J., Shaw, G.R., and Webb, P.M. (2007). Health effects of recreational exposure to Moreton Bay, Australia waters during a *Lyngbya majuscula* bloom. Environ. Internat. 33, 309-314.

Padisak, J. (1997). *Cylindrospermopsis raciborskii* (Woloszynska) Seenayya et Subba Raju, an expanding, highly adaptive cyanobacterium: worldwide distribution and review of its ecology. Archiv für Hydrobiologie, Suppl. 107, 563-593.

Paerl, H.W. (1982). Chapter 17. Interactions with bacteria. In The Biology of Cyanobacteria, N.G. Carr and B.A. Whitton, eds. (Oxford: Blackwell Scientific Publications Ltd.), pp. 441461.

Paerl, H.W. (1986). Growth and reproductive strategies of freshwater bluegreen algae (cyanobacteria), In Growth and reproductive strategies of freshwater phytoplankton, C.D. Sandgren, ed. (Cambridge, UK: Cambridge University Press).

Paerl, H.W. (1988). Nuisance phytoplankton blooms in coastal, estuarine, and inland waters. Limnol. Oceanogr. 33, 823847.

Paerl, H.W. (2012). Marine Plankton. In Ecology of Cyanobacteria II: Their Diversity in Space and Time, B.A. Whitton, ed. (Dordrecht, The Netherlands: Springer). pp. 127-153.

Paerl, H.W. (2013). Combating the global proliferation of harmful cyanobacterial blooms by integrating conceptual and technological advances in an accessible water management toolbox. Environ. Microbiol. Rep. 5, 12-14.

Paerl, H.W. (2017). Controlling harmful cyanobacterial blooms in a climatically more extreme world: Management options and research needs. J. Plankt. Res. DOI: 10.1093/plankt/fbx042

Paerl, H.W., Tucker, J., and Bland, P.T. (1983). Carotenoid enhancement and its role in maintaining bluegreen algal (*Microcystis aeruginosa*) surface blooms. Limnol. Oceanogr. 28, 847857.

Paerl, H.W., and Millie, D.F. (1996). Physiological ecology of toxic cyanobacteria. Phycologia 35, 160-167.

Paerl, H.W., Dennis, R.L., and Whitall, D.R. (2002). Atmospheric deposition of nitrogen: implications for nutrient over-enrichment of coastal waters. Estuaries 25, 677-693.

Paerl, H.W., and Fulton, R.S. III. (2006). Ecology of harmful cyanobacteria. In Ecology of harmful marine algae, E. Graneli and J. Turner, eds. (Berlin: Springer-Verlag). pp. 95-107.

Paerl, H.W., and Huisman J. (2008). Blooms like it hot. Science 320, 57-58.

Paerl, H.W., and Huisman J. (2009). Climate change: a catalyst for global expansion of harmful cyanobacterial blooms. Environ. Microbiol. Rep. 1, 27-37.

Paerl, H.W., and Scott, J.T. (2010). Throwing fuel on the fire: synergistic effects of excessive nitrogen inputs and global warming on harmful algal blooms. Environ. Sci. Technol. 44, 7756-7758.

Paerl, H.W., Hall, N.S., and Calandrino, E.S. (2011a). Controlling harmful cyanobacterial blooms in a world experiencing anthropogenic and climatic-induced change. Sci. Total Environ. 409, 1739-1745.

Paerl, H.W., and Paul, V. (2011). Climate change: links to global expansion of harmful cyanobacteria. Water Res. 46, 1349-1363.

Paerl, H.W., Xu, H., McCarthy, M.J., Zhu, G., Qin, B., Li, Y., and Gardner, W.S. (2011b). Controlling harmful cyanobacterial blooms in a hyper-eutrophic lake (Lake Taihu, China): the need for a dual nutrient (N & P) management strategy. Water Res. 45, 1973-1983.

Paerl, H.W., and Otten, T.G. (2013a). Harmful Cyanobacterial Blooms: Causes, Consequences and Controls. Microb. Ecol. 65, 995-1010.

Paerl, H.W., and Otten, T.G. (2013b). Blooms bite the hand that feeds them. Science 342, 433-434.

Paerl, H.W., Gardner, W.S., Havens, K.E., Joyner, A.R., McCarthy, M.J., Newell, S.E., Qin, B., and Scott, J.T. (2016a). Mitigating cyanobacterial harmful algal blooms in aquatic ecosystems impacted by climate change and anthropogenic nutrients. Harmful Algae 54, 213-222.

Paerl, H. W., Scott, J.T., McCarthy, M.J., Newell, S.E., Gardner, W.S., Havens, K.E., Hoffman, D.K., Wilhelm, S.W., and Wurtsbaugh, W.A. (2016b). It takes two to tango: When and where dual nutrient (N & P) reductions are needed to protect lakes and downstream ecosystems. Environ. Sci. & Technol. 50, 10805−10813.

Paerl, H.W., Crosswell, J.R., Van Dam, B., Hall, N.S., Rossignol, K.L., Osburn, C.L., Hounshell, A.G., Sloup, R.S., and Harding L.W. Jr. (2018a). Two decades of tropical cyclone impacts on North Carolina's estuarine carbon, nutrient and phytoplankton dynamics: Implications for biogeochemical cycling and water quality in a stormier world. Biogeochem. 141, 307–332. DOI: 10.1007/s10533-018-0438-x

Paerl, H.W., Otten, T.G., and Kudela, R. (2018b). Mitigating the expansion of harmful algal blooms across the freshwater-to-marine continuum. Environ. Sci. & Technol. 52, 5519−5529. DOI: 10.1021/acs.est.7b05950

Paerl, H.W., Havens, K.E., Hall, N.S., Otten, T.G., Zhu, M., Xu, H., Zhu, G., and Qin, B. (2019a). Mitigating a global expansion of toxic cyanobacterial blooms: confounding effects and challenges posed by climate change. Mar. Freshwat. Res. DOI: 10.1071/MF18392

Paerl, H.W., Havens, K.E., Xu, H., Zhu, G., McCarthy, M.J., Newell, S.E., Scott, J.T., Hall, N.S., Otten, T.G., and Qin, B. (2019b). Mitigating eutrophication and toxic cyanobacterial blooms in large lakes: The evolution of a dual nutrient (N and P) reduction paradigm. Hydrobiologia DOI: 10.1007/s10750-019-04087-y

Paerl, H.W., Hall, N.S., Hounshell, A.G., Luettich, R.A., Jr, Rossignol, K..L, Osburn, C.L., and Bales, J. (2019c). Recent increase in catastrophic tropical cyclone flooding in coastal North Carolina, USA: Long-term observations suggest a regime shift. Nature Sci. Rep. 9, 10620. DOI: 10.1038/s41598-019-46928-9

Paul, V.J. (2008). Global warming and cyanobacterial harmful algal blooms. In Cyanobacterial harmful algal blooms: state of the science and research needs, H.K. Hudnell, ed., Advances in Experimental Medicine and Biology, Vol. 619 (Springer), pp. 239-257.

Peeters, F., Straile, D., Lorke, A., and Livingstone, D.M. (2007). Earlier onset of the spring phytoplankton bloom in lakes of the temperate zone in a warmer climate. Global Change Biol. 13, 1898-1909.

Petersen, S.A. (1982). Lake restoration by sediment removal. Journal of the American Water Resources Association 18, 423-435.

Potts, M., and Whitton, B.A. (2000). The Biology and Ecology of Cyanobacteria (Oxford: Blackwell Scientific Publications).

Priscu, J.C. (1998). Ecosystem Dynamics in a Polar Desert: The McMurdo Dry Valleys, Antarctica. American Geophysical Union Publications 72 (Washington, DC).

Priscu, J.C., Fritsen, C.H., Adams, E.E., Giovannoni, S.J., Paerl, H.W., McKay, C.P., Doran, T., Lanoil, B.D., and Pinckney, J.L. (1998). Perennial Antarctic lake ice: An oasis for life in a polar desert. Science 280, 2095-2098.

Qin, B., Zhu, G., Gao, G., Zhang, Y., Li, W., Paerl, H.W., and Carmichael, WW. (2010). A Drinking Water Crisis in Lake Taihu, China: Linkage to Climatic Variability and Lake Management. Environ. Manag. 45, 105-112.

Qin, B, Paerl, H.W., Brookes, J.D., Liu, J., Jeppesen, E., Zhu, G., Zhang, Y., Xu, H., Shi, K., and Deng, J. (2019). Why Lake Taihu continues to be plagued with cyanobacterial blooms through 10 years (2007–2017) efforts. Science Bulletin. DOI: 10.1016/j.scib.2019.02.008

Reynolds, C.S. (2006). Ecology of Phytoplankton (Ecology, Biodiversity and Conservation). (Cambridge, UK: Cambridge University Press).

Robarts, R.D., and Zohary, T. (1987). Temperature effects on photosynthetic capacity, respiration, and growth rates of bloom-forming cyanobacteria. New Zeal. J. Mar. Fresh. 21, 391-399.

Robb, M., Greenop, B., Goss, Z., Douglas, G., and Adeney, J. (2003). Application of Phoslock, an innovative phosphorous binding clay, to two Western Australian waterways: preliminary findings. Hydrobiologia 494, 237-243.

Sánchez-Baracaldo, P., Handley, B.A., and Hayes, P.K. (2008). Picocyanobacterial community structure of freshwater lakes and the Baltic Sea revealed by phylogenetic analyses and clade-specific quantitative PCR. Microbiol. 11, 3347-3357.

Scheffer, M. (1998). Ecology of shallow lakes (London: Chapman and Hall).

Schopf, J.W. (2000). The fossil record: tracing the roots of the cyanobacterial lineage. In The ecology of cyanobacteria, B.A. Whitton and M. Potts, eds. (Dordrecht: Kluwer Academic Publishers), pp. 13-35.

Scott, J.T., McCarthy, M.J., Otten, T.G., Steffen, M.M., Baker, B.C., Grantz, E.M., Wilhelm, S.W., and Paerl, H.W. (2013). Comment: An alternative interpretation of the relationship between TN:TP and microcystins in Canadian lakes. Can. J. Fish. Aqua. Sci. 70, 1-4. DOI: 10.1139/cjfas-2012-0490

Scott, J.T., McCarthy, M.J., and Paerl, H.W. (2019). Nitrogen transformations differentially affect nutrient-limited primary production in lakes of varying trophic state. Limnol. Oceanogr. Lett. DOI: 10.1002/lol2.10109

Sharpley, A.N., Daniel, T., Sims, T., Lemunyon, T.J., Stevens, R., and Parry, R. (2010). Agricultural phosphorus and eutrophication, Second edition. (University Park, PA: USDA-ARS, Pasture Systems & Watershed Management Research Unit).

Smith, V.H. (1983). Low nitrogen to phosphorus ratios favor dominance by blue-green algae in lake phytoplankton. Science 221, 669–671.

Smith, V.H., and Schindler, D.W. (2009). Eutrophication science: Where do we go from here? TREE 24, 201-207.

Stüken, A., Rücker, J., Endrulat, T., Preussel, K., Hemm, M., Nixdorf, B., Karsten, U., and Wiedner, C. (2006). Distribution of three alien cyanobacterial species (Nostocales) in northeast Germany: *Cylindrospermopsis raciborskii*, *Anabaena bergii* and *Aphanizomenon aphanizomenoides*. Phycologia 45, 696-703.

Suikkanen, S., Laamanen, M., and Huttunen, M. (2007). Long-term changes in summer phytoplankton communities of the open northern Baltic Sea. Estuar. Coast. S. Sci. 71, 580-592.

Tonk, L., Bosch, K., Visser, P.M., and Huisman, J. (2007). Salt tolerance of the harmful cyanobacterium *Microcystis aeruginosa*. Aquat. Microb. Ecol. 46,117-123.

Trenberth, K.E. (2005). The impact of climate change and variability on heavy precipitation, floods, and droughts. In Encyclopedia of Hydrological Sciences, M.G. Anderson, ed. (John Wiley and Sons). DOI: 10.1002/0470848944.hsa211

U.S. Environmental Protection Agency. (2011). Reactive nitrogen in the United States: an analysis of inputs, flows, consequences, and management options. (Washington, DC: EPA Scientific Advisory Board publication EPA-SAB-11-013).

Van Vuuren, S.J., and Kriel, G.P. (2008). *Cylindrospermopsis raciborskii*, a toxic invasive cyanobacterium in South African fresh waters. Afr. J. Aquat. Sci. 1, 17-26.

Vincent, W.F. (1987). Dominance of bloom forming cyanobacteria (Blue-green algae). New Zeal. J. of Mar. and Fresh. 21, 361-542.

Vincent, W.F. (1988). Microbial Ecosystems of Antarctica. (Cambridge, UK: Cambridge University Press).

Vincent, W.F. (2000). Cyanobacterial dominance in the polar regions. In The Ecology of Cyanobacteria, B.A. Whitton and M. Potts, eds. (The Netherlands: Kluwers Academic Press), pp. 321-340.

Vincent, W.F., and Quesada, A. (2012). Cyanobacteria in high latitude lakes, rivers and seas. In Ecology of Cyanobacteria II: Their Diversity in Space and Time, B.A. Whitton, ed. (New York: Springer), pp. 371-385.

Visser, P.M., Ibelings, B.W., Van der Veer, B., Koedood, J., and Mur, L.R. (1996). Artificial mixing prevents nuisance blooms of the cyanobacterium *Microcystis* in Lake Nieuwe Meer, the Netherlands. Freshwater Biol. 36, 435-450.

Wagner, C., and Adrian, R. (2009). Cyanobacteria dominance: quantifying the effects of climate change. Limnol. Oceanogr. 54, 2460-2468.

Walsby, A.E., Hayes, P.K., Boje, R., and Stal, L.J. (1997). The selective advantage of buoyancy provided by gas vesicles for planktonic cyanobacteria in the Baltic Sea. New Phytol. 136, 407-417.

Watkinson, A.J., O'Neil, J.M., and Dennison, W.C. (2005). Ecophysiology of the marine cyanobacterium *Lyngbya majuscula* (Oscillatoriaceae) in Moreton Bay, Australia. Harmful Algae 4, 697-715.

Webster, P.J., Holland, G.J., Curry, J.A., and Chang, H.R. (2005). Changes in tropical cyclone number, duration, and intensity in a warming environment. Science 309, 1844-1846.

Weithoff, G., Taube, A., and Bolius, S. (2017). The invasion success of the cyanobacterium *Cylindrospermopsis raciborskii* in experimental mesocosms: genetic identity, grazing loss, competition and biotic resistance. Aquatic Invasions 12: 333–341. DOI: 10.3391/ai.2017.12.3.07

Wells, M.L. et al. (2019). Future HAB science: Directions and challenges in a changing climate. Harmful Algae, DOI: 10.1016/j.hal.2019.101632

Wetzel, R.G. (2001). Limnology: lake and river ecosystems, 3rd edition (San Diego, CA: Academic Press).

Weyhenmeyer, G.A. (2001). Warmer winters: are planktonic algal populations in Sweden's largest lakes affected? Ambio 30, 565-571.

Whitton, B.A. (2012). Ecology of Cyanobacteria II: Their Diversity in Space and Time (Dordrecht, The Netherlands: Springer).

Wiedner, C., Rücker, J., Brüggemann, R., and Nixdorf, B. (2007). Climate change affects timing and size of populations of an invasive cyanobacterium in temperate regions. Oecologia 152, 473-484.

Wilhelm, S.W., and Trick, C.G. (1994). Iron-limited growth of cyanobacteria: multiple siderophore production is a common response. Limnol. Oceanogr. 39, 1979-1984.

Wolk, C.P. (1996). Heterocyst Formation. Annu. Rev. Genet. 30, 59-78.

Wuebbles, D., Meehl, G., Hayhoe, K., et al. (2014). CMIP5 climate model analyses: Climate extremes in the United States. Bull. Am. Meteorol. Soc. 95, 571–583.

Wurtsbaugh, W.A., Paerl, H.W., and Dodds, W.K. (2019). Nutrients, Eutrophication and Harmful Algal Blooms Along the Freshwater to Marine Continuum. Wiley Interdisc. Rev. 6 DOI: 10.1002/wat2.1373

Zakhia, F., Jungblut, A.-D., Taton, A., Vincent, W.F., and Wilmotte, A. (2008). Cyanobacteria in Cold Ecosystems. In Psychrophiles: from Biodiversity to Biotechnology, R. Margesin et al., eds. (Berlin, Heidelberg: Springer-Verlag) pp. 121-135.

Chapter 2

Climate Change Effects on Planktonic Bacterial Communities in the Ocean: From Structure and Function to Long-term and Large-scale Observations

Ingrid Brettar[1]*, Manfred G. Höfle[1], Carla Pruzzo[2] and Luigi Vezzulli[2]

[1] Helmholtz Centre for Infection Research, Department of Vaccinology and Applied Microbiology, D-38124 Braunschweig, Germany; [2] Department of Earth, Environment and Life Sciences (DISTAV), University of Genoa, Corso Europa, 26, 16132 Genoa, Italy; * corresponding author

Email: IBrettar@web.de, manfred.hoefle@helmholtz-hzi.de, carla.pruzzo@unige.it, luigi.vezzulli@unige.it

DOI: https://doi.org/10.21775/9781913652579.02

Abstract

Planktonic bacterial communities of marine environments can be considered to be affected by climate change through a set of direct and indirect effects. As direct effects of climate change, elevated temperature and ambient CO_2 levels have to be taken into account. As indirect effects a large spectrum of impacts ranging from increased stratification of surface water, deoxygenation of subsurface water, increased occurrence of extreme weather events, and a changed food chain and nutrient regime resulting in a changed top down and bottom up control for bacterioplankton. All these direct and indirect effects will affect bacterial communities in a multifaceted way. Changes of the bacterial communities due to climate change impacts can be expected on all different taxonomic levels, i.e. from the clonal intraspecies level to the phylum level. The focus of the article will be on the evaluation of bacterioplankton observations over time and space and along climate gradients in the frame of environmental parameters allowing modelling with respect to climate change scenarios. A specific emphasis will be on bacterioplankton analysis based on analysis of samples of the Continuous Plankton Recorder Archive. This sample archive allowed insights into particle associated bacteria of coastal environments over more than the last 60 years. Beside effects of climate change effectors on bacterial growth and community composition, effects of climate change on bacteria-mediated marine biogeochemical cycling and potential hazards by increased abundance of pathogenic bacteria such as *Vibrio cholerae* will be considered.

Introduction

Marine bacterioplankton is affected by the effects of climate change in a multifactorial way, with different impact on coastal and open ocean environments. Global warming had noticeable effects on sea surface temperature (SST), and is considered to affect in the future decades and centuries the deep water (Deser et al., 2010). The warming of the sea surface generates enhanced stratification with a broad set of effects, among others a reduction of nutrient transfer from deeper waters with the potential to impact the bottom up control of the bacterioplankton (Gruber, 2011; Rees, 2012). Temperature increase might affect the higher trophic levels to a larger extent than the bacterioplankton. Quantitative and qualitative changes of the trophic cascades with impact on the grazing regime have to be taken into account for the top down control of bacterioplankton (Beaugrand et al., 2002; Lewandowska et al., 2014; Sarmento et al., 2010; Tittensor et al., 2010; Voigt et al., 2003). The increased atmospheric CO_2-level has an impact on the pH and the bicarbonate availability, especially in the surface water, with far-reaching influences on the respective abiotic and biotic processes (Gruber, 2011; Krause et al., 2012, The MerMex Group, 2011).

Coastal and open ocean bacterioplankton are differently impacted by sea surface warming, stratification, deoxygenation and acidification (Duarte et al., 2013; Gruber, 2011; Rees, 2012). The effect on the respective bacterioplankton may vary substantially, mainly due to different nutrient and bicarbonate availabilities. Extreme weather events are a consequence of global warming. Heavy storm and rain events, and the resulting erosion and floods, increase the discharge of nutrients and suspended solids to coastal environments. Open ocean environments are far less impacted by extreme weather events; long distance transfer of Saharan dust and Hurricanes are examples for such rare events (Jickels et al., 2005; Maranon et al., 2010; Schlosser et al., 2014).

Bacterial communities of all water bodies, i.e. from freshwater to marine ecosystems, can be considered to be affected by climate change effects. The smaller the water body the more the impact of climate change on the surrounding or connected terrestrial ecosystem will affect the aquatic bacterial communities. Rivers, lakes and groundwater will be affected in different ways and to a different extent, also depending on the climate zone. There is a wealth of work performed on freshwater ecosystems showing a diverse impact and response (Kernan et al., 2009; Jeppesen et al., 2009). Marine ecosystems are more buffered towards the small scale, highly heterogeneous and patchy effects of global warming occurring in terrestrial ecosystems that are always in the vicinity of freshwater ecosystems. Especially changed nutrient fluxes are highly impacting freshwater ecosystem and these make distinction between eutrophication and climate change effects very difficult. Therefore, marine ecosystems allow a better insight into major climate change effects on bacterioplankton such as temperature increase, stratification and acidification. In addition, the large surface of the oceans gives them a major role for CO_2 cycling and makes them an important buffer for human driven climate change.

Marine bacterioplankton plays an essential role for the biogeochemical cycling of the ocean (Herndl and Reinthaler, 2013; Höfle et al., 2008; Jiao et al., 2010). The climatic impact on bacterioplankton is closely linked to the ocean's impact on climate change because it is a key driver of the marine carbon cycle with respect to mineralisation and carbon sequestration on the one hand, and its consumption and production of climate relevant trace gases on the other hand. Thus, the bacterioplankton is not only the recipient of the climatic effects, but also a driver of climatic changes. Despite the important role of the bacterioplankton, there is a severe lack of data that allow an assessment of the "reaction" of bacterioplankton with respect to climate change effects (Ladau et al., 2012).

To obtain insights into climate change effects, long-term observations are needed, that allow correlating the climate change effects to the respective bacterioplankton community. There were many experimental approaches achieved in order to extrapolate the effects of global warming (Moran et al., 2006; Sarmento et al., 2010; White et al., 1991). Though experimental approaches improved our mechanistic understanding of biogeochemical processes to a large extent, they were not evaluated in this article due to the following considerations: The assessment of bacterioplankton composition and turnover rates and an extrapolation for increased temperature or more complex climate change effects suffer from two basic drawbacks: i) the difficulty to obtain and measure in an undisturbed sample, and ii) the impossibility to predict adaptation and evolution of the analyzed bacterioplankton to the respective climate change impacts during an interval long enough to observe relevant changes in the marine ecosystem.

The focus of this review will therefore, be on the evaluation of bacterioplankton observations over time and space and along climate gradients. A specific emphasis will be on bacterioplankton analysis based on the use of samples of the Continuous Plankton Recorder Archive (Reid et al., 2011; Vezzulli et al., 2012) allowing retrospective insights into particle associated bacteria of coastal environments for over more than the last 60 years. After giving an overview on climate change effects on marine ecosystems and its bacterioplankton, we will try to delineate basic features of bacterioplankton that may lead to an estimate on the functioning and the potential reactivity towards climate change. In detail, we try to tackle a set of basic questions with respect to climate change effects, such as impacts on bacterial growth, e.g. oxidative stress, the taxonomic level of relevance for observation of climate change effects on bacterioplankton, and the role of r- and K-strategists. We will address the potential of the now increasing data sets on molecular analysis of marine bacterioplankton of the ocean and the potential to deviate models for future climate change scenarios by linking today's environmental parameters with the existing bacterioplankton community. The increasing risk of health relevant bacteria in coastal waters is discussed and its tight link to climate change. The potential of satellite-based monitoring for bacterioplankton and pathogens, the use of retrospect analyses of particle attached bacterioplankton using the CPR archive, and the modelling of bacterioplankton assemblages is perceived as an integrative set of measures that have the potential to support a more thorough understanding of the dynamics of marine bacterioplankton and may allow predictions and prevention measures with respect to climate change impacts.

Planktonic bacteria and biogeochemical cycling of the ocean

Marine planktonic bacteria make major contributions to the biogeochemical cycling in marine ecosystems. Marine bacterioplankton is closely linked to the carbon cycling of the ocean and have thus a tight link to climate change (Herndl and Reinthaler, 2013; Höfle et al., 2008; Jiao et al., 2010). Planktonic bacteria are major drivers of degradation and mineralization of particulate and dissolved organic matter of allochthonous and autochthonous sources. On the other hand, they contribute to carbon sequestration. The main impact on carbon sequestration is their contribution to the production of the largest marine pool of organic carbon, i.e. refractory dissolved organic carbon (rDOC). However, a detailed knowledge on the processes and dynamics of rDOC production is still missing (Hansell, 2013; Herndl and Reinthaler, 2013; Jiao et al., 2010). Another contribution to carbon sequestration is the calcite formation by bacteria that contribute to carbon binding and gravity transfer of marine particulate organic matter to deeper layers (Armstrong et al., 2002).

A second main climate related bacterial activity is the production and consumption of climate related trace gases such as methane, N_2O, and dimethyl sulfide (DMS) compounds. Planktonic bacteria contribute largely to the consumption of methane, N_2O and algae-derived DMS-compounds in the water column (Brettar and Rheinheimer, 1992; Brettar et al., 2006; Howard et al., 2006; Mou et al., 2008; Rees 2012; Ward et al., 1989). While the production of methane is mainly restricted to sediments and particles, this is different for N_2O whose production and consumption can occur by free-living and particle attached planktonic bacteria as well as in sediments. A strong increase of climate related trace gases can be observed especially in coastal environments because trace gas production is closely linked to the rate of primary and secondary production. Therefore, climate change related nutrient transfer after extreme weather events can increase trace gas production substantially. However, in total it is difficult to attribute this production to climate change impacts for areas with high eutrophication background.

Planktonic bacteria are thus not only driven by the impacts of climate change but are also impacting the climate by their activities. Though the development of the earth's atmosphere over the last 3.5 billion years is the most impressive example, the interactions of bacterial activity and climate during the more recent glacial and warming periods indicate this tight interaction and its potential impact for the current global warming (Rees, 2012). A good knowledge of the planktonic bacterial community and its biogeochemical activities is therefore considered as an important issue for a thorough understanding of the ocean's role and interaction in climate change scenarios.

Different impact of climate change effects on coastal and open ocean bacterial communities and regional variability

Coastal and open ocean bacterial communities are affected differently by climate change effects (Duarte et al., 2013). While oligotrophic open ocean environments are mainly affected by temperature and CO_2 increase, the coastal environments are more impacted by the effects of extreme weather events such as storms, heavy rains and floods. In addition, coastal environments are generally more affected by human activity such as eutrophication, making discrimination between climate change effects and general anthropogenic impact more difficult. A brief listing of climate change effects on coastal and open ocean environments is given in Table 2.1.

A comprehensive view on impacts of climate change on oceanic environments according to different scenarios for human CO_2 emission was elaborated by Gruber et al. (2011). In general, it can be considered that surface water bacterial communities are most affected by climate change due to their higher exposition towards changing environmental parameters such as temperature, acidification, UV-regime, stratification and the resulting changed nutrient regime and food chain. Therefore, the main impact on the bacterioplankton in coastal and offshore environments is considered for water masses above the thermocline.

Stratification is considered to affect all oceanic environments but with different effects. At low latitudes, stratification is estimated to reduce primary production (PP) due to more pronounced nutrient limitation, while it is assumed to enhance PP at high latitudes due to lower light limitation of phytoplankton. Acidification is more affecting open ocean than coastal regions (Duarte et al., 2013). In coastal water discharge from rivers and groundwater may compensate for bicarbonate loss due to CO_2 absorption, and eutrophication and the resulting high PP may cause an increase of surface water pH. For the global mean surface ocean and for an increase in atmospheric CO_2 by 100 ppm, pH drops by about 0.07 pH units and the aragonite saturation is lowered by 0.3%. Open ocean surface water is more affected at low to mid latitudes than at the high latitudes; this discrepancy is especially pronounced with respect to future CO_2 scenarios (Grube, 2011).

Deoxygenation is another climate change impact that is considered to play a major role in the future. Deoxygenation is assumed to be triggered by reduced oxygen solubility in water at increased temperature, reduced oxygen transfer due to stratification and increased metabolic rates of oxygen consumption. The main oxygen decrease of the open ocean was observed for the thermocline (200-600 m) of the Eastern boundary upwelling areas. Coastal waters as well as low oxygen water in the thermocline at low latitudes (Eastern Tropical North and South Pacific, Arabian Sea) are considered to be most affected in the future due to critical low oxygen concentrations affecting all biota including bacterioplankton.

Table 2.1 Expected direct and indirect climate change effects on coastal and open ocean environments and their expected impact on planktonic bacterial communities.
Legend: o, neutral; +, relevant

Climate Change effect	Coastal	Open ocean
Temperature increase: increased growth & metabolic turnover rates	+	+ (growth limitation by nutrients, increased starvation & decay)
Increased stratification of surface water	+	+ (increased primary production at high latitudes)
Decreased nutrient transfer from deeper water layers due to enhanced stratification -> **change of bottom up control**	o/+	+ (negative impact on primary production at low latitudes)
Acidification : -> change of carbonate dependent higher trophic levels, less calcite production by bacteria, -> support of cyanobacteria over algae	o/+ (land based carbonate supply)	+
Change of higher trophic levels, changed top down control	+	+
De-oxygenation	+	+ (water masses at thermocline (200-600 m) mostly affected)
Increased impact from extreme weather events	++ (nutrient and suspended solids supply from heavy rains, floods, erosion)	o/+ (lower & occasional impact, e.g. Saharan dust, hurricane)

Difficulties in the assessment of temperature effects on bacterioplankton

Sea surface temperature (SST) has already increased by 0.7°C over the last 100 years, whereas the deep water is hardly affected; the overall temperature increase of the ocean is estimated as less than 0.04°C (Gruber, 2011). Depending on the scenarios, the future SST is expected to increase by 1.4 to 3.1°C by the end of the century. SST warming is considered to mostly affect tropical and Northern high latitudes especially Arctic areas due to disappearance of sea ice. Ocean warming will affect all marine ecosystems and their biogeochemical processes directly due to their temperature dependence. Assuming a Q_{10} factor of 1.5 - 4.0, i.e. a measure of how much a rate increases per 10°C increase, and a predicted SST warming by 2°C, an acceleration of most physiological processes by 7-30% was estimated.

Temperature effects of climate change could be assessed at three major levels of relevance for marine bacteria: i) community, ii) species and iii) strain. The universal Arrhenius equation relates temperature and chemical reactions in an exponential way. This equation allows a precise prediction of the bacterial growth rate as long as the temperature range of a given bacterium is in its optimal range. In this range, increasing temperature result in higher metabolic turnover and growth rate (Mohr and Krawiec, 1980). Even more important is that most physiological processes have Q10 factors in the range from 1.5 to 4. The temperature effects on species and communities are more difficult to predict. Commonly, a bacterial species has a fixed temperature range at which it can operate with some but little variation among individual strains. This results in a certain buffering capacity within a natural bacterial community in terms of temperature effects. If this buffering capacity is exceeded, the species will be replaced by another species better adapted to the respective temperature. The above mentioned processes can be considered as "adaptation" at the strain to the community level; for climate change scenarios, however, evolutionary processes have to be taken additionally into account (Lavergne et al., 2010).

Experimental approaches with marine bacterial communities aiming to elucidate the impact of temperature in the range of the global warming scenarios show unequivocally an increase in metabolic turnover rates (Gruber, 2011; Lopez-Urruttia and Moran, 2006; Nydahl et al., 2013; Vaquer-Sunyer and Duarte, 2013) with an expected higher impact in cold environments and high latitudes. Since the time and space for evolution and adaption is not given in experiments, an extrapolation for the small temperature changes over decades are difficult to be deduced. This is considered valid for the metabolic turnover, but also for temperature induced changes of the bacterial community composition (Lavergne et al., 2010).

With respect to an extrapolation of the measured turnover rates, a second issue has to be addressed: precision of the measurements and representativeness of the analyzed samples. Though many methods for the assessment of bacterial growth and activities are well established, the measurement itself can still be considered to create too much of an artifact in a way that it could be considered critical to extrapolate to large areas and to climate change scenarios (Gasol et al., 2008). As recently demonstrated the microstructure of planktonic samples is very delicate and directly linked to the bacterial growth and activity rates (Stocker, 2012). Sampling (due to the inherent shear stress, and changes of the microstructure of organisms, compounds and partial gas pressures) and especially fractionation of water samples often disrupt the delicate microstructure resulting in changed growth and turnover rates, and are often reflected by a change of the bacterial community composition during follow-up incubation. However, the experimental assessment of in situ rates is in general a difficult task and highly debated within the scientific community (Gasol et al., 2008); thus, it is also affecting the assessment of all other bacterial processes including the ones impacted by climate change.

In addition to the problem to experimentally assess the temperature effect on bacterioplankton under natural conditions, the mere temperature effect is also often masked by additional factors such as nutrient limitations (Lopez-Urruttia and Moran, 2006; Kirchman et al., 2009). However, factors interfering with the mere temperature effect are real and finally highly important for the outcome of global change on bacterioplankton C-turnover in the ocean.

Oxidative stress and its potential role in aquatic environments

Oxidative stress is of relevance in all oxic environments and could play a - so far neglected - role for bacterioplankton due to climate change. Though many studies on the role of oxidative stress have been performed in medical microbiology, little attention has been attributed to oxidative stress in aquatic microbial ecology (Brettar et al., 2012; Hassett and Cohn, 1989; Imlay 2008). Oxygen is an excellent electron acceptor for the use of organic substrates but life in an oxic environment is not free of costs due to the inherent oxidative damages caused by reactive oxygen species (ROS) (Hassett and Cohn, 1989; Imlay, 2008; Touati, 2000). The production of intracellular

ROS is highly dependent on the oxygen content in the environment because oxygen passes easily the cell membrane. In addition, external sources, such as ROS produced by photochemical reactions (UV, visible light), protozoan grazing, virus infection, and competing or predatory bacteria, cause oxidative stress for bacterial cells. These ROS cause a large variety of damages to nucleic acids, lipids and proteins. Based on genome analyses of marine bacteria, detailed insights into the genes enabling bacteria to cope with oxidative stress were gained (Math et al., 2012; Médigue et al., 2005; Wright et al., 2014). To cope with oxidative stress basically three different strategies were observed: i) avoidance of the generation of ROS/RNS (reactive oxygen and nitrogen species), ii) scavenging of ROS/RNS by antioxidants (e.g. pigments), and iii) repair/replacement mechanisms for damaged molecules. For all these mechanisms a broad and diversified set of responses was observed. A set of regulator genes (*OxyR, PerR, SoxR*) is inducible by ROS and known to drive a rapid and comprehensive response to protect the bacterial cell from ROS damage (Imlay 2008). However, for all the measures to cope with oxidative stress, such as antioxidants, detoxifying enzymes and repair systems, the needed energy and nutrients are critical under conditions of nutrient limitations. Considering energetic aspects, most interesting are mechanisms to avoid the generation of ROS/RNS. These mechanisms comprise export of metals (out of the cell) reactive with oxygen and ROS such as Cu and Fe, avoidance of metabolic pathways with high ROS generation such as the molybdoterine pathways (e.g. TMAO reductase, oxido-reductase YedY), and avoidance of highly reactive metal containing enzymes and co-factors (Médigue et al., 2005). In addition, low oxic environments like particles and low oxic water provide sites of reduced oxidative stress (Brettar et al., 2012; Roslev and King, 1995; Roslev et al., 2004). In general, a strong decline of bacterial biomass was observed in many studies under conditions of nutrient limitation and oxic starvation of bacteria (del Giorgio and Cole, 1998; Matin et al., 1989). Based on their studies on oxic starvation, McDougald et al. (2002) assumed that oxidative stress is a phenomenon most relevant to survival of bacteria especially under conditions of nutrient limitation. Oxidative stress could be of special relevance for oligotrophic environments. For starving bacteria not able to compensate the oxidative stress damages by using repair mechanisms a stronger decay due to damages and enhanced virus susceptibility could be assumed. Swan et al. (2013) hypothesized a lack of repair of the genomic DNA as causing the observed low GC content of bacteria representative for marine oligotrophic planktonic environments. Oxidative stress, in turn, could be at least to a certain extent causing these damages to the genomic DNA.

Due to climate change a changed oxidative stress regime for bacterioplankton is conceivable. The increase of sea water temperature and stratification are considered to decrease the oxygen content on the one hand (Gruber, 2011), and thus reduce the oxidative stress; on the other hand, higher metabolic rates due to increased water temperature would increase oxidative stress. In addition, de-oxygenation of oceanic water masses around the thermocline and in coastal environments will reduce oxidative stress and improve bacterial survival under starving conditions at reduced energetic costs as evidenced for the Baltic Sea deep water (Brettar et al., 2012).

Oxidative stress can be considered to increase the maintenance cost for bacterial growth and survival, and thus reduce bacterial growth efficiency. Therefore, a change of the oxidative stress regime has to be taken into account with respect to carbon-allocation in marine ecosystems by bacterioplankton. However, for an assessment of the impact of oxidative stress due to climate change on bacterial growth efficiency, vitality, abundance and C-remineralisation in coastal and open ocean environments, more targeted studies are needed.

Taxonomic levels of changes in bacterial communities in response to climate change

Bacterial community changes due to climate change impacts can be expected at all taxonomic levels from the clone to the phylum level. However, the mere temperature effect might remain with more subtle taxonomic changes, i.e. clonal replacement of a better adapted strain of the same species. In a recent study, based on single amplified genomes (SAGs) of individual bacterial cells representing different climatic regions, Swan et al. (2013) showed that the distinction among the main phylogenetic lineages of marine bacterioplankton has to be substantially lower than the species level, i.e. at a sequence similarity of genomic DNA of 95% or higher. Below 80% genomic DNA sequence similarity, corresponding to about 97% similarity of the SSU rRNA gene, the genotypes within a "species" adapted to the different climate zones could not be distinguished. The authors concluded that the climatic adaptation cannot be distinguished on the widely used "species" level, but needs a higher degree of genetic similarity (best matched at or above 95% genomic DNA similarity) that can be tentatively addressed as subspecies or even clonal level. Based on these observations for adaptations to different climate zones, changes of bacterial communities in response to the mere temperature increase could be restricted to the clonal level thus being not "visible" on the species level.

Clonal richness, changes over time and metabolic versatility

The clonal structure of populations of marine bacteria was mainly studied in isolates due to the better access to strains obtained by classical cultivation than to single cells obtained directly form a specific marine environment. A comprehensive study conducted on the gammaproteobacterium *Shewanella baltica* based on a large set of about 150 strains, showed a highly diverse clonal structure with rapid evolution during the study period of 12 years (Caro-Quintero et al., 2011, 2012; Deng et al., 2014; Höfle et al., 2000; Ziemke et al., 1997). By comparing fully aligned whole genome sequences, horizontal gene transfer (HGT) was shown as an important gene exchange mechanism for evolution within the species, with different gene exchange rates for intracladal and extracladal exchange. The *S. baltica* clades, defined by comparison of whole genome sequences and high resolution molecular genotyping, were assigned different metabolic features that allowed the distinction of different ecotypes that might fill different niches in the ecosystem, due to different metabolic features such as different capabilities with respect to sulphur metabolism and a rather diverse use of a broad set of carbon compounds. Taken together, the genome and the metabolic analysis and the conditions and abundance at the site of isolation, indicate a rather diverse and partially complementary consortium of the *S. baltica* clades with a high gene exchange rate, a broad spectrum of diverse niches and rapid evolutionary changes under the "umbrella" of a single species.

Rapid evolution at the clonal level was also demonstrated in the marine environment as well as in serial transfer experiments for *Vibrio parahaemolyticus* using multi-locus variable number of tandem repeat analysis (MLVA) (Garcia et al., 2012; Hart-Chu et al., 2010). This high rate of microevolution was able to explain the world wide rapid evolution of highly pathogenic strains of *V. parahaemolyticus* and corresponding outbreaks during the last decade in Chile. The rapid evolution of novel ecotypes may allow a rapid adaptation to the changing conditions induced by climate change, especially in less oligotrophic environments with occasional nutrient pulses. Overall, these versatile species, non-pathogenic as well as pathogenic species, are characterized by a large pan-genome and a substantial accessory genome providing high flexibility to react to environmental changes including climate change effects within a rather short evolutionary time frame meaning years to decades. This is in

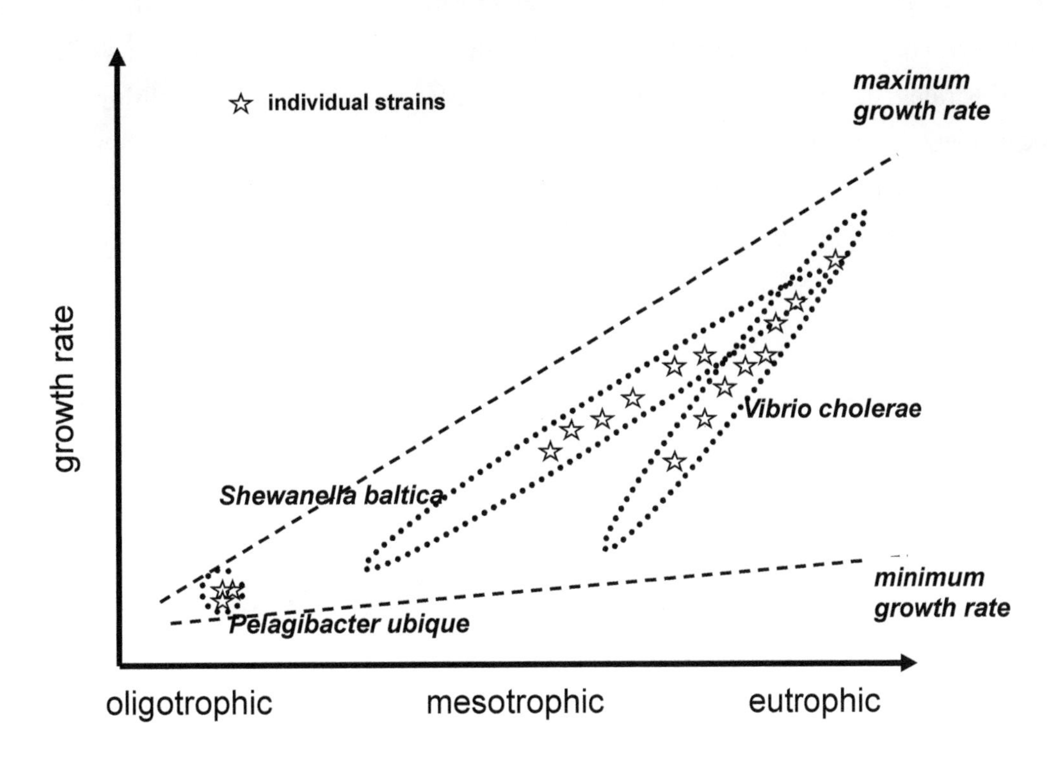

Figure 2.1 Conceptual model of the r- & k-strategist continuum for pelagic marine bacteria with respect to the trophic status of the marine environment and growth rate. Species given are only current examples of marine species better understood in terms of clonal population structure than many other members of the bacterioplankton community. Stars indicate the optimum of the respective strains; the trophy vs. growth rate-spectrum per strain may cover a strain-dependent part of the indicated area of the respective species.

contrast to the small and highly streamlined genomes of oligotrophic bacteria, such as *Pelagibacter* and *Synechococcus* species, providing less metabolic versatility and might need far longer time spans for evolutionary adaptation to climate change effects (Giovannoni et al., 2014).

r- vs. K-strategists - different responses to climate change impacts

The recent understanding of marine bacteria, communities as well as populations, based on genomic and metagenomic data can be transferred in the concept of r- and

K-strategists adapted to bacteria in the 1970s to characterize their physiological functioning and competitive abilities in the marine environment (Andrews and Harris, 1986). r-strategists are bacteria with high growth rates at nutrient rich environments that are well adapted to rapid changes of types and concentrations of nutrients (Egli, 1995). K-strategists are bacteria with low growth rates at low nutrient conditions and are best adapted to stable environmental conditions and correspond to oligothrophs (Egli, 2010). In reality this concept of r- and K-strategists describes a continuum where r- and K-strategist represent the two extremes and each bacterial species can be placed according to its metabolic capabilities. In this continuum we see the r-

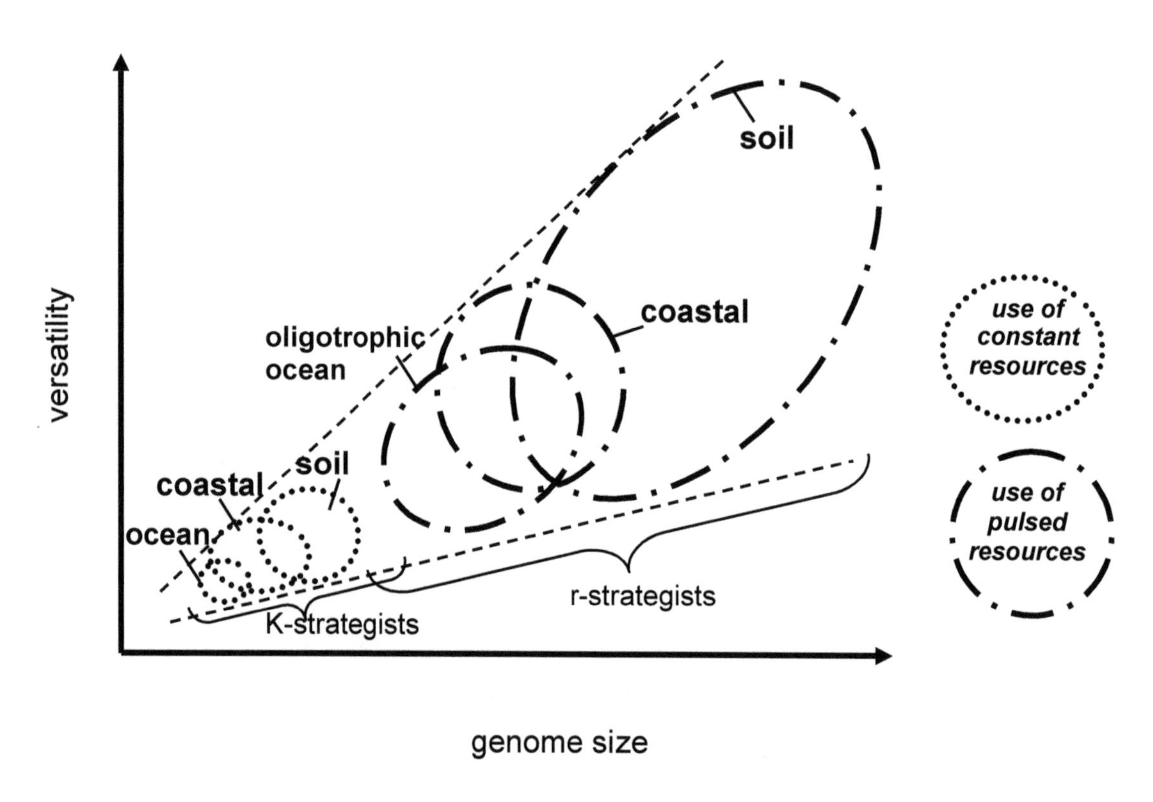

Figure 2.2 Conceptual model of the r- & k-strategist continuum for pelagic marine bacteria with respect to genome size and versatility. As "versatility" all features from metabolism/anabolism to phage, grazing and oxidative stress resistance are considered. As K-strategists we address slow growing bacteria using resources that are available at a relatively constant but low concentration; as r-strategist fast-growing bacteria using mainly resources available at highly variable concentrations. This scheme mainly refers to predominantly heterotrophic bacteria; primarily autotrophic bacteria are considered to escape this scheme (e.g. sulphur-oxidizing *Sulfurimonas* species).

strategists as the more versatile group operating under (temporarily) meso- to eutrophic conditions (Fig. 2.1). The position of an individual bacterial species can be identified, as first approximation, by the absolute size of its genome, its size variation within the species (pan-genome versus core-genome) and the number of gene duplications (abundance of paralog genes) (Swan et al., 2013) (Fig. 2.2). A more complex scheme for the assessment of the trophic strategy of marine bacteria based on genome analysis was suggested by Lauro et al. (2009).

Many studies have demonstrated the global distribution of oligotrophs among marine bacterioplankton as slow growing bacteria with low versatility but with high abundances in pelagic communities (Giovannoni et al., 2014; Swan et al., 2013). Thus, typical K-strategists are in general assumed for oligotrophic off-shore environments but are also considered of relevance for coastal environments (Swan et al., 2013). However, bacteria with a higher metabolic versatility, a larger genome and the capability to grow fast should not be considered as an artifact of cultivation. These r-strategists can be considered to play an important role under conditions of temporarily or spatially increased nutrient concentrations, such as nutrient pulses after a plankton bloom or in the vicinity of a nutrient rich particle. Gradients and microstructure also present in pelagic samples as shown by Stocker (2012) may provide a wealth of habitats for r-strategists - though at a microscale. Of special relevance might be their capability to react to gradients and sudden nutrient pulses, to grow, take up and metabolize rapidly a broad set of compounds. Their role has been nicely demonstrated by experimental metagenome studies (Mou et al., 2008) as well by in situ studies of phytoplankton blooms (Tada et al., 2011; Teeling et al., 2012). These studies demonstrate the rapid growth and organic matter degradation of typical r-strategists, especially *Flavobacteria* and *Gammaproteobacteria* of different genera, e.g. the gammabacterial genera *Alteromonas, Vibrio,* and *Reineckea,* and clade SAR92. Patchiness and the short term growth and decay of populations of r-strategists, may contribute to its underestimation in marine metagenome studies due to their inherent limitation with respect to sample numbers. However, the contribution of r-strategists to carbon and nutrient cycling in coastal environments might be substantial.

The world's ocean can be considered as an n-dimensional space of bacterial niches. Major factors (dimensions) structuring this space with respect to bacterial niches are: i) physical factors such as temperature, light, pH, redox potential, etc.; ii) chemical factors such as inorganic nutrients (N, P, Fe, Co, Mo, etc.), electron donors (glucose, leucine, H_2S, etc.), and electron acceptors (O_2, NO_3, Mn, etc.); and iii) biological factors such as bacterial abundances, microhabitats (particle bound vs. free living), grazing resistance, virus susceptibility, genome size and organization. During the last decade our understanding of the genomes and metagenomes of marine bacterioplankton communities and its individual members has increased enormously by the application of advanced molecular analyses (Glöckner et al., 2012; Rinke et al., 2013; Swan et al., 2013). This advancement has led to a genome enabled biogeochemistry of the world oceans based on a set of essential genes orchestrating the global biogeochemical cycles (Falkowski et al., 2008). About 200 of these essential genes are substantial to run the ocean's biogeochemistry. For a more detailed understanding of the effects of climate on bacterioplankton we have to move to a more detailed model (systems analysis) combining ecological theory (n-dimensional niche space) and single ecotypes as described above. This meta-analysis has to be tested with long-term observations using the available long-term research (LTR) sites and global plankton archives. Overall, we think there is a need for integration of conceptual models with large global data sets on bacterioplankton composition and functioning to enable predictions of climate change effects (Giovannoni and Vergin, 2012; Glöckner et al., 2012; Ladau et al., 2012).

Modelling worldwide distribution patterns of marine bacteria and predictions in response to climate change

Fortunately, large-scale sampling and molecular analysis of marine bacterial biomass in the last decade has enabled a more thorough insight into the world-wide distribution of marine bacteria with most data available for the layer most susceptible to climate change, i.e. the surface water (Amaral-Zettler et al., 2010; Fuhrman et al., 2008; Pommier et al., 2007; Rusch et al., 2007; Schattenhofer et al., 2009; Wietz et al., 2010). Based on such data sets and the respective environmental metadata, statistical modelling approaches were successful in modelling of the worldwide distribution of marine bacterial groups and its relationship with environmental parameters (Ladau et al., 2012; Larsen et al., 2012). Modelling was based on species distribution modelling (SDM), a fundamental tool for predicting the diversity patterns of macroorganisms (Franklin and Miller, 2009). Marine bacteria are considered as being well-suited to SDM due to their strong environmental sorting (Tamames et al., 2010) and their low limitation in dispersal (Caropaso et al., 2012).

Modelling by Ladau et al. (2012) showed the world-wide distribution, diversity and richness of major bacterial groups, and revealed unexpected maxima of bacterial richness in winter at high latitudes. While acidification and sea surface temperature increase did not show a specific impact on bacterial communities, an increase of bacterial diversity was observed for areas of high human impact on marine ecosystems (Halpern et al., 2008), both for coastal and open ocean environments. By Ladau et al. (2012)'s comprehensive analysis of relationships with environmental factors, a strong positive correlation was revealed for bacterial diversity and richness with P-availability and distance from the thermocline. Their findings on the impact of human disturbed areas may indicate that the "direct" human disturbance factor including eutrophication largely interferes with the detection of mere global warming impacts on the one hand. On the other hand, the findings that P-availability and distance from the thermocline correlate with increased richness may indicate that climate change-induced increased stratification will affect bacterial communities towards a reduced diversity.

In addition, the modelling of worldwide bacterial distribution revealed biases introduced by sampling and recommended season-independent sampling at higher latitudes. Though the modelling and predictive power can still be considered very limited, future input by the scientific community may contribute to a more thorough understanding in the near future. For this purpose, a website was established by Ladau et al. (2012) (http://doc.pollard.org/marine_diversity) as a publicly available interactive tool to be used for hypothesis testing and future data input. Future large scale modelling using an artificial neural network approach - so far applied only at a regional scale - is promising to achieve even better insight into the relationship of marine bacterial communities and changes triggered by climate change (Larsen et al., 2012).

Long-term observation of planktonic bacterial communities

Reliable data and sample sets are needed covering the period of significant increase of the sea surface temperature of the respective marine ecosystem to make science based predictions on climate change effects on marine bacterioplankton communities and their ecological functioning (Glöckner et al., 2012). Ideally, the last one to two hundred years would be an appropriate time span for such predictions. Unfortunately, true long-term observations of planktonic bacterial communities do not exist. However, with the recent development of molecular biology techniques the retrospective study of microbial populations is becoming possible by the analysis of historical (archived) biological samples. For instance, some insight into the planktonic bacteria of coastal environments has been obtained by the analysis of formalin-fixed plankton samples collected in the last 50 years by the Continuous Plankton Recorder (CPR) survey (http://www.sahfos.ac.uk/sahfos-home.aspx). The CPR is a high-speed plankton sampler designed to be towed from ships of opportunity over long distances (Reid et al., 2003). Sampling takes place in the surface layer (about 7 metres) and plankton is collected on a band of silk (mesh-size 270 μm) that moves across the sampling aperture at a rate proportional to the speed of the towing ship. The CPR technology has been used since 1931 for plankton research on a global scale and this methodology was recently applied and successfully tested for the first time in a large scale study of the ecology and molecular epidemiology of vibrios in coastal water (Vezzulli et al., 2012). In their study Vezzulli et al. (2012) exploited the well known association between vibrios and plankton to assess possible linkage between *Vibrio* occurrence in the sea and environmental variables during the last 50 years. Using advanced molecular techniques to analyze the samples, they observed that the genus *Vibrio*, including the human pathogen *Vibrio cholerae*, has increased in prevalence in the last 44 years in the coastal Southern North Sea and that this increase is correlated significantly, during the same period, with warming sea surface temperature. Fig. 2.3 summarizes the findings of an increasing fraction of *Gammaproteobacteriaceae, Vibrionaceaea* and *V. cholerae* as analyzed by pyrosequencing, and furthermore supported by real-time PCR. These findings are in agreement with laboratory studies showing that increasing temperature (up to 30°C) has a pronounced effect on the multiplication of *Vibrio* species in the aquatic environment including those pathogenic for humans (Huq et al., 1984; Vezzulli et al., 2013). These data are also consistent with the current observations of increased *Vibrio*-related human diseases in this region (Baker-Austin et al., 2012) and provide support for the view that global warming is having a strong impact on the composition of marine prokaryotic communities, with potential important implications for animal and human health. Retrospective analysis such as the one described above may be possibly applied in the future on a wider range of historical archived samples including, if available, those collected by long-term ecological surveys at sites such as LTER (http://www.lternet.edu/), HOTS (http://

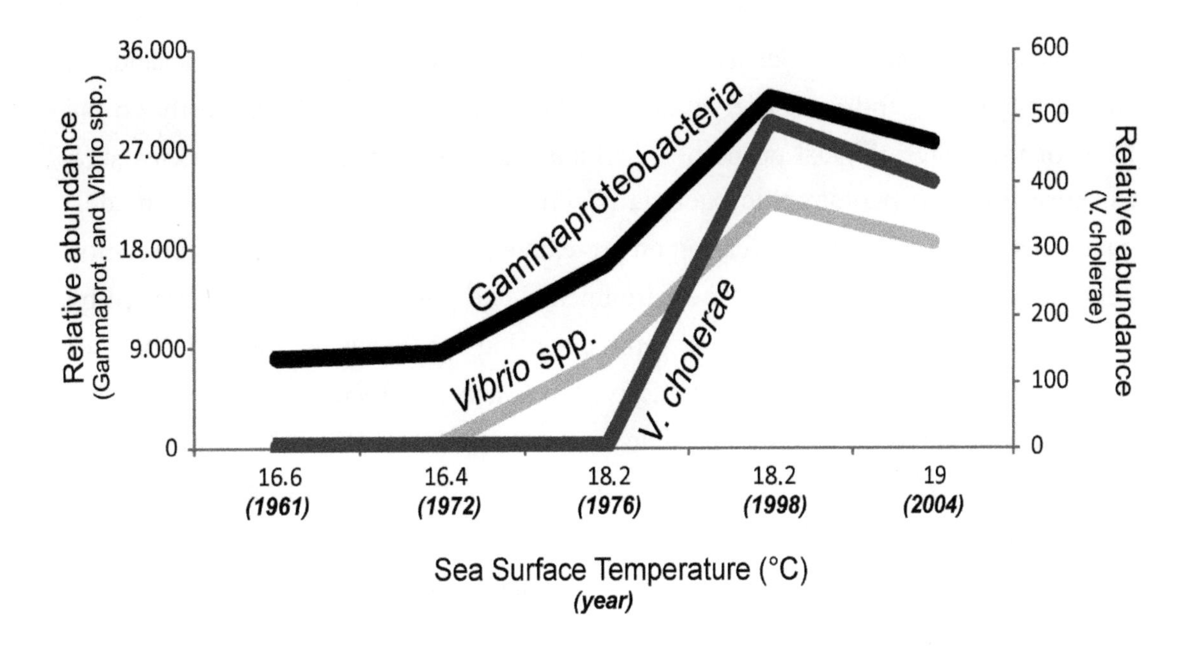

Figure 2.3 Long-term relationship between Sea Surface Temperature (at sampling time in August) and relative abundance (numbers of normalised reads) of *Gammaproteobacteria, Vibrio spp.* and *V. cholerae* calculated by 16S rDNA pyrosequencing analysis of CPR samples collected in 1961, 1972, 1976, 1998 and 2004 off the Rhine Estuary (total normalized read numbers per sample were 100.000 reads; the relative abundance was thus ranging between aprox. 0 to 33% for *Gammaproteobacteriaceae* and *Vibrionaceae*, and between 0 to 0.5 permil for *V. cholerae*) (Modified from Vezzulli et al., 2012).

hahana.soest.hawaii.edu/hot), BATS (http://bats.bios.edu) and DYFAMED (http://www.obs-vlfr.fr/dyfBase).

Large-scale satellite based observation of planktonic bacterial communities

Satellite-based remote sensing of biota has wide application to study large scale distribution of microorganisms in the ocean, especially in the light of ongoing climate change, also providing a valuable tool in predicting human health risks associated with pathogenic microorganisms (Grimes et al., 2014; Lleo et al., 2008). Early applications in this direction were focused on harmful algal blooms, but more recently methods have been developed to interrogate the ocean for abundances of specific bacteria. The modes of satellite observation can be classified into direct, indirect, and inference or model based. Direct observation identify bacteria exclusively through their optical properties; mapping of bacterial mass (i.e. *Cyanobacteria*) that can cover large areas of the oceans in the range of thousands of square kilometres. Indirect observations are those identifying bacteria mostly on the basis of their physiological properties, including optical properties of pigments such as those found in phototrophic bacteria (Grimes et al., 2014). Inference or model-based observations are finally relying on pre-existing knowledge about the correlation between target bacteria counts and environmental and physical variables measured by satellites. An example of such a remote-sensing approach is the detection of *Enterococcus* where a correlation is derived between *Enterococcus spp.* counts and particle concentration and the particle concentration is estimated from satellite measurements (Zhang et al., 2010). Among the main applications in ocean sciences, model-based satellite observation of marine microorganisms has become a useful tool to develop predictive models for disease surveillance also in consideration of ongoing climate change. Climate change induced effects including flooding, extreme rainfall causing increased runoff and washing of faecal material into water bodies, and warmer water temperatures are associated with increased transmission of waterborne diseases such as cholera, cryptosporidiosis, leptospirosis and the spread of harmful algal blooms (Brettar et al., 2007; Harvell et al., 2002). In the context of waterborne diseases, the possibility that satellites can monitor hydrological and oceanographic parameters is facilitating the study of environmental factors favouring or affecting the emergence, concentration and distribution of different types of aquatic organisms from viruses to bacteria, and algae. A notable example of this has been the applications of satellite-based environmental monitoring to characterize and map environmental variables in the Bay of Bengal that are associated with the temporal patterns of cholera cases. In particular, satellite-based studies and monitoring of key environmental parameters allowed researchers to demonstrate that increases in sea surface temperature and sea surface height preceded cholera outbreaks in Bangladesh and that by monitoring these two parameters from space it is possible to predict epidemics (Jutla et al., 2010). In general, recent advances in our knowledge of pathogens, vectors, reservoirs and host ecology have allowed the assessment of a broader range of environmental factors that promote disease transmission, vector production and the emergence and maintenance of disease foci. Monitoring of such factors on a large scale from space offers the opportunity to predict and thus prevent occurrence of diseases (Grimes at al., 2014; Jutla et al., 2010; Lleo et al., 2008).

Pathogenic vibrios as sentinels of climate change impact on bacterial communities

The record of disease incidence due to pathogenic vibrios, such as *V. cholerae*, *V. parahaemolyticus* and *V. vulnificus,* is a matter of worldwide concern especially in the light of global warming scenarios (Baker-Austin et al., 2010; Martinez-Urtaza et al., 2010). Vibrios are considered as indigenous members of marine bacterial communities and are generally considered as copiotrophs. A common feature among vibrios is the presence of multiple lifestyles: a planktonic, free swimming state and a sessile existence on a surface (Pruzzo et al., 2005). These bacteria are spread in coastal aquatic environments throughout the world, being more common in warmer waters, as shown by a large number of studies (Baffone et al., 2006; Huq et al., 1984). The effects of climate change on *V. cholerae* interaction with plankton (mainly copepods) merit special attention because mesoplankton plays a role not only as environmental reservoir of vibrios but also as a vector for human infections (Colwell, 1996). The chitin exoskeleton of a single colonized copepod has been shown to contain up to 10^4 cells of *V. cholerae* (Colwell, 1996), thus providing the required infectious dose for clinical cholera, ranging from 10^4 to 10^{11} bacteria depending on the strain and the infected host (Nelson et al., 2009). Laboratory experiments showed that temperatures ranging from 25°C to 30°C significantly promoted the attachment of *V. cholerae* to copepods (Huq et al., 1983). This may be due, at least in part, to an enhanced expression, under warmer conditions, of the two main adhesins involved in *V. cholerae* binding to chitin surfaces, namely N-acetylglucosamine-binding protein A (GbpA) and mannose-sensitive haemaglutinine (MSHA) (Stauder et al., 2010, 2012).

Analysis of cholera outbreaks showed that meso-plankton concentration, in addition to temperature and other variables (e.g., salinity), influenced seasonal transmission of cholera in those regions of the world where inhabitants rely on untreated water as a source of drinking water (De Magny and Colwell, 2009). In India during the monsoon, the appearance of cholera cases in coastal human communities correlated with variations in phyto- and zooplankton abundance in the marine environment induced by variations in rainfall (Colwell, 1996; Faruque et al., 2005; Sedas 2007). Cholera dynamics in Peru and the Bay of Bengal seemed to be particularly affected by El Nino Southern Oscillation-driven anomalies (Lipp et al.,

2002; Pascual et al., 2000). Warm waters along the coast, coupled with plankton blooms driven by El Niño rains, showed an increased population of *V. cholerae* and were also associated with increased numbers of diarrheal cases, including cholera (Lipp et al., 2002).

Besides increased SST, eutrophication of coastal waters that is partially increased by climate change effects improves the survival of pathogenic bacteria by i) increasing the availability of organic nutrients, ii) increased particle abundance and iii) increased abundance of phyto- and zooplankton. Therefore, the direct and indirect effects of climate change vs. the general eutrophication on pathogen abundance are difficult to be distinguished and deserve more detailed studies. Pathogenic vibrios are valuable reflectors of the multifaceted effects of climate change and of high health relevance. Therefore, vibrios can be considered as prime targets of detailed studies, e.g. by CPR record analyses as highly valuable approach for retrospective long-term studies. Due to the better attention of the public to pathogenic microorganisms their data base is larger than those for the general marine bacterial community. Taken together with the observations based on CPR by Vezzulli et al. (2012), there incidence could be evaluated and used as an indicator of climate change for marine coastal ecosystems.

Conclusions and future challenges

There is a wealth of knowledge on bacterioplankton assemblages achieved in the last decade with major advances due to single amplified genome (SAG) analyses. Modelling based on the available data is promising in order to increase our understanding with respect to environmental drivers of bacterioplankton assemblages that are highly needed to cope with climate change impacts. Complementation of the bacterioplankton databases and improved modelling tools will greatly contribute to deepen our understanding in the future.

There is still a major lack of understanding in order to link bacterioplankton composition to the biogeochemical processes and more over to deviate the turn over rates under given environmental conditions. Metagenome libraries have built a link to the biogeochemical potential, but do not allow deviation of the turnover rates that are the real important issues for the driver/reactor role of bacterioplankton vs. climate change.

There are still a lot of knowledge gaps on the effects of climate change on the level of bacterial growth and physiology. Oxidative stress response could be highly important for the bacterial growth efficiency. Though a wealth of different genes are known from marine bacterial genomes and metagenome data sets and different mitigation strategies could be assumed, it is still a large and unknown field in bacterioplankton ecology. Targeted studies could help to elucidate the effect of oxidative stress on marine bacteria and, more specifically, its effects due to climate change on the overall bacterial communities and their carbon allocation.

The analysis of the Continuous Plankton Recorder (CPR) archive is unique in a way that it is an important access to bacterioplankton communities of a climate-wise highly relevant past period of global warming. It may teach us about resilience of bacterioplankton communities and the future potential of *Vibrio* communities under climate change effects.

Satellite based analyses of bacterioplankton are promising for future observation and may contribute to monitoring, prevention and also to a better understanding of bacterioplankton and pathogens under climate-induced pressures.

Pathogens such as vibrios are part of the bacterioplankton and could be used as sentinels for bacterioplankton changes due to global warming. In addition, with help of a more thorough understanding of *Vibrio* dynamics with respect to climate change impacts a better prevention could be achieved. Thus, improved study and monitoring systems for vibrios could serve as health prevention and indicate climate-change related impacts on bacterioplankton.

Taken all together, increasing data bases on bacterioplankton biogeography, improved modelling with respect to environmental and climate change-relevant parameters, lessons learnt from CPR-archive analyses, and the use of pathogens as sentinels by using early warning systems, all this may lead to an improved understanding of bacterioplankton and its reaction potential towards climate change in the near future. This may allow improved prevention measures and help to overcome and, hypothetically, to avoid the most negative effects of climate change on marine ecosystems.

References

Amaral-Zettler, L., Artigas, L.F., Baross, J., Bharati, L., Boetius, A., Chandramohan, D., Herndl, G., Kogure, K., Neal, P., Pedros-Alios, C., et al. (2010). A global census of marine microbes, In Life in the World's Oceans, Diversity, Distribution and Abundance. A.D. McIntyre, ed. (Oxford, UK: Blackwell Publishing), pp. 223-245.

Andrews, J.H., and Harris, R.F. (1986). r- and k-selection and microbial ecology. Adv. Microb. Ecol. *9*, 99-147.

Armstrong, R.A., Lee, C., Hedges, J., Honjo, S., and Wakeham, S.G. (2002).A new, mechanistic model for organic carbon fluxes in the ocean based on the quantitative association of POC with ballast minerals. Deep Sea Res. II *49*, 219-236.

Baffone, W., Tarsi, R., Pane, L., Campana, R., Repetto, B., Mariottini, G.L., and Pruzzo C. (2006). Detection of free-living and plankton-bound vibrios in coastal waters of the Adriatic Sea (Italy) and study of their pathogenicity-associated properties. Environ. Microbiol. *28*, 1299-1305.

Baker-Austin, C., Stockley, L., Rangdale, R., and Martinez-Urtaza, J. (2010). Environmental occurrence and clinical impact of *Vibrio vulnificus* and *Vibrio parahaemolyticus*, a European perspective. Environ. Microbiol. Rep. *2*, 7–18.

Baker-Austin, C., Trinanes, J.A., Taylor, N.G.H., Hartnell, R., Siitonen, A., and Martinez-Urtaza, J. (2012). Emerging *Vibrio* risk at high latitudes in response to ocean warming. *3*, 73-77. DOI: 10.1038/nclimate1628

Beaugrand, G., Reid, P.C., Ibañez, F., Alistair Lindley J., and Edwards, M. (2002). Reorganization of North Atlantic marine copepod biodiversity and climate. Science *296*, 1692-1694.

Brettar, I., and Rheinheimer, G. (1992). Influence of carbon availability on denitrification in the central Baltic. Limnol. Oceanogr. *37*, 1146-1163.

Brettar, I., Labrenz, M., Flavier, S., Bötel, J., Kuosa, H., Christen, R., and Höfle, M.G. (2006). Identification of a *Thiomicrospira denitrificans*-like epsilonproteobacterium as a catalyst for autotrophic denitrification in the central Baltic Sea. Appl. Environ. Microbiol. *72*, 1364-1372.

Brettar, I., Guzman, C., and Höfle, M.G. (2007). Human pathogenic micro-organisms in the marine environment - an ecological perspective. CIESM Reports No. 31: Marine Sciences and public health - some major issues. Geneva, September 2006, p. 59-68.

Brettar, I., Christen, R., and Höfle, M.G. (2012). Analysis of bacterial core communities in the central Baltic by comparative RNA-DNA-based fingerprinting provides links to structure-function relationships. ISME J. *6*, 195-212. DOI: 10.10038/ismej.2011.80

Caporaso, J.G., Paszkiewicz, K., Field, D., Knight, R., and Gilbert, J.A. (2012). The western English channel contains a persistent microbial seed bank. ISME J. *6*, 1089–1093.

Caro-Quintero, A., Deng, J., Auchtung, J., Klappenbach, J., Brettar, I., Höfle, M.G., and Konstantinidis, K.T. (2011). Unprecedented levels of horizontal gene transfer among spatially co-occurring *Shewanella* bacteria from the Baltic Sea. ISME J. *5*,131-140. DOI: 10.1038/ismej.2010.93

Caro-Quintero, A., Auchtung, J., Deng, J., Brettar, I., Höfle, M.G., Tiedje, J.M., and Konstantinidis, K.T. (2012). Genome sequencing of five *Shewanella baltica* strains recovered from the oxic-anoxic interface of the Baltic Sea. J. Bacteriology *194*, 1236.

Colwell, R.R. (1996). Global climate and infectious disease, the cholera paradigm. Science. *274*, 2025-2031.

Deng, J., Brettar, I., Luo, C., Auchtung, J., Konstantinidis, K.T., Rodrigues, J.L.M., Höfle, M.G., and Tiedje, J.M.. (2014). Stability, genotypic and phenotypic diversity of *Shewanella baltica* in the redox transition zone of the Baltic Sea. Environ. Microbiol. *16*, 1854-1866. DOI: 10.1111/1462-2920.12344

Deser, C., Alexander, M.A., Xie, S.P., and Phillips, A.S. (2010). Sea surface temperature variability, patterns and mechanisms. Ann. Rev. Mar. Sci. *2*, 115-143.

Duarte, C.M., Hendriks, I.E., Moore, T.S., et al. (2013). Is ocean acidification an open-ocean syndrome? Understanding anthropogenic impacts on seawater pH. Estuaries Coasts *36*, 221–236.

Egli, T. (1995). The ecology and physiological significance of the growth of heterotrophic microorganisms with mixtures of substrates. Adv. Microb. Ecol. *14*, 305-386.

Egli, T. (2010). How to live at low substrate concentration. Water Res *44*, 4826-4837.

Falkowski, P.G., Fenchel, T., and Delong, E.F. (2008). The microbial engines that drive Earth's biogeochemical cycles. Science *320*, 1034-1039.

Faruque, S.M., Naser, I.B., Islam, M.J., Faruque, A.S., Ghosh, A.N., Nair, G.B., Sack, D.A., and Mekalanos, J.J. (2005). Seasonal epidemics of cholera inversely correlate with the prevalence of environmental cholera phages. Proc. Natl. Acad. Sci. USA *102*, 1702–1707.

Franklin, J., and Miller, J.A. (2009). Mapping species distributions, spatial inference and prediction (Cambridge, UK: Cambridge University Press).

Fuhrman, J.A., Steele, J.A., Hewson, I., Schwalbach, M.S., Brown, M.V., Green, J.L., and Brown, J.H., (2008). A latitudinal diversity gradient in planktonic marine bacteria. Proc. Natl. Acad. Sci. USA *105*, 7774–7778.

García, K., Gavilán, R.G., Höfle, M.G., Martínez-Urtaza, J., and Espejo R.T. (2012). Microevolution of pandemic *V. parahaemolyticus* assessed by the number of repeat units in short sequence tandem repeats regions. PLOS ONE *7*, e30873.

Gasol, J.M., Pinhassi, J., Alonso-Sáez, L., Ducklow, H., Herndl, G.J., Koblizek, M., Labrenz, M., Luo, Y., Morán, X.A.G., Reinthaler, T., et al. (2008). Towards a better understanding of microbial carbon flux in the sea. Aquat. Microb. Ecol. *53*, 21-38.

del Giorgio, P.A., and Cole, J.J. (1998). Bacterial growth efficiency in natural aquatic systems. Ann. Rev. Ecol. Syst. *29*, 503-541.

Giovannoni, S.J., and Vergin, K.L. (2012). Seasonality in ocean microbial communities. Science *335*, 671–676.

Giovannoni, S.J., Thrash, J.C., and Temperton, B. (2014). Implications of streamlining theory for microbial ecology. ISME J. *8*, 1553–1565. DOI: 10.1038/ismej.2014.60

Glöckner, F.O., Stal, L.J., Sandaa, R-A., Gasol, J.M., O'Gara, F., Hernandez, F., Labrenz, M., Stoica, E., Varela Rozados, M., Bordalo, A., et al. (2012). Marine microbial diversity and its role in ecosystem functioning and environmental change. Marine Board Position Paper 17. J.B. Calewaert and N. McDonough, eds. (Ostend, Belgium: Marine Board, European Science Foundation).

Grimes, D.J., Ford, T.E., Colwell, R.R., Austin, C.B., Martinez-Urtaza, J., Subramaniam, A., and Capone, D.G. (2014). Viewing marine bacteria, their activity and response to environmental drivers from orbit satellite remote sensing of bacteria. Microb. Ecol. *67*, 489-500.

Gruber, N. (2011). Warming up, turning sour, losing breath, ocean biogeochemistry under global change. Philos. Trans. A Math. Phys. Eng. Sci. *369*, 1980-1996.

Hansell, D. (2013). Recalcitrant dissolved organic carbon fractions. Ann. Rev. Sci. *5*, 421-45.

Halpern, B.S., Walbridge, S., Selkoe, K.A., Kappel, C.V., Micheli, F., D'Agrosa, C., Bruno, J.F., Casey, K.S., Ebert, C., Fox, H.E., et al. (2008). A global map of human impact on marine ecosystems. Science *319*, 948–952.

Hassett, D.J., and Cohen, M.S. (1989). Bacterial adaptation to oxidative stress, implications for pathogenesis and interaction with phagocytic cells. FASEB J. *3*, 2574-2582.

Harth-Chu, E., Especio, R.T., Christen, R., Guzman, C.A., and Höfle, M.G. (2009). Multiple-locus variable-number of tandem-repeats analysis for clonal identification of *Vibrio parahaemolyticus* isolates using capillary electrophoresis. Appl. Environ. Microbiol. *75*, 4079-4088.

Harvell, C.D., Mitchell, C.E., Ward, J.R., Altizer, S., Dobson, A.P., et al. (2002). Ecology -climate warming and disease risks for terrestrial and marine biota. Science *296*, 2158–2162.

Herndl, G.J., and Reinthaler, T. (2013). Microbial control of the dark end of the biological pump. Nature Geoscience *6*, 718-724.

Höfle, M.G., Ziemke, F., and Brettar, I. (2000). Niche differentiation of *Shewanella putrefaciens* populations from the Baltic as revealed by molecular and metabolic fingerprinting. In Microbial Biosystems, New Frontiers, C.R. Bell, M. Brylinsky and P. Johnson-Green, eds, Proceedings of ISME8, Halifax, Canada, pp. 135-143.

Höfle, M.G., Kirchman, D.L., Christen, R., and Brettar, I. (2008). Molecular diversity of bacterioplankton - link to a predictive biogeochemistry of pelagic ecosystems? Aquat. Microb. Ecol. *53*, 39-58.

Howard, E.C., Henriksen, J.R., Buchan, A., Reisch, C.R., Bürgmann, H., Welsh, R., Ye, W., González, J.M., Mace, K., Joye, S.B., et al. (2006). Bacterial taxa that limit sulfur flux from the ocean. Science *314*, 649-651.

Huq, A, Small, E.B., West, P.A., Huq, M.I., Rahman, R., and Colwell, R.R. (1983). Ecological relationship between *Vibrio cholerae* and planktonic copepods. Appl. Environ. Microbiol. *45*, 275–283.

Huq, A., West, P.A., Small, E.B., Huq, M.I., and Colwell, R.R. (1984). Influence of water temperature, salinity, and pH on survival and growth of toxigenic *Vibrio cholerae* serovar 01 associated with live copepods in laboratory microcosms. Appl. Environ. Microbiol. *48*, 420–424.

Imlay, JA. (2008). Cellular defenses against superoxide and hydrogen peroxide. Ann. Rev. Biochem. *77*, 755-776.

Jeppesen, E., Kronvang, B., Meerhoff, M., Søndergaard, M., et al. (2009). Climate change effects on runoff , catchment phosphorus loading and lake ecological state, and potential adaptations. J. Environ. Qual. *38*, 1930–1941.

Jiao, N., Herndl, G.J., Hansell, D.A., Benner, R., Kattner, G., Wilhelm, S.W., Kirchman, D.L., Weinbauer, M.G., Luo, T., Chen, F., et al. (2010). Microbial production of recalcitrant dissolved organic matter, long-term carbon storage in the global ocean. Nat. Rev. Microbiol. *8*, 593–599.

Jickells, T.D., An, Z.S., Andersen, K.K., Baker, A.R., et al. (2005). Global iron connections between desert dust, ocean biogeochemistry, and climate. Science *308*, 67–71.

Jutla, A.S., Akanda, A.S., and Islam, S. (2010). Tracking Cholera in Coastal Regions using Satellite. J. Am. Water Resour. Assoc. *464*, 651–662.

Kernan, M., Battarbee, R., and Moss, B., eds. (2009). Changing climate and changing freshwaters: a European perspective. (Oxford, UK: Blackwell).

Kirchman, D.L., Moran, X.A.G., and Ducklow, H. (2009). Microbial growth in the polar oceans - role of temperature and potential impact of climate change. Nature Rev. Microbiol. *7*, 451-459.

Krause, E., Wichels, A., Gimenez, L., Lumau, M,. Schilhabel, M.B., and Gerdts, G. (2012). Small changes in pH have direct effects on marine bacterial community composition, a microcoam approach. PLOS ONE *7*, e47035.

Ladau, J., Sharpton, T.J., Finucane, M.M., Jospin, G., et al. (2013). Global marine diversity peaks at high latitudes in winter. ISME J. *7*, 1669-1677.

Larsen, P.E., Field, D., and Glibert, J.A. (2012). Predicting bacterial community assemblages using an artificial neural network approach. Nature Methods *9*, 621-625.

Lauro, F.M., McDougald, D., Thomas, T., Williams, T.J., et al. (2009). The genomic basis of trophic strategy in marine bacteria. Proc. Natl. Acad. Sci. U.S.A. *106*, 15527–15533.

Lavergne, S., Mouquet, N., Thuiller, W., and Ronce, O. (2010). Biodiversity and climate change, Integrating evolutionary and ecological responses of species and communities. Ann. Rev. Ecol. Syst. *41*, 321-350.

Lewandowska, A.M., Boyce, D.G., Hofmann, M., Matthiessen, B., Sommer, U., and Worm, B. (2014). Effects of sea surface warming on marine plankton. Ecol. Lett. *17*, 614–623.

Lipp, E.K., Huq, A., and Colwell, RR. (2002). Effects of global climate on infectious disease, the cholera model. Clin. Microbiol. Rev. *15*, 757-770.

Lleo, M.M., Lafaye, M., and Guell, A. (2008). Application of space technologies to the surveillance and modelling of waterborne diseases. Curr. Opin. Biotechnol. *19*, 307–312.

López-Urrutia, A., and Morán, X.A.G. (2007). Resource limitation of bacterial production distorts the temperature dependence of oceanic carbon cycling. Ecology *88*, 817–822.

McDougald, D., Gong, L., Srinivasan, S., Hild, E., Thompson, L., Takayama, K., Rice, S.A., and Kjelleberg, S., (2002). Defenses against oxidative stress during starvation in bacteria. Antonie Van Leeuwenhoek *81*, 3–13.

de Magny, G.C., and Colwell, R.R.. (2009). Cholera and climate: a demonstrated relationship. Trans. Am. Clin. Climatol. Assoc. *120*, 119–128.

Maranon, E., Fernandez, A., Mouriño-Carballido, B., and Martínez-García, S., (2010). Degree of oligotrophy controls the response of microbial plankton to Saharan dust. Limnol. Oceanogr. *55*, 2339-2352.

Martinez-Urtaza, J,, Bowers, J.C., Trinanes, J., and DePaola, A. (2010). Climate anomalies and the increasing risk of *Vibrio parahaemolyticus* and *Vibrio vulnificus* illnesses. Food Res. Internat. *43*, 1780–1790.

Math, R.K., Jin, H.M., Kim, J.M., Hahn, Y., Park, W., Madsen, E.L., and Jeon, C.O., (2012). Comparative genomics reveals adaptation by *Alteromonas* sp, SN2 to marine tidal-flat conditions. PLOS ONE *7*, e35784.

Matin, A., Auger, E.A., Blum, P.H., and Schultz, J.E. (1989). Genetic basis of starvation survival in nondifferentiating bacteria. Ann. Rev. Microbiol. *43*, 293–314.

Médique, C., Krin, E., Pascal, G., Barbe, V., Bernsel, A., Bertin, P.N., Cheung, F., Cruveiller, S., D'Amico, S., Duilio, A., et al. (2005). Coping with the cold, the genome of the versatile marine Antarctica bacterium *Pseudoalteromonas haloplanktis* TAC125. Genome Res. *15*, 1325-1335.

The MerMex Group (2011). Review marine ecosystem's response to climatic and anthropogenic forcings in the Mediterranean. Progr. Oceanogr. *91*, 97-166.

Mohr, P.W., and Krawiec, S. (1980). Temperature characteristics and Arrhenius plots for nominal psychrophiles, mesophiles and thermophiles. J. Gen. Microbiol. *121*, 311-317.

Moran, X.A.G., Sebastian, M., Pedros-Alio, C., and Estrada, M. (2006). Response of Southern Ocean phytoplankton and bacterioplankton production to short-term experimental warming. Limnol. Oceanogr. *51*, 1791-1800.

Mou, X., Sun, S., Edwards, R.A., Hodson, R.E., and Moran, M.A. (2008). Bacterial carbon processing by generalist species in the ocean. Nature *451*, 708-711.

Nelson, E.J., Harris, J.B., Morris, J.G.Jr., Calderwood, S.B., and Camilli, A. (2009). Cholera transmission, the host, pathogen and bacteriophage dynamic. Nat. Rev. Microbiol. *7*, 693-702.

Nydahl, A., Panigrahi, S., and Wikner, J. (2013). Increased microbial activity in a warmer and wetter climate enhances the risk of coastal hypoxia. FEMS Microbiol. Ecol. *85*, 338-347.

Pascual, M., Rodo, X., Ellner, S.P., Colwell, R.R., and Bouma, M.J. (2000). Cholera dynamics and El Niño-southern oscillation. Science *289*, 1766-1769.

Pommier, T., Canback, B., Riemann, L., Boström, K.H., Simu, K., Lundberg, P., Tunlid, A., and Hagström, A., (2007). Global patterns of diversity and community structure in marine bacterioplankton. Mol. Ecol. *16***,** 867–880.

Pruzzo, C., Huq, A., Colwell, R.R., and Donelli, G. (2005). Pathogenic *Vibrio* species in marine and estuarine environment. In Oceans and Health: Pathogens in the Marine Environment, R. Colwell and S. Belkin, eds. (Dordrecht, The Netherlands: Kluwer Academic/Plenum Publishers), pp. 217-252.

Rees, A.P. (2012). Pressures on the marine environment and the changing climate on ocean biogeochemistry. Philos. Trans. A Math. Phys. Eng. Sci. *370*, 5613-5635.

Reid, P.C., Colebrook, J.M., Matthews, J.B.L., and Aiken, J. (2003). The Continuous Plankton Recorder, concepts and history, from Plankton Indicator to undulating recorders. Progr. Oceanogr. *58*, 117–173.

Reid, P.C., Gorick, G., and Edwards, M. (2011). Climate change and European Marine Ecosystem Research (Plymouth, UK: Sir Alister Hardy Foundation for Ocean Science).

Rinke, C., Schwientek, P., Scyrba, A., Ivanova, N.N., Anderson, I.J., Cheng, J.F., Darling, A., Malfatti, S., Swan, B.K., Gies, E.A., et al. (2013). Insights into the phylogeny and coding potential of microbial dark matter. Nature *499*, 431-437.

Roslev, P., Bjergbaek, L., and Hesselsoe, M. (2004). Effect of oxygen on survival of faecal pollution indicators in drinking water. J. Appl. Microbiol. *96*, 938-945.

Roslev, P., and King, G.M., (1995). Aerobic and anaerobic starvation metabolism in methanotrophic bacteria. Appl. Environ. Microbiol. *61*, 1563-1570.

Rusch, D.B., Halpern, A.L., Sutton, G., Heidelberg, K.B., Williamson, S., et al. (2007). The Sorcerer II global ocean sampling expedition, northwest Atlantic through eastern tropical pacific. PLOS Biol. *5*, e77.

Sarmento, H., Montoya, J.M., Vazquez-Dominguez, E., Vaque, D., and Gasol, J.M. (2010). Warming effects on marine microbial food web processes, how far can we go when it comes to predictions? Philos. Trans. R. Soc. Lond. B Biol. Sci. *365*, 2137-2149.

Schattenhofer, M., Fuchs, B.M., Amann, R., Zubkov, M.V., et al. (2009). Latitudinal distribution of prokaryotic picoplankton populations in the Atlantic Ocean. Environ. Microbiol. *11*, 2078–2093.

Schlosser, C., Klar, J. K., Wake, B.D., Snow, J.T., Honey, D.J., Woodward, E.M.S., Lohan, M.C., Achterberg, E.P., and Moore, C.M. (2014). Seasonal ITCZ migration dynamically controls the location of the subtropical Atlantic biogeochemical divide. Proc. Natl. Acad. Sci. U.S.A. *111*, 1438-1442.

Sedas, V.T. (2007). Influence of environmental factors on the presence of *Vibrio cholerae* in the marine environment, a climate link. J. Infect. Dev. Ctries. *1*, 224-241.

Stauder, M., Huq, A,, Pezzati, E., Grim, C.J., Ramoino, P., Pane, L., Colwell, R.R., Pruzzo, C., and Vezzulli, L. (2012). Role of GbpA protein, an important virulence-related colonization factor, for *Vibrio cholerae*'s survival in the aquatic environment. Environ. Microbiol. Rep. ***4*,** 439-445.

Stauder, M., Vezzulli, L., Pezzati, E., Repetto, B., and Pruzzo, C. (2010). Temperature affects *Vibrio cholerae* O1 El Tor persistence in the aquatic environment via an enhanced expression of GbpA and MSHA adhesins. Env. Microbiol. Rep. *2*, 140–144.

Stoecker, R. (2012). Marine microbes see a sea of gradients. Science *338*, 628.

Swan, B.K., Tupper, B., Sczyrba, A., Lauro, F.M., Martinez-Garcia, M., Gonzalez, J.M., Luo, H., Wright, J.J., Landry, Z.C., Hanson, N.W., et al. (2013). Prevalent genome streamlining and latitudinal divergence of planktonic bacteria in the surface Ocean. Proc. Natl. Acad. Sci. U.S.A. *28*, 11483.

Tada, Y., Taniguchi, A., Nagao, I., Miki, T., Uematsu, M., Tsuda, A., and Hamasaki, K. (2011). Differing growth responses of major phylogenetic groups of marine bacteria to natural phytoplankton blooms in the western North Pacific Ocean. Appl. Environ. Microbiol. *77*, 4055–4065.

Tamames, T., Abellan, J.J., Pignatelli, M., Camacho, A., and Moya, A. (2010). Environmental distribution of prokaryotic taxa. BMC Microbiol. *10*, 85.

Teeling, H., Fuchs, B.M., Becher, D., Klockow, C., Gardebrecht, A., Bennke, C.M., Kassabgy, M., Huang, S., Mann, A.J., Waldmann, J., et al. (2012). Substrate-controlled succession of marine bacterioplankton populations. Science *336*, 608-611.

Tittensor, D.P., Mora, C., Jetz, W., Lotze, H.K., Ricard, D., Berghe, E.V., and Worm, B. (2010). Global patterns and predictors of marine biodiversity across taxa. Nature *466*, 1098–1101.

Touati D. (2000). Iron and oxidative stress in bacteria. Arch. Biochem. Biophys. *373*, 1-6.

Vaquer-Sunyer, R., and Duarte, C. (2013). Experimental evaluation of the response of coastal Mediterranean planktonic and benthic metabolism to warming. Estuaries Coasts *36*, 697-707.

Vezzulli, L., Brettar, I., Pezzati, E., Reid, P.C., Colwell, R.R., Höfle, M.G., and Pruzzo, C. (2012). Long-term effects of ocean warming on the prokaryotic community, evidence from the vibrios. ISME J. *6*, 21-30. DOI: 10.10038/ismej.2011.89

Vezzulli, L., Colwell, R.R., and Pruzzo, C. (2013). Ocean warming and spread of pathogenic vibrios in the aquatic environment. Microb. Ecol. *654*, 817-825.

Voigt, W., Perner, J., Davis, A.J., Eggers, T., Schumacher, J., Bährmann, R., Fabian, B., Heinrich, W., Köhler, G., Lichter, D., et al. (2003). Trophic levels are differentially sensitive to climate. Ecology *84*, 2444–2453.

Wietz, M., Gram, L., Jorgensen, B., and Schramm, A. (2010). Latitudinal patterns in the abundance of major marine bacterioplankton groups. Aquat. Microb. Ecol. *61*, 179–189.

White, P.A., Kalff, J., Rasmusssen, J.B., and Gasol, J.M. (1991). The effect of temperature and algal biomass on bacterial production and specific growth rates in freshwater and marine habitats. Microb. Ecol. *21*, 99-118.

Ward, B.B., Kilpatrick, K.A., Wopat, A.E., Minnich, E.C., and Lindstrom, M.E. (1989). Methane oxidation in Saanich Inlet. Cont. Shelf Res. *9*, 65–75.

Wright, J.J., Mewis, K., Hanson, N.W., Konwar, K.M., Maas, K.R., and Hallam, S.J. (2014). Genomic properties of Marine Group A bacteria indicate a role in the marine sulfur cycle. ISME J. *8*, 455-468.

Zhang, Z., Deng, Z-Q., Rusch, K., Gutierrez, W.M., and Chenier, K. (2010). Remote sensing algorithms for estimating enterococcus concentration in coastal Louisiana Beaches, The 5th International Conference on Environmental Science and Technology, Houston, Texas.

Ziemke, F., Brettar, I., and Höfle, M.G. (1997). Stability and diversity of the genetic structure of a *Shewanella putrefaciens* population in the water column of the central Baltic. Aquat. Microb. Ecol. *13*, 63-74.

Chapter 3

Climate Change, Microbial Communities and Agriculture in Semiarid and Arid Ecosystems

Felipe Bastida[1,*], Alfonso Vera[1], Marta Díaz, Carlos García, Antonio Ruíz-Navarro and José Luis Moreno

CEBAS-CSIC, Department of Soil and Water Conservation, Campus Universitario de Espinardo, Murcia, Spain; * corresponding author

[1] these authors contributed equally

Email: fbastida@cebas.csic.es, avera@cebas.csic.es, mdiaz@cebas.csic.es, cgarizq@cebas.csic.es, ruiznavarro@cebas.csic.es, jlmoreno@cebas.csic.es

DOI: https://doi.org/10.21775/9781913652579.03

Abstract

Soil microbial communities perform critical functions for the maintenance of ecosystem services and planet's sustainability. Climate change strongly impacts soil microbial diversity and their ecosystem functions due to changes in the physical and chemical environment in soil. In arid and semiarid ecosystems, which are commonly water and nutrient limited, agricultural activities may enhance the effects of climate change. Thus, the maintenance of agricultural productivity under arid and semiarid climate often requires the application of nutrients in the form of organic amendments and fertilizers, as well as the utilization of alternative water sources for irrigation. In some cases, these practices can be accompanied by contamination of soils with heavy metals and pesticides. Overall, these practices have strong impacts in the composition, biomass and activity of the soil microbial communities. However, the knowledge on the effects of desalinated seawaters and wastewaters in the soil microbial community is still limited. Future studies should focus on the impacts of emerging contaminants in the soil microbial community and its mediated functions.

Introduction

In this chapter, the aim is to provide an overview of the effects of climate change adaptations in semiarid and arid agricultural systems and in soil microbes. However, before doing so, it is important to provide a brief overview on the impacts of climate change. One of the most notable impacts in soil microbial communities is related to the alteration of carbon turnover and the capacity of soils to act as climate regulators. For example, in Southern Europe, a reduction in precipitation is expected in the coming decades (Kirtman et al., 2013) and this is likely to affect soil microbiota. Overall, climate change can alter soil microbial communities via effects on plant communities (Bardgett et al., 2008). Drought may affect the biomass, activity, and composition of soil microbial communities (Bastida et al., 2017; Ochoa-Hueso et al., 2018) and some studies have noted that fungi can be considered as more resistant to drought than bacteria (De Vries and Shade, 2013; Evans and Wallenstein, 2014). Nevertheless, the wide spectrum of life-history strategies of soil microorganisms also implies a tremendous variety of responses to climate change factors. Moreover, many studies have focused on only one or two climate change factors. However, global change is driven by multiple factors that coincide in soil microbial communities (Rillig et al., 2019); so, more multi-factorial studies are needed in order to assess and predict how microbial communities will evolve in the coming decades.

As mentioned above, we will focus this chapter on the soil microbial ecology of semiarid ecosystems; particularly, those subjected to agriculture. The global population is expected to increase in the coming decades and agriculture will need to provide it with adequate levels of food and fibre. Agricultural ecosystems provide humans with food, forage (for livestock), bioenergy, and pharmaceuticals and are essential to human wellbeing. Moreover, agroecosystems also produce a variety of ecosystem services, such as regulation of soil and water quality, carbon sequestration, support for biodiversity, and cultural services. However, the change and adaptation of agroecosystems constitute a serious concern as a consequence of the effects of global change.

Agriculture will need to withstand the impact of changes in anthropogenic activities derived from a growing human population and, particularly, climate change. This impact will not only affect crop production; ultimately, it will also alter soil ecosystem services. For this reason, in this chapter, we will focus on how the human adaptations to climate change (in particular, drought) in Mediterranean agroecosystems may affect soil microbial communities.

Relationships between of soil microbiota, climate change, and agriculture in a changing world

Soil is a natural resource, non-renewable on the human timescale, which carries out fundamental services for all living beings on the planet. The immediate consequence of this is that we need to protect, conserve, and, if necessary, restore soils. Soils are the foundation for agriculture and food security. They are the basis for food, feed, fuel, and fibre production and for many ecological services, including climate regulation, plant debris decomposition, and pollutants degradation (Van Der Heijden et al., 2008; Delgado-Baquerizo et al., 2016). Soil degradation, which is being exacerbated by climate change and inadequate land uses through their influence on organic matter reductions and losses of biodiversity, can compromise the sustainability of agriculture in a changing world. Thus, the importance of soil is starting to be recognized beyond its scientific context by different organisms. For instance, 2015 was appointed the International Year of Soils by the FAO. Moreover, soils are vital to the Sustainable Development Goals of the United Nations, having a fundamental importance in those of Zero Hunger, Good Health and Well-being, Life on Land, and Climate Action.

Soil is a living system capable of performing functions that are key functions from ecological and human perspectives. The agents responsible for such functions are the millions of bacterial, fungal, and invertebrate species which live in soil. For instance, the total number of species of microorganisms in terrestrial ecosystems is about 10^{29} and, in turn, microorganisms support the existence of all higher trophic life forms (Cavicchioli et al., 2019). Soil organisms comprise 25% of global terrestrial diversity and are known to regulate important ecosystem functions and services such as plant productivity, nutrient cycling, water filtration, organic matter (OM) decomposition, pollutant degradation, climatic regulation (e.g., greenhouse gas fluxes), pest control, and human health (Van Der Heijden et al., 2008; Bastida et al., 2016; Delgado-Baquerizo et al., 2016, 2018).

Soil is the largest terrestrial repository of OM, storing nearly 1500 Gt of carbon, and soil organisms are mainly responsible for the turnover of this OM (Crowther et al., 2019). Microbes, through their respiration, release vast amounts of CO_2 into the atmosphere and this may contribute to climate change (Bastida et al., 2019a; Singh et al., 2010), but they also participate in carbon immobilization in soil - within their biomass, which is subsequently complexed in organo-mineral fractions (Miltner et al., 2012). Hence, a deep understanding of the physiology and dynamics of microbial communities is essential to extend our knowledge of the mechanisms that regulate greenhouse gas fluxes in soil (Allison et al., 2010; Singh et al., 2010; Miltner et al., 2012). For instance, some recent studies have permitted a simplistic classification of soil bacteria on the basis of carbon mineralization, highlighting that

Acidobacteria are usually oligotrophs while Betaproteobacteria and Bacteroidetes are copiotrophs (Fierer et al., 2007).

Soil nutrient limitation in semiarid agroecosystems: impact of climate change on the soil biological phosphorus cycle

Biogeochemical cycles of nutrients in dryland agroecosystems are key drivers of ecosystem functioning (i.e., primary production, respiration, and decomposition) and services (i.e., food production and carbon storage) (Delgado-Baquerizo et al., 2013; Castellanos et al., 2018). The aridity increase predicted by climate change models would affect severely nutrient cycles in terrestrial ecosystems, by alteration of soil moisture and nutrient availability and soil–plant–microbial interactions (Emmett et al., 2004). Consequently, changes in soil fertility due to climate change could trigger dramatic effects on ecosystem functions and services in a warmer and drier world (Sardans et al., 2004; Fernández-Martínez et al., 2014; Luo et al., 2016; Delgado-Baquerizo et al., 2018). In the agronomic context, crop yield might be constrained by global warming and this would condition the utilization of fertilizers (Erbas and Solakoglu, 2017). Increased fertilizer consumption might help farmers to maintain crop yields under global warming, but some fertilizers also contribute to global greenhouse gas emissions, as those containing nitrogen (N) are one of the major anthropogenic sources of nitrous oxide (Vitousek et al., 1997). Nevertheless, application of fertilizers may have little impact on the soil organic carbon (SOC) concentration and microbial activity unless it occurs in conjunction with the application of biosolids (Lal, 2004). However, given the wide variety of soil conditions and agricultural practices, monitoring of the effects of fertilizers in soils might be adequate for assessing the sustainability of agricultural soils.

The soil nutrients availability in dryland agroecosystems is strongly determined by the soil OM and the activity of the soil microbial community (Lal, 2004), although the physicochemical properties (pH, texture, cation exchange capacity, carbonates content, salinity, etc.) also play important roles (Luo et al., 2016; Augusto et al., 2017). Dryland soils usually have a low OM content and high pH (Valdecantos et al., 2006; Albaladejo et al., 2013; Sardans and Peñuelas, 2015), resulting in low nutrients availability (Lajtha and Bloomer, 1988; Cross and Schlesinger, 2001; Rashid and Ryan, 2004).

Nitrogen (N) has been extensively considered as the major soil nutrient limiting primary productivity in dryland ecosystems (West, 1991; Austin, 2011). However, more recent studies have shown that phosphorus (P) can be at least as limiting as N, or even more so, in many dryland ecosystems, especially in calcareous soils (Rashid and Ryan, 2004; Luo et al., 2016; Ruiz-Navarro et al., 2019). Soil P is present in different forms, which differ greatly in their solubility in water and in their chemical reactivity. They include inorganic forms, such as crystalline apatites, amorphous phosphates of calcium, potassium, iron, and aluminium, other phosphates, inorganic polyphosphate, and orthophosphate. In addition, P is also an important element in most organic compounds present in the soil, such as nucleic acids, phospholipids, inositol phosphates, and many metabolic intermediates. Generally, more than 50% of the P in dryland soils is inorganic P, in contrast to more mesic soils, which often contain a higher proportion of organic P.

In spite of the diversity of P forms present in the soil, most of it is biologically unavailable. According to Pierzynski et al. (2005), the main soil processes that affect the availability of soil P are: dissolution–precipitation (mineral equilibria), sorption–desorption (interactions between P in solution and soil solid surfaces), and mineralization–immobilization (biologically mediated conversions of P between inorganic and organic forms). In drylands, the environmental processes that drive the soil P cycle and determine the amount of biologically available soil P are mainly controlled by the timing and amount of precipitation, since soil moisture directly affects the rates of geochemical reactions, ion diffusion, and biotic activity. As Belnap (2011) pointed out, the soil wetting and drying influences the biological availability of P in several ways. In dry periods, P could accumulate in surface soil due to the effects of desiccation and radiation on microbial mortality and OM degradation. After the rewetting of dry soils, the soil microbial biomass increases quickly, using mainly the surface soil P released by the death and decomposition of soil microbiota that has not flowed down to the subsurface soil profile. Much larger events that move P down to the rhizosphere depth in the soil are necessary to produce plant responses, resulting in a temporal decoupling of microbial and plant responses to rises in soil P availability. Moreover, independent of microbial activity, wetting can chemically increase organic P solubility by disrupting OM coatings and detaching and mobilizing soil colloids, but also by controlling the rate of formation of H_2CO_3 - which decreases soil pH, dissolving carbonates, and increases the transition of P between solid and liquid phases, in both cases increasing the presence of P in the soil solution.

Water is a strong limiting factor for plant and microbial development in arid and semiarid ecosystems, and greatly influences the soil P cycle. When soil moisture increases, the microbial community becomes active and influences soil P availability in several ways (Belnap, 2011). Most soil microbes create polyfunctional metal-binding sites by the secretion of exopolymers and these can adsorb cations and anions containing P; these adsorbed ions are not absorbed by the cell, thus remaining available to plants while the losses by leaching to the subsurface soil are reduced (Poole, 1990). Soil microorganisms also increase the availability of P by the release of protons (H^+) during respiration, which decreases soil pH and frees P from synthetic hydroxyapatite (Meyer et al., 2019). Many soil microbes secrete organic acids - such as citrate, malate, acetate, pyruvate, lactate, or fumarate - that could solubilize bound P. Some also secrete metal chelators, such as siderochromes, that increase the availability of soil solution P to plants and/or other microbes (Meena et al., 2017). The previously cited mechanisms that release P and make it available in soil are often specific to particular species or genera, and they act particularly to increase P solubility from inorganic forms. However, most of the microbes existing in soils have the ability to release extracellular phosphatase enzymes into the surrounding soil; these hydrolyse organic phosphates, releasing P (Nannipieri et al., 2011).

For dryland regions, the IPCC predicts a climate scenario in which temperatures will be around 2.5 °C higher and the average annual precipitation will be drastically decreased, with more severe rainfall events. These factors have a strong influence on the availability of P through both direct and indirect mechanisms, so climate change will have a substantial impact on soil P cycling in drylands due to multiple factors (Belnap, 2011). Thus, although much research is required on this topic, the strong decline in soil moisture could slow down all abiotic processes that release soil P, but it will also decrease the microbial abundance and activity that is responsible for the release of bound P, through soil acidification by respiration activities, production of enzymes, or excretion of acidic compounds and chelators. Moreover, drought and warming will increase plant retention of P, resulting in the input to the soil of more recalcitrant litter with a lower P content.

Phosphate rock is a non-renewable resource that will be exhausted 70–100 years from now (Cordell et al., 2009) and there is no substitute source of P in nature (Kim et al., 2018). In consequence, it is expected that P limitation will play a pivotal role in the agricultural productivity in the coming decades. The P-dependence of agriculture is increasing because P is a finite resource that, by scarcity or price, will limit future food production; this will be particularly problematic in dryland regions according to future climate change scenarios. Therefore, it is clear that P use will have to be accompanied by greater efforts towards recycling and strategic, targeted applications (Oberson et al., 2011). In this regard, an important P-rich waste stream is being produced - originating from effluents of municipal and industrial wastewater treatment systems, slaughterhouse refuse, or manure from livestock production - that can be used as recycled organic P directly on agricultural land (van Dijk et al., 2016). Also, more eco-friendly, efficient and food-safe P fertilizers derived from secondary raw materials, such as struvite-like materials, have been evaluated for their use by European agricultural sectors within a circular economy context (Huygens and Saveyn, 2018). There is some scientific evidence of how struvite can impact crop yield and the activity and biomass of the soil microbial community. Struvite is an ammonium magnesium phosphate mineral that can be considered an alternative source of P in soil (Plaza et al., 2007). However, it is quite stable and insoluble in soil. Several studies observed that struvite can increase plant production (dry matter) and plant P uptake (Plaza et al., 2007). Recent studies on barley suggest that the application of struvite, alone or preferentially in combination with sludge, can increase the availability of P in soil, plant uptake, and yield with significant changes in the biomass of some actinobacterial and verrucomicrobial populations, as revealed by metaproteomics (Bastida et al., 2019b).

The use of organic amendments for agro-ecological and sustainable soil management in arid and semiarid areas

The soil carbon (C) pool to a depth of 1 m has been calculated to be 2500 Pg (1 Pg = 1 billion tonnes) and represents more than 3 times the atmospheric pool (760 Pg) and 4.5 times the biotic pool (560 Pg) (Singh et al., 2018). Increasing the soil organic C (SOC) content in agricultural systems, by the fixation of atmospheric CO_2 in soil organic matter (soil C sequestration), has been considered a way to mitigate climate change. On the other hand, small losses of SOC constitute a feedback to boost atmospheric CO_2 and CH_4 levels. Thus, agricultural soils may function as a C sink or source of greenhouse effect gases (GEG) depending on the farming practices used (Smith, 2004). In fact, article 3.4 of the Kyoto Protocol emphasizes the role of agriculture in CO_2 sequestration in tilled soils and advocates sustainable cropping techniques for this purpose. An important fraction of the C contained in organic amendments originates from the atmosphere and thus the recycling of these organic materials in soil helps to increase the sequestered C pool and may constitute one way in which to mitigate climate change.

The SOC is subjected to a continuous turnover and its content reflects a net balance between organic C losses and inputs. Some farming practices that enhance SOC losses are fallowing, cultivation, and stubble burning or removal (Singh et al., 2018). On the other hand, agricultural management practices that increase plant productivity and diversity or involve the application of organic amendments to soil can increase

SOC storage. To maintain the equilibrium between C inputs and outputs, a number of sustainable agricultural management practices have been recommended to increase SOC and to maintain it at optimum levels. Depending on the local conditions, traditional and emerging management practices are efficient for improving soil C sequestration. These include: cover crops, green manures, crop rotations, integrated application of organic amendments and inorganic nutrients, agroforestry, integrated crop-livestock mixed farming systems, biochar application, and addition of clay to sandy soil (FAO, 2017). Among these management practices, application of organic amendments - such as manures, composts, biosolids, biochar, and crop residues - is proving to be the most suitable way to increase SOC in arid and semiarid degraded soils, which are characterized by scarce soil organic matter (SOM), in order to restore it for sustainable agricultural use and improve soil quality (Tejada et al., 2007; Hernández et al., 2014; Moreno et al., 2017). Further, the addition of some organic materials - such as biochar, compost, and biosolids - has been shown to be a highly effective practice with regard to increasing SOC storage and achieving other environmental benefits (Martinez-Blanco et al., 2013), in addition to improving soil quality and crop yield in sustainable agrosystems (Larney and Angers, 2012).

It is assumed that by 2050 agricultural production will have to have increased by 50% to feed the projected global population of over 9 billion (Alexandratos and Bruinsma, 2012). Organic farming is one concrete, but controversial, suggestion for improving the sustainability of agrosystems. It avoids the use of synthetic fertilizers and pesticides, promotes crop rotation, and can contribute to the provision of sufficient food and the reduction of environmental impacts. However, organic systems produce lower yields and thus require larger land areas to produce the same output as conventional production systems (Muller et al., 2017). Intensive agriculture negatively affects soil fertility, principally because of a loss of SOM. Sustainable practices such as the addition of organic amendments could be a useful tool to maintain or increase the organic matter content in agricultural soils, preserving or improving soil fertility in intensive farming systems (Scotti et al., 2015). It has been shown that the application of organic amendments such as compost is a reliable and effective tool to improve the soil structure and both the chemical (Ayuso et al., 1997) and biological fertility of soils (Bastida et al., 2012) as well as to suppress soil-borne pathogens.

The increasing human population has directly resulted in the worldwide generation of huge amounts of diverse solid wastes. The most important solid wastes generated are municipal solid waste (MSW) and biosolids (sewage sludge), regarding the high amounts produced annually worldwide (Sharma et al., 2017). Unsuitable disposal of biosolids and MSW results in several environmental issues such as surface and groundwater contamination, degradation of land, and food chain contamination. Agricultural recycling of biosolids is the environmentally preferential option, rather than the traditional disposal methods (Sharma et al., 2017). The application of biosolids to soil recycles valuable plant nutrients and is considered an effective practice which, in turn, helps the sustainable management of this waste and minimizes the reliance on chemical fertilizers. However, biosolids might contain toxic heavy metals that may limit their usage on croplands. Heavy metals at concentrations higher than the permissible limits may lead to food chain contamination (Moreno et al., 1996).

Healthy soil, as part of the soil quality concept, is defined as a stable soil system with high levels of biological diversity and activity, internal nutrient cycling, and resilience to disturbance. A large number of different physical, chemical, and biological properties of soil are employed as quantitative indicators to define soil quality. In comparison to the rapid shifts in biochemical and biological properties that occur after soil disturbance, changes in physical properties may occur relatively less quickly. Among the biological properties, soil microorganisms are very sensitive to external perturbations and can act as a sensor for monitoring the soil response and, more generally, the soil quality (Bastida et al., 2006). Indeed, microorganisms drive many fundamental nutrient cycling processes, soil structural dynamics, degradation of pollutants, and various other services and respond quickly to natural perturbations and environmental stress. Particularly, the application of organic amendments to agricultural soil has an important effect on the soil microbial community: thus, several studies have revealed an improvement in microbial activity (Moreno et al., 1999) and shifts in the composition, abundance, and structure of the soil microbial community (Bastida et al., 2008; Zhen et al., 2014). Saha's review (Saha, 2010) indicates that organic farming could play a significant role in increasing biodiversity. Other authors showed how compost amendments act as a microbial inoculum in the soil and revealed the long-term effects of different composts and doses on soil microbial diversity (Knapp et al., 2010).

Soils support biological activities, such as the decomposition and recycling of dead organic matter, and play a major role in the mitigation of climate change through the sequestration of C. Soil enzyme activities, which are produced by the microbial community, are reliable indicators of different soil management practices as they rapidly respond to different environmental conditions. Dehydrogenase activity is usually considered as an indicator of the average activity of viable microbial populations in soil. It was suggested that enzymes related to microbial activity, such as dehydrogenases, may be less suitable for the prediction of long-term changes in soil because they respond to recent and transitory management or seasonal/climatic effects. In contrast, extracellular enzyme activities closely related to the organic matter content may be better indicators of permanent changes in soil quality because such enzymes are probably associated with surviving microorganisms and/or are complexed and protected against inactivation by soil humic or clay complexes (Laudicina et al., 2012). Thus, hydrolytic extracellular enzymes are generally used as indices of soil fertility and quality. In this regard, the most widely assayed enzyme activities are those involved in the degradation of cellulose and lignin, those catalysing the hydrolysis of reservoir compounds of N (proteins, chitin, and peptidoglycan), and phosphatases, for their role in the mineralization of organic P. The activities of β-glucosidase, urease, proteases, phosphomonoesterases, and arylsulfatase have been assayed by many authors in order to investigate soil fertility and quality changes following the implementation of different management practices (Moreno et al., 2019). Some authors reported that alkaline and acid phosphatase, arylsulfatase, β-glucosidase, and dehydrogenase activities discriminated organic and conventional management systems, as they were higher in organic management systems than in conventional ones (Bastida et al., 2012; Bowles et al., 2014).

The decline in the organic matter content is becoming a major process of degradation for many soils, particularly in semiarid Mediterranean regions of Europe, as a consequence of the climatic conditions and land use intensification (Garcia et al., 2005). Soil management strategies for sustainable agriculture should focus not only on increasing the organic matter in the soil, but also on the storing of soil residual nutrients in a way that prevents leaching of excess nutrients into the groundwater. Indeed, with inappropriate use, organic amendments can also be a source of environmental pollution. The cropland application of immature organic wastes can also produce environmental and agronomic problems (Diacono and Montemurro, 2010). Hence, organic products of high quality must be produced and their stability determined. As reported previously (Diacono and Montemurro, 2010), long-lasting and repeated applications of organic amendments in agroecosystems improve C sequestration and soil physical properties. This soil management practice increases soil biological activity, organic C, and soil fertility but it may also lead to the build-up of toxic elements. Crop yields are increased by long-term application of high rates of compost and, with its appropriate use, there is no evidence of negative impacts of heavy metals in soil when high-quality compost is used for long periods (Bastida et al., 2008). Repeated application of composted materials significantly increases soil organic nitrogen, storing it for gradual mineralization without inducing nitrate leaching to groundwater (Diacono and Montemurro, 2010).

Climate change and water demands: effects on the soil microbial community

The climatic conditions have changed over recent decades. Beginning in the second half of the 20th century, several groups of scientific researchers have predicted a concerning scenario for the next decades (Giorgi and Lionello, 2008). Mainly caused by humans, climate change is one of the most important concerns today and it influences each political and administrative decision. The possible scenarios predicted for the coming decades are still uncertain since they are extracted from probabilistic climate models (Beniston et al., 2007).

Nevertheless, the fact is that the majority of these models shows the same trends, with a series of predictions as follows: i) an increase in the surface-air temperature on a global scale; ii) an increase in the rainfall distribution variability over the globe, with an increase in mean precipitation values in wet and mountainous regions and a decrease in dry regions; iii) changes in the current pattern of the atmospheric circulation; iv) an increase in extreme climatic events (i.e., droughts, floods, heatwaves); and v) a change in the sea level (Beniston et al., 2007).

Of all these climate change effects, those related to changes in the hydrological cycle are the most concerning for soil science, in particular within the agricultural context. Soil is one of the most important pieces in the environmental system. As part of an ecosystem, soil represents an important recipient containing a high diversity of different macro and microorganisms. It is the habitat where many chemical, biochemical, physical, and biological changes happen. Soil is the physical, but also the chemical and biological, support for plant growth and development. Plants are one of the key factors that make life possible on the Earth´s surface. This is the level of importance that can be attached to a shallow layer little more than a few inches deep in some instances (Yaalon, 1997).

Soil is usually sensitive to environmental changes, including variations in the water content. Climate change is affecting the hydrological system and that includes the relationship between water and soil (Ragab and Prudhomme, 2002). In some terrestrial regions, especially arid and semiarid ones, the water availability is already restricted or might be in future scenarios due to climate change (García-Ruiz et al., 2011). For these regions, dryer conditions in the coming years are predicted, with lower annual precipitation, hotter summers (with an increase in heatwaves), warmer winters, etc. All this leads to the following two conclusions: i) a lower water input to the terrestrial system; and ii) higher evapotranspiration due to the rise in temperature.

These conclusions imply a worrying future scenario where water scarcity episodes will be more extreme and durable. This will be the case for the majority of the regions that have Mediterranean climate conditions: the Mediterranean basin and Californian coast, central Chile, and some small areas in Australia and South Africa. In general, regions with these climatological conditions have important agricultural systems (Giorgi and Lionello, 2008).

Agriculture is an economic activity that provides income and employment for a high number of people around the globe. Nevertheless, it also has a high impact on the environment and the agroecosystems it comprises. Currently, farming is the human activity that consumes the greatest amount of water. Besides, during the crop growth period, farmers add to the soil different materials (manure, fertilizers, and pesticides) with the aim of achieving the best conditions for growth and development and thus the maximum yield (Beniston et al., 2007).

According to the high volume consumed and the future shortages in water supply predicted, society faces severe water deficits. For this reason, in recent decades, scientists and farmers have been working to find other sources of water and to use them more efficiently. The alternative water sources used most frequently are desalinated seawater, desalinated brackish water, and reclaimed water (Downward and Taylor, 2007; Qadir et al., 2007; Martinez-Alvarez et al., 2016; Maestre-Valero et al., 2019). However, the use of water of different quality and different origin can disturb the balance in the soil system: changes in irrigation can alter different soil chemical, physical, and biological properties. Among these, biological properties have a fundamental role in the development of the plant-soil system.

In the last few decades, some agricultural areas in the Mediterranean basin have been subjected to important pressure by farmers in drought periods. For this reason, several desalination plants have been built near to the coast, to obtain water from the sea for human and agricultural consumption. The first countries to develop such systems to make seawater potable and suitable for irrigation were Israel, Spain, and Malta (Martinez-Alvarez et al., 2016). The desalination process consists of the removal of salts from seawater with a membrane system. It starts with a pre-treatment phase in which large particles are removed and continues with a Reverse Osmosis (RO) phase, during which seawater passes through the tiny pores in the membrane that retain the microorganisms, salts, and other ionic elements (Hilal et al., 2011).

However, this process has two disadvantages: the high energy consumption of RO and the lack of some minerals in the desalinated seawater (Jacob, 2007). Due to the latter, it is mandatory to add a final phase of enrichment, to correct the lack of minerals, and adjust the pH up to normal levels (Martinez-Alvarez et al., 2016).

From an agricultural and environmental perspective, the desalinated seawater presents some problems when used for irrigation purposes (Fig. 3.1). Of these, the most vexing is the high boron (B) concentration in desalinated seawater. At the same time, as we have commented in the above paragraph, the desalinated seawater can have a lack of nutrients that may injure plant growth (Martinez-Alvarez et al., 2017). Desalinated brackish water is the other type of water used after a desalination process. Brackish water is obtained from aquifers. It can be obtained close to the coast, like desalinated seawater, and in inland areas (Valdes-Abellan et al., 2013). The toxic effect of the presence of high concentration of B in the desalinated seawater used for irrigation has been widely studied by scientists and agronomists for many years. This research has always been more focused on the plant system (Eaton, 1940; Hu and Brown, 1997; Nable et al., 1997; Brown et al., 2002; Zhu, 2002; Tester and Davenport, 2003; Munns and Tester, 2008) than on the soil effects. However, this has changed with the passage of time. Now, there is an important volume of work addressing soil salinity effects on soil properties, even microbiological properties. In the case of boron, its impact on the soil physicochemical properties has been studied by several research groups (Goldberg and Forster, 1991; Yermiyahu et al., 1995; Goldberg, 1997; Goldberg et al., 2000), whereas its impact on the microbial community is poorly understood. The fact that the toxic effects of boron in soil are the least studied of those arising from the use of desalinated seawater is surprising since it is one of the chemical elements with the thinnest range between deficiency and toxicity for plants (Brown et al., 2002).

Soil microbial communities suffer the impact of elevated boron levels in soil. Recent research (Vera et al., 2019) demonstrated that boron could accumulate in soil when continuous irrigation with desalinated water took place. In this case, boron accumulation caused an important decrease in microbial activity (via its effects on microbial respiration) and it influenced the bacterial and fungal composition. This research also demonstrated that fungal communities were more sensitive to boron than bacterial communities. It is possible that changes in microbial communities related to boron are due to the presence of bacteria capable of living in soil with high levels of this element, since a few species of bacteria have been found in soils polluted with boron in recent years (Ahmed et al., 2007a, b; Ahmed and Fujiwara, 2010). The enzyme activity

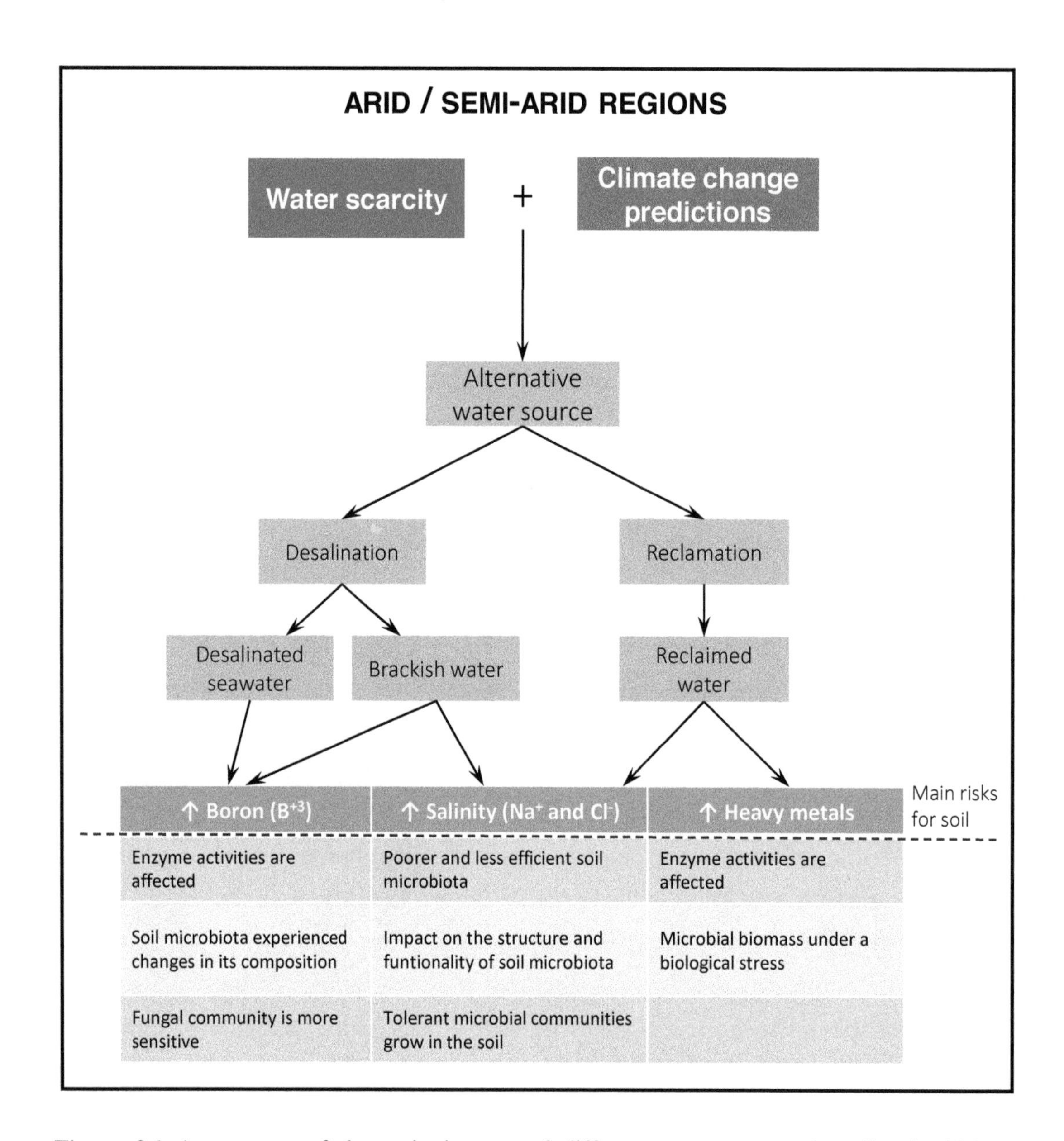

Figure 3.1 A summary of the main impacts of different water sources in soil microbial communities.

in soils can be affected by boron, as demonstrated by Kolesnikov et al. (2008) and Vera et al. (2019). The lack of information about the effects of boron on soil microbiota is important and further research is needed to increase our knowledge about its role in soil microbiology.

Wastewater has been used in agricultural irrigation for centuries. However, in recent decades its use has increased a lot in countries with water availability problems

(Adrover et al., 2012). As we have commented before, agriculture is the largest consumer of water in the world and its demand for water may be a concern in some regions, such as the Mediterranean area. There, an important number of countries have decades of experience in the use of wastewater for irrigation (Angelakis et al., 1999; Pedrero et al., 2010).

Overall, there are two types of wastewater: one is produced in the domestic environment and the other comes from industrial activity (Carolin et al., 2017). The composition of wastewater depends on its origin. Nevertheless, wastewaters always contain both organic and inorganic fractions. Some chemicals (pesticides, metals, and ionic salts) can be present at important concentrations, whereas others (like emerging pollutants; i.e., drugs) are present in trace amounts (Becerra-Castro et al., 2015). Together with these chemicals, microorganisms are also found in wastewater. To sum up, pollution due to wastewater has both chemical and microbiological dimensions. From a biological perspective, bacteria, viruses, and nematodes are the pathogens of most concern in wastewater. The EC and the presence of metals are the main chemical concerns (Salgot et al., 2006). Among the heavy metals, Carolin et al. (2017) highlighted that certain ionic forms can react with bioparticles and cause toxicity problems. Thus, the reuse of treated wastewater entails some health and environmental risks.

Overall, the use of reclaimed water as an irrigation alternative can have effects on the soil biodiversity. Several investigations have demonstrated that microbial communities suffer changes in their structure and functionality upon irrigation with reclaimed water (Adrover et al., 2012; Bastida et al., 2018; Gupta et al., 2019). However, the response of soil microbial communities to such irrigation is not uniform. Adrover et al. (2012) found that this irrigation enhanced the soil microbial biomass and its activity, whereas Gupta et al. (2019) found that an increase in soil heavy metals caused a biological stress that affected the microbial biomass and enzyme activities. Besides, the use of reclaimed water also affects the root-associated microbiome and causes important changes in the microbial community structure that has a key role in the soil-plant system (Zolti et al., 2019).

Besides, the increase in salinity in soils makes it likely that salinity-tolerant microbial communities will remain in the soil. The use of reclaimed water during regulated deficit irrigation promoted a more-resilient community that translated into a recovery of microbial biomass and enzyme activities after the water restriction ended. These results imply potential ecological benefits of the irrigation with reclaimed water that should be considered in relation to the water supply limitation predicted for

Mediterranean areas in climate change models. Bastida et al. (2018) found that the annual use of reclaimed water or the combined irrigation with fresh water benefited the microbial biomass and enzyme activities of grapefruit crop soil. In contrast, the microbial community of mandarin crop soil seemed more affected by the annual irrigation with reclaimed water.

Emerging pollutants of concern in agricultural soils: impacts in the microbial community and strategies of remediation.

In many Mediterranean regions agriculture is the principal economic activity and, in the last few decades, it has been intensified in order to improve crop yield and quality. Agricultural practices have changed, to supply the rising demand for agricultural products, with increased pesticide application to crops and greater use of irrigation. Recent studies have focused on the determination of contaminants in soil, due to the rising concern about soil quality and health risks derived from contamination in the food chain (Bastida et al., 2008; Fantke et al., 2012; Hodson and Lewis, 2016). Many contaminants can be found in soil and their persistence depends on many factors driven by a wide network of physical, chemical, and biological interactions. In this complex network, the chemical structure of the contaminant and the soil composition stand out, since they determine the contaminant-soil interaction (Fig. 3.2). Before reviewing the main effects of contaminants on the soil microbial community, it is important to describe the contaminants and their origins.

Besides pests, water scarcity is one of the main problems in arid and semiarid regions, where irrigation is widely practiced. This makes necessary the use of alternative water sources, and wastewater and regenerated water treatments seem to be a solution. However, the use of these types of water has advantages and disadvantages. On the one hand, they provide fertilizer elements and low molecular weight organic compounds, which usually are part of the macro and micronutrients required by plants and the soil microbial community (Wang et al., 2019). On the other hand, contaminants such as heavy metals and contaminants of emerging concern (CEC) are also present in this kind of water source (Hurtado et al., 2016; Welikala et al., 2018) and their presence in agricultural soils is even more common (Elgallal et al., 2016). Many studies have evaluated heavy metals contents in soils, resulting in the detection of contaminating levels of As, Cd, Co, Cr, Cu, Hg, Ni, Pb, Sb, and Zn. A study in the Jordan Valley concluded that the use of inorganic fertilizers and treated wastewater may have increased the heavy metal contents in soil due to intensive agricultural practices (Gharaibeh et al., 2020). Similar results were found in 132

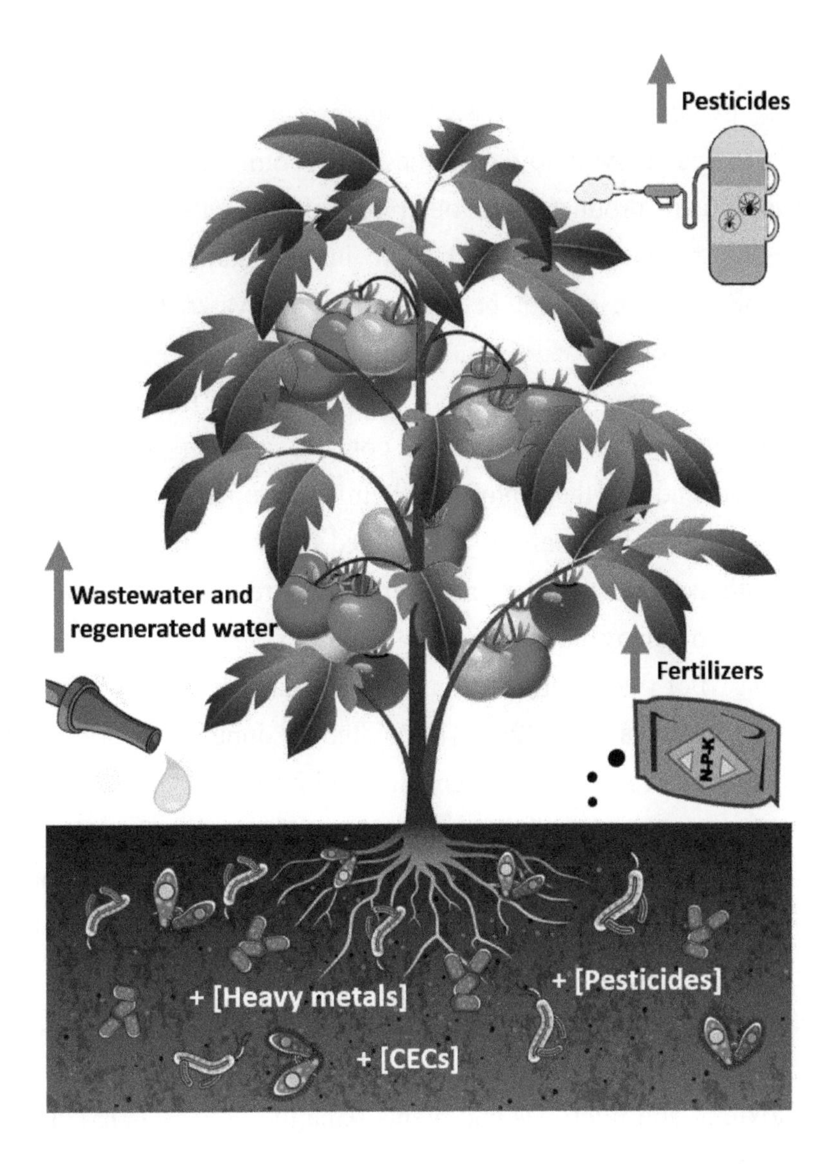

Figure 3.2 Main sources of soil contaminants from wastewaters and agricultural activities.

citrus grove soils from the Argolida basin: the abundances of specific heavy metals (Cu, Zn, Cd, Pb, As) were directly correlated with an anthropogenic origin - the uncontrolled application of fertilizers and pesticides (Kelepertzis, 2014). However, agricultural practices differ among geographical regions because of crop demands. Thus, a study conducted on soils throughout the European Union revealed diverse patterns of accumulation of heavy metals, caused by both geographic variability and the distribution of samples. Western-European and Mediterranean regions showed concentrations above the permissible threshold for at least one element (Tóth et al.,

2016). This scenario is alarming, so monitoring of soils around the EU has been established.

The CEC are also present in wastewaters, but in lesser amounts than the contaminants described above. Some of the most common CEC are surfactants, antibiotics, personal care products, and other pharmaceuticals. Compounds derived from drugs usually reach urban water through human and animal excretions, but unfortunately some of them cannot be removed from water. Crop irrigation with polluted water and fertilization with polluted biosolids make possible their presence in agricultural soils (Gomes et al., 2017). Sorption, desorption, and degradation patterns determine the persistence of these contaminants in soil; in the case of antibiotics, these depend on their physico-chemical properties (Boy-Roura et al., 2018). A recent review (Snow et al., 2019) pointed out the current situation of these compounds in both soils and water, regarding the presence, concentrations, and degradation patterns of many CEC. Several studies in Spain determined somewhat high doses (300 ng/L) of fluoroquinolones and sulphonamides in surface water (Boy-Roura et al., 2018). Ivanová et al. (2018) estimated an annual addition to Slovakian agricultural soils of up to 120 kg of fexofenadine and 29 kg of verapamil. Assays with triclosan and carbamazepine highlighted the role of the soil carbon content in the dissipation of contaminants. Thus, increasing the total organic carbon content in soil could promote sorption, thereby decreasing the bioavailability of such chemicals and restricting their potential degradation (Shao et al., 2018).

The non-biodegradability of contaminants, due to their synthetic nature or their intrinsic toxicity, might increase their persistence in soil and they may reach concentrations toxic to soil microbes and plants (Chagnon et al., 2015). Therefore, it is necessary to perform an evaluation of the microbial community in soils that have suffered human activities, including agriculture. This is critical because soil microbes play a key role in ecosystem services such as biogeochemical cycles and soil fertility (Muñoz-Leoz et al., 2012; Bowles et al., 2014; Moreno et al., 2019). Generally, studies have focused on the impact on the soil microbial community of a single or a couple of pesticides. We can observe differential effects on soil enzyme activities and microbial biomass and biodiversity depending on the pesticide studied (Imfeld and Vuilleumier, 2012; Riah et al., 2014). García-Delgado et al. (2019) reported that triasulfuron and prosulfocarb had a negative effect on microbial biomass. Azoxystrobin and pirimicarb drastically reduced basal respiration, but dehydrogenase activity –a general indicator of soil microbial activity- was only reduced by pirimicarb (Álvarez-Martín et al., 2016). Imidacloprid, which has already been banned, produced a reduction in dehydrogenase activity and altered the N cycle (Cycoń and Piotrowska-Seget, 2015). In similar experiments with imidacloprid and acetamiprid (analysed separately), a significant decrease in the activities of nearly all the enzymes tested (dehydrogenase, phosphomonoesterase, and urease) was observed; the exception was arginine deaminase, whose activity increased (Wang et al., 2014).

These results suggest that emerging contaminants have an impact on the activity of soil microbial communities, and also on the soil microbial biomass. Furthermore, the microbial biomass can be determined by analysis of the phospholipid fatty acid (PLFA) content and sequencing approaches have provided new insights into the diversity and composition of soil microbial communities. Cycoń et al. (2013) observed that napropramide increased the content of Gram-negative bacteria and fungi in soil, whereas Zhang et al. (2009) found that cypermethrin increased Gram-negative bacteria, but decreased the relative abundance of Firmicutes. Nonetheless, the shift in the soil microbial community differs extremely depending on the pollutant.

A similar situation seems to exist in soils contaminated by heavy metals. These contaminants usually have a negative effect on terrestrial microbial communities because they are environmental stressors. Many investigations have evaluated the presence of heavy metals in soil, but few have focused on the effects on the soil microbial community. However, microorganisms seem to be far more sensitive to heavy metals than other living organisms in the same habitat (Giller et al., 1998). One study found a huge range of sensitivity in the microbial community, from relatively sensitive (rhizobia) to insensitive (nitrifiers and organisms involved in carbon compounds degradation) populations (Giller et al., 2009). Recent work showed that the resistance to heavy metals may be led by multiple heavy metal oxidase genes encoded by Proteobacteria (Wang et al., 2019). Nonetheless, the overall microbial activity is usually affected by the presence of heavy metals (Moreno et al., 2001, 2006). For instance, in a study of 145 vineyard soils in Spain, Cu-contaminated soils had lower enzyme activities (Fernández-Calviño et al., 2010).

Much less is known about the effects of CEC on the soil microbial community in comparison with those of pesticides and heavy metals. Evaluations of these effects have been conducted in recent years, obtaining different results. Liu et al. (2009) analysed the effects of six antibiotics (chlortetracycline, tetracycline, tylosin, sulfamethoxazole, sulfamethazine, and trimethoprim) on the soil microbial community and found that they inhibited acid~~ic~~ phosphatase activity and soil respiration. However, the deleterious effects in soil depend greatly on the nature of the antibiotic. The persistence and accumulation of antibiotics in soil have several effects, such as the development of resistance genes and changes in the soil microbial population structure due to inhibition of the growth of certain types of microbe (Engelhardt et al., 2015). Similar results were described by Du and Liu (2012) and Gomes et al. (2017), who concluded that CEC related to antibiotics reduced microbial activity and readjusted the microbial community structure. Nonetheless, the estimation of ecotoxicological parameters is very difficult because experimental designs are not always carefully addressed. Much more research needs to be done in order to understand the toxicity thresholds and impacts of CEC in soil microbial communities, the extent to which soil microbes are able to degrade them and their interactions in plant-soil systems.

As we have previously highlighted, the persistence of pollutants in soil is a serious environmental concern. The wide spectrum of persistence shown by these chemical compounds makes necessary the development of strategies that can remove them, partially or totally, from soils. These strategies have been implemented from different perspectives, attending to the chemical, physical, and/or biological properties of the contaminants. Remediation techniques are carried out in three ways: 1) *in-situ*, the contaminant is treated in the same place without disturbing the soil; 2) soil is excavated and treated on the spot; and 3) *ex-situ*, polluted soil is excavated and treated in another location. Depending on the characteristics of the contaminant, different strategies can be applied for its removal. Hence, we can find specific and broad spectrum strategies for soil contaminants.

Bioremediation is a strategy that is broadly distributed, not only in the case of pesticides, but also for heavy metals in agricultural soils. Some reports have documented certain microorganisms capable of decompose pesticides, with Actinobacteria being highlighted (Alvarez et al., 2016, 2017). A combined degradation by root exudates and rhizosphere microorganisms also could represent a more efficient tool for the biodegradation of some pollutants in soils. For instance, a direct relationship between degradation and microbial activity was found in a deltamethrin-contaminated soil (Sun et al., 2018). Removal of Cu, As, and Cr, among other elements, has been achieved by this strategy. However, phytoremediation usually requires long periods of time to be effective and is limited to surface contamination (Guemiza et al., 2017).

New removal strategies have been developed in recent years. Biosurfactants are able to desorb and dissolve contaminants, facilitating their biodegradation. Moreover, they have good environmental compatibility and their combination with other agents can further improve the efficiency of the decontamination. The removal of pesticides and heavy metals has been dealt with using this strategy, obtaining promising results (Mao et al., 2015). However, there is a lack of knowledge in the case of CEC, owing to their recent appearance in soil as contaminants. Solarisation and biosolarisation have been demonstrated to reduce pesticides contents in agricultural soils (Fenoll et al., 2017). Further, biochar application can be useful in the removal of heavy metals; but, as far as we know, the use of this approach for pesticides and CEC removal has not been sufficiently studied (Yaashikaa et al., 2019).

In summary, a great amount of contaminants reaches agricultural soils, directly or indirectly, and can remain there even for years. The maintenance of the ecosystem services and good soil health can be considered in two complex steps. The first is concerned with how contaminants could affect living organisms and the agricultural and exosystemic services they provide. The second involves the development of sustainable agricultural practices, to reduce the amount of these contaminants applied, and strategies that allow their removal from agricultural soils. For this reason, monitoring of the application of these compounds is necessary.

Acknowledgements

The authors are grateful to Spanish Ministry of Science, Innovation and Universities, and FEDER funds, through the project AGL2017-85755-R; the project LIFE17 ENV/ES/000203 with the contribution of the LIFE Programme of the European Union; the project BIOFORG from the ComFuturo program (Fundación General CSIC), and Fundación Séneca (19896/GERM/15).

References

Adrover, M., Farrús, E., Moyà, G., and Vadell, J. (2012). Chemical properties and biological activity in soils of Mallorca following twenty years of treated wastewater irrigation. J. Environ. Manage. 95, S188-S192. DOI: 10.1016/j.jenvman.2010.08.017

Ahmed, I., and Fujiwara, T. (2010). Mechanism of boron tolerance in soil bacteria. Can. J. Microbiol. 56, 22–26. DOI: 10.1139/W09-106

Ahmed, I., Yokota, A., and Fujiwara, T. (2007a). A novel highly boron tolerant bacterium, *Bacillus boroniphilus* sp. nov., isolated from soil, that requires boron for its growth. Extremophiles 11, 217–224. DOI: 10.1007/s00792-006-0027-0

Ahmed, I., Yokota, A., and Fujiwara, T. (2007b). *Gracilibacillus boraciitolerans* sp. nov., a highly boron-tolerant and moderately halotolerant bacterium isolated from soil. Int. J. Syst. Evol. Microbiol. 57, 796–802. DOI: 10.1099/ijs.0.64284-0

Albaladejo, J., Ortiz, R., Garcia-Franco, N., Navarro, A.R., Almagro, M., Pintado, J.G., and Martínez-Mena, M. (2013). Land use and climate change impacts on soil organic carbon stocks in semi-arid Spain. J. Soils Sediments 13, 265–277. DOI: 10.1007/s11368-012-0617-7

Alexandratos, N., and Bruinsma, J. (2012). World Agriculture Towards 2030/2015: The 2012 Revision (Rome).

Allison, S.D., Wallenstein, M.D., and Bradford, M.A. (2010). Soil-carbon response to warming dependent on microbial physiology. Nat. Geosci. 3, 336–340. DOI: 10.1038/ngeo846

Álvarez-Martín, A., Hilton, S.L., Bending, G.D., Rodríguez-Cruz, M.S., and Sánchez-Martín, M.J. (2016). Changes in activity and structure of the soil microbial community after application of azoxystrobin or pirimicarb and an organic amendment to an agricultural soil. Appl. Soil Ecol. 106, 47–57. DOI: 10.1016/j.apsoil.2016.05.005

Alvarez, A., Saez, J.M., Davila Costa, J.S., Colin, V.L., Fuentes, M.S., Cuozzo, S.A., Benimeli, C.S., Polti, M.A., and Amoroso, M.J. (2017). Actinobacteria: Current research and perspectives for bioremediation of pesticides and heavy metals. Chemosphere 166, 41–62. DOI: 10.1016/j.chemosphere.2016.09.070

Angelakis, A.N., Marecos Do Monte, M.H.F., Bontoux, L., and Asano, T. (1999). The status of wastewater reuse practice in the Mediterranean basin: Need for guidelines. Water Res. 33, 2201–2217. DOI: 10.1016/S0043-1354(98)00465-5

Augusto, L., Achat, D.L., Jonard, M., Vidal, D., and Ringeval, B. (2017). Soil parent material—A major driver of plant nutrient limitations in terrestrial ecosystems. Glob. Chang. Biol. 23, 3808–3824. DOI: 10.1111/gcb.13691

Austin, A.T. (2011). Has water limited our imagination for aridland biogeochemistry. Trends Ecol. Evol. 26, 229–235. DOI: 10.1016/j.tree.2011.02.003

Ayuso, M., Moreno, J.L., Hernández, T., and García, C. (1997). Characterisation and evaluation of humic acids extracted from urban waste as liquid fertilisers. J. Sci. Food Agric. 75, 481–488. DOI: 10.1002/(SICI)1097-0010(199712)75:4<481::AID-JSFA901>3.0.CO;2-K

Bardgett, R.D., Freeman, C., and Ostle, N.J. (2008). Microbial contributions to climate change through carbon cycle feedbacks. ISME J. 2, 805–814. DOI: 10.1038/ismej.2008.58

Bastida, F., Luis Moreno, J., Hernández, T., and García, C. (2006). Microbiological degradation index of soils in a semiarid climate. Soil Biol. Biochem. 38, 3463–3473. DOI: 10.1016/j.soilbio.2006.06.001

Bastida, F., Kandeler, E., Moreno, J.L., Ros, M., García, C., and Hernández, T. (2008). Application of fresh and composted organic wastes modifies structure, size and activity of soil microbial community under semiarid climate. Appl. Soil Ecol. 40, 318–329. DOI: 10.1016/j.apsoil.2008.05.007

Bastida, F., Jindo, K., Moreno, J.L., Hernández, T., and García, C. (2012). Effects of organic amendments on soil carbon fractions, enzyme activity and humus-enzyme complexes under semi-arid conditions. Eur. J. Soil Biol. 53, 94–102. DOI: 10.1016/j.ejsobi.2012.09.003

Bastida, F., Torres, I.F., Moreno, J.L., Baldrian, P., Ondoño, S., Ruiz-Navarro, A., Hernández, T., Richnow, H.H., Starke, R., García, C., and Jehmlich, N. (2016). The active microbial diversity drives ecosystem multifunctionality and is physiologically related to carbon availability in Mediterranean semi-arid soils. Mol. Ecol. 25, 4660–4673. DOI: 10.1111/mec.13783

Bastida, F., Torres, I.F., Andrés-Abellán, M., Baldrian, P., López-Mondéjar, R., Větrovský, T., Richnow, H.H., Starke, R., Ondoño, S., García, C., Lopez-Serrano, F.R., and Jehmlich, N. (2017). Differential sensitivity of total and active soil microbial communities to drought and forest management. Glob. Chang. Biol. 23, 4185–4203. DOI: 10.1111/gcb.13790

Bastida, F., Torres, I.F., Abadía, J., Romero-Trigueros, C., Ruiz-Navarro, A., Alarcón, J.J., García, C., and Nicolás, E. (2018). Comparing the impacts of drip irrigation by freshwater and reclaimed wastewater on the soil microbial community of two citrus species. Agric. Water Manag. 203, 53–62. DOI: 10.1016/j.agwat.2018.03.001

Bastida, F., Jehmlich, N., Martínez-Navarro, J., Bayona, V., García, C., and Moreno, J.L. (2019a). The effects of struvite and sewage sludge on plant yield and the microbial community of a semiarid Mediterranean soil. Geoderma 337, 1051–1057. DOI: 10.1016/j.geoderma.2018.10.046

Bastida, F., García, C., Fierer, N., Eldridge, D.J., Bowker, M.A., Abades, S., Alfaro, F.D., Asefaw Berhe, A., Cutler, N.A., Gallardo, A., García-Velázquez, L., Hart, S.C., Hayes, P.E., Hernández, T., Hseu, Z.Y., Jehmlich, N., Kirchmair, M., Lambers, H., Neuhauser, S., Peña-Ramírez, V.M., Pérez, C.A., Reed, S.C., Santos, F., Siebe, C., Sullivan, B.W., Trivedi, P., Vera, A., Williams, M.A., Luis Moreno, J., and Delgado-Baquerizo, M. (2019b). Global ecological predictors of the soil priming effect. Nat. Commun. 10, 3481. DOI: 10.1038/s41467-019-11472-7

Becerra-Castro, C., Lopes, A.R., Vaz-Moreira, I., Silva, E.F., Manaia, C.M., and Nunes, O.C. (2015). Wastewater reuse in irrigation: A microbiological perspective on implications in soil fertility and human and environmental health. Environ. Int. 75, 117–135. DOI: 10.1016/j.envint.2014.11.001

Belnap, J. (2011). Biological Phosphorus Cycling in Dryland Regions. In Phosphorus in Action: Biological Processes in Soil Phosphorus Cycling, E. Bünemann, A. Oberson and E. Frossard, eds. (Berlin, Heidelberg; Springer), pp. 371–406.

Beniston, M., Stephenson, D.B., Christensen, O.B., Ferro, C.A.T., Frei, C., Goyette, S., Halsnaes, K., Holt, T., Jylhä, K., Koffi, B., Palutikof, J., Schöll, R., Semmler, T., and Woth, K. (2007). Future extreme events in European climate: An exploration of regional climate model projections. Clim. Change 81, 71–95. DOI: 10.1007/s10584-006-9226-z

Bowles, T.M., Acosta-Martínez, V., Calderón, F., and Jackson, L.E. (2014). Soil enzyme activities, microbial communities, and carbon and nitrogen availability in organic agroecosystems across an intensively-managed agricultural landscape. Soil Biol. Biochem. 68, 252–262. DOI: 10.1016/j.soilbio.2013.10.004

Boy-Roura, M., Mas-Pla, J., Petrovic, M., Gros, M., Soler, D., Brusi, D., and Menció, A. (2018). Towards the understanding of antibiotic occurrence and transport in groundwater: Findings from the Baix Fluvià alluvial aquifer (NE Catalonia, Spain). Sci. Total Environ. 612, 1387–1406. DOI: 10.1016/j.scitotenv.2017.09.012

Brown, P.H., Bellaloui, N., Wimmer, M.A., Bassil, E.S., Ruiz, J., Hu, H., Pfeffer, H., Dannel, F., and Römheld, V. (2002). Boron in plant biology. Plant Biol. 4, 205–223. DOI: 10.1055/s-2002-25740

Carolin, C.F., Kumar, P.S., Saravanan, A., Joshiba, G.J., and Naushad, M. (2017). Efficient techniques for the removal of toxic heavy metals from aquatic environment: A review. J. Environ. Chem. Eng. 5, 2782–2799. DOI: 10.1016/j.jece.2017.05.029

Castellanos, A.E., Llano-Sotelo, J.M., Machado-Encinas, L.I., López-Piña, J.E., Romo-Leon, J.R., Sardans, J., and Peñuelas, J. (2018). Foliar C, N, and P stoichiometry characterize successful plant ecological strategies in the Sonoran Desert. Plant Ecol. 219, 775–788. DOI: 10.1007/s11258-018-0833-3

Cavicchioli, R., Ripple, W.J., Timmis, K.N., Azam, F., Bakken, L.R., Baylis, M., Behrenfeld, M.J., Boetius, A., Boyd, P.W., Classen, A.T., Crowther, T.W., Danovaro, R., Foreman, C.M., Huisman, J., Hutchins, D.A., Jansson, J.K., Karl, D.M., Koskella, B., Mark Welch, D.B., Martiny, J.B.H., Moran, M.A., Orphan, V.J., Reay, D.S., Remais, J.V., Rich, V.I., Singh, B.K., Stein, L.Y., Stewart, F.J., Sullivan,

M.B., van Oppen, M.J.H., Weaver, S.C., Webb, E.A., and Webster, N.S. (2019). Scientists' warning to humanity: microorganisms and climate change. Nat. Rev. Microbiol. 17, 569–586. DOI: 10.1038/s41579-019-0222-5

Chagnon, M., Kreutzweiser, D., Mitchell, E.A.D., Morrissey, C.A., Noome, D.A., and Van Der Sluijs, J.P. (2015). Risks of large-scale use of systemic insecticides to ecosystem functioning and services. Environ. Sci. Pollut. Res. 22, 119–134. DOI: 10.1007/s11356-014-3277-x

Cordell, D., Drangert, J.O., and White, S. (2009). The story of phosphorus: Global food security and food for thought. Glob. Environ. Chang. 19, 292–305. DOI: 10.1016/j.gloenvcha.2008.10.009

Cross, A.F., and Schlesinger, W.H. (2001). Biological and geochemical controls on phosphorus fractions in semiarid soils. Biogeochemistry 52, 155–172. DOI: 10.1023/A:1006437504494

Crowther, T.W., van den Hoogen, J., Wan, J., Mayes, M.A., Keiser, A.D., Mo, L., Averill, C., and Maynard, D.S. (2019). The global soil community and its influence on biogeochemistry. Science 365 (6455), eaav0550. DOI: 10.1126/science.aav0550

Cycoń, M., and Piotrowska-Seget, Z. (2015). Biochemical and microbial soil functioning after application of the insecticide imidacloprid. J. Environ. Sci. (China) 27, 147–158. DOI: 10.1016/j.jes.2014.05.034

Cycoń, M., Markowicz, A., Borymski, S., Wójcik, M., and Piotrowska-Seget, Z. (2013). Imidacloprid induces changes in the structure, genetic diversity and catabolic activity of soil microbial communities. J. Environ. Manage. 131, 55–65. DOI: 10.1016/j.jenvman.2013.09.041

Delgado-Baquerizo, M., Maestre, F.T., Gallardo, A., Bowker, M.A., Wallenstein, M.D., Quero, J.L., Ochoa, V., Gozalo, B., García-Gómez, M., Soliveres, S., García-Palacios, P., Berdugo, M., Valencia, E., Escolar, C., Arredondo, T., Barraza-Zepeda, C., Bran, D., Carreira, J.A., Chaieb, M., Conceicao, A.A., Derak, M., Eldridge, D.J., Escudero, A., Espinosa, C.I., Gaitán, J., Gatica, M.G., Gómez-González, S., Guzman, E., Gutiérrez, J.R., Florentino, A., Hepper, E., Hernández, R.M., Huber-Sannwald, E., Jankju, M., Liu, J., Mau, R.L., Miriti, M., Monerris, J., Naseri, K., Noumi, Z., Polo, V., Prina, A., Pucheta, E., Ramírez, E., Ramírez-Collantes, D.A., Romao, R., Tighe, M., Torres, D., Torres-Díaz, C., D. Ungar, E., Val, J., Wamiti, W., Wang, D., and Zaady, E. (2013). Decoupling of soil nutrient cycles as a function of aridity in global drylands. Nature 502, 672–676. DOI: 10.1038/nature12670

Delgado-Baquerizo, M., Maestre, F.T., Reich, P.B., Jeffries, T.C., Gaitan, J.J., Encinar, D., Berdugo, M., Campbell, C.D., and Singh, B.K. (2016). Microbial diversity drives multifunctionality in terrestrial ecosystems. Nat. Commun. 7, 10541. DOI: 10.1038/ncomms10541

Delgado-Baquerizo, M., Eldridge, D.J., Maestre, F.T., Ochoa, V., Gozalo, B., Reich, P.B., and Singh, B.K. (2018). Aridity Decouples C:N:P Stoichiometry Across Multiple Trophic Levels in Terrestrial Ecosystems. Ecosystems 21, 459–468. DOI: 10.1007/s10021-017-0161-9

Diacono, M., and Montemurro, F. (2010). Long-term effects of organic amendments on soil fertility. A review. Agron. Sustain. Dev. 30, 401–422. DOI: 10.1051/agro/2009040

van Dijk, K.C., Lesschen, J.P., and Oenema, O. (2016). Phosphorus flows and balances of the European Union Member States. Sci. Total Environ. 542, 1078–1093. DOI: 10.1016/j.scitotenv.2015.08.048

Downward, S.R., and Taylor, R. (2007). An assessment of Spain's Programa AGUA and its implications for sustainable water management in the province of Almería, southeast Spain. J. Environ. Manage. 82, 277–289. DOI: 10.1016/j.jenvman.2005.12.015

Du, L., and Liu, W. (2012). Occurrence, fate, and ecotoxicity of antibiotics in agro-ecosystems. A review. Agron. Sustain. Dev. 32, 309–327. DOI: 10.1007/s13593-011-0062-9

Eaton, S. V. (1940). Effects of boron deficiency and excess on plants. Plant Physiol. 15, 95–107. DOI: 10.1104/pp.15.1.95

Elgallal, M., Fletcher, L., and Evans, B. (2016). Assessment of potential risks associated with chemicals in wastewater used for irrigation in arid and semiarid zones: A review. Agric. Water Manag. 177, 419–431. DOI: 10.1016/j.agwat.2016.08.027

Emmett, B.A., Beier, C., Estiarte, M., Tietema, A., Kristensen, H.L., Williams, D., Peñuelas, J., Schmidt, I., and Sowerby, A. (2004). The response of soil processes to climate change: Results from manipulation studies of shrublands across an environmental gradient. Ecosystems 7, 625–637. DOI: 10.1007/s10021-004-0220-x

Engelhardt, I., Sittig, S., Šimůnek, J., Groeneweg, J., Pütz, T., and Vereecken, H. (2015). Fate of the antibiotic sulfadiazine in natural soils: Experimental and numerical investigations. J. Contam. Hydrol. 177–178, 30–42. DOI: 10.1016/j.jconhyd.2015.02.006

Erbas, B.C., and Solakoglu, E.G. (2017). In the presence of climate change, the use of fertilizers and the effect of income on agricultural emissions. Sustain. 9, 1989. DOI: 10.3390/su9111989

Evans, S.E., and Wallenstein, M.D. (2014). Climate change alters ecological strategies of soil bacteria. Ecol. Lett. 17, 155–164. DOI: 10.1111/ele.12206

Fantke, P., Friedrich, R., and Jolliet, O. (2012). Health impact and damage cost assessment of pesticides in Europe. Environ. Int. 49, 9–17. DOI: 10.1016/j.envint. 2012.08.001

FAO (2017). Soil Organic Carbon: The Hidden Potential. Food and Agriculture Organization of the United Nations.

Fenoll, J., Garrido, I., Vela, N., Ros, C., and Navarro, S. (2017). Enhanced degradation of spiro-insecticides and their leacher enol derivatives in soil by solarization and biosolarization techniques. Environ. Sci. Pollut. Res. 24, 9278–9285. DOI: 10.1007/s11356-017-8589-1

Fernández-Calviño, D., Soler-Rovira, P., Polo, A., Díaz-Raviña, M., Arias-Estévez, M., and Plaza, C. (2010). Enzyme activities in vineyard soils long-term treated with copper-based fungicides. Soil Biol. Biochem. 42, 2119–2127. DOI: 10.1016/ j.soilbio.2010.08.007

Fernández-Martínez, M., Vicca, S., Janssens, I.A., Sardans, J., Luyssaert, S., Campioli, M., Chapin, F.S., Ciais, P., Malhi, Y., Obersteiner, M., Papale, D., Piao, S.L., Reichstein, M., Rodà, F., and Peñuelas, J. (2014). Nutrient availability as the key regulator of global forest carbon balance. Nat. Clim. Chang. 4, 471–476. DOI: 10.1038/nclimate2177

Fierer, N., Bradford, M.A., and Jackson, R.B. (2007). Toward an ecological classification of soil bacteria. Ecology 88, 1354–1364. DOI: 10.1890/05-1839

García-Delgado, C., Barba-Vicente, V., Marín-Benito, J.M., Mariano Igual, J., Sánchez-Martín, M.J., and Sonia Rodríguez-Cruz, M. (2019). Influence of different agricultural management practices on soil microbial community over dissipation time of two herbicides. Sci. Total Environ. 646, 1478–1488. DOI: 10.1016/ j.scitotenv.2018.07.395

García-Ruiz, J.M., López-Moreno, I.I., Vicente-Serrano, S.M., Lasanta-Martínez, T., and Beguería, S. (2011). Mediterranean water resources in a global change scenario. Earth-Science Rev. 105, 121–139. DOI: 10.1016/j.earscirev.2011.01.006

Garcia, C., Moreno, J.L., Hernandez, T., Cano, A.F., Silla, R.O., and Mermut, A.R. (2005). Microbial activity as an indicator of soil degradation in some soils of Murcia, SE Spain. Sustain. Use Manag. Soils - Arid Semiarid Reg. 36, 291-302.

Gharaibeh, M.A., Marschner, B., Heinze, S., and Moos, N. (2020). Spatial distribution of metals in soils under agriculture in the Jordan Valley. Geoderma Reg. 20, e00245. DOI: 10.1016/j.geodrs.2019.e00245

Giller, K.E., Witter, E., and Mcgrath, S.P. (1998). Toxicity of heavy metals to microorganisms and microbial processes in agricultural soils: A review. Soil Biol. Biochem. 30, 1389–1414. DOI: 10.1016/S0038-0717(97)00270-8

Giller, K.E., Witter, E., and McGrath, S.P. (2009). Heavy metals and soil microbes. Soil Biol. Biochem. 41, 2031-2037. DOI: 10.1016/j.soilbio.2009.04.026

Giorgi, F., and Lionello, P. (2008). Climate change projections for the Mediterranean region. Glob. Planet. Change 63, 90–104. DOI: 10.1016/j.gloplacha.2007.09.005

Goldberg, S. (1997). Reactions of boron with soils. Plant and Soil 193, 35–48.

Goldberg, S., and Forster, H.S. (1991). Boron sorption on calcareous soils and reference calcites. Soil Sci. 152, 304–310. DOI: 10.1097/00010694-199110000-00009

Goldberg, S., Lesch, S.M., and Suarez, D.L. (2000). Predicting boron adsorption by soils using soil chemical parameters in the constant capacitance model. Soil Sci. Soc. Am. J. 64, 1356–1363. DOI: 10.2136/sssaj2000.6441356x

Gomes, A.R., Justino, C., Rocha-Santos, T., Freitas, A.C., Duarte, A.C., and Pereira, R. (2017). Review of the ecotoxicological effects of emerging contaminants to soil biota. J. Environ. Sci. Heal. - Part A Toxic/Hazardous Subst. Environ. Eng. 52, 992–1007. DOI: 10.1080/10934529.2017.1328946

Guemiza, K., Coudert, L., Metahni, S., Mercier, G., Besner, S., and Blais, J.F. (2017). Treatment technologies used for the removal of As, Cr, Cu, PCP and/or PCDD/F from contaminated soil: A review. J. Hazard. Mater. 333, 194–214. DOI: 10.1016/j.jhazmat.2017.03.021

Gupta, S.K., Roy, S., Chabukdhara, M., Hussain, J., and Kumar, M. (2019). Risk of metal contamination in agriculture crops by reuse of wastewater: An ecological and human health risk perspective. In Water Conservation, Recycling and Reuse: Issues and Challenges, pp. 55–79.

Van Der Heijden, M.G.A., Bardgett, R.D., and Van Straalen, N.M. (2008). The unseen majority: Soil microbes as drivers of plant diversity and productivity in terrestrial ecosystems. Ecol. Lett. 11, 296–310. DOI: 10.1111/j.1461-0248.2007.01139.x

Hernández, T., Chocano, C., Moreno, J.L., and García, C. (2014). Organic wastes as alternative to inorganic fertilizers in crop cultivation. Acta Horticulturae 1028, 371–376.

Hilal, N., Kim, G.J., and Somerfield, C. (2011). Boron removal from saline water: A comprehensive review. Desalination 273, 23–35. DOI: 10.1016/j.desal.2010.05.012

Hodson, A., and Lewis, E. (2016). Managing for soil health can suppress pests. Calif. Agric. 70, 137–141. DOI: 10.3733/ca.2016a0005

Hu, H., and Brown, P.H. (1997). Absorption of boron by plant roots. Plant and Soil 193, 49–58.

Hurtado, C., Domínguez, C., Pérez-Babace, L., Cañameras, N., Comas, J., and Bayona, J.M. (2016). Estimate of uptake and translocation of emerging organic contaminants from irrigation water concentration in lettuce grown under controlled conditions. J. Hazard. Mater. 305, 139–148. DOI: 10.1016/j.jhazmat.2015.11.039

Huygens, D., and Saveyn, H.G.M. (2018). Agronomic efficiency of selected phosphorus fertilisers derived from secondary raw materials for European agriculture. A meta-analysis. Agron. Sustain. Dev. 38, 52. DOI: 10.1007/s13593-018-0527-1

Imfeld, G., and Vuilleumier, S. (2012). Measuring the effects of pesticides on bacterial communities in soil: A critical review. Eur. J. Soil Biol. 49, 22–30. DOI: 10.1016/j.ejsobi.2011.11.010

Ivanová, L., Mackuľak, T., Grabic, R., Golovko, O., Koba, O., Staňová, A.V., Szabová, P., Grenčíková, A., and Bodík, I. (2018). Pharmaceuticals and illicit drugs – A new threat to the application of sewage sludge in agriculture. Sci. Total Environ. 634, 606–615. DOI: 10.1016/j.scitotenv.2018.04.001

Jacob, C. (2007). Seawater desalination: Boron removal by ion exchange technology. Desalination 205, 47–52. DOI: 10.1016/j.desal.2006.06.007

Kelepertzis, E. (2014). Accumulation of heavy metals in agricultural soils of Mediterranean: Insights from Argolida basin, Peloponnese, Greece. Geoderma 221–222, 82–90. DOI: 10.1016/j.geoderma.2014.01.007

Kim, J.H., An, B. min, Lim, D.H., and Park, J.Y. (2018). Electricity production and phosphorous recovery as struvite from synthetic wastewater using magnesium-air fuel cell electrocoagulation. Water Res. 132, 200–210. DOI: 10.1016/j.watres.2018.01.003

Kirtman, B., Power, S.B., Adedoyin, A.J., Boer, G.J., Bojariu, R., Camilloni, I., Doblas-Reyes, F., Fiore, A.M., Kimoto, M., Meehl, G., et al. (2013). Near-term climate change: Projections and predictability. In Climate Change 2013 the Physical Science Basis: Working Group I Contribution to the Fifth Assessment Report of the Intergovernmental Panel on Climate Change, T.F. Stocker, D. Qin, G.-K. Plattner, M.M.B. Tignor, S.K. Allen, J. Boschung, A. Nauels, and Y. Xia, eds. (Cambridge, UK, New York, NY, USA: Cambridge University Press), pp. 953–1028.

Knapp, B.A., Ros, M., Insam, H., Goberna, M. (2010). Do composts affect the soil microbial community? In Microbes at Work. H. Insam, I. Franke-Whittle and M. Goberna, eds. (Berlin, Heidelberg: Springer), pp. 271-291.

Kolesnikov, S.I., Popovich, A.A., Kazeev, K.S., and Val'kov, V.F. (2008). The influence of fluorine, boron, selenium, and arsenic pollution on the biological properties of ordinary chernozems. Eurasian Soil Sci. 41, 400–404. DOI: 10.1134/S1064229308040066

Lajtha, K., and Bloomer, S.H. (1988). Factors affecting phosphate sorption and phosphate retention in a desert ecosystem. Soil Sci. 146, 160–167. DOI: 10.1097/00010694-198809000-00003

Lal, R. (2004). Carbon sequestration in dryland ecosystems. Environ. Manage. 33, 528–544.

Larney, F.J., and Angers, D.A. (2012). The role of organic amendments in soil reclamation: A review. Can. J. Soil Sci. 92, 19–38. DOI: 10.4141/CJSS2010-064

Laudicina, V.A., Dennis, P.C., Palazzolo, E., and Badalucco, L. (2012). Key Biochemical Attributes to Assess Soil Ecosystem Sustainability. In Environmental Protection Strategies for Sustainable Development. A. Malik and E. Grohmann, eds. (Dordrecht, The Netherlands: Springer), pp. 193-227.

Liu, F., Ying, G.G., Tao, R., Zhao, J.L., Yang, J.F., and Zhao, L.F. (2009). Effects of six selected antibiotics on plant growth and soil microbial and enzymatic activities. Environ. Pollut. 157, 1636–1642. DOI: 10.1016/j.envpol.2008.12.021

Luo, W., Sardans, J., Dijkstra, F.A., Peñuelas, J., Lü, X.T., Wu, H., Li, M.H., Bai, E., Wang, Z., Han, X., and Jiang, Y. (2016). Thresholds in decoupled soil-plant elements under changing climatic conditions. Plant Soil 409, 159–173. DOI: 10.1007/s11104-016-2955-5

Maestre-Valero, J.F., González-Ortega, M.J., Martínez-Álvarez, V., and Martin-Gorriz, B. (2019). The role of reclaimed water for crop irrigation in southeast Spain. Water Sci. Technol. Water Supply 19, 1555–1562. DOI: 10.2166/ws.2019.024

Mao, X., Jiang, R., Xiao, W., and Yu, J. (2015). Use of surfactants for the remediation of contaminated soils: A review. J. Hazard. Mater. 285, 419–435. DOI: 10.1016/j.jhazmat.2014.12.009

Martínez-Alvarez, V., Martin-Gorriz, B., and Soto-García, M. (2016). Seawater desalination for crop irrigation - A review of current experiences and revealed key issues. Desalination 381, 58–70. DOI: 10.1016/j.desal.2015.11.032

Martínez-Alvarez, V., González-Ortega, M.J., Martin-Gorriz, B., Soto-García, M., and Maestre-Valero, J.F. (2017). The use of desalinated seawater for crop irrigation in the Segura River Basin (south-eastern Spain). Desalination 422, 153–164. DOI: 10.1016/j.desal.2017.08.022

Martínez-Blanco, J., Lazcano, C., Christensen, T.H., Muñoz, P., Rieradevall, J., Møller, J., Antón, A., and Boldrin, A. (2013). Compost benefits for agriculture evaluated by life cycle assessment. A review. Agron. Sustain. Dev. 33, 721–732. DOI: 10.1007/s13593-013-0148-7

Meena, V.S., Meena, S.K., Verma, J.P., Kumar, A., Aeron, A., Mishra, P.K., Bisht, J.K., Pattanayak, A., Naveed, M., and Dotaniya, M.L. (2017). Plant beneficial rhizospheric microorganism (PBRM) strategies to improve nutrients use efficiency: A review. Ecol. Eng. 107, 8–32. DOI: 10.1016/j.ecoleng.2017.06.058

Meyer, G., Maurhofer, M., Frossard, E., Gamper, H.A., Mäder, P., Mészáros, Schönholzer-Mauclaire, L., Symanczik, S., and Oberson, A. (2019). Pseudomonas protegens CHA0 does not increase phosphorus uptake from 33 P labeled synthetic hydroxyapatite by wheat grown on calcareous soil. Soil Biol. Biochem. 131, 217–228. DOI: 10.1016/j.soilbio.2019.01.015

Miltner, A., Bombach, P., Schmidt-Brücken, B., and Kästner, M. (2012). SOM genesis: Microbial biomass as a significant source. Biogeochemistry 111, 41–55. DOI: 10.1007/s10533-011-9658-z

Moreno, J.L., García, C., Hernández, T., and Pascual, J.A. (1996). Transference of heavy metals from a calcareous soil amended with sewage-sludge compost to barley plants. Bioresour. Technol. 55, 251–258. DOI: 10.1016/0960-8524(96)00009-0

Moreno, J.L., Hernández, T., and Garcia, C. (1999). Effects of a cadmium-contaminated sewage sludge compost on dynamics of organic matter and microbial activity in an arid soil. Biol. Fertil. Soils 28, 230–237. DOI: 10.1007/s003740050487

Moreno, J.L., García, C., Landi, L., Falchini, L., Pietramellara, G., and Nannipieri, P. (2001). The ecological dose value (ED50) for assessing Cd toxicity on ATP content and dehydrogenase and urease activities of soil. Soil Biol. Biochem. 33, 483–489. DOI: 10.1016/S0038-0717(00)00189-9

Moreno, J.L., Sanchez-Marín, A., Hernández, T., and García, C. (2006). Effect of cadmium on microbial activity and a ryegrass crop in two semiarid soils. Environ. Manage. 37, 626–633. DOI: 10.1007/s00267-004-5006-6

Moreno, J.L., Ondoño, S., Torres, I., and Bastida, F. (2017). Compost, leonardite, and zeolite impacts on soil microbial community under barley crops. J. Soil Sci. Plant Nutr. 17, 214–230. DOI: 10.4067/S0718-95162017005000017

Moreno, J.L., Torres, I.F., García, C., López-Mondéjar, R., and Bastida, F. (2019). Land use shapes the resistance of the soil microbial community and the C cycling response to drought in a semi-arid area. Sci. Total Environ. 648, 1018–1030. DOI: 10.1016/j.scitotenv.2018.08.214

Muller, A., Schader, C., El-Hage Scialabba, N., Brüggemann, J., Isensee, A., Erb, K.H., Smith, P., Klocke, P., Leiber, F., Stolze, M., and Niggli, U. (2017). Strategies for feeding the world more sustainably with organic agriculture. Nat. Commun. 8, 1290. DOI: 10.1038/s41467-017-01410-w

Munns, R., and Tester, M. (2008). Mechanisms of Salinity Tolerance. Annu. Rev. Plant Biol. 59, 651–681. DOI: 10.1146/annurev.arplant.59.032607.092911

Muñoz-Leoz, B., Garbisu, C., Antigüedad, I., and Ruiz-Romera, E. (2012). Fertilization can modify the non-target effects of pesticides on soil microbial communities. Soil Biol. Biochem. 48, 125–134. DOI: 10.1016/j.soilbio.2012.01.021

Nable, R.O., Bañuelos, G.S., and Paull, J.G. (1997). Boron toxicity. Plant and Soil 193, 181–198.

Nannipieri, P., Giagnoni, L., Landi, L., and Renella, G. (2011). Role of Phosphatase Enzymes in Soil. In Phosphorus in Action, F.E. Bünemann E., Oberson A., ed. (Springer, Berlin, Heidelberg), pp. 215–243.

Oberson, A., Pypers, P., Bünemann, E.K., and Frossard, E. (2011). Management Impacts on Biological Phosphorus Cycling in Cropped Soils. In Phosphorus in Action, F.E. Bünemann and A. Oberson, eds. (Berlin, Heidelberg: Springer), pp. 431–458.

Ochoa-Hueso, R., Collins, S.L., Delgado-Baquerizo, M., Hamonts, K., Pockman, W.T., Sinsabaugh, R.L., Smith, M.D., Knapp, A.K., and Power, S.A. (2018). Drought consistently alters the composition of soil fungal and bacterial communities in grasslands from two continents. Glob. Chang. Biol. 24, 2818–2827. DOI: 10.1111/gcb.14113

Pedrero, F., Kalavrouziotis, I., Alarcón, J.J., Koukoulakis, P., and Asano, T. (2010). Use of treated municipal wastewater in irrigated agriculture-Review of some practices in Spain and Greece. Agric. Water Manag. 97, 1233–1241. DOI: 10.1016/j.agwat.2010.03.003

Pierzynski, G.M., Vance, G.F., and Sims, J.T. (2005). Soils and Environmental Quality, 3rd edition (Boca Raton, CRC Press).

Plaza, C., Sanz, R., Clemente, C., Fernández, J.M., González, R., Polo, A., and Colmenarejo, M.F. (2007). Greenhouse evaluation of struvite and sludges from municipal wastewater treatment works as phosphorus sources for plants. J. Agric. Food Chem. 55, 8206–8212. DOI: 10.1021/jf071563y

Poole, R. (1990). Microbial mineral recovery. Trends Biotechnol. 8, 370–370. DOI: 10.1016/0167-7799(90)90235-p

Qadir, M., Sharma, B.R., Bruggeman, A., Choukr-Allah, R., and Karajeh, F. (2007). Non-conventional water resources and opportunities for water augmentation to achieve food security in water scarce countries. Agric. Water Manag. 87, 2–22. DOI: 10.1016/j.agwat.2006.03.018

Ragab, R., and Prudhomme, C. (2002). Climate change and water resources management in arid and semi-arid regions: Prospective and challenges for the 21st century. Biosyst. Eng. 81, 3-34. DOI: 3–34 10.1006/bioe.2001.0013

Rashid, A., and Ryan, J. (2004). Micronutrient constraints to crop production in soils with mediterranean-type characteristics: A review. J. Plant Nutr. 27, 959–975. DOI: 10.1081/PLN-120037530

Riah, W., Laval, K., Laroche-Ajzenberg, E., Mougin, C., Latour, X., and Trinsoutrot-Gattin, I. (2014). Effects of pesticides on soil enzymes: A review. Environ. Chem. Lett. 12, 257–273. DOI: 10.1007/s10311-014-0458-2

Rillig, M.C., Ryo, M., Lehmann, A., Aguilar-Trigueros, C.A., Buchert, S., Wulf, A., Iwasaki, A., Roy, J., and Yang, G. (2019). The role of multiple global change factors in driving soil functions and microbial biodiversity. Science 366 (6467), 886–890. DOI: 10.1126/science.aay2832

Ruiz-Navarro, A., Fernández, V., Abadía, J., Abadía, A., Querejeta, J.I., Albaladejo, J., and Barberá, G.G. (2019). Foliar fertilization of two dominant species in a semiarid ecosystem improves their ecophysiological status and the use efficiency of a water pulse. Environ. Exp. Bot. 167, 103854. DOI: 10.1016/j.envexpbot.2019.103854

Saha, S. (2010). Soil Functions and Diversity in Organic and Conventional Farming. In Sociology, Organic Farming, Climate Change and Soil Science. E. Lichtfouse, ed. (Dordrecht, The Netherlands: Springer), pp.275-301.

Salgot, M., Huertas, E., Weber, S., Dott, W., and Hollender, J. (2006). Wastewater reuse and risk: Definition of key objectives. Desalination 187, 29–40. DOI: 10.1016/j.desal.2005.04.065

Sardans, J., and Peñuelas, J. (2015). Potassium: A neglected nutrient in global change. Glob. Ecol. Biogeogr. 24, 261–275. DOI: 10.1111/geb.12259

Sardans, J., Rodà, F., and Peñuelas, J. (2004). Phosphorus limitation and competitive capacities of *Pinus halepensis* and *Quercus ilex* subsp. *rotundifolia* on different soils. Plant Ecol. 174, 307–319. DOI: 10.1023/B:VEGE.0000049110.88127.a0

Scotti, R., Bonanomi, G., Scelza, R., Zoina, A., and Rao, M.A. (2015). Organic amendments as sustainable tool to recovery fertility in intensive agricultural systems. J. Soil Sci. Plant Nutr. 15, 333–352. DOI: 10.4067/s0718-95162015005000031

Shao, Y., Yang, K., Jia, R., Tian, C., and Zhu, Y. (2018). Degradation of triclosan and carbamazepine in two agricultural and garden soils with different textures amended with composted sewage sludge. Int. J. Environ. Res. Public Health 15, 1-14. DOI: 10.3390/ijerph15112557

Sharma, B., Sarkar, A., Singh, P., and Singh, R.P. (2017). Agricultural utilization of biosolids: A review on potential effects on soil and plant grown. Waste Manag. 64, 117–132. DOI: 10.1016/j.wasman.2017.03.002

Singh, B.K., Bardgett, R.D., Smith, P., and Reay, D.S. (2010). Microorganisms and climate change: Terrestrial feedbacks and mitigation options. Nat. Rev. Microbiol. 8, 779–790. DOI: 10.1038/nrmicro2439

Singh, B.P., Setia, R., Wiesmeier, M., and Kunhikrishnan, A. (2018). Agricultural Management Practices and Soil Organic Carbon Storage. In Soil Carbon Storage: Modulators, Mechanisms and Modelling, B.K. Singh, ed. (London: Academic Press), pp.207-244.

Smith, P. (2004). Carbon sequestration in croplands: The potential in Europe and the global context. Eur. J. Agron. 20, 229–236. DOI: 10.1016/j.eja.2003.08.002

Snow, D.D., Cassada, D.A., Biswas, S., Malakar, A., D'Alessio, M., Carter, L.J., Johnson, R.D., and Sallach, J.B. (2019). Detection, occurrence, and fate of emerging contaminants in agricultural environments (2019). Water Environ. Res. 91, 1103–1113. DOI: 10.1002/wer.1204

Sun, S., Sidhu, V., Rong, Y., and Zheng, Y. (2018). Pesticide Pollution in Agricultural Soils and Sustainable Remediation Methods: a Review. Curr. Pollut. Reports 4, 240–250. DOI: 10.1007/s40726-018-0092-x

Tejada, M., Moreno, J.L., Hernandez, M.T., and Garcia, C. (2007). Application of two beet vinasse forms in soil restoration: Effects on soil properties in an arid environment in southern Spain. Agric. Ecosyst. Environ. 119, 289–298. DOI: 10.1016/j.agee.2006.07.019

Tester, M., and Davenport, R. (2003). Na^+ tolerance and Na^+ transport in higher plants. Ann. Bot. 91, 503–527. DOI: 10.1093/aob/mcg058

Tóth, G., Hermann, T., Da Silva, M.R., and Montanarella, L. (2016). Heavy metals in agricultural soils of the European Union with implications for food safety. Environ. Int. 88, 299–309. DOI: 10.1016/j.envint.2015.12.017

Valdecantos, A., Cortina, J., and Vallejo, V.R. (2006). Nutrient status and field performance of tree seedlings planted in Mediterranean degraded areas. Ann. For. Sci. 63, 249–256. DOI: 10.1051/forest:2006003

Valdes-Abellan, J., Candela, L., Jiménez-Martínez, J., and Saval-Pérez, J.M. (2013). Brackish groundwater desalination by reverse osmosis in southeastern Spain. Presence of emerging contaminants and potential impacts on soil-aquifer media. Desalin. Water Treat. 51, 2431–2444. DOI: 10.1080/19443994.2012.747506

Vera, A., Moreno, J.L., García, C., Morais, D., and Bastida, F. (2019). Boron in soil: The impacts on the biomass, composition and activity of the soil microbial community. Sci. Total Environ. 685, 564–573. DOI: 10.1016/j.scitotenv.2019.05.375

Vitousek, P.M., Aber, J.D., Howarth, R.W., Likens, G.E., Matson, P.A., Schindler, D.W., Schlesinger, W.H., and Tilman, D.G. (1997). Human alteration of the global nitrogen cycle: Sources and consequences. Ecol. Appl. 7, 737–750. DOI: 10.1890/1051-0761(1997)007[0737:HAOTGN]2.0.CO;2

De Vries, F.T., and Shade, A. (2013). Controls on soil microbial community stability under climate change. Front. Microbiol. 4, 265. DOI: 10.3389/fmicb.2013.00265

Wang, F., Yao, J., Chen, H., Yi, Z., and Choi, M.M.F. (2014). Influence of short-time imidacloprid and acetamiprid application on soil microbial metabolic activity and enzymatic activity. Environ. Sci. Pollut. Res. 21, 10129–10138. DOI: 10.1007/s11356-014-2991-8

Wang, M., Chen, S., Chen, L., Wang, D., and Zhao, C. (2019). The responses of a soil bacterial community under saline stress are associated with Cd availability in long-term wastewater-irrigated field soil. Chemosphere 236, 124372. DOI: 10.1016/j.chemosphere.2019.124372

Welikala, D., Hucker, C., Hartland, A., Robinson, B.H., and Lehto, N.J. (2018). Trace metal mobilization by organic soil amendments: insights gained from analyses of solid and solution phase complexation of cadmium, nickel and zinc. Chemosphere 199, 684–693. DOI: 10.1016/j.chemosphere.2018.02.069

West, N.E. (1991). Nutrient Cycling in Soils of Semiarid and Arid Regions. In Semiarid Lands and Deserts, J. Skujins, ed. (New York: Marcel Dekker Inc.), pp. 295-332.

Yaalon, D.H. (1997). Soils in the Mediterranean region: What makes them different? Catena 28, 157–169. DOI: 10.1016/S0341-8162(96)00035-5

Yaashikaa, P.R., Senthil Kumar, P., Varjani, S.J., and Saravanan, A. (2019). Advances in production and application of biochar from lignocellulosic feedstocks for remediation of environmental pollutants. Bioresour. Technol. 292, 122030. DOI: 10.1016/j.biortech.2019.122030

Yermiyahu, U., Keren, R., and Chen, Y. (1995). Boron Sorption by Soil in the Presence of Composted Organic Matter. Soil Sci. Soc. Am. J. 59, 405–409. DOI: 10.2136/sssaj1995.03615995005900020019x

Zhang, B., Bai, Z., Hoefel, D., Tang, L., Wang, X., Li, B., Li, Z., and Zhuang, G. (2009). The impacts of cypermethrin pesticide application on the non-target microbial community of the pepper plant phyllosphere. Sci. Total Environ. 407, 1915–1922. DOI: 10.1016/j.scitotenv.2008.11.049

Zhen, Z., Liu, H., Wang, N., Guo, L., Meng, J., Ding, N., Wu, G., and Jiang, G. (2014). Effects of manure compost application on soil microbial community diversity and soil microenvironments in a temperate cropland in China. PLoS One 9, e108555. DOI: 10.1371/journal.pone.0108555

Zhu, J. (2002). Salt and Drought Stress Signal Transduction in Plants. Annu Rev Plant Biol 53, 247–273. DOI: 10.1146/annurev.arplant.53.091401.143329.Salt

Zolti, A., Green, S.J., Ben Mordechay, E., Hadar, Y., and Minz, D. (2019). Root microbiome response to treated wastewater irrigation. Sci. Total Environ. 655, 899–907. DOI: 10.1016/j.scitotenv.2018.11.251

Chapter 4

Responses of Aquatic Protozoans to Climate Change

Hartmut Arndt* and Mar Monsonís Nomdedeu

University of Cologne, Institute of Zoology, Department of General Ecology, Zülpicher Str. 47b, 50674 Köln (Cologne), Germany; * corresponding author

Email: Hartmut.Arndt@uni-koeln.de

DOI: https://doi.org/10.21775/9781913652579.04

Abstract

Heterotrophic protists (protozoans) play a key role in marine as well as in freshwater microbial food webs. They influence the abundance and taxonomic composition as well as size structure of bacteria and archaeans and thus regulate nutrient and carbon transfer in aquatic systems. Interactions between the different trophic levels of protists and different metazoan trophic levels are much more complex than generally believed. Laboratory experiments indicated a high capacity of genetic and non-genetic adaptation of heterotrophic protists to local temperature variations. Global warming does not only increase metabolic activity of protists and their impact on other microbes but since sensitivity to temperatures is potentially very different for each protist population unforeseeable impacts of warming have to be expected. Even if only one species in a system is temperature sensitive, the whole community is affected due to a close network of interactions. Warming has also a lot of important indirect effects on protists. Most important are changes in vertical stratifications and mixing processes in freshwater as well as in marine systems with fundamental effects on microbial food webs. Some of the major effects of temperature increases on free-living heterotrophic protist communities will be summarized on the background of complex microbial interactions.

Introduction

Protozoans - heterotrophic protists - are generally small sized unicellular organisms; their size ranges between 1 µm (some tiny heterotrophic flagellates) and more than 1 mm (some large ciliates and rhizarians (e.g. foraminiferans, radiolarians). Free-living heterotrophic protists include also many mixotrophic eukaryotes (Hausmann et al., 2003). Half of oceanic protists may contain autotrophic symbionts or kleptoplasts (e.g., Stoecker et al., 2009). Thus, protists comprise a large number of functions in food webs ranging from facultative or obligate primary producers, predators on viruses, bacteria and archaeans, as well as on algae, and on other heterotrophic protists and even on metazoans. Originally, the function of protists in aquatic systems was considered as a "loop" transferring dissolved organic matter (DOC) released from heterotrophic bacteria and autotrophic microbes and larger phototrophs to heterotrophic bacteria, heterotrophic flagellates and ciliates to metazoans, thus making DOC-carbon available to larger organisms (Sorokin and Paveljeva, 1972; Azam et al., 1983). This "loop"-concept is replaced today by considering protists as part of a complex food web, occupying not only one or two trophic levels, but due to their functional (and phylogenetic) diversity of protists, they may occupy a large number of trophic levels depending on the structure and functioning of the whole food web. Metazoans and protozoans interact at many different size and trophic categories far more complex than considered within the original loop-concept. All these trophic direct and indirect interactions as well as nutrient recycling and other feed-back mechanisms are influenced by climate changes as illustrated in Fig. 4.1 based on the original figure of Landry (2001).

The protistan "players"

The fundamental importance of protists in aquatic ecosystems was first mainly considered regarding the transfer of carbon to metazoan food webs and nutrient remineralization (e.g., Caron, 1982; Fenchel, 1987), many recent studies have indicated their specific role also in structuring bacteria and archaean communities and thereby influencing functioning of microbial food webs by selecting specific strains of bacteria etc. (Bonkowski, 2004; Jürgens and Massana, 2008; Jousset, 2012). Today, microbial food webs are considered to act as very important and essential components in all types of ecosystems (Arndt et al. 2000; Bever et al., 2012). The major players among protists in marine and freshwater habitats are phylogenetically and functionally extremely diverse. Fig. 4.2 should illustrate the quantitatively most important groups (up to our present knowledge) of aquatic free-living protists. It is evident that protists comprise a large variety of phylogenetically very distant groups (see also Simpson et al., 2017; Adl et al., 2019). Heterotrophic nanoflagellates as the major bacterivores in most aquatic environments range mainly in size from 1-15μm

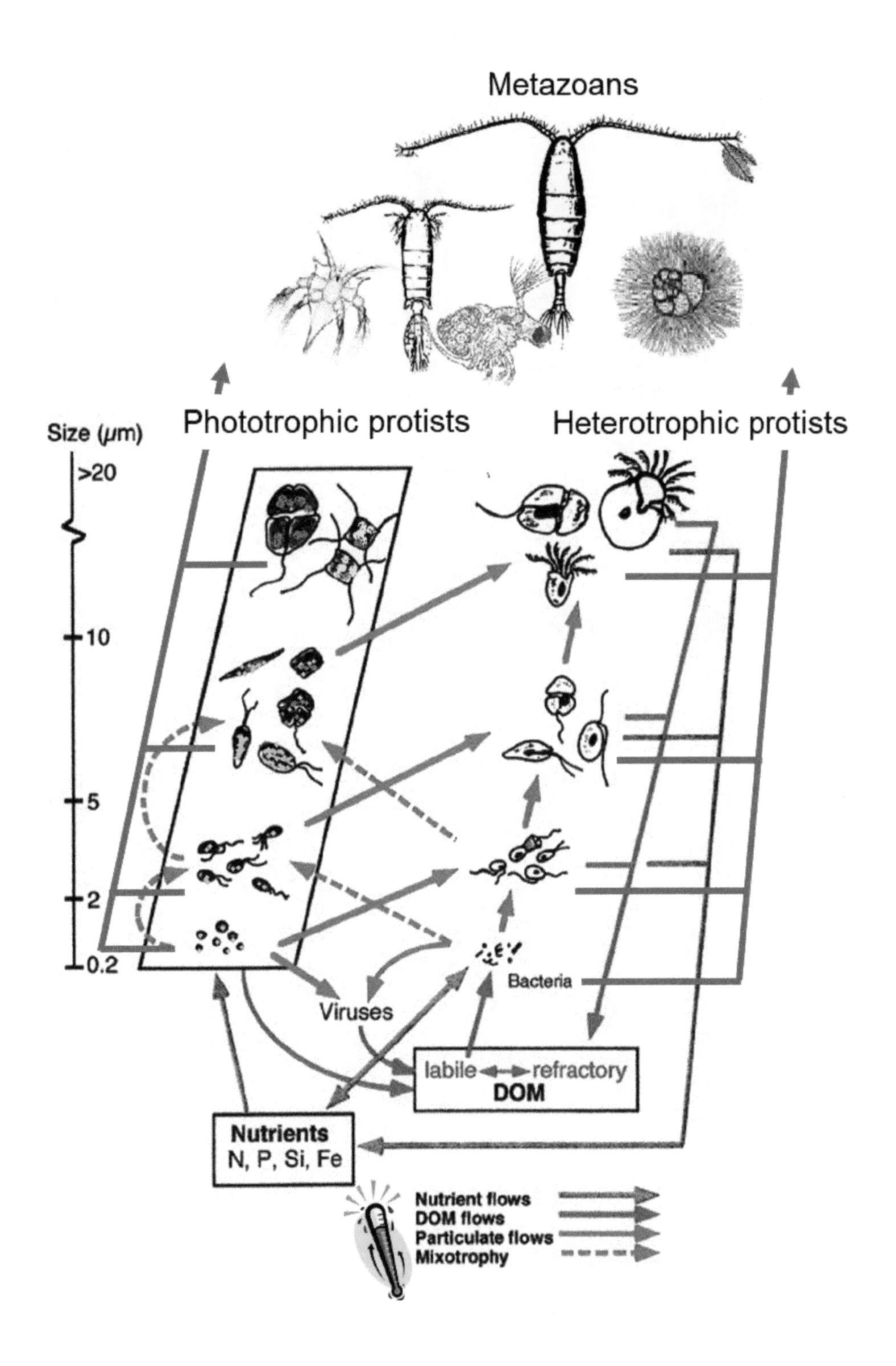

Figure 4.1 Protists as an integrated part of the food web. Flows of mineral nutrients and dissolved organics and interactions among various size classes of bacteria and protists are shown. Modified from Landry (2001).

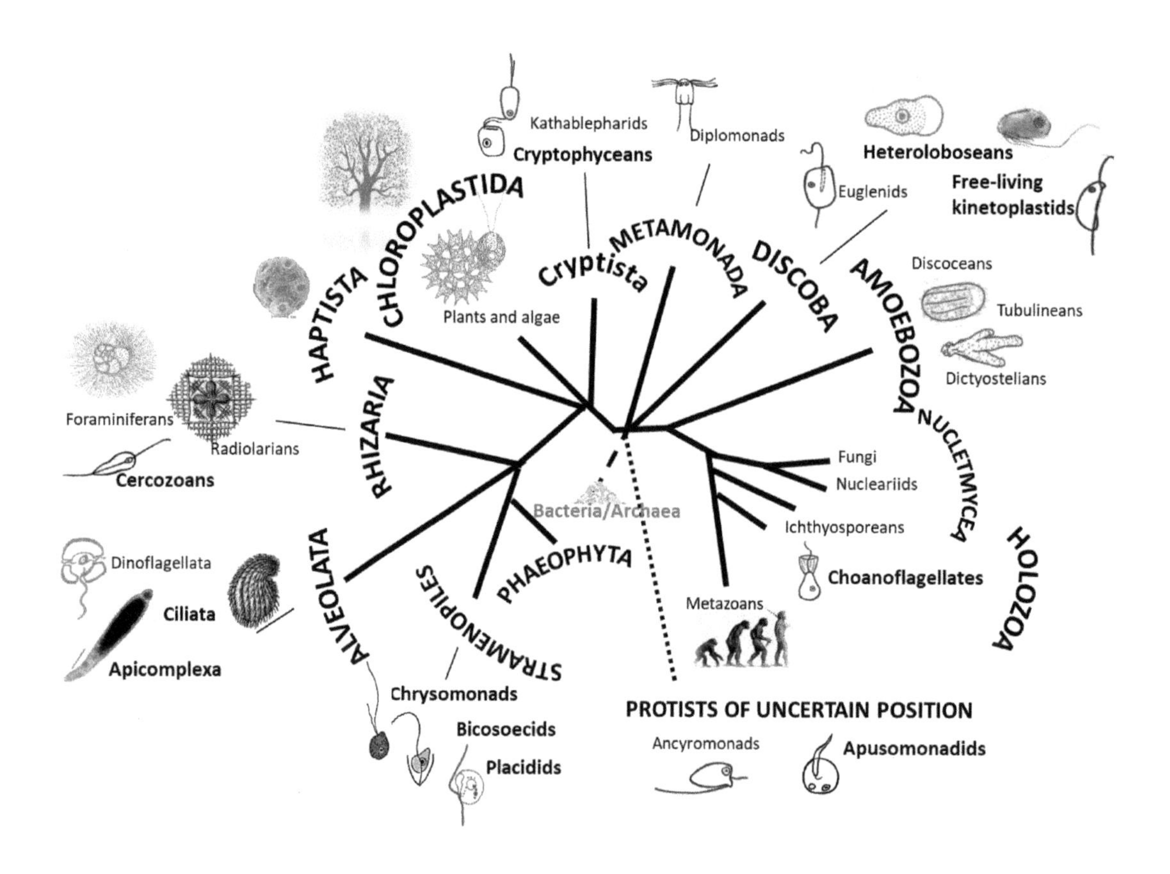

Figure 4.2 Current ideas of the systematic position of heterotrophic protists (after Simpson et al., 2017, and Adl et al., 2019) most common in environmental aquatic samples.

(as nanoflagellates, "HNF"). In addition, there are large heterotrophic flagellates ranging in size between 15μm and up to 450μm which can also consume algae and other nano- and even microplankton (Arndt et al., 2000). The dominant taxonomic groups within different marine and freshwater pelagic flagellate communities are stramenopile taxa, dinoflagellates, choanoflagellates, and kathablepharids, while benthic communities are dominated by euglenids, free-living kinetoplastids, cercozoans and others. Naked amoebae, especially small forms in the size range of 3-6μm mainly belonging to amoebozoans and heteroloboseans, can be important protists in benthic as well as pelagic systems (e.g., Arndt, 1993). In addition, there are large rhizarian forms (though quantitatively less important than nanoprotists), like foraminiferans and radiolarians (polycystineans) in oceanic environments (e.g., de Vargas et al., 2015). Ciliates are very diverse microprotists in pelagic as well as benthic habitats with a large variety of morphological and functional groups being important in all types of habitats (Foissner et al., 2008; de Vargas et al., 2015; Schoenle et al., 2017). A functional group which is often neglected in ecological studies are parasitic forms, which are known from nearly all phylogenetic lineages and may contribute a substantial part to protistan diversity in pelagic and benthic habitats. They can infect metazoans as well as protozoans and channel organic carbon back from higher trophic levels. Molecular studies of the last years have indicated that these principal components are typical for all kinds of aquatic environments ranging from pelagial to benthic and from shallow water to deep-sea environments (e.g., de Vargas et al., 2015; Grossmann et al., 2016; Gooday et al., 2020).

Present estimates indicate that protists are the genetically most diverse group of organisms overriding bacteria, plants and metazoans. The relatively small genetic distance between choanoflagellates and man (Fig. 4.2) might give an idea regarding the large genetic distance among the different groups of free-living protists. Naturally, this diversity is also connected with a diverse evolutionary adaption to their environment: The diversity of functional roles in ecosystems is much more complex than it can be anticipated from Fig. 4.1.

The thermal niche of protists

Regarding the ongoing climate change (e.g., IPCC, 2014), the question of how protists react to temperature is of fundamental importance in order to predict the consequences of global change for the functioning of microbial food webs. The ectothermal nature of single-celled organisms causes a direct impact of temperature on all metabolic processes in protists. Population growth, feeding activity and respiration rates are strongly dependent on temperature (e.g., Gillooly et al., 2001; Savage et al., 2004; Krenek et al., 2011). The unidirectional impact of temperature on the performances of a protist is described by its temperature reaction norm (Fig. 4.3). Growth rates of protists usually respond linearly to temperature (Montagnes et al., 2003). The thermal niche of an organisms/population is defined as the range of temperature in which moving trophont stages may occur. In many free-living protozoans, the phenotype remains fairly constant during stable environmental conditions, except for changes in cell size in the course of growth processes. Many protists are able to change their phenotypic appearance as a response to environmental factors. The most striking feature is the ability to form resting stages, which may protect the organism from unfavourable environmental conditions (e.g.,

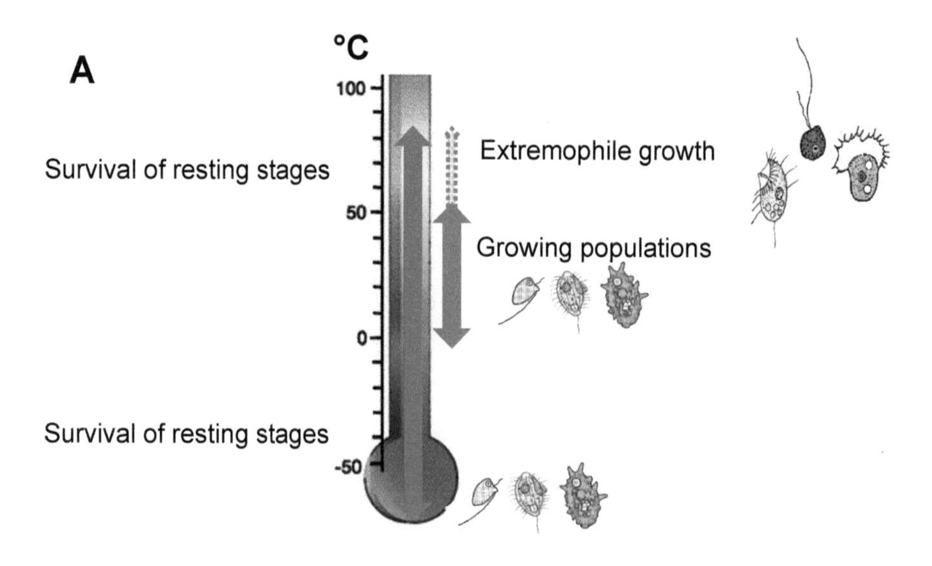

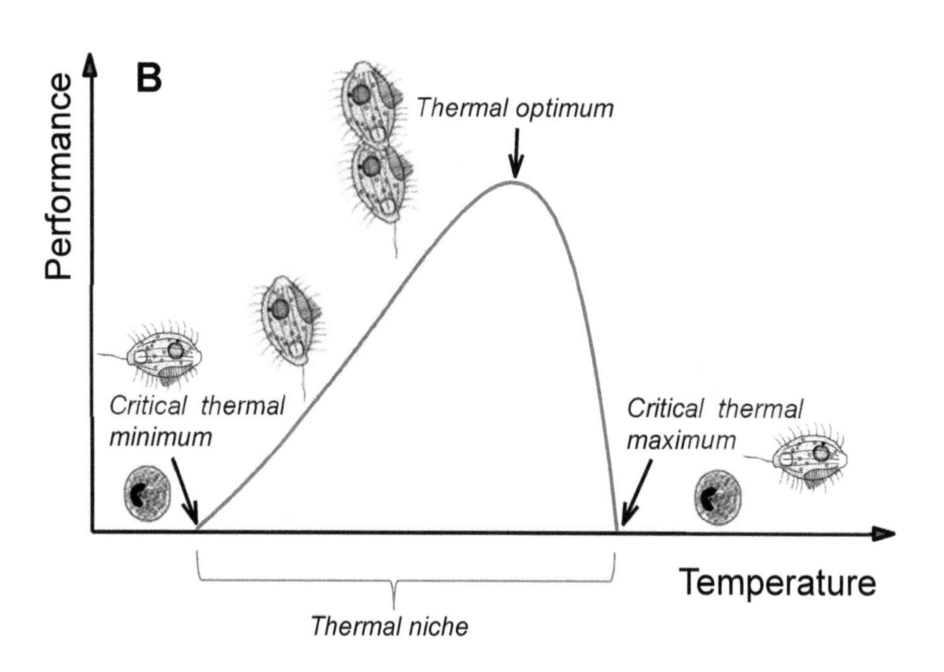

Figure 4.3 A) Temperature limits of protistan life. Many protist might survive low even extremely low temperatures in form of resting stages (shock freezing even in form of trophonts); there are representatives of all functional groups which might live up to temperatures of up to 50-70°C (for more details see text). B) Thermal niche of heterotrophic protists comprising the critical thermal minimum below which protists might survive only in dormant stages, the thermal optimum at which maximum performance (reproduction etc.) is realized, and the critical thermal maximum at which protists cannot be active any more (modified after Krenek et al., 2011).

low or high temperatures) and aid in dispersal to occupy favourable habitats. The formation of cysts is called cryptobiosis. This genetically defined process involves the formation of a resting stage, and consists of a reversible cytodifferentiation (Verni and Rosati, 2011). It is generally assumed that the temperature reaction norms were already determined for the majority of species (or at least the most habitual in laboratory investigations), and are available in literature. However, there are only a very few species (examples are e.g., *Tetrahymena pyriformis* (Schmid, 1967), *Paraphysomonas imperforata* (Caron et al., 1986), *Tetrahymena geleii* (Phelps, 1946), *Astasia sp.* (Schoenborn, 1947) for which the thermal niche is known. Montagnes et al. (2003) gave a comprehensive summary of available data sets; this list was not extended very much during the last years. Most dominant species in ecosystems have not been studied yet and data on tolerances regarding extreme temperatures are mostly missing.

Temperature creates a series of problems for protists such as structural devastation by ice crystals on the one extreme side to the denaturation of proteins and DNA on the other extreme side. Additionally, the solubility of oxygen is reduced at high temperatures and fluidity of membranes is dramatically increased (Rothschild and Mancinelli, 2001). Generally, protists may occur as (dehydrated) dormant stages below the freezing point and above temperatures of 35-40°C. In marine waters, especially in brine ice, it is known that protist trophonts may occur at temperatures even below -2°C, often physiologically adapted by the presence of freezing point reducing metabolites or salts (e.g., Garrison et al., 2005). Microbes with maximum growth rates at temperatures below 15°C are called psychrophilic. On the other hand, there are thermophile species among parasitic protists (e.g., some naked amoebae, Brown et al., 1983; Tyndall et al., 1989) which may live at temperatures of endothermic animals (37°C) and above.

Starting in the 1970s, extremophiles received increasing attention with the discovery of deep-sea hydrothermal vents. And due to PCR techniques, the presence of even non-cultivable forms became evident in hot springs (e.g. Reysenbach et al., 2001; Bown and Wolfe, 2006). However, up to now, only a very few species have been reported as trophonts at temperatures above 40°C. Among these exceptions are some ciliates, heterotrophic flagellates and amoebae. E.g. *Echinamoeba thermarum* isolated from hot springs grows only at temperatures between 33 and 57°C (Baumgartner et al., 2003), and *Trimyema minutum*, an anaerobic ciliate from hydrothermal vents, grows only at temperatures between 28 and 52°C (Baumgartner et al., 2002).

As mentioned above, species or population specific reaction norms are not well understood. And microbial biogeography with regard to temperature preferences, especially that of eukaryotic microbes, is still poorly studied (Dolan, 2005; Foissner et al., 2009; Weisse, 2008). The main reasons for this lack of investigations are diverse. On one hand, some scientists thought that microbes might not be subjected to geographical barriers and therefore all species may be found everywhere (Fenchel and Finlay, 2004), on the other hand, the population characteristics inherent to all microorganisms: small sizes, short generation times and high abundances make the biogeography question difficult to asset as a general concept. Meanwhile, molecular studies have shown that there are clear differences between locally separated protist populations and that speciation processes in protists are comparable to that known for metazoans (de Vargas et al. 1999; Arndt et al., 2020). Within the last years, several studies have focused on protistan microbial biogeography and found that within a species, geographically distant clones show differences in their temperature reaction norms (e.g., Gächter and Weisse, 2006; Krenek et al., 2011) indicating that heterotrophic protists may adapt to the thermal conditions of their environment and physiological differences do exist between organisms of a single species (Fig. 4.4). Species specific parameterization of mathematical models leads to better predictions (Yang et al., 2013; Brown and Gillooly, 2003; Krenek et al., 2011), especially if, in addition to single species temperature reaction norms, temperature dependent interaction strength is considered (Davis et al., 1998).

One can conclude from the temperature reaction norms known for protists that temperatures above about 35°C may be critical for most protists, meaning that free-living tropical communities may be under severe thermal stress upon further global warming. However, temperature increases of 2-3°C may not substantially change the performance of most protist species in temperate systems. Nevertheless, protists are a central component of carbon and energy flux in most aquatic systems and slight changes in their performance compared to other components of the food web might alter their relative function completely (e.g., Viergutz et al., 2007; Weitere et al., 2008).

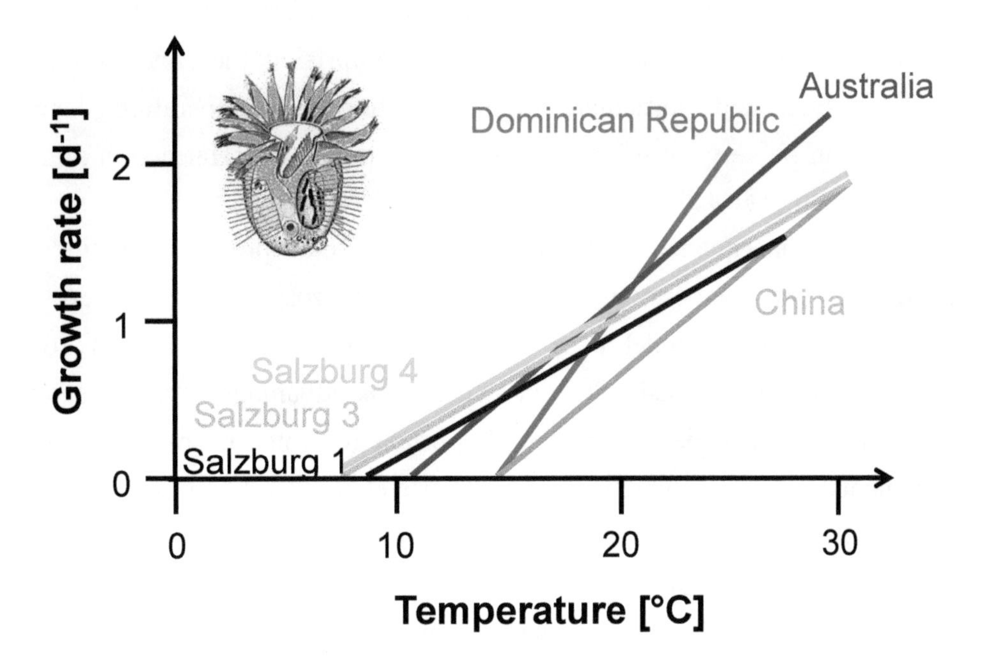

Figure 4.4 Growth rate of 6 different clones for the range of tolerated temperatures for the ciliate *Meseres corlissi* isolated from the Dominican Republic, Australia, China, and Austria near Salzburg (3 clones) (linear regressions of growth rate over the temperature range from the minimum temperature tolerated to the temperature at which maximum growth rates were observed; modified from Gaechter and Weisse, 2006).

Food web effects of temperature responses by protists

The knowledge on the function of protists in food webs has significantly increased in the course of the last decades and the old concept of the microbial loop is replaced by considering the activity of protists as being embedded in a network of complex trophic interactions (Fig. 4.1). There is an enormous variability in habitats, food preferences and trophic interactions within eukaryote microbes. Depending on habitat and season, the different trophic interactions may vary regarding its importance. Protozoans play a fundamental role in the transfer of organic matter and energy from bacteria and algal/plant producers to higher food web levels (Arndt et al., 2000; Azam et al., 1983; Bonkowski, 2004). Up to our present knowledge, protists are the genetically most diverse group of organisms (e.g., de Vargas et al., 2015). As mentioned above, warming affects single species directly through changes in the metabolic rate (Gillooly et al., 2001; Brown et al., 2004; Fig. 4.3), but indirect effects (e.g. food web mediated) may be as important as direct effects. All facets of the food web indicated by different colours in Fig. 4.1 comprising the flow of inorganic nutrients, dissolved and particulate organic matter are temperature dependent processes and can be very sensitive to changes in temperature. Harmon et al. (2009) showed how ecological and evolutionary complexities should be incorporated into predictions of the consequences of environmental change for protist populations.

Due to their short generation times and their microscopic size, heterotrophic protists are very useful model organisms for basic research on such complex interactions (Jessup et al., 2004). Several studies took advantage of these features of microbial food webs and delivered some important insights on how warming may affect organisms and their interactions (e.g., Aberle et al., 2012; Monsonís Nomdedeu et al., 2012; Fussmann et al., 2014). Hoekman (2010) showed that changes in temperature may cause changes in top-down and bottom-up effects in protozoan pitcher plant communities. Microcosm experiments run under well-controlled conditions may allow the analysis of isolated processes and to derive general ecosystem responses to climate change. Many investigations predict that climate change affects food web structure and ecosystem function (e.g., Petchey et al., 1999), although climate change models of ecosystem functioning usually ignore protistan microbial communities (e.g., Allison and Martiny, 2008).

One can derive from Fig. 4.4 that population specific reaction norms of protists (Gaechter and Weisse, 2006) matter when considering the impact of temperature on ecosystem functioning. Jiang and Morin (2004) focused on the different competition capacity of two protozoan species (*Colpidium* and *Paramecium*) and analysed how the interaction strength varied with temperature. At a constant experimental temperature, this variation drove one species or the other species to extinction depending on the experimental temperature. When temperature was set as a variable parameter, coexistence of both species was more likely to occur (Jiang and Morin, 2007).

Predictions of future temperature patterns on earth have fairly well advanced (e.g. IPCC, 2014), though the knowledge on the impact of temperature changes on populations, communities and ecosystems is still far behind (e.g. Walther et al., 2002; McKee et al., 2002) and this is especially true for reactions of food web interactions (Norf et al., 2007; Viergutz et al., 2007). Temperature changes can generate mismatches in the interactions of the components of a microbial food web. Variable predator-prey interactions due to temperature variations may have strong influences on population stability and could drive them to extinctions (e.g., Rall et al., 2010). Though, many mathematical models associated with simple predator-prey experiments predict species extinctions derived from climate change (e.g., Clements et al., 2014), others have shown that warming may increase food web stability (in the sense of oscillations with smaller amplitude) near to a temperature threshold (Fussmann et al., 2014). On the other hand, mathematical models show and experimental data indicated rapid evolution (genetically fixed adaptation) to environmental conditions (e.g., Ellner et al., 2011) and it has to be expected that heterotrophic protists also show a rapid evolution capacity that may reduce warming effects.

Match/mismatches of processes within food webs have been indicated as one of the principal negative effects of climate change (Walther et al., 2002; Durant et al., 2007; Ovaskainen et al., 2013). Investigations on the structure and functioning of microbial communities have shown strong temperature effects on phytoplankton and zooplankton production (Aberle et al., 2012) and on protistan community and population parameters like grazing rates (Norf et al., 2007; Rall et al., 2010), nutrient recycling (Norf and Weitere, 2010) or growth rates (Savage et al., 2004). Several recent studies have indicated that temperature effects on microbial protists have to be considered in concert with other factors (e.g., Kimmance et al., 2006). Two-factorial experiments with natural biofilm communities with manipulated temperature and either mechanical disturbance or food availability have shown that both disturbance and food supply can strongly alter the response of protistan communities towards warming (Norf and Weitere, 2010; Marcus et al., 2013). Thus, temperature effects can only be estimated when the communities and all environmental parameters are considered in its complex interactions. Investigations of microbial mesocosms allow addressing complex ecological questions at the community level, and help disentangling how indirect food web effects influence ecosystem responses to climate change (e.g., Andrushchyshyn et al., 2009).

As discussed above, there is empirical evidence that protist species and even populations show significant changes in temperature reaction norms, thus climate change does not affect all species and all processes in the same way. A mismatch between originally synchronized processes (e.g., life cycle strategies, changes in food concentrations, competitors and predators) may change the ecosystem functioning in future in an unforeseeable way. Recently, we were able to establish a controlled one-stage-chemostat system consisting of a bacterivorous ciliate (*Tetrahymena*) and two bacterial species, where the bacterium preferred by the ciliate predator is the superior competitor. Only one bacterium species (*Acinetobacter*) was temperature sensitive in the tested temperature range (20-25°C). Fig. 4.5 shows that temperature changes may

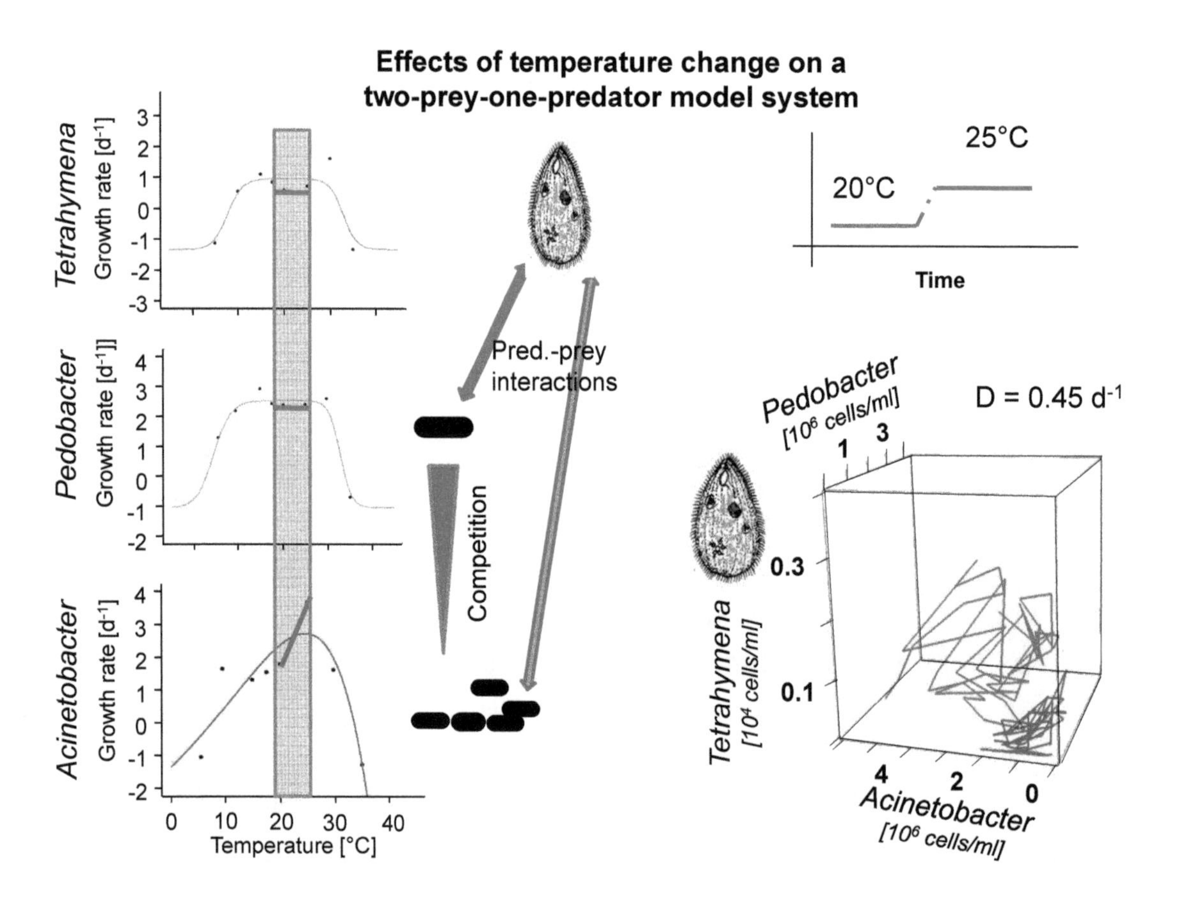

Figure 4.5 Effects of temperature increase on a model food web consisting of a predatory ciliate *Tetrahymena pyriformis* and two bacteria *Acinetobacter johnsonii* and *Pedobacter sp. Acinetobacter* is temperature sensitive and doubles its growth rate with an increase of temperature from 20°C to 25 °C (left panel), while the predatory ciliate and the other competing bacterium *Pedobacter* did not change growth rates. Indirect food-web effects drastically change interactions of all three species as indicated in the phase space diagram in the lower right panel. Blue line indicates the succession of abundances of all three species at low temperature (20°C) and red line after a rise of temperature to 25°C. Although experiments have shown no significant changes in the temperature reaction norm between 20 and 25°C for the ciliate, it changes its abundances significantly in response to the altered dynamics of the temperature sensitive food bacterium *Acinetobacter.*

drive the microbial system into different basins of attraction having fundamental effects also on species which are not sensitive to the offered temperatures due to species interactions. Such phenomena are difficult to derive from field data. Food web mediated effects may be of essential relevance for population dynamics and cannot be predicted directly from the single species temperature response (Monsonís Nomdedeu et al., 2012).

Temperature effects on complex protistan field communities

Multifactorial interactions of environmental factors and multi-species interactions make forecasts of temperature effects very difficult for natural communities. In order to estimate the effect of increasing temperatures on communities, several investigations took advantage of the analysis of large historical data sets (e.g., Walther et al., 2002; Mysterud et al., 2001; Ovaskainen et al., 2013, Wiltshire et al., 2010). Such studies give an idea about long-term effects of global warming, for example the uncoupling of trophic interactions in the spring peak of primary producers and zooplankton (Winder and Schindler, 2004; Wiltshire et al. 2010). Nevertheless, long-term data sets are rare especially regarding microorganisms.

In pelagic systems, either marine or freshwater, mesocosm studies were used to observe responses to warming mediated through different food web effects. Studies on the combined effects on the plankton community in Canada revealed no food web reaction at an increase in ultraviolet radiation but higher temperatures drove changes in the structure of the planktonic food web to smaller sized organisms (Rae and Vincent, 1998). Two-factorial experiments manipulating nutrients and temperature in a global change scenario of freshwater mesocosms in Denmark showed that picoplankton and heterotrophic nanoflagellates changed in a similar manner over time with lower abundances in winter than in summer. Warming itself had no effects on abundances, albeit it significantly modified the positive effect of the nutrients. Overall direct effects of warming were far less important than the nutrient effects (Christoffersen et al. 2006).

During the last ten years, several deep lakes exhibited significant changes in thermal stratification in the course of global warming. One of the best studied lakes in this respect is Lake Zürich (e.g., Posch et al., 2012, 2015). The lack of complete mixing or reduced mixing has led to a dominance of toxic blue-greens (*Planktothrix rubescens*) favoured by increased N:P ratios. Gas vacuoles formed by *P. rubescens* allow for buoyancy to accumulate within the depth of optimal irradiance and outcompetition of other algae. These changes in algal dominances due to the absence of holomixis had also dramatic effects on the protistan communities. Though the cyanobacterial dominance is also partly attributed to its grazing resistance towards grazers, Dirren et al. (2017) could show that species of the protistan genus *Nuclearia* (Nucleariidae, Nucletmycea; see Fig. 4.2) may feed on the blue-green filaments by phagocytosing small fragments or by breaking up cells during feeding. In connection with bacterial degradation this process contributes to a destruction and detoxification of blue-greens.

Also, significant changes in stratification and seasonal succession have already been observed in marine pelagic systems, which are especially dramatic in polar regions (e.g., Deppeler and Davidson, 2017). In marine mesocosm experiments, Aberle et al. (2012) reported that ciliate peak biomass was reached by 2.5 days earlier and occurred almost synchronously with biomass peaks of phytoplankton in the warm mesocosms (6°C higher temperatures compared to controls). Interestingly, these studies indicated that instead of a mismatch, warming might lead to a stronger match between protist grazers and their phytoplankton food altering the transfer of matter and energy towards higher trophic levels.

For benthic communities, Dossena et al. (2012) reported an altered size structure in response to warming and altered rates of key ecosystem processes they mediate. A 4°C rise led to steeper size spectra driven by declines in total community biomass and the proportion of large organisms in spring and to shallower size spectra driven by elevated total community biomass and a greater proportion of large organisms in autumn. The response of benthic microbial food webs has also intensively been studied in different biofilm communities consisting of complex communities of algae, bacteria, heterotrophic flagellates, ciliates, amoebae and small metazoans embedded in a matrix of exopolymeric substances (Arndt et al., 2003; Hall-Stoodley et al., 2004; Fleming and Wingeder, 2010). Biofilms play a key role in aquatic systems (Battin et al., 2003) which may be altered by temperature (Bouletreau et al., 2014; Ferreira and Canhoto, 2014). Several investigations took advantage of the natural complex structure of biofilms in order to study warming effects on food webs. Several studies combined *in-situ* measurements with *in-situ* manipulations of natural biofilm communities of the River Rhine (e.g., Norf et al., 2007; Norf and Weitere, 2010; Marcus et al., 2013). Temperature increases similar to the climate change model predictions of 2, 4, and 6 °C combined with manipulations of resource availability resulted in strong effects on the biofilm associated ciliate community. Those effects were strongly seasonal for example, higher temperatures in winter affected the early development of the biofilm reducing the carrying capacity (Norf et al., 2007). Furthermore, seasonal effects were observed in biofilms having reached their maximal carrying capacity; warming in winter positively affected both biomass and community structure, while in summer biomass was negatively affected and the community structure did not show changes (Norf and Weitere, 2010). Nevertheless, those negative effects in summer could be neutralized by increasing the resource availability, showing the transcendental role of complex interplays in microbial communities. Such interplays were also analysed regarding mechanical disturbances which led to reduced ciliate abundance in winter. This effect could be partially compensated by warming, whereas no effects could be observed in summer, neither due to disturbance nor due to warming (Marcus et al., 2013).

Temperature may also affect the transfer of matter and energy from biofilms (microbial community) to higher trophic levels in aquatic ecosystems. For example, grazing rates of invasive river clams *Corbicula* and *Dreissena* decreased with increasing temperature while growth rates of the microbial community increased indicating unbalanced predator-prey interactions (Kathol et al., 2009; Viergutz et al., 2007; Weitere et al., 2008; Weitere et al., 2009). Protozoans react faster to temperature increase than macrofaunal filter-feeders. Therefore their role controlling the benthic-pelagic coupling might be equal or become even more important with increasing temperatures (e.g., Kathol et al., 2011).

An additional factor which became important in recent years is the increasing speed of permafrost melting which might change protist communities. Stoupin et al. (2012) and Shatilovich et al. (2015) showed that protists cysts of flagellates and ciliates, respectively, have survived for several thousands of years in permafrost. A release of these genetically relict forms might influence evolutionary processes in protist communities in the near future in arctic habitats.

Conclusions

All in all, relatively little is known about the direct effects of climate change on microbial food webs although several empirical and theoretical studies indicate an important role of heterotrophic protists in food webs and for the resistance and resilience of food webs to climate change. Protists may occur in a very wide range of temperatures comprising environments ranging from such extremes as the vicinity of hot springs and ice and the temperature reaction norm of individual protist populations may be quite different. Only a few of the dominant protistan species have been studied regarding the full range of tolerated temperatures. Even if only one species in a system is temperature sensitive, the whole community is affected due to a close network of interactions. The complexity of impacts of global warming on the functioning of ecosystems has been largely overlooked in the past. Such complex features including indirect consequences of warming on food webs are difficult to analyse. Experimental studies on protist communities at semi-natural conditions including several trophic levels which contribute the major part of carbon transfer among eukaryotes in most ecosystems offer a very potent tool to understand even complex warming effects. The number of studies which consider the effect of warming on heterotrophic unicellular eukaryotes, in contrast to their fundamental role in the matter flux of aquatic ecosystems (Allison and Martiny, 2008; Shurin et al., 2012; Dossena et al., 2012), is still very low. The main advantage of investigations of warming effects on microbial food webs is that one reagent tube might contain already a large variety of trophic links and several general conclusions regarding the impact of global change on the structure and function of ecosystems can be drawn (Petchey et al., 1999; Dossena et al., 2012; Bouletreau et al., 2014) providing a close cooperation between experimentalists and theoreticians. Such studies might include also biodiversity responses (Eisenhauer et al., 2012; Finlay et al., 1997; Hooper et al., 2005). Dramatic changes in thermal stratification and seasonal dynamics in freshwater as well as in marine systems observed during the last 10 years indicate the urgent necessity of considering global climate change and its impact on microbial and especially protistan communities which comprise the majority of organismic genotypes.

Acknowledgements

This study was supported by grants from the German Research Foundation (DFG) to H.A. project numbers 268236062 (SFB 1211; B03, B02), AR288/17, 21, 23).

References

Aberle, N., Bauer, B., Lewandowska, A., Gaedke, U., and Sommer, U. (2012). Warming induces shifts in microzooplankton phenology and reduces time-lags between phytoplankton and protozoan production. Mar. Biol. 159, 2441-2453.

Adl, S.M., Simpson, A.G.B., Lane, C.E., Lukes, J., Bass, D., Bowser, S.S., Brown, M.W., Burki, F., Dunthorn, M., Hampl, V., Heiss, A., Hoppenrath, M., Lara, E., Le Gall, L., Lynn, D.H., McManus, H., Mitchell, E.A.D., Mozley-Stanridge, S.E., Parfrey, L.W., Pawlowski, J., Rueckert, S., Shadwick, L., Schoch, C.L., Smirnov, A., and Spiegel, F. (2012). The revised classification of eukaryotes. J. Eukaryot. Microbiol. 59, 429–493.

Allison, S.D., and Martiny, J.B.H. (2008). Resistance, resilience, and redundancy in microbial communities. Proc. Natl. Acad. Sci. U.S.A. 105, 11512-11519.

Andrushchyshyn, O.P., Wislon, K.P., and Williams, D.D. (2009). Climate change-predicted shifts in the temperature regime of shallow groundwater produce rapid responses in ciliate communities. Global Change Biol. 15, 2518-2538.

Arndt, H. (1993). A critical review of the importance of rhizopods (naked and testate amoebae) and actinopods (heliozoa) in lake plankton. Mar. Microb. Food Webs 7, 3-29.

Arndt, H., Dietrich, D., Auer, B., Cleven, E.-J., Gräfenhan, T., Weitere, M., and Mylnikov, A.P. (2000): Functional diversity of heterotrophic flagellates in aquatic ecosystems. In The Flagellates, B.S.C. Leadbeater and J. C. Green, eds. (London, UK: Taylor and Francis Ltd.), pp. 240-268.

Arndt, H., Schmidt-Denter, K., Auer, B., and Weitere, M. (2003). Protozoans and biofilms. In Fossil and Recent Biofilms, W.E. Krumbein, D.M. Paterson and G.A. Zavarzin, eds. (Dordrecht, Netherlands: Kluwer Academic Publ.), pp. 173-189.

Arndt, H., Ritter, B., Rybarski, A., Schiwitza, S., Dunai, T., and Nitsche, F. (2020). Mirroring the effect of geological evolution: Protist divergence in the Atacama Desert. Glob. Planet. Change 190, 103193. DOI: 10.1016/j.gloplacha.2020.103193

Azam, F., Fenchel, T., Field, J. G., Gray, J. S., Meyer-Reil, L. A., and Thingstad, F. (1983). The ecological role of water-column microbes in the sea. Mar. Ecol. Progr. Ser. 10, 257-263.

Battin, T.J., Kaplan, L.A., Newbold, J.D., and Hansen, C.M.E. (2003). Contributions of microbial biofilms to ecosystem processes in stream mesocosms. Nature 426, 439-442.

Baumgartner, M., Stetter, K.O., and Foissner, W. (2002). Morphological, small subunit rRNA, and physiological characterization of *Trimyema minutum* (Kahl, 1931), an anaerobic ciliate from submarine hydrothermal vents growing from 28°C to 52°C. J. Eukaryot. Microbiol. 49, 227-238.

Baumgartner, M., Yapi, A., Gröbner-Ferreira, R., and Stetter, K.O. (2003). Cultivation and properties of *Echinamoeba thermarum* n. sp., an extremely thermophilic amoeba thriving in hot springs. Extremophiles 7, 267-274.

Bever, J.D., Platt, T.G., and Morton, E.R. (2012). Microbial population and community dynamics on plant roots and their feedbacks on plant communities. Ann. Rev. Microbiol. 66, 265-283.

Bonkowski, M. (2004). Protozoa and plant growth: the microbial loop in soil revisited. New Phytol. 162, 617-631.

Bouletreau, S., Lyautey, E., Dubois, S., Compin, A., Delattre, C., Touron-Bodilis, A., Mastrorillo, S., and Garabetian, F. (2014). Warming-induced changes in denitrifier community structure modulate the ability of phototrophic river biofilms to denitrify. Sci. Total Environ. 466, 856-863.

Brown, J.H., and Gillooly, J.F. (2003). Ecological food webs: High-quality data facilitate theoretical unification. Proc. Natl. Acad. Sci. USA 100, 1467-1468.

Brown, P.B., and Wolfe, G.V. (2006). Protist genetic diversity in the acidic hydrothermal environments of Lassen Volcanic National Park, USA. J. Eukaryot. Microbiol. 53, 420-431.

Brown, J.H., Gillooly, J.F., Allen, A.P., Savage, V.M., and West, G.B. (2004). Toward a metabolic theory of ecology. Ecology 85, 1771-1789.

Brown, T.J., Cursons, R.T., Keys, E.A., Marks, M., and Miles, M. (1983). The occurrence and distribution of pathogenic free-living amoebae in thermal areas of the North Island of New Zealand. N.Z. J. Mar. Freshw. Res. 17, 59-69.

Caron, D.A., Davis, P.G., Madin, L.P., and Sieburth, J.McN. (1982). Heterotrophic bacteria and bacterivorous protozoa in oceanic macroaggregates. Science 218, 795-797.

Caron, D.A., Goldman, J.C., and Dennett, M.R. (1986). Effect of temperature on growth, respiration, and nutrient regeneration by an omnivorous microflagellate. Appl. Environ. Microbiol. 52, 1340-1347.

Christoffersen, K., Andersen, N., Søndergaard, M., Liboriussen, L., and Jeppesen, E. (2006). Implications of climate-enforced temperature increases on freshwater pico- and nanoplankton populations studied in artificial ponds during 16 months. Hydrobiologia 560, 259-266.

Clements, C.F., Collen, B., Blackburn, T.M., and Petchey, O.L. (2014). Effects of directional environmental change on extinction dynamics in experimental microbial communities are predicted by a simple model. Oikos 123, 141-150.

Davis, A.J., Jenkinson, L.S., Lawton, J.H., Shorrocks, B., and Wood, S. (1998). Making mistakes when predicting shifts in species range in response to global warming. Nature 391, 783-786.

Deppeler, S.L., and Davidson, A.T. (2017). Southern ocean phytoplankton in a changing climate. Front. Mar. Sci. 4, 40. DOI: 10.3389/fmars.2017.00040

De Vargas, C., Norris, R., Zaninetti, L., Gibb, S.W., and Pawlowski, J. (1999). Molecular evidence of cryptic speciation in planktonic foraminifers and their relation to oceanic provinces. Proc. Natl. Acad. Sci. U.S.A. 96, 2864-2868.

De Vargas, C., Audic, S., Henry, N., Decelle, J., Mahé, F., Logares, R., Lara, E., Berney, C., Le Bescot, N., Probert, I., Carmichael, M., Poulain, J., Romac, S., Colin, S., Aury, J-M., Bittner, L., Chaffron, S., Dunthorn, M., Engelen, S., Flegontova, O., Guidi, L., Horák, A., Jaillon, O., Lima-Mendez, G., Lukeš, J., Malviya, S., Morard, R., Mulot, M., Scalco, E., Siano, R., Vincent, F., Zingone, A., Dimier, C., Picheral, M., Searson, S., Kandels-Lewis, S., Tara Oceans Coordinators, Acinas, S.G., Bork, P., Bowler, C., Gorsky, G., Grimsley, N., Hingamp, P., Iudicone, D., Not, F., Ogata, H., Pesant, S., Raes, J., Sieracki, M.E., Speich, S., Stemmann, L., Sunagawa, S., Weissenbach, J., Wincker, P., and Karsenti, E. (2015). Eukaryotic plankton diversity in the sunlit ocean. Science 348(6237), 1261605. DOI: 10.1126/science.1261605

Dirren, S., Salcher, M.M., Blom, J.F., Schweikert, M., and Posch, T. (2014). *Ménage-à-trois:* The amoeba *Nuclearia* sp. from Lake Zürich with its ecto- and endosymbiotic bacteria. Protist 165, 745–758.

Dolan, J.R. (2005). An introduction to the biogeography of aquatic microbes. Aquat. Microb. Ecol. 41, 39-48.

Dossena, M., Yvon-Durocher, G., Grey, J., Montoya, J.M., Perkins, D.M., Trimmer, M., and Woodward, G. (2012). Warming alters community size structure and ecosystem functioning. Proc. R. Soc. B 279, 3011-3019.

Durant, J. M., Hermann, D.O., Ottersen, G., and Stenseth, N.C. (2007). Climate and the match or mismatch between predator requirements and resource availability. Climate Res. 33, 271-283.

Eisenhauer, N., Cesarz, S., Koller, R., Worm, K., and Reich, P.B. (2012). Global change belowground: impacts of elevated CO2, nitrogen, and summer drought on soil food webs and biodiversity. Global Change Biol. 18, 435-447.

Ellner, S.P., Geber, M.A., and Hairston, N.G. (2011). Does rapid evolution matter? Measuring the rate of contemporary evolution and its impacts on ecological dynamics. Ecol. Lett. 14, 603-614.

Fenchel, T. (1987). Ecology of Protozoa. The Biology of Free-living Phagotrophic Protists. (Madison/Wisconsin and Berlin, Germany: Science Tech Publishers/ Springer Verlag).

Fenchel, T., and Finlay, B.J. (2004). Here and there or everywhere? Response from Fenchel and Finlay. Bioscience, 54, 884-885.

Ferreira, V., and Canhoto, C. (2014). Effect of experimental and seasonal warming on litter decomposition in a temperate stream. Aquat. Sci. 76, 155-163.

Finlay, B.J., Maberly, S.C., and Cooper, J.I. (1997). Microbial diversity and ecosystem function. Oikos 80, 209-213.

Flemming, H.C., and Wingender, J. (2010). The biofilm matrix. Nature Rev. Microbiol. 8, 623-633.

Foissner, W. (1999). Protist diversity: Estimates of the near-imponderable. Protist 150, 363-368.

Foissner, W., Chao, A., and Katz, L.A. (2008). Diversity and geographic distribution of ciliates (Protista: Ciliophora). Biodiv. Conserv. 17, 345-363.

Fussmann, K.E., Schwarzmueller, F., Brose, U., Jousset, A., and Rall, B.C. (2014). Ecological stability in response to warming. Nature Climate Change 4, 206-210.

Gächter, E., and Weisse, T. (2006). Local adaptation among geographically distant clones of the cosmopolitan freshwater ciliate *Meseres corlissi*. I. Temperature response. Aquat. Microb. Ecol. 45, 291-300.

Garrison, D.L., Gibson, A., Coale, S.L., Gowing, M.M., Okolodkov, Y.B., Fritsen, C.H., and Jeffries, M.O. (2005). Sea-ice microbial communities in the Ross Sea: autumn and summer biota. Mar. Ecol. Progr. Ser. 300, 39-52.

Gillolly, J.F., Brown, J.H., West, G.B., Savage, V.M., and Charnov, E.L. (2001). Effects of size and temperature on metabolic rate. Science 293, 2248-2251.

Gillolly, J.F., Charnov, E.L., West, G.B., Savage, V.M., and Brown, J.H. (2002). Effects of size and temperature on developmental time. Nature 417, 70-73.

Gooday, A.J., Schoenle, A., Dolan, J., and Arndt, H. (2020). Protist diversity and function in the dark ocean - challenging the paradigms of deep-sea ecology with special emphasis on foraminiferans and naked protists. Europ. J. Protistol. 75, 125721. DOI: 10.1016/j.ejop.2020.125721

Grossmann, L., Jensen, M., Heider, D., Jost, S., Glücksman, E., Hartikainen, H., Mahamdallie, S.S., Gardner, M., Hoffmann, D., Bass, D., and Boenigk, J. (2016) Protistan community analysis: key findings of a large-scale molecular sampling. ISME J. 10, 2269–2279.

Hall-Stoodley, L., Costerton, J.W., and Stoodley, P. (2004). Bacterial biofilms: from the natural environment to infectious diseases. Nature Rev. Microbiol. 2, 95-108.

Harmon, J.P., Moran, N.A., and Ives, A.R. (2009). Species response to environmental change: Impacts of food web interactions and evolution. Science 323, 1347-1350.

Hausmann, K., Hülsmann, N., and Radek, R. (2003). Protistology. 3rd ed. (Stuttgart, Germany: Schweizerbart Science Publishers).

Hillebrand, H., Kahlert, M., Haglund, A.L., Beringer, U.G., Nagel, S., and Wickham, S. (2002). Control of microbenthic communities by grazing and nutrient supply. Ecology 83, 2205-2219.

Hoekman, D. (2010). Turning up the heat: Temperature influences the relative importance of top-down and bottom-up effects. Ecology 91, 2819-2825.

Hooper, D.U., Chapin, F.S., Ewel, J.J., Hector, A., Inchausti, P., Lavorel, S., Lawton, J.H., Lodge, D.M., Loreau, M., Naeem, S., Schmid, B., Setala, H., Symstad, A.J., Vandermeer, J., and Wardele, D.A. (2005). Effects of biodiversity on ecosystem functioning: A consensus of current knowledge. Ecol. Monogr. 75, 3-35.

IPCC (2014). Climate Change 2014: Synthesis Report. Contribution of Working Groups I, II and III to the Fifth Assessment Report of the Intergovernmental Panel on Climate Change [Core Writing Team, R.K. Pachauri and L.A. Meyer (eds.)], (Geneva, Switzerland: IPCC).

Jeuck, A., and Arndt, H. (2014). A short guide to common heterotrophic flagellates of freshwater habitats based on the morphology of living organisms. Protist 164, 842-860.

Jessup, C.M., Kassen, R., Forde, S.E., Kerr, B., Buckling, A., Rainey, P.B., and Bohannan, B. J.M. (2004). Big questions, small worlds: microbial model systems in ecology. Trends Ecol. Evol. 19, 189-197.

Jiang, L., and Morin, P.J. (2004). Temperature-dependent interactions explain unexpected responses to environmental warming in communities of competitors. J. Anim. Ecol. 73, 569-576.

Jiang, L., and Morin, P.J. (2007). Temperature fluctuation facilitates coexistence of competing species in experimental microbial communities. J. Anim. Ecol. 76, 660-668.

Jousset, A. (2012). Ecological and evolutive implications of bacterial defences against predators. Environ. Microbiol. 14, 1830-1843.

Jürgens, K., and Massana, R. (2008). Protistan Grazing on Marine Bacterioplankton. In Microbial Ecology of the Oceans, 2nd ed., D.L. Kirchman, ed. (Hoboken, NJ, USA: John Wiley and Sons, Inc.), pp. 383-441.

Kathol, M., Fischer, H., and Weitere, M. (2011). Contribution of biofilm-dwelling consumers to pelagic-benthic coupling in a large river. Freshw. Biol. 56, 1160-1172.

Kathol, M., Norf, H., Arndt, H., and Weitere, M. (2009). Effects of temperature increase on the grazing of planktonic bacteria by biofilm-dwelling consumers. Aquat. Microb. Ecol. 55, 65-79.

Kimmance, S.A., Atkinson, D., and Montagnes, D.J.S. (2006). Do temperature–food interactions matter? Responses of production and its components in the model heterotrophic flagellate *Oxyrrhis* marina. Aquat. Microb. Ecol. 42, 63–73.

Krenek, S., Berendonk, T.U., and Petzoldt, T. (2011). Thermal performance curves of *Paramecium caudatum*: A model selection approach. Europ. J. Protistol. 47, 124-137.

Landry, M.R. (2001). Microbial loops. In J.H. Steele, S. Thorpe and K. Turekian, eds., Encyclopedia of Ocean Sciences (London: Academic Press,), pp. 1763–1770.

Marcus, H., Wey, J.K., Norf, H., and Weitere, M. (2013). Disturbance alters the response of consumer communities towards warming: a mesocosm study with biofilm-dwelling ciliates. Ecosphere 5,10. DOI: 10.1890/ES13-00170.1

McKee, D., Atkinson, D., Collings, S., Eaton, J., Harvey, I., Heyes, T., Hatton, K., Wilson, D., and Moss, B. (2002). Macro-zooplankter responses to simulated climate warming in experimental freshwater microcosms. Freshw. Biol. 47, 1557-1570.

Monsonís Nomdedeu, M., Willen, C., Schieffer, A., and Arndt, H. (2012). Temperature-dependent ranges of coexistence in a model of a two-prey-one-predator microbial food web. Mar. Biol. 159, 2423-2430.

Montagnes, D.J.S., Kimmance, S.A., and Atkinson, D. (2003). Using Q10: Can growth rates increase linearly with temperature? Aquat. Microb. Ecol. 32, 307-313.

Mysterud, A., Stenseth, N.C., Yoccoz, N.G., Langvatn, R., and Steinheim, G. (2001). Nonlinear effects of large-scale climatic variability on wild and domestic herbivores. Nature 410, 1096-1099.

Norf, H., Arndt, H., and Weitere, M. (2007). Impact of local temperature increase on the early development of biofilm-associated ciliate communities. Oecologia 151, 341-350.

Norf, H., and Weitere, M. (2010). Resource quantity and seasonal background alter warming effects on communities of biofilm ciliates. FEMS Microb. Ecol. 74, 361-370.

Ovaskainen, O., Skorokhodova, S., Yakovleva, M., Sukhov, A., Kutenkov, A., Kutenkova, N., Shcherbakov, A., Meyke, E., and Delgado, M.D.M. (2013). Community-level phenological response to climate change. Proc. Natl. Acad. Sci. U.S.A. 110, 13434-13439.

Petchey, O.L., Mcphearson, P.T., Casey, T.M., and Morin, P.J. (1999). Environmental warming alters food-web structure and ecosystem function. Nature 402, 69-72.

Phelps, A. (1946). Growth of protozoa in pure culture. 3. Effect of temperature upon the division rate. J. Exp. Zool. 102, 277-292.

Pielou, E.C. (1981). Crosscurrents in biogeography. Science 213, 324-325.

Posch, T., Köster, O., Salcher, M.M., and Pernthaler, J. (2012). Harmful filamentous cyanobacteria favoured by reduced water turnover with lake warming. Nature Climate Change 2, 809-813.

Posch, T., Eugster, B., Pomati, F., Pernthaler, J., Pitsch, G., and Eckert, E.M. (2015). Network of interactions between ciliates and phytoplankton during spring. Front. Microbiol. 6, 1289. DOI: 10.3389/fmicb.2015.01289

Pounds, J.A., Bustamante, M.R., Coloma, L.A., Consuegra, J.A., Fogden, M.P.L., Foster, P.N., La Marca, E., Masters, K.L., Merino-Viteri, A., Puschendorf, R., Ron, S.R., Sanchez-Azofeifa, G.A., Still, C.J., and Young, B.E. (2006). Widespread amphibian extinctions from epidemic disease driven by global warming. Nature 439, 161-167.

Rae, R., and Vincent, W.F. (1998). Effects of temperature and ultraviolet radiation on microbial foodweb structure: potential responses to global change. Freshw. Biol. 40, 747-758.

Rall, B.C., Vucic-Pestic, O., Ehnes, R.B., Emmerson, M., and Brose, U. (2010). Temperature, predator-prey interaction strength and population stability. Global Change Biol. 16, 2145-2157.

Reysenbach, A.L., Voytek, M., and Mancinelli, R., eds. (2001). Thermophiles: Biodiversity, Ecology, and Evolution. (New York, USA: Kluwer Academic/Plenum Publ.)

Rothschild, L.J., and Mancinelli, R.L. (2001). Life in extreme environments. Nature 409, 1092-1101.

Savage, V.M., Gilloly, J.F., Brown, J.H., West, G.B., and Charnov, E.L. (2004). Effects of body size and temperature on population growth. Am. Nat. 163, 429-441.

Schmid, P. (1967). Temperature adaptation of growth and division process of *Tetrahymena pyriformis*. I. Adaptation phase. Exp. Cell Res. 45, 460-470.

Schoenle, A., Nitsche, F., Werner, J., and Arndt, H. (2017). Deep-sea ciliates: recorded diversity and experimental studies on pressure tolerance. Deep-Sea Res I, 128, 55-66.

Schoenborn, H.W. (1947). The relation of temperature to growth of *Astasia* (Protozoa) in pure culture. J. Exp. Zool. 105, 269-277.

Shatilovich, A., Stoupina, D., and Rivkina, E. (2015). Ciliates from ancient permafrost: Assessment of cold resistance of the resting cysts. Europ. J. Protistol. 51, 230–240.

Shurin, J.B., Clasen, J.L., Greig, H.S., Kratina, P., and Thompson, P.L. (2012). Warming shifts top-down and bottom-up control of pond food web structure and function. Phil. Trans. Roy. Soc. B 367, 3008-3017.

Simpson, A.G.B., Slamovits, C.H., and Archibald, J.M. (2017). Protist diversity and eukaryote phylogeny. In Handbook of the Protists, J.M. Archibald, A.G.B. Simpson and C.H. Slamovits, eds. (New York: Springer), 2nd ed., pp. 1-22.

Sorokin, Y.I., and Paveljeva, E.B. (1972). On the quantitative characteristics of the pelagic ecosystems of Dalnee Lake (Kamchatka). Hydrobiologia 40, 519-552.

Stoecker, D.K., Johnson, M.D., de Vargas, C., and Not, F. (2009). Acquired phototrophy in aquatic protists. Aquat. Microb. Ecol. 57, 279–310.

Stoupin, D., Kiss, A.K., Arndt, H., Shatilovich, A.V., Gilichinsky, D.A., and Nitsche, F. (2012). Cryptic diversity within the choanoflagellate morphospecies complex *Codosiga botrytis* – Phylogeny and morphology of ancient and modern isolates. Europ. J. Protistol. 48, 263-273.

Tyndall, R.L., Ironside, K.S., Metler, P.L., Tan, E.L., Hazen, T.C., and Fliermans, C.B. (1989). Effect of thermal additions on the density and distribution of thermophilic amoebae and pathogenic *Naegleria fowleri* in a newly created cooling lake. Appl. Environ. Microbiol. 55, 722-732.

Verni, F., and Rosati, G. (2011). Resting cysts: A survival strategy in Protozoa Ciliophora. Ital. J. Zool. 78, 134-145.

Viergutz, C., Kathol, M., Norf, H., Arndt, H., and Weitere, M. (2007). Control of microbial communities by the macrofauna: A sensitive interaction in the context of extreme summer temperatures? Oecologia 151, 115-124.

Walther, G.R., Post, E., Convey, P., Menzel, A., Parmesan, C., Beebee, T.J.C., Fromentin, J.-M., Hoegh-Guldberg, O., and Bairlein, F. (2002). Ecological responses to recent climate change. Nature 416, 389-395.

Weisse, T. (2008). Distribution and diversity of aquatic protists: an evolutionary and ecological perspective. Biodiv. Conserv. 17, 243-259.

Weitere, M., Dahlmann, J., Viergutz, C., and Arndt, H. (2008). Differential grazer-mediated effects of high summer temperatures on pico- and nanoplankton communities. Limnol. Oceanogr. 53, 477-486.

Weitere, M., Vohmann, A., Schulz, N., Linn, C., Dietrich, D., and Arndt, H. (2009). Linking environmental warming to the fitness of the invasive clam *Corbicula fluminea*. Global Change Biol. 15, 2838-2851.

Wiltshire, K.H., Kraberg, A., Bartsch, I., Boersma, M., Franke, H.-D., Freund, J., Gebuehr, C., Gerdts, G., Stockmann, K., and Wichels, A. (2010). Helgoland Roads, North Sea: 45 Years of Change. Estuar. Coasts 33, 295-310.

Winder, M., and Schindler, D.E. (2004). Climate change uncouples trophic interactions in an aquatic ecosystem. Ecology 85, 2100-2106.

Yang, Z., Lowe, C.D., Crowther, W., Fenton, A., Watts, P.C., and Montagnes, D.J.S. (2013). Strain-specific functional and numerical responses are required to evaluate impacts on predator-prey dynamics. ISME J. 7, 405-416.

Chapter 5

Terrestrial Fungi and Global Climate Change

Irina Sidorova and Elena Voronina*

Lomonosov Moscow State University, Moscow, Russia; * corresponding author

Email: irsidor2008@yandex.ru, mvsadnik@list.ru

DOI: https://doi.org/10.21775/9781913652579.05

Anecdotal evidence, like "when I was a kid, the sulphur tuft never fruited before the first of October, now I find it at the end of August," is nice, but does not convince me, nor do the politicians or scientists, that there is any change. Data, hard core data, long term data, preferably collected in a standard way —this is what is needed. Fluctuations from year to year, caused by changes in weather, have to be filtered out before we can say anything about the long-term, about the trends caused by climate changes.

Else Vellinga. MycoDigest: More Mushrooms Thanks to Climate Change?

Abstract

Global change is expected to affect fungi both directly and through associated organisms. Due to ubiquity and many-sided ecosystem roles, fungi can largely contribute to ecosystem resilience under negative impacts, thus their responses to warming, extreme weather events, carbon dioxide and nitrogen-elevated concentrations are essential. Fungal responses to climate change factors are hard to discriminate from non-climatic ones because of retaining gaps in fungal ecology and geography and huge variety across taxa and functional guilds. Here we present a review of recent data on different groups of terrestrial saprotrophic, mycorrhizal and pathogenic fungi perspectives under climate change with discussing possible mechanisms underlying effects observed or predicted.

Introduction

Global climate change became recently a ground for debates and discussion both for world policymakers and common citizens and surely got a keen interest of scientists. Such effects as global warming, changes in precipitation and sea level, greenhouse gases emission and change of albedo assessed in regular Intergovernmental Panel on Climate Change (IPCC) reports can create considerable risks for both terrestrial and water ecosystems (IPCC, 2019). Thorough investigation of all known biota groups' responses to climate change is a pressing issue to make a precise evaluation of its global impacts and potential danger.

Fungi are widely acknowledged ecosystem "drivers" and "engineers" shaping and stabilizing ecosystems (Read et al., 2004; Berke, 2010). It is hard to overestimate role of fungi in global nitrogen (N) and carbon (C) cycling, soil maintenance and fertility, and plant and animal performance (Dighton, 2016). But due to fungal predominantly hidden mode of life, microscopic size, and tremendous ecological and species diversity, the knowledge on their biodiversity, geography and ecology, not to mention responses to climate change, remains scarce as compared to plant and animal study. Despite of a number of objective difficulties in studying fungi in the light of global climate change, these organisms already deserve attention from the researchers. As follows from Fig. 5.1 representing search results from Google Scholar accessed on January 27, 2020, number of papers mentioning both "fungi" and "global climate change" increases visibly. This estimation is quite rough, but it is obvious that two last decades gave a noticeable augment to fungal climate responses research.

The chapter focuses on effects of global climate change on various ecological guilds of terrestrial non-lichenized fungi to predict responses of land ecosystems. It covers possible patterns of both observed and tentative climate change impact on fungal distribution, numbers and diversity with emphasis on mechanisms revealed, ecological consequences and debatable problems waiting to be solved.

Fungal ecology: an outline

Fungi represent one of the most ancient and diverse groups of living organisms. Its species number is an ambiguous issue and strongly depends on an estimation

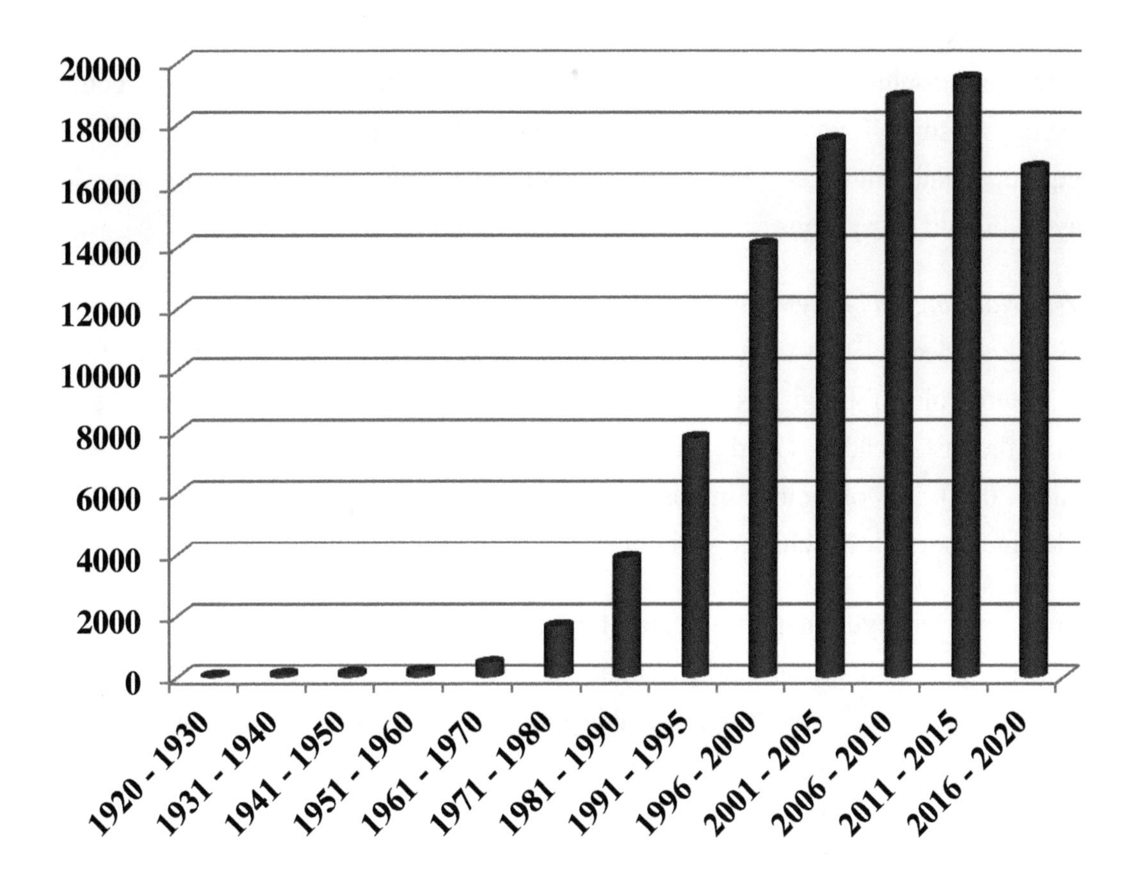

Figure 5.1 Dynamics of research on fungi under global climate change for last 100-yr period (based on Google Scholar search results, see explanation in text). Note: The time-scale step was reduced for period from 1991 until now to stress the sharpness of increase.

approach and species concept, but according to recent data there are about 120,000 already described fungal species and the total number (including hidden diversity) reaches 3.8 or even 6 million species (Blackwell and Dega, 2018). The terrestrialization of fungi became a key-stone event both in fungal evolution and land ecosystem shaping. There is a striking discrepancy between dates estimated either by fossil records or by molecular clocks, but even the most moderate, fossil-based, dating shows the presence of terrestrial fungi already in Ordovician (485–443 Mya) (Berbee et al., 2017). The loss of flagellar motility and development of filamentous organization are considered to be the main fungal adaptations for land environments (Naranjo-Ortiz and Gabaldón, 2019). The hyphal fast and in theory endless apical growth and developmental plasticity and extracellular enzymes` production perfectly suited osmotrophic life style and along with high reproduction rate and genetic plasticity ensured fungal success in occupation of terrestrial habitats making fungi ubiquitous and numerous. Shortly after their terrestrialization fungi became key players in land ecosystems.

Terrestrial fungi in ecosystems and fungal guilds

It is acknowledged that there is no contemporary land ecosystems free from fungi. Ecosystem role of fungi is considered to be great and underwent extensive study during the last decades (Gadd, 2006; Gadd et al., 2007; Dighton, 2016; Dighton and White, 2017). It is obviously impossible to predict ecosystems` dynamics and fitness or decline under global change without exploration of fungal responses.

Fungi in nutrient cycling and soil detoxification

Being the predominant terrestrial primary decomposers with extremely diverse and potent chemical activities fungi are able to utilize a wide range of complex both organic and inorganic substrates as a nutritional source. They are in charge for plant-derived complex compounds (including lignocellulose) cleavage (litter mineralization and wood biodegradation) which results in nutrient mobilization and nutrient cycles` facilitation (Dighton, 2016). According to comparative phylogenomics and molecular clock data, the fungi began consume lignocellulose just the first woody plants appeared on Earth (Watkinson, 2016). With the assistance of bacteria attracted by hyphal exudation and inhabiting mycelium-derived unique microniche called hyphosphere (Voronina and Sidorova, 2017), fungi can play an important role in bioweathering. "Rock-eating" fungi can extract nutrients from apatite, silicates of feldspar and mica groups, and other minerals (Rosling et al., 2004; Rosenstock, 2009; Römheld and Kirkby, 2010). Thus nutrients from recalcitrant substrates become available for primary producers and fungi can be considered as "the eye of the needle through which virtually all nutrients must pass" (Jenkinson, 1990). The back side of high enzymatic activity coupled with short generation time and high mutation speed in fungi is their ability to degrade toxic chemicals invading soil due to human activities thus remediating polluted environment (Dighton, 2016; Walker and White, 2018). High tolerance to toxic metals or arsenic compounds and ability to sequester them in inactive forms in some mycorrhizal fungi is crucial for plants and is applied for reforestation of highly disturbed areas (Smith and Read, 2008).

Fungi and soil formation and maintenance

Another prominent ecosystem role of fungi is multi-faceted soil mediation. For "soils do not just occur" (Dighton, 2016) it requires a number of abiotic and biotic effects to transform parent rock into fertile habitat for plants and other associated organisms. Fungi help to create and maintain soil both by physical and chemical means. Mycelia largely contribute to initial weathering and subsequent soil aggregation, which is of great importance for soil fertility (Johnson et al., 2017). Chemical transformation of plant and animal remnants enriches soil in nutrients available for other soil biota groups. Fungal biomass in soils estimated as 500 – 5000 kg per ha and becoming necromass can noticeably contribute to soil organic matter (Hays and Watson, 2019). Ectomycorrhizal fungi mycelium considered as one of the most significant soil organic C reservoir comprising about 20–50% of plant photosynthates (Fernandez et al., 2016). Basidiomycete saprotrophs equipped with peroxidases required for lignin break-down serve as N repository (Watkinson, 2016). Hyphal activity strongly modifies soil environment (humidity, pH, chemistry, etc.), especially the locations of robust aggregates such as perennial mycelial mats. They act as a locus both for acidification-based bioweathering and associative N_2 fixation (Griffiths et al., 1994).

Fungi in food web and biotic interactions of fungi

Fungi are linked to a wide range of other biota groups through direct or indirect interactions. Besides associated hyphosphere bacteria fungi can serve as mutualistic symbionts, pathogens or food source for numerous eukaryotes. Mycorrhizal symbioses with plants are as ubiquitous as terrestrial fungi themselves comprising more than 90% of vascular plants species around the world (Brundrett and Tedersoo, 2018). Most boreal and temperate tree species are obligatory mycorrhizal, so this fungal symbiosis is necessary for wood ecosystems` formation and functioning. Mycorrhizal fungi are involved in nutrient cycling by creating a mycelial network linking a number of plants with a common nutrient transfer stabilizing ecosystem and preventing nutrient loss. They determine plant community composition and drive plant successions by selecting their hosts and increasing its fitness (Smith and Read, 2008). Pathogenic fungi are commonly closely and selectively connected to their hosts. Mostly such fungi are biotrophs and normally do not devastate their hosts` populations maintaining equilibrium with the hosts and serving rather as regulators of their numbers in multispecies communities. Yet crop monocultures are subjected to severe epidemics which result in great financial losses and even food security breach

(Agrios, 2005). At last fungi themselves can serve as food source for a number of animal groups both invertebrate and vertebrate. Fungi are rich in proteins due to their ability to synthesize amino acids by conversion of substrates with high C: N ratio (Watkinson, 2016). Large fruitbodies of macromycete fungi comprise the diet of small mammals, slugs and dipteran larvae. Microarthropods and nematodes feed upon mycelia of various fungal species making fungi a significant component of soil and dead wood food webs (Dighton, 2016).

Functional guilds of fungi

Terrestrial fungi evolved a number of nutritional strategies to obtain organic C from a range of sources. Fungal functional guilds are recognized according to the food source. The most significant terrestrial fungi guilds addressed in the chapter are listed in Table 1.

Factors affecting terrestrial fungi fitness, biodiversity and distribution

In order to recognize the potential responses of terrestrial fungi to global climate change, it is necessary to realize the factors affecting fungal life in the natural environment. It is "key to the development of predictive models and to our general understanding of the resilience and resistance of fungal communities to environmental perturbations such as climate change" (Walker and White, 2018), but our knowledge on fungal requirements under non-laboratory conditions is insufficient. It is generally considered that filamentous fungi can grow near endlessly presuming the availability of nutrition and the absence of lethal stresses. Main abiotic factors affecting fungal survival are temperature, moisture, pH, nutrient forms and amounts, oxygen supply, and radiation (light and ultraviolet (UV)). For the outline of current knowledge on abiotic factors affecting fungal growth and reproduction see Watkinson (2016) and Walker and White (2018). Biotic interactions are also essential, especially for symbionts and pathogens depending in its organic C on their hosts. For example, range shifts in host plants under changing environment will lead to correspondent shifts in their fungal suite.

The fungal responses to global climate change display three main patterns:

1) increase / decrease of fungal fitness — mycelial growth, colonization rate, sporulation, or fruitbody density;
2) range shifts in fungi — changing the areas due to climatic factors, or habitat disturbance, or following hosts` shifts;

Table 1 The main ecological guilds of terrestrial fungi. The system of fungi follows Spatafora et al. (2017).

Fungal guild	Organic carbon supply	Main ecosystem functions	Main fungal lineages
Soil saprotrophic fungi (SSF)	Soil organic matter, fallen and buried plant debris	Litter mineralization, contribution to C and N global cycles	Many lineages within Mucoromycota, Ascomycota, Basidiomycota
Wood-decaying saprotrophic fungi (WSF)	Dead wood	Wood decomposition	Agaricomycetes (Basidiomycota), some Ascomycota
Arbuscular mycorrhizal fungi (AMF)	Living host plants, predominantly herbaceous, typically harmless for hosts	Host plant growth facilitation and protection, plant succession driving, soil aggregation, preventing nutrients from leaching	Glomeromycetes (Mucoromycota)
Ectomycorrhizal fungi (EcMF)	Living host plants, predominantly woody and shrubs, typically harmless for hosts	Host plant growth facilitation and protection, plant succession driving, litter mineralization, contribution to C and N global cycles and soil organic matter, bioweathering, soil detoxification, preventing nutrients from leaching	Many lineages within Agaricomycetes (Basidiomycota) and Pezizomycetes (Ascomycota)
Ericoid mycorrhizal fungi (ErMF)	Living host plants (Ericaceae), typically harmless for hosts	Host plant growth facilitation and protection, plant succession driving, litter mineralization, soil detoxification	Leotiomycetes (Ascomycota), some Basidiomycota
Plant pathogenic fungi (PPF)	Living or dead host plants, typically deleterious for hosts	Impact on plant community composition and structure, host population dynamics	Many lineages within Ascomycota, Pucciniomycetes and Ustilaginomycetes (Basidiomycota). Pseudofungi: Peronosporales (Oomycota)
Animal and human pathogenic fungi (APF)	Living animal and human hosts, typically deleterious for hosts	Impact on food web, host population dynamics	Entomophthoromycetes and Zoopagomycetes (Zoopagomycota), Saccharomycetes, Laboulbeniomycetes, Sordariomycetes and Eurotiomycetes (Ascomycota), Malasseziomycetes (Basidiomycota)

3) temporal shifts in fungi — changes in fruiting phenology, or disease symptoms development.

All these effects can be observed under laboratory or natural conditions, but it is always a task to clarify the mechanisms underpinned and main factors involved due to complexity of impact. At first, many effects are mediated by biotic interactions. Stimulation of fungal "foes" (e.g. mycelia grazers) or negative effect on "friends" (host plants) can be both deleterious for fungi even without direct influence on their own life parameters. Secondly, besides the climatic changes *per se* there are many human-induced predominantly negative effects such as destroying fungal habitats by land use and polluting the environment with fertilizers, pesticides, fungicides, herbicides etc. (Boddy, 2016a). The main obstacles in fungal responses to climate change factors research are summarized below.

1) Increase / decrease of fungal fitness. The main functional part of a fungus, mycelium, is hidden and hard to study under natural conditions. The most data at our disposal were obtained in the laboratory and must be expanded to natural conditions with much precaution. More obvious part is fruitbodies produced by so called macromycetes from Ascomycota and Basidiomycota. The fruitbody development and density was shown to be influenced predominantly by required combination of temperature and moisture (see in Ferreira and Voronina, 2016). But due to sharp interseasonal fluctuations in fungal fruiting it is always hard to discriminate between indigenous and exogenous factors` contribution and long-term monitoring required to reveal global change trends concealed by short-term climatic responses. Besides, even a decade halt in fruiting in some macromycete species do not ensure its total elimination from the community and regular and long-term application of molecular research approach (e.g. environmental DNA or metabarcoding surveys) to clarify fungal community dynamics (Dickie and St John, 2017).
2) Range shifts in fungi. Biological invasions are considered as global phenomenon threatening biodiversity and in some part driven by climate change (Catford et al., 2012). There are numerous data on alien species impact on indigenous communities accumulated during the last years, but it is acknowledged that underestimate of complexity of responses will entail "unrealistically simplistic answers to often more complicated problems" (Töpper et al., 2018). This is the case for vegetation, but it is related to fungi even more. Microbial biogeography is

critically behind plant and animal ones and due to difficulties of observation and lack of prominent morphological discrimination within fungi they were until recently considered as ubiquists lacking distinct geographical patterns (Peay and Matheny, 2017). Wide application of molecular approaches to fungal biodiversity study resulted in revealing enormous numbers of cryptic species thus debunking the previous concepts of "pan-global" fungi. For example, a model species in ectomycorrhizal studies, basidiomycete *Pisolithus tinctorius*, was of particular interest for Australian researchers serving as a mycobiont for economically important *Eucalyptus*. Subsequently the complexity of the taxon was shown, and *P. tinctorius sensu stricto* turned out to be a species non-native for Australia at all and non-compatible with *Eucalyptus* under natural conditions (Cairney, 2002; Martin et al., 2002). With paucity of mycogeographical data available we have little chance to make a precise evaluation of range shifts in fungi, especially those that have no obvious and specific associations with other groups of biota. Plant pathogens and mycorrhizal symbionts are easier to predict by their hosts range shifts, but it should be recognized that vegetation responses are not only climate-depending (Töpper et al., 2018).

3) Temporal shifts in fungi. The most obvious changes can be demonstrated through shifts in fruiting of macromycetes and pathogen-caused disease development. The problem is the same as for fungi fruiting phenomenon at all: the scarcity of long-term monitoring data. Macrofungi demonstrate enormous inter- and intra-seasonal fluctuations in their fruiting caused by either environmental or indigenous factors. This may produce considerable obstacles in studying phenology and discriminating global change impacts from short-term ordinary fluctuations. For more details on factors driving fruiting phenology shifts see Ferreira and Voronina (2016).

Impact of global climate change on terrestrial fungi

The main global climate change factors influencing living organisms including fungi directly or through associated species are: elevated concentrations of carbon dioxide (CO_2) and ozone (O_3), increasing N deposition, enlarging of UV radiation, global warming (both in soil and air), and altered precipitation (summary of effects see in Rillig et al., 2002; Ghini et al., 2012). No one of these factors acts apart of others, thus making hard to reveal the mechanism underlying. Gases (including CO_2) emission facilitate warming, greening trends (increasing of photosynthetic activity in vegetation) are caused by a complex action of CO_2 elevation, N deposition, increase of diffuse radiation and warming, etc. (IPCC, 2019). More than that, near all factors produces non-uniform effects region- and species depending. Some trends can be interfered with by side effects, i.e. plants` benefit from enriched CO_2 facilitating photosynthesis can be overweighed by increasing activity of pathogens and grazers. Some climatic changes, as altered precipitation, can lead both to drought and excessive moisture thus leading to community reorganization being stressful for one species and favourable for another. And, indeed, effects from climatic changes are near impossible to put apart from other global changes related to increasing land use and consequent land degradation with habitat destruction or fragmentation and intensive soil and air pollution (González-Cortés et al., 2012). One of the most important and the most hard to predict risks to ecosystem is disruption across trophic levels (Bell et al., 2019). To prevent it takes at least detailed knowledge on not only individual taxa responses, but on the ways various biotic interactions mediate the effects observed or modelled under natural, *in situ*, conditions.

The effects of global change factors on different functional groups of fungi reported up to date are summarized in Table 2. There are two main empirical approaches to study fungal responses to global change factors. The first one is represented by experimental studies modelling the effects contributing to global change. Fungi are exposed to increasing temperature, elevated CO_2 etc. to predict their response to future changes. The most common simulation are experiments with artificial CO_2 elevation (FACE — free-air CO_2 enrichment and OTC — open top chamber) (Staddon et al., 2002). The benefit of such experimental design is the accuracy of the assessment for the factor studied is predetermined and the response can be further visualized as a predicting model. The weak point is short time duration and impossibility to evaluate more than one, or few, factors which under environmental conditions can be outweighed by some neglected side effects. Thus elevated temperature stimulating decomposer fungi activity can also increase mycelium-grazing pressure by collembolans (A'Bear et al., 2014). The critical appraisal of plant disease causal fungi research under controlled environment facilities is provided by Gullino et al. (2018).

The second approach is based on visible changes in fungal community composition and spatio-temporal shifts observed worldwide and correlated with known changes in temperature and CO_2 elevation, precipitation and N deposition

Table 2 Responses of main fungal guilds and plants to different global change factors. Fungal guilds: see abbreviation expansions in Table 1. Responses detected: + = positive; - = negative; 0 = absent. Few data means insufficient number of studies. When more than one type of response was detected, it is marked with '/'.

Fungal parameter / response	Fungal guilds							Plant growth	References
	SSF	WSF	AMF	EcMF	ErMF	PPF	APF		
Elevated CO_2									
% colonization			+/0	+/0				+	Gorissen and Kuyper (2000), Rillig et al. (2002), Albertson et al. (2005), Ghini et al. (2008), Andrew and Lilleskov (2009), Hall et al. (2010), Melloy et al. (2010), Luck et al. (2011), Pickles et al. (2012), Boddy (2016a), Cotton (2018), Maček et al. (2019), Juroszek et al.(2020)
Hyphal length	+/0/-		+/0	+/0		-	+		
Sporulation						+			
Pathogenicity						+/0/-	+		
Nitrogen deposition									
% colonization			-	+/0/-				+/0/-	Arnolds (1991), Termorshuizen (1993), Lilleskov et al. (2001), Mack et al. (2004), Moore et al. (2008), Deslippe et al. (2011), Chen et al. (2017b), Purahong et al. (2018), Maaroufi et al. (2019)
Hyphal length	0		-	+/0/-	+/0/-				
Fruit bodies	- (?, few data)	0		0/-					
Altered precipitation									
Hyphal length			+/0/-					+/-	Eveling et al. (1990), O'Dell et al. (1999), Moore et al. (2008), Büntgen et al. (2012), Cotton (2018), Corredor-Moreno and Saunders (2020)
Sporulation and fruit bodies	+			+/0		+			
Warming									
% colonization				0	-			+	Bret-Harte et al. (2001), Clemmensen et al. (2006), Malcolm et al. (2008), Moore et al. (2008), Garcia-Solache and Casadevall (2010), Hall et al. (2010), Deslippe et al. (2011), Luck et al. (2011), Ghini et al. (2012), Fukasawa and Matsuoka (2015), Boddy (2016a), Ferreira and Voronina (2016), Lorberau et al. (2017), Yan et al (2017), Cowden et al. (2019), DeAngelis et al. (2019), Corredor-Moreno and Saunders (2020)
Hyphal length	+			+					
Sporulation and fruit bodies	+	+/-		0/-		-			
Pathogenicity						+	+		
Range shifts	+			+	+	+	+		
Phenology shifts	+	+		+		+			
Increase in O_3 concentration									
% colonization			- (few data)	0/- (few data)				0/-	Cairney and Meharg (1999), Rillig et al. (2002), Ghini et al. (2008), Andrew and Lilleskov (2009), Cotton (2018), Juroszek et al.(2020)
Hyphal length	0 (few data)			- (few data)					
Pathogenicity						+/0/-			
Increase in UV radiation									
% colonization				0 (few data)				0/-	Rillig et al. (2002), Ghini et al. (2012), Juroszek et al.(2020)
Sporulation						-			
Pathogenicity						0			

alteration. Mycorrhizal, pathogenic and litter saprotroph fungi are migrating northward concomitant with host / substrate tree species (Boddy, 2016a) and many macromycete species are proved to fruit earlier in Europe, North America and China (see references in Ferreira and Voronina, 2016). The evidences are plural, but the exact mechanisms underpinning and established relation to global climate changes are not always presented. The data available contain much controversy and uncertainty impeding generalization of fungal responses and there are yet large gaps in our knowledge of factors governing fungal life even in the unchanged environment. The large data sets of long-term collections-based records can be of great value as it proved by Andrew et al. (2019). The analyses revealed significant correlations of both saprotrophic and ectomycorrhizal fungi species richness with climatic conditions and allowed to predict global change factors affecting fungal richness patterns at large scales. Besides there are numerous examples of non-empiric "*in silico*" research approaches including meta-analysis of data-sets, published data reviews and generation of predicting models (Juroszek et al., 2020). Further the most recent insights in different fungal functional guilds responses to global change factors are overviewed and discussed.

Saprotrophic fungi

This fungal guild is serving as main decomposers in land ecosystems charged with mineralization processes and C and N cycling. Many fungi possess polymer hydrolysing enzymes and are able to degrade both natural (cellulose, chitin, cutin, hemicellulose, lignin, nucleic acids, proteins, starch, etc.) and artificial polymers including plastics (Watkinson, 2016).

Soil saprotrophs (SSF)

Near all major lineages of terrestrial fungi contain SSF (Table 1). According to research approaches they can be divided into macromycetes (Ascomycota and Basidiomycota with macroscopic fruitbodies) and micromycetes (sporulating without fruitbodies predominantly anamorphic fungi of ascomycete affinity). The global change effects on the first group are assessed largely by observations on fruiting phenology represented in long-term monitoring datasets. According to data summarized in Boddy (2016a) and Ferreira and Voronina (2016), the changes in fruiting patterns of basidiomycete species for the period since 1950s up now are rather prominent. The responses were expectedly species-specific, region and habitat-dependent, but the noticeable trend detected for fungal community in the south of the UK was the doubling of the length of the fruiting period. The mean start date of fruiting was significantly earlier and the date of its final was significantly later (for grassland species 13 and 48% correspondingly). The correlation with late summer temperatures and precipitation were shown (Boddy, 2016a). The second change revealed was the spring fruiting adding to normal autumn one for some species, and spring species tended to fruit earlier. Such drastic changes may have significant large-scale ecosystem consequences. Presuming that more prolonged fruiting season mirrors enriching of C budget of SSF (fruitbodies are rather "costly") and spring fruiting points at increased winter activity of fungi, the rate of decomposition should elevate too. The outcome for ecosystem will depend on vegetation response. If primary production fails to keep up with decomposition rate, it can result in greenhouse effect enhancement, especially in soils enriched in organic matter (Boddy, 2016a). On climate-driven SSF migrations assessed by fruitbody surveys see in Ferreira and Voronina (2016).

Nitrogen is a crucial macronutrient for fungi as well as other organisms, being a main limiting factor on boreal ecosystems, but excessive N is unfavourable for most fungal species. The most important ecosystem outcome of N deposition increase in boreal and temperate forest soils is the negative effect on decomposition rate because of C sequestration. The long-term N addition was shown to impede litter, but not humus decomposition, and was significant only with high-dosage inputs. High N concentrations were shown to decrease SSF biomass and cause community shifts, but it is suggested that current N deposition rates in boreal regions (≤ 12 kg N ha^{-1} $year^{-1}$) would not noticeably affect SSF community and humus and litter decomposition (Maaroufi et al., 2019).

The studies of SSF responses to global change not based on fruiting dynamics are few. They commonly discuss SSF within all soil fungi entity without discriminating from other fungal ecological guilds or even coupled with soil prokaryotic microorganisms. The responses observed are various and depend on experimental design and region. One of possible outcomes in response to warming and altered precipitation (shown under experimental conditions), was compositional resilience to predicted environmental change (Jumpponen and Jones, 2014). More than 40 000 fungal amplicons analysed from soil samples (tallgrass prairie in eastern Kansas, USA) were not restricted to SSF, but the data on Eurotiomycetes and

Dothideomycetes (representing most common soil decomposers) show no significant changes in response to climate alterations at community level, but taxa-specific responses are suggested. These data are consistent with a number of previous studies, but opposed by others (see references in Jumpponen and Jones, 2014), but it can be concluded that SSF communities have enough plasticity to adapt to environmental extremes, at least, created experimentally. The same absence of SSF decrease in richness and diversity was reported for Mediterranean shrubland (Birnbaum et al., 2019). More "taxa-oriented" research provides less optimistic for fungal-driven decomposition processes projections. With a model organism *Neurospora discreta* (Ascomycota) the potential for adaptation of soil fungi to warming was assessed (see references in DeAngelis et al., 2019). The adaptation providing fungal survival under elevated temperature resulted in prioritized sporulation rather than growth, thus supporting predicted decrease of microbial biomass in soil thus affecting C global cycling. However, linking SSF responses to soil C dynamics under global change is rather speculative, for there is great uncertainty in predicting of soil organic C future alterations (Sulman et al., 2018). Contrary, the hyphal growth increase in cord-forming basidiomycete *Phallus impudicus* was shown in response to CO_2 and temperature elevation, but these conditions were demonstrated to favour mycelia grazers too (A'Bear et al., 2014). The significance of considering trophic interactions in soil for correct prediction of soil ecosystem response to global change and magnitude of C cycle–climate feedbacks is also stressed by Crowther et al. (2015).

Wood-decaying saprotrophs (WSF)

This functional guild is represented by Basidiomycota and Ascomycota with basidiomycete fungi as key players. Wood decomposition is essential for global C cycling for cellulose became available for utilization by various animal groups due to fungal activity on recalcitrant lignin coat removing (Watkinson, 2016). As C sequestered in woody biomass is estimated approximately 80 Tg annually, it is obvious that environmental changes` impact on WSF will be crucial for modelling forest communities response to climate change (see references in Hiscox et al., 2018). Wood decomposition is a complex process accomplished by a wide range of different organisms and wood community dynamics is crucial for modelling of ecosystem responses to environmental change. WSF species differ in preferable wood-derived resources (lignin or cellulose) and in decomposing rate. Their competitive interactions drive community dynamics and its potential changes are crucial for future ecosystem resilience or vulnerability (Hiscox et al., 2018). This field of research is a priority for climate change effects on wood decomposition assessment, but there is a strong need for accumulation of ecologically relevant field studies data rather than the results of artificial laboratory experiments.

Similarly to SSF, fruiting shifts driven by climatic factors are recognized for WSF too. The main pattern is elongation of the fruiting season with changes of both start and final points. According to data for south UK fungal community, earlier start of fruitbody formation was detected for 53% species, and 20% had later end of the fruiting season (Boddy, 2016a). But, again, it would be irrelevant simplification to claim potential warming is beneficial for WSF in total. A large-scale field study aimed at coarse wood decomposition (*Pinus densiflora* dead logs) was accomplished at 12 sites in Japan (Fukasawa and Matsuoka, 2015). By pyrosequencing of fungal ITS rDNA authors analysed WSF community structure to reveal probable association with a number of factors including mean annual temperature. The results confirmed significant correlation with climatic conditions, but preferences towards lower or higher temperature were species specific. The first group of species (e.g., *Sistotrema brinkmannii* and *Trichaptum abietinum*) was suggested to eliminate locally from decomposer communities under global warming. The second (*Rigidoporus* sp., *Xeromphalina campanella* and *Skeletocutis odora*) was presumed to expand its range. But the considerations on significant correlation of any functional group within WSF (as white or brown rot fungi, see in Fukasawa and Matsuoka, 2015) with some environmental factor got no reliable supporting evidence.

However, trying to estimate and predict wood decomposition rate in changed environment we should bear in mind other than climatic factors interfering with this process. The impact of substrate traits on decomposition rates at a large scale was recognized and statistically confirmed as significant. Fukasawa and Matsuoka (2015) revealed the association of WSF with log diameter along with mean annual temperature. Hu et al. (2018) claimed the wood traits (namely diameter and N content) "were stronger predictors of variation in wood decomposition rates than climate variables" which explain about a half of the global variation in decomposition rates while climate parameters (temperature and precipitation) a fifth only. Another research performed under laboratory conditions with *Pinus*-associated WSF revealed the species specificity in relative contribution of climatic factors and wood traits to different WSF species activity (Venugopal et al., 2016). Fast decaying species such as

Dichomitus squalens and *Fomitopsis pinicola* were affected by wood quality only whereas slow decaying *Antrodia xantha* and *Gloeophyllum protractum* responded both to wood quality and climatic effects. Meanwhile, the wood traits are neglected by most global decomposition models and its incorporation into models seems to be necessary to link wood degradation to C cycling to predict global changes` ecosystem effects. Considering the role of substrate traits in decomposition processes the most prominent effect on WSF expected is underpinned by changes in tree biology thus creating altering niches for decomposers. Atmospheric CO_2 elevation promotes plant tissue lignification, and increase in phenolic content and C:N ratio makes wood substrates more recalcitrant for fungi (Boddy, 2016a). Another side of climatic effects needing further investigation is potentially contrasting responses to the same factors in different groups of decomposers sharing dead wood substrates as it was recently shown for WSF and saproxylic beetles (Thorn et al., 2018).

There is a large uncertainty in evaluation of increasing N deposition effect on WSF community and wood decomposition process. One opinion is suppression of lignin decomposition by inhibition of correspondent fungal enzymes (Boddy, 2016a), but another one claims that additive N has no significant effect neither on WSF composition, nor on decomposition rate (Purahong et al., 2018). An experiment with N addition was carried out in Central Germany at 4 forest sites with both broadleaf and conifer trees, and WSF biodiversity in wood was assessed directly by high-throughput sequencing (Purahong et al., 2018). Authors detected some changes in WSF community structure due to N concentration (presence/absence of species) similarly to results obtained for terrestrial SSF and EcMF under N fertilization (e.g., Lilleskov et al., 2002a). OTUs ascribed to a number of basidiomycete (*Amylostereum, Bjerkandera, Dentipellis, Ischnoderma, Phanerochaete, Phlebia, Polyporus, Ramaria, Skeletocutis*) and ascomycete (*Hymenoscyphus, Hypoxylon, Nemania*) genera eliminated from community after N concentration increase. Nevertheless, some WSF genera (e.g., *Hyphoderma, Mycena, Stereum*) being absent from decomposer assemblage of one tree species after N addition appeared under high-N conditions in the wood of another tree species (Purahong et al., 2018). The absence of N increase effect on overall community and its ecological functions was contrary to observed for SSF communities (Purahong et al., 2016 cited in Purahong et al., 2018) and was explained by N-limitation typically higher for wood compared to leaf litter. However, authors acknowledge that such experiments should be performed for other habitats and other N concentrations to make generalizations. Fungal plasticity and capacity to alleviate stressful changes in environment by prompt adaptation is well known, but it is not boundless, and more high N inputs probably can cause more drastic changes on WSF communities.

Warming effects on WSF communities are rarely reported, but if it goes to Polar Regions, it is obvious that tree vegetation shifts poleward and increase in woody area and substrate availability should benefit wood decayers. According to data obtained for Polar Urals, the complex impact of warming and precipitation elevation facilitates tree growth and decomposition processes and resulted in two-fold increase in WSF species richness over 60 years. Appearance of large-sized deadwood favoured aphyllophoroid poroid fungi which were revealed to increase in richness by 12% (Shiryaev et al., 2019).

The potential alteration of WSF community by direct effects of global changes factors or mediated by altered antagonistic interactions between decomposers or by altered substrate traits might result in noteworthy alterations in carbon-use efficiency and further in amounts of wood-sequestered CO_2 input into global cycle (Hiscox and Boddy, 2017). There is an urgent need of multi-factorial and multi-faceted studies of fungal-driven wood decomposition and concomitant WSF community interactions and dynamics *in situ* under ecologically realistic conditions.

Mycorrhizal fungi

Mycorrhizal fungi are crucial for terrestrial ecosystem being as widely distributed as plants themselves and presenting in near all types of plant communities (Smith and Read, 2008). Plant beneficial mycobionts have many-sided impact on plant growth and performance providing nutrient supply, protection from pathogens and alleviating plant abiotic stresses such as drought and intoxication. Mostly mycorrhizal fungi co-evolved with their hosts and have reciprocal interdependence on them (Brundrett, 2002). Mycorrhizal fungi responses to environmental changes are largely mediated by its hosts but concomitantly determine plant communities` future under climatic change conditions. Several types of mycorrhiza are recognized at the base of symbionts` taxonomy and symbiotic interface ultrastructure (Smith and Read, 2008). Here we address three main types distinguished for the most prominent ecological roles, namely arbuscular mycorrhiza, ectomycorrhiza, and ericoid mycorrhiza. Mycorrhizal fungi are considered as plant successions drivers (Read et al., 2004) and should by no means be neglected in predicting the vegetation responses to future changes. For different mycorrhizal types presume variety in host-fungus interactions which result in different ecosystem outcomes, the most prominent changes are expected in the case of alteration of dominant mycorrhizal type in a plant community.

The ways global change factors can affect mycorrhizal fungi are numerous. Roughly all the mechanisms can be divided into direct (impact on fungi) and indirect (impact on host plants). According to Soudzilovskaia et al. (2015), ca. 50% of the colonization rate variability is explained by global climate and soil patterns. CO_2 and ozone elevation acts indirectly, while effects of N deposition increase, altered precipitation and warming are considered to be more complex, affecting both symbionts concomitantly (Cotton, 2018). The future of mycorrhizal fungi under global change is crucial for terrestrial ecosystems as they are expected to enhance ecosystems` resilience and alleviate negative effects such as nutrient loss caused by increased precipitation (Martínez-García et al., 2017).

For all mycorrhizal fungi are dependent on host plant assimilation products, it is important to consider potential changes in photosynthesis caused by CO_2 elevation and it is the most studied effect related to mycorrhizas. The predictions of its rate are rather controversial due to different scenarios existing, but significant increase of CO_2 return to atmosphere during the last two centuries and its acceleration are acknowledged (Chen et al., 2017a). Increased photosynthesis should result in mycorrhizal abundance grow, but it was shown not for all species and turned out to depend on mycorrhizal type (Boddy, 2016a). Soil warming, according to experimental data, can act in the same way resulting in higher plant productivity and correspondent increase in mycorrhizal colonization due to plant higher nutrient require. But correspondent N deposition is suggested to interfere with temperature effect *per se* leading to diminishing of mycorrhizas significance in plant life thereby reducing its fitness and affecting nutrient cycling and losses at ecosystem scale (Cowden et al., 2019).

The variety of mycorrhizal fungi responses to different global change factors are outlined below, but in sum the most probable prerequisite of their future prosperity is C allocation dynamics. The positive response of mycorrhizal fungi to any environmental change might be expected only if it favours sufficient C influx from their hosts.

Arbuscular mycorrhizal fungi (AMF)

Arbuscular mycorrhiza is the most ancient and widespread plant symbiosis known to determine the plant terrestrialization event and further evolution of terrestrial plant communities (van der Heijden et al., 2015). AMF comprise a monophyletic group Glomeromycetes within Mucoromycota and are mandatory associated with mostly herbaceous plants from near all major plant lineages (Table 1). AMF unable to exist without host plant due to lack of a number of primary metabolism genes and are totally dependent on their hosts in C supply (Morin et al., 2019). The most prominent ecological roles of AMF are host plant fitness and competitiveness increase by nutrient fortification (phosphorus (P) and N supply), protection from intoxication and pathogens and mediating soil fertility (Johnson et al., 2017). AMF are essential component of soil biota worldwide and their responses to climatic changes worth thorough study (Treseder, 2016), but recognizing of direct environmental impact on AMF is hampered by their obligate biotrophy.

The great disparity between species richness of AMF (< 2,000) and their hosts (> 200,000) led to an erroneous view of AMF as non-specific and functionally identical symbionts (see in Smith and Read, 2008). The strict specificity of symbiosis is not shown but there are growing evidences of some selectivity towards host plants. The main problem is poor species concept in AMF caused by irrelevance of morphological traits for species delimitation and high intraspecies genetic variation (van der Heijden et al., 2015). AMF biodiversity is obviously understudied and underestimated thus hampering elucidation of particular species functionality, contribution to the community and response to any environmental factors under natural conditions. Only recently novel research tools such as stable isotope probing combined with nucleic acids labelling disclaimed the presumed equal contribution of different AMF species to ecosystem processes and allowed to distinguish the metabolically active part of community (Taylor et al., 2017). Some cue to observed variety of AMF responses to global change factors is provided by patterns of fungal biomass allocation (external hyphae / intraradical mycelium / spores) (Maherali and Klironomos, 2007 cited in Weber et al., 2019). Three functional AMF guilds were recognized by Weber et al. (2019): edaphophilic, rhizophilic and ancestral, different in their responses to environmental factors and interactions with host plants. The first group (e.g., Gigasporaceae) — fungi with more biomass allocated to external compared to intraradical hyphae, considered to facilitate plant nutrition by intensive mining for nutrients with vast mycelia which is particularly important for host plants with coarse roots with low ability to direct nutrient uptake. The second group (e.g., Glomeraceae) has limited external mycelia and main biomass locates in root tissues. These fungi are suggested to serve mostly protective function suppressing root pathogens and are crucial for plants with fine roots more susceptible to pathogen attack. "Ancestral" AMF differ in lower biomass and absence of allocation preference and presumed to be an ancestral state for other groups with exact functional specificity unclear (Powell et al., 2009 cited in Weber et al., 2019). These groups were shown to have different ecological preferences and limitations, so their differing responses to climate change should be considered.

Atmospheric CO_2 increase is expected to influence AMF through the impact on hosts` photosynthesis rate, but the actual effect until recently was unknown. It was considered that the more effective hosts` photosynthesis will result in higher C allocation to roots and supply to mycobiont and further to increase in AMF root colonization under elevated CO_2 conditions (e.g., Albertson et al., 2005), but this response is not uniform (Maček et al., 2019). However, the alteration or stability of mycorrhizal abundance alone could hardly provide ecologically relevant information on community-level response to any change if AMF diversity and composition remains obscure. The studies of AMF communities changes in richness and diversity are rather few and two contrasting types of response to CO_2 increase were expected (see details in Cotton, 2018). If CO_2 elevation would actually increase plant C supply to AMF, the removing of C limitation would probably promote competition thus leading to elimination of some species from community and diversity decline. By contrary opinion, the absence of C limitation would be compensated by soil nutrient limitation thus increasing the role of mycobionts for plant fitness and facilitating AMF biodiversity rise. But despite the plausibility of both premises, until recently the most research results indicated no significant impact of CO_2 elevation on AMF overall richness (Cotton, 2018). However, there are data demonstrating the effect on AMF biodiversity. Some insight into indigenous AMF community dynamics under CO_2 increase was provided by FACE long-term (more than 15 years) experiment in old-growth semi-natural grasslands in Northern Hemisphere (Maček et al., 2019). The exposure to 20% increased CO_2 concentration did not affect the root colonization significantly, but led to increase in taxa (OTU) richness. No change in AMF community evenness was detected under elevated CO_2 conditions. While alteration in population densities (both growth and decline depending on taxa) was revealed, the overall community-level changes were inconsistent and depend on sampling time and experimental treatment. These results confirm the fallibility of extrapolation of single or few species responses to a community level and prove the necessity of long-term research for recognizing global change effects. Besides, suggesting the recently revealed functional difference within AMF such local fluctuations can help to specify plant community responses to climatic factors. Another experiment with OTC-provided 8 years` exposure of paddy soil (Cuttack, India) to elevated CO_2 demonstrated contrasting responses of two AMF taxa (Panneerselvam et al., 2020). AMF community in flooded paddy soil analysed through Illumina MiSeq platform revealed drastic reduction of common species of Glomeraceae (Glomerales) and increase in *Scutellospora* spp. (Gigasporaceae, Diversisporales). For the first group is considered rhizophilic and r-strategists while the second — edaphophilic and k-strategists (Panneerselvam et al., 2020 and references therein) the selection favouring k-strategists against r-strategists might have outcome in preferential allocation of nutrients and organic C. The last experiment mentioned is of particular interest because the results obtained are inconsistent with the pattern presumed to be common based on a number of researches (see in Cotton, 2018): the selective benefit of Glomeraceae along with decrease of Gigasporaceae under the CO_2 enrichment.

It can be concluded that the majority of data indicates drastic AMF community shifts independent of region studied and experimental design which may have great outcome for plant communities. As for inconsistent numerical data regarded either AMF abundance or biodiversity, the probable explanation is the complexity of factors involved and the absence of common experimental protocols and approaches impeding comparing of results. The only unequivocal conclusion emerged is that the abrupt CO_2 elevation affects AMF (as well as other fungi) communities more drastic than gradual increase.

The main nutritional role of AMF known for a long time is P supply to the host with small intervention into N transformation and transfer for they inhabit soils with main limitation in P. However, recent studies stress the importance of AMF in plant N nutrition too (Mohan et al., 2014; Johnson et al., 2017). Besides, the popularity of this particular global change factor research can be explained by its relative simplicity for long-term investigations of fertilizer effect on soil biota provide study sites varying in level of N addition (Cotton et al., 2018). The N deposition increase may lessen the plant dependence on its symbionts with outcome in critical shifts in AMF communities due to obligate biotrophy of the fungi and often only facultative mycotrophy of their plant partners (Cowden et al., 2019). There are experimental data on N enrichment-caused AMF general suppression demonstrated both as species diversity and abundance decrease under the conditions of semi-arid steppe (Chen et al., 2017b). However, authors regard the results as variable and site-dependent with the possible explanation by decreased host C allocation to AMF under particular soil nutrient status. More prominent AMF-inhibiting effect on intraradical abundance coupled with species richness decrease was shown in the case of combined treatment of N-P addition (Camenzind et al., 2014). Another interesting effect observed in tropical montane rainforest in southern Ecuador was AMF community shifts resulted in rare species elimination and variation in response between different taxonomic groups. Diversisporales appeared to be affected mainly be N enrichment while Glomerales were highly sensitive only to P additions. There are many other research results consistent with mentioned, but opposite, stimulating, effect of N deposition increase on AMF was reported too (see in Cotton, 2018). At least partly the discrepancy can be underpinned by different research approaches and initial nutrient status of soils studied. Due to diversity of experimental design it is hard to carry out a correct comparison of data available, but the view of AMF community shifts with Diversisporales more vulnerable and Glomerales more resistant to elevated N concentrations is rather constant.

Precipitation alteration as a factor affecting AMF communities is not quite well studied yet. Most research revealed only small changes (e.g., only one species showed noticeable response), impact varied greatly across plant communities and environmental conditions, and no ecologically realistic patterns could be established on the base of information available (see references in Cotton, 2018). Implying the recognized AMF role in alleviating drought stress in plants this aspect deserves further investigation.

The direct effect of warming on AMF growth and development is hard to discriminate from the impact on host plant. However studies with compartmentalized chambers available only for hyphae but not to plant roots elucidate AMF growth responses. As always they were inconsistent, but some species were obviously stimulated by elevated temperature (see references in Cotton, 2018). The temperature-induced lag time of spore germination can be particularly significant for the order of root colonization is known to shape intraradical AMF community by prevalence of the first colonist (Werner and Kiers, 2015 cited in Cotton, 2018).

Elevated O_3 concentrations are presumed to be unfavourable for plant C fixation thus lessening C allocation to AMF. According to data available, it takes rather long time for noticeable changes to emerge due to cumulative nature of the effect. Therefore the short-term experiments or research of young plants might not reveal any significant changes (Cotton, 2018).

For no global change factor acts in isolation, there is a pressing need in multifactorial experiments allowing obtaining more realistic data to predict future alterations in fungal community structure and functions. This type of research is yet rare but a seven-year experiment assessing concomitantly effects of aridity, N deposition, and plant invasions demonstrated changes both in AMF community composition and the interaction of fungi with hosts (Weber et al., 2019). The factors studied showed multidirectional effect. The aridity caused decrease in total AMF abundance, N enrichment — increase in colonization while reduction of external mycelia. Invasive grasses differed from indigenous ones in tendency to host rhizophilic rather than edaphophilic AMF. The responses of different functional AMF guilds were not uniform, thus stressing again the necessity of considering difference of ecological strategies in fungi for their community composition and diversity might significantly affect their contribution to ecosystems` functioning under climate changes.

Ectomycorrhizal fungi (EcMF)

EcMF is essential fungal guild for boreal and temperate forest biomes being mutualistic symbionts of tree species. Most EcMF belong to Agaricomycetes (Basidiomycota), but there are some ascomycete (Pezizomycetes) and even "zygomycete" (Endogonales) species too. Most ectomycorrhizal symbioses are mandatory for both partners, and EcMF are expected to respond to global changes affecting them indirectly by influence on hosts. Temperature changes drives tree migration to the poles under warming and backward under temperature decrease as it follows from paleoecological data (Boddy, 2016a). Plant success in their new habitats will depend on co-migration with EcMF or ability to adapt to new aboriginal symbionts, and *vice versa*. The most fungi produce air-dispersed spores which facilitate migration, but those who have any special requirements for dispersion, such as animal vectors, are vulnerable at the face of drastic vegetation changes.

EcMF are predominantly macromycetes with conspicuous fruitbodies, so their responses to climatic factors can be estimated on the basics of fruitbody density and fruiting phenology, the same as for saprotrophs. Long-term monitoring data suggest the temporal shifts in EcMF fruiting but the patterns differ from SSF due to another C source (Boddy, 2016a; Ferreira and Voronina, 2016). Contrary to SSF, EcMF have regular organic supply from host tree and are disposed of C mining, but the reverse of the coin is risk created by negative environmental impacts on the hosts without affecting fungi *per se*. For the UK forests the different phenology responses were detected for deciduous (later fruiting in 59% of species) and conifer (no significant changes) trees` mycobionts. It can be probably caused by elongation of vegetation period of season-green trees allowing them to share its photosynthates with EcMF much longer thus lasting fungal activity (Boddy, 2016a). Such deciduous tree-associated EcMF prolonged vigil can influence the forest ecosystems considerably for there are data accumulating on mycorrhizal fungi role in litter mineralization. Albert Bernhard Frank who coined the term "mycorrhiza" in 19th century and started ectomycorrhizal research was the first to claim EcMF as soil organic matter decomposers (Trappe, 2005; Tunlid et al., 2017). But it took about a century for those "decomposers in disguise" (Talbot et al., 2008) to become widely acknowledged. It is known that ectomycorrhizal life-style evolved in multiple (more than 60) SSF and WSF asco- and basidiomycete lineages through losses of genes required for plant cell wall polysaccharides degradation, but the magnitude of these losses varied greatly within lineages (Martin et al., 2016). It is obvious that EcMF have no need in foraging for C as energy source and material for fruitbody formation as true saprotrophs do, but they can break down complex organic compounds from litter and soil to obtain macronutrients such as P and N and probably deliver it to hosts (Nicolás et al., 2019). Such "ecological saprotrophy" significantly contributes to forest ecosystems nutrient cycling and should be considered while predicting litter decomposition rate future changes. By the way, a negative effect of experimental drought on EcMF laccase activity leading to a decrease in phenolic compounds degrading capacity was recently revealed for beech-spruce forest ecosystem (Nickel et al., 2017).More attention should be paid to EcMF with different foraging strategies (Agerer, 2001) mirroring their nutrient mining pattern to specify the community responses.

In forest ecosystems soil response to increased C availability is largely mediated by EcMF (Fransson, 2012). CO_2 elevated concentrations effect on EcMF is discussed in numerous papers (see references in Ferreira and Voronina, 2016). Again, responses varied within experimental designs and species studied, but the main pattern revealed was stimulation of fungal growth under elevated CO_2 (both under field and laboratory conditions) and community shifts. The discrepancies can be caused by a number of EcMF peculiarities such as exploration strategies, successional stages, C requirements (Fransson, 2012), and this should be regarded to obtain the realistic picture of EcMF response to CO_2 elevation. Besides, the limited number of model species in EcMF laboratory research with prevalence of producing large hyphal biomass and high C-demanding can shift data to the point of overestimation of fungal responsiveness to C additional concentrations. The large-scale field-based experiments can largely contribute to understanding of EcMF responses and ecosystem functions under predicted further increase in atmospheric CO_2.

Atmospheric N deposition increased through human activities has profound impact on forest growth under natural conditions. Its complex action with CO_2 elevation, warming and precipitation affects C allocation and forest ecosystem carbon sequestration (Chen et al., 2017a). One of the possible mechanisms of negative effect of N inputs on soil biota is reducing both autotrophic and heterotrophic soil respiration (Hasselquist et al., 2012). For facilitation of plant N nutrition is one of the vital EcMF functions, especially in naturally N-limited boreal biomes (Smith and Read, 2008), their responses to N enrichment are widely studied. EcMF fruitbody surveys at N-fertilized sites and along the gradients of N pollution in Europe demonstrated decline both in numbers and species richness (Arnolds, 1991; Termorshuizen, 1993). The observed effects on fruiting are likely based on N impacts on below-ground patterns of C allocation to EcMF mycelia (Smith and Read, 2008). With the example of N deposition gradient in Alaska Lilleskov et al. (2001; 2002a) demonstrated both above- and belowground EcMF community changes. Industrial ammonia inputs increased N soil availability which had multiple ecosystem outcomes in soil acidification, enhanced tree growth and vegetation changes. EcMF was generally shown to decline both in species richness and fruitbody abundance, but the responses were species-specific. Two ecological N-related EcMF groups, namely "nitrophobic" (*Cortinarius*, *Hebeloma*, *Lactarius*, *Russula*, *Tricholoma* species) and "nitrophilic" (*Hygrophorus olivaceoalbus*, *Laccaria* spp., *Lactarius theiogalus*,

Paxillus involutus) were emerged (Lilleskov et al., 2001). The first group was suppressed by mineral N inputs into organic horizon, while the second was not affected or responded slightly positive. Partly this separation mirrors the discrimination between "early" and "late stage" EcMF (for details see Smith and Read, 2008), but not always in the expected way. "Early stage" EcMF colonize hosts at initial succession stage or in disturbed habitats while "late stage" fungi integrate seedlings in already established mycorrhizal network requiring adult trees and mature forest. The first group occupies habitats typically enriched in N, while the second one appears when humus is already accumulated, and mineralization rate slowed down. According to Lilleskov et al. (2001) an "early" EcMF *Laccaria bicolor* unsurprisingly turned out to be a dominant nitrophilic species, and "late" *Cortinarius*, *Russula*, and *Tricholoma* decreased under elevated N concentrations. But abundance of "early" *Hebeloma* (especially, *H. mesophaeum*) known to inhabit fertile soils and even greenhouses negatively correlated with N increase. When the belowground picture of EcMF community was clarified by molecular and morphological tools applied to the same study site, the main trend was consistent with above-ground survey data (Lilleskov et al., 2002a). EcMF species richness sharply decreased (more than 30 species in low-N sites vs 9 in high-N). The community structure changed rather noticeable too with complete replacement of dominants under N increase. The suggested EcMF community succession direction — from species adapted to low-N conditions (*Cortinarius*, *Piloderma*) through species requiring high nutrient availability (*Tomentella sublilacina*, *Thelephora terrestris*) to fungi adapted high-N, low-P and low pH habitats (*Paxillus involutus*, *Lactarius theiogalus*). Besides, the data obtained by pure culture method showed that the significance of organic N sources lessened under increased N availability (Lilleskov et al., 2002b). Thus N deposition can impede decomposition processes for N scavenging is acknowledged main reason for saprotrophic activity in EcMF (Nicolás et al., 2019).

Despite of the drastic EcMF community shifts regularly reported, total amount of mycorrhizas can remain unchanged. Eighteen-year long experiment on fertilizing in arctic tundra *Betula nana* community did not alter the proportion of root tips colonized by EcMF contrary to 3-year long resulting in near 50% decrease (Deslippe et al., 2011). It can be explained by short-term inhibition of mycorrhization by N elevated input which subsequently recovers with N-tolerant species coming to the foreground. But again it was demonstrated that despite of visible mycorrhizal persistency the EcMF diversity declined, and the effect on its functionality is not clear. Not only can the longevity of inputs, but amount of N addition also influence the response of EcMF community. Slightly greater EcMF fruitbody production was reported for Scots pine forest (northern Sweden) under low N addition, but fruiting was nearly stopped and mycorrhizal respiration critically decreased with high N amounts (Hasselquist et al., 2012). There is an evidence of host plants C allocation to EcMF fungi is linked with N retention in fungal fruitbodies. The disruption of this link by high N concentrations or rather long exposure to elevated N can affect greatly forest ecosystems C and N dynamics (Hasselquist and Högberg, 2014). Despite of the obvious capacity of EcMF community to recover after stress elimination, it can be concluded that long-term or high N inputs may result in EcMF diversity loss and these fungi "should be considered in the establishment of critical loads of N" (Lilleskov et al., 2001).

Soil warming in boreal forests is known to increase microbial respiration and C utilization while which can affect C balance (see references in Cowden et al., 2019). Similar to AMF, both host and mycobiont response to elevated temperature should be considered to predict ecosystem outcome. Observed EcMF hyphal growth under elevated temperature with advance of species rich in extraradical mycelium such as Cortinariaceae (Deslippe et al., 2011) may alleviate the stress in host plants. However, the outcome can be opposed by another global change effects. Vast mycelial networks are often suited for active nutrient foraging and organic N compounds cleavage (as it was shown for *Cortinarius* spp.), so increased inorganic N concentrations would be suppressive (Lilleskov et al., 2002a, b). The most prominent effects are expected in ecotonal habitats, so artificial warming experiments were recently carried out at boreal-temperate ecotonal boundary in northern Minnesota. Community shifts and the selectivity of EcMF response were revealed again. The benefit to short-contact exploration strategists provided by warming under condition of suppressed host photosynthesis were revealed which can be explained with a selection towards less C demanding EcMF (Fernandez et al., 2017). The observed advance of any EcMF exploration strategy is inconsistent across different studies, e.g., medium distance exploration type turned out to prevail in Arctic Alaskan tundra under temperature elevation (Morgado et al., 2015), but the warming-driven shifts in EcMF extraradical part are reported regularly. These changes affecting soil nutrient mobilization greatly might be caused by alterations in host plant fitness under changed environment and depend on plant nutrient requirements and assimilation rate and further ability to afford more or less C costly EcMF assemblage. Another study with increase of air and soil temperature by 1.7 and 3.4 °C above ambient level, respectively, did not reveal changes in EcMF species richness but showed community alterations too and no putative fungal switch from boreal to temperate tree species (Mucha et al., 2018). Fungal plasticity was shown too for only few EcMF species turned out to be highly sensitive to warming, which is consistent with other studies` results (e.g., Morgado et al., 2015). However, another aspect, a niche breadth, should not be neglected, for species with narrow niches are presumed to be in greater danger under environmental changes. Investigation of temperature preferences and niche breadth for EcMF taxa with high occurrence in Japan with application of nested theory and weighted-randomization null model for analysis demonstrated restriction to low-temperature habitats. These fungi might be limited by global warming, and, moreover, colder-site dwelling EcMF exhibited narrower temperature breadths than expected (Miyamoto et al., 2018), thus indicating that local EcMF diversity decline can be caused by warming.

Ericoid mycorrhizal fungi (ErMF)

Ericoid mycorrhiza is restricted to rather small but widespread plant group Ericaceae (Ericales) (Table 1) and is vital for these plants predominantly dwelling rather harsh environment. The main ErMF functions are N supply to plants under N-limited oligotrophic conditions and alleviation of environmental stresses such as extremely high or low temperatures and high acidity enhancing metal cations mobility endangering plant with intoxication (Smith and Read, 2008). Moreover, Ericaceae are prevalent in the biomes considered as the most vulnerable under climatic changes: arcto-alpine and circumpolar areas, and thus ErMF responses are of great value to predict potential vegetation shifts. Ericoid plants are known to be mycorrhizal under natural conditions, but ErMF can be present as free-living saprotrophs (Van der Heijden et al., 2015). Mostly ascomycete fungi, often anamorphic, they have strong decomposing ability and are capable to degrade complex polymer compounds — pectin, cellulose, cellobiose, hemicellulose, polyphenols, lignin and proteins (Read et al., 2004; Grelet et al., 2017). Another recently recognized ErMF ecosystem role is building the deep organic humus layers in the mor soils of heathlands and some boreal forests, the biomes mostly abundant in ericoid plants (Lindahl and Clemmensen, 2017). Besides, some strains of recently revised complex taxon

"*Rhizoscyphus ericae*" known as the most wide-spread symbiont of Ericaceae are capable to colonize neighbouring trees in boreal forests thus creating a network with nutrients shared between tree and shrub connected by common mycobiont (Fehrer et al., 2019).

The factors affecting performance, distribution and the biogeography of ErMF are poorly known for the data on their host-specificity is lacking (Grelet et al., 2017). Generally the studies of ErMF response to climate change are yet few, and the predictions could be mostly speculative. The expected effect of N deposition increase on ErMF is the same as for EcMF communities. High input of the inorganic N might lessen the role of mycobiont in plant nutrition and lead to ErMF abundance decline. The changes might be less sharp than for EcMF communities for ErMF do not depend on plant C entirely.

Elevation gradients are considered to be a good model for climatic changes with expected migration of species under warming conditions upward in elevation and north in latitude. The study of ErMF community associated with shrub *Vaccinium membranaceum* shifts was carried out along the altitudinal gradient in Canadian Rocky Mountains (Gorzelak et al, 2012). High elevation ErMF communities differed from lower elevation ones: the first were dominated by *Rhizoscyphus ericae*, while the second ones — by *Phialocephala fortinii*. Ubiquitous nature of both leading symbiont species which can also represent free-living saprotrophs or common endophytes of non-ericoid plants suggests that the plant-fungal interactions would be persistent under global warming scenario. Interesting data was obtained for circumpolar ericoid plant *Cassiope tetragona* reported to be involved both in ecto- and ericoid mycorrhizal symbiosis (Lorberau et al., 2017). OTC-based warming experiment revealed potential shift in plant-fungal interactions for only a minor part of the root-associated fungal community could be ascribed to ErMF with the greatest proportion of the OTUs identified as either saprotrophs (e.g., *Mycena*) or EcMF (e.g., *Cortinarius*). Authors hypothesized the possibility of *C. tetragona* involved in symbiosis looking like ericoid mycorrhiza but with non-ErMF species or had non-mycorrhizal status on the study site in the High Arctic. By now we have definitely insufficient data to obtain a clear picture of ErMF responses to climate change. It is promising field for future research for ErMF are common dwellers of boreal and heathland ecosystems involved in regulation of prominent terrestrial C stocks (Leopold, 2016).

Pathogenic fungi

Due to broad enzymatic activities and ability to invade tissues and cells with specific hyphae adjusted to obtain nutrition through intracellular contact zones, haustoria, fungi evolved the pathogenic life style numerously within all major lineages. Fungal hosts are represented by other fungi, plants and animals, including humans, and two latter groups are addressed further.

Plant pathogens (PPF)

PPF are extremely diverse in their taxonomy, life style and disease symptoms (Table 1). Most pathogens are specifically attack a narrow group of hosts, but some are broad-range generalists such as *Botrytis cinerea* (Leotiomycetes, Ascomycota) destroying various dicotyledonous plants. Due to great annual yield losses endangering food security, the predicting food crop diseases` development in the changed environment is essential. The most economically important and causing the greatest expenses PPF are predominantly associated with food grains (*Magnaporthe oryzae*, *Puccinia* spp., *Fusarium* spp., *Blumeria graminis* etc.) (Dean et al., 2012). The PPF-caused threat is not restricted to yield losses because of some species` toxigenity. *Fusarium* is particularly notorious for production of trichothecene toxins causing serious, even lethal damage to human health (McCormick et al., 2011). Plant protection suggests not only chemical treatment (fungicides) and crop rotation, but also breeding of new resistant plant varieties for the main determinants of plant disease initiation and development are PPF compatibility and virulence, host susceptibility and environmental conditions favourable for pathogen (Doohan and Zhou, 2018). Breeding procedure itself takes years, and the plant resistance to PPF might come into conflict with acclimation to environment. Due to complexity of the pathogen-host-environment interactions, it is always hard to attribute disease emergency and severity to climate change and to distinguish climatic effects from non-climatic ones (Ghini et al., 2012).

The data on PPF under global change are restricted to causal agents of crop plants with obvious lack of information on unmanaged plant communities (Juroszek et al., 2020, but see Zhan et al., 2018). This ignorance of PPF dynamics in the wild is unsafe for a range of non-cultivated plant species is known to harbour economically important pathogens (Frederick et al., 2017; Kokaeva et al., 2019). According to the recent data, there is noticeable uncertainty in predicting global change impact on PPF-caused diseases and its consequences for future food security (Table 3). Three comprehensive reviews, Luck et al. (2011) on food crops, Ghini et al. (2011) on tropical and plantation crops, and Sturrock et al. (2011) on forest diseases demonstrated all possible outcomes: positive, negative and neutral. Besides the difference in PPF environmental requirements, this variety of responses can be explained both by complexity of factors influencing and non-uniform manifestation of global changes across the regions, such as altered precipitation resulting in drought leading to desertification in some regions, while in excessive rainfalls and flood in other ones (IPCC, 2019). Global warming and elevated CO_2 can result in increasing of plant biomass and productivity, but it can be easily outweighed both by reducing of cropping area and growing severity of plant diseases. Elevation in O_3 concentrations decreases plant biomass and provokes tissue necrosis, but concomitant CO_2 increase neutralizes the effect (Ghini et al., 2012). Different PPF taxa may vary in their temperature and humidity optima and require some specific climatic factors` combinations for their development, so it is hard to make any generalizations, but "dormant" PPF may be locally "awaken" by milder and wetter climate conditions. Local environment means too, so the variety at regional scale is rather large (Hýsek et al., 2017). It should be kept in mind that crop management can influence plant diseases along with climatic conditions and climate and non-climate dependent PPF range shifts can drastically change plant disease distribution and incidence.

Juroszek et al. (2020) took a titanic task to summarize more than 100 review articles on potential climate change effects on plant disease risks published in 1988 – 2019 (July) and highlight research gaps. This overview comprises reviewed data on different crop diseases` causal agents, but PPF (including oomycete pseudofungi) obtained a leading position. The authors did not aim at predicting the future plant disease risks, but the outline of current study approaches, advances and gaps are contribute largely to PPF response recognizing and should be mentioned. Temperature is the most often discussed parameter regarding PPF followed by water availability and CO_2 elevation (Juroszek et al., 2020). Operating review papers authors revealed some strange discrepancies between experimental and modelling studies: the former more often consider combination of temperature and CO_2 elevation to simulate future climatic conditions, while the latter are mainly based on temperature / water availability-dependent potential responses. In total, similarly to non-pathogenic fungal guilds research, the current experimental design in PPF studies does not allow to simulate ecologically realistic conditions when the effect of one

Table 3 Summary of global change effects on PPF-caused crop plant diseases (modified from Juroszek et al., 2020, based on data from review articles* for 1988 – 2019)

Effect type	Hosts and pathogens included	Putative risk increase	Putative not changing risk	Putative risk decrease
Climate change in general	Field, vegetable and horticultural crops; PPF and other plant pathogens	36	7	73
	Tropical plantation crops; PPF and other plant pathogens	12	5	9
	Maize; PPF, viruses and a bacterium	31	2	12
	Agricultural and horticultural crops; PPF, viruses and bacteria	43	7	22
	Wheat; PPF and viruses	28	16	15
	Field crops; mainly PPF	27	11	15
Elevated CO_2	Horticultural crops; PPF	11	8	0
	Mainly crops; PPF and a virus	15	4	10
	Agricultural and horticultural crops; PPF and a virus	8	4	5
Increase in O_3 concentration	Mainly crops; PPF and bacteria	25	10	14
Increase in UV radiation	Mainly crops; PPF	10	2	5

*See the references in Juroszek et al. (2020). The numbers refer to experimental studies which demonstrated the mentioned effect quantity.

factor might be easily neutralized or overweighed by the impact of another one. Another misleading issue is data extrapolation. If some region is claimed to be affected (or unaffected) by future plant disease increase, the similar outcome for a neighbouring one cannot be ensured. Generally, tropical regions are understudied compared to temperate and subtropical ones. Moreover, some PPF (e.g., rust fungi) are pleomorphic, and it is important to be sure to address all the recognized stages in the research, for fungal response (e.g., to elevated temperature) may vary across stages (Ghini et al., 2012; Zhan et al., 2018). Another gap ought to be filled is the lack of agronomical parameters incorporated in the disease progress models. This is important research prospect for future changes may demand to adapt current practices to altered environment (Barnes et al., 2010). Despite of the fact that the most significant outcome of PPF responses to climate changes are potential yield alterations (losses or gains), the PPF regarding models are largely neglect this aspect, so pathogens interference with yield dynamics under ongoing or future changes cannot be estimated.

Attempting to predict crop diseases progress under changed climate, one should kept in mind all four main components determining the outcome: 1) plant response to changes, 2) PPF response to changes, 3) plant-fungus interaction alterations in response to changes, and 4) crop management aspects related to changed environment. Unfortunately, such significant issues as PPF adaptive potential and climate impact on plant defence systems are yet poorly known, not to mention plant-pathogen interaction. For both increase and decrease (with the former prevailing) in PPF-caused diseases are expected by different researchers (see references in Juroszek et al., 2020) under climate change conditions, the both scenarios` underlying mechanisms are briefly outlined below.

PPF-caused plant diseases increase under global change

Plant increased susceptibility:

1. Climatic-driven morphological changes in plants favouring PPF infection. Altered precipitation resulting in drought can change leaf stomatal aperture thus facilitating PPF invasion (Qi et al., 2018 cited in Juroszek et al., 2020).
2. Extreme weather events stressing plants thus making them more prone to infection (Bidzinski et al., 2016 cited in Juroszek et al., 2020). Decreased precipitation may be especially suppressive for trees in urban areas due to

limited soil volume occupied by roots compared to natural environment. Caused by drought reduced photosynthesis may limit plant`s ability to produce defensive compounds (see references in Ghini et al., 2012).

3. Enhanced plant biomass, sugar production, root exudation and more rapid leaf senescence under CO_2 elevation facilitating PPF growth and aggressiveness, especially in rusts and powdery mildews (Váry et al., 2015; Gullino et al., 2018).

4. Biotic stress defence genes suppression under CO_2 elevated conditions. For wheat and maize it was shown to destroy the jasmonate-dependent defence thus making plants susceptible to *Fusarium* (see references in Váry et al., 2015).

5. Biotic stress defence genes suppression under increased temperature. Stem rust resistant wheat cultivars were shown to lack resistance at 27°C, the same pattern was observed for *Brassica napus* stem canker (*Leptosphaeria maculans*) (see references in Ghini et al., 2012).

6. Lack of native plant resistance to invasive PPF immigrated due to climatic changes (Corredor-Moreno and Saunders, 2020).

7. Caused by O_3 elevation acceleration in plant tissues necrosis favourable for necrotrophic PPF (Ghini et al., 2012).

8. Elevated UV radiation making already susceptible plant cultivars more vulnerable to PPF infection, especially to attack of melanised UV-resistant species (e.g., *Cladosporium cucumerinum*, Finckh et al., 1995 cited in Ghini et al., 2012).

PPF increased benefits:

1. Increased PPF growth under elevated temperatures in temperate regions and earlier infection start (Kremer et al., 2016)

2. Reduction of frosty day's number resulting in more successful overwintering of most pathogens in high latitudes (Harvell et al., 2002 cited in Corredor-Moreno and Saunders, 2020; Levitin, 2015b).

3. Increased precipitation favouring PPF development, especially, for late blight causing hydrophilic oomycetes (e.g., *Phytophthora infestans*, see in Ghini et al., 2012). For modelling-based projections regarding precipitation dynamics see Madgwick et al., 2011.

4. PPF dispersing via insect vectors in temperate regions (e.g., *Ophiostoma ulmi*, causal agent of Dutch elm disease) benefiting from warming favourable for insects (Boland et al., 2004).
5. Climate-driven weed migration or growth stimulation creating an infection reservoir for crop PPF (Frederick et al., 2017; Kokaeva et al., 2019).
6. PPF climate-driven migration accompanied with switch to new, non-resistant host species or varieties (Corredor-Moreno and Saunders, 2020).
7. Caused by elevated CO_2 increase in PPF virulence (e.g., *Fusarium*, see in Velasquez et al., 2018).
8. Temperature elevation caused emergence and wide distribution of new more aggressive PPF races (e.g., *Puccinia striiformis*, see in Velasquez et al., 2018).
9. More high resistance to deleterious global change factors in disease agents compared to fungal biocontrol agents (e.g., *Trichoderma* spp. are more sensitive to UV radiation increase than melanin-containing target PPF *Botrytis cinerea*, see in Ghini et al., 2012).

Crop-management aspects:
1. New crop varieties established to accommodate elevated CO_2 or temperature concomitantly turning out to be less resistant to PPF due to stomata number increase (increase in PPF entry points) and higher leaf trichome density (facilitating fungal spore adhesion and survival) (Váry et al., 2015).
2. Fungicides efficiency decrease under raised temperature through PPF growing adaptation (Delcour et al., 2015; Greiner et al., 2019 cited in Juroszek et al., 2020).

PPF-caused plant diseases decrease under global change

Plant increased resistance:
1. Caused by elevated CO_2 reduction in stomatal opening impeding PPF invasion (rusts, downy mildews etc.) (Ghini et al., 2008), but see the opposite effect above.
2. Caused by elevated CO_2 acceleration of plant life cycle decreasing PPF infection period (Ghini et al., 2008).
3. Caused by elevated CO_2 increase in root exudation favourable for rhizosphere microbiota antagonistic to PPF (Ghini et al., 2008).

4. Higher efficiency of fungal biocontrol agents under elevated CO_2 and temperature (e.g., *Ampelomyces quisqualis* against zucchini powdery mildew, see in Gullino et al., 2018), but see the opposite effect above.
5. Caused by elevated UV radiation resistance due to accumulation of phenolic and other defensive compounds such as chitinases and glucanases, but the effect is local and context-depending (Ghini et al., 2012), and see the opposite effect above.

PPF suppression:
1. Reduced precipitation in spring-summer period discouraging infection and inhibiting disease progress (Boland et al., 2004). The more drastic effect can be expected regarding to more hydrophilic oomycete pseudofungi causative for the late blight (Ghini et al., 2008).
2. Hotter and drier summer weather in temperate regions leading to decrease in PPF sporulation and inocula for next season infection, especially in rust fungi and oomycetes (Boland et al., 2004; Ghini et al., 2012; Zhan et al., 2018).
3. Warming in humid tropical and humid warm subtropical zones pushing PPF beyond the climatic optima (Ghini et al., 2008). However, the attention should be paid to the possible discrepancies between PPF responses to the same temperature ranges under the laboratory vs. field conditions, and to possibility of tropical PPF movement to the temperate regions (Velasquez et al., 2018).
4. Caused by elevated CO_2 extended incubation period and decrease in conidia germination (Hibberd et al., 1996a, b cited in Ghini et al., 2008).
5. Increase in UV radiation reducing PPF spore germination. Not all PPF species are vulnerable, but the effect was shown for causal agents of blister blight disease of tea (*Exobasidium vexans*) and grape powdery mildew (*Uncinula necator*) (see references in Ghini et al., 2012).

The mechanisms and ways of disease dynamics presented are by no means uniform and universal; they operate at local scale and context-depending, and outcomes are often contradictive as it obvious from summarized data (see Table 3).

Another widely discussed PPF response to global change factors is spatial shifts. However, here the climate effects *per se* especially hard to discriminate from non-climate ones. The migration of fungal pathogens, as well as beneficial symbionts, often accompanies its host range shifts but is not restricted to them. The most destructive PPF (causal agents of rust, smut, powdery mildew etc.) are air-dispersed and their spores can travel long distances with the wind or use animal or human vectors. The most deleterious migrations are well-known from history and have no relation to drastic climatic change. Thus coffee rust *Hemileia vastatrix* totally destroyed coffee plantations and caused tremendous economic losses to British colonists on Ceylon in 1870s arriving probably from the Horn of Africa with the southwest monsoons (Money, 2007). Active people migration and travelling, plants and plant material intensive trade and introduction enhanced during 20th century create new ways for PPF entrance the new habitats that should be considered in pathogen migration along with habitats` changes and hosts` range shifts (Corredor-Moreno and Saunders, 2020). Such invasions may result in unexpected sudden and devastating plant epidemics caused by the native plants` lack of resistance to newly arrived PPF, or invasive PPF switching to new host species. West et al. (2012) reported wheat fusarium ear blight causative agent *Fusarium graminearum* became more common in UK than *F. culmorum* since 2002, but considered the survey period insufficient to attribute this shift to changing climate only, especially with the concomitant changes in farming practices.

Examples of climate-related migration can be provided by hurricane-driven PPF spore dispersal and warming-dependent shifts. The causal agent of Asian soybean rust, *Phakopsora pachyrhizi*, was carried to USA by hurricane Ivan from South America and succeeds in occupying southeastern United States and Mexico (Stokstad, 2004 cited in Corredor-Moreno and Saunders, 2020). The clearer picture of PPF actually climate-depending migrations can be obtained with the temperature optima analysis in the light of warming. According to large dataset of long-term records (since 1960) analysed by Bebber et al. (2013), the estimated average rate of PPF poleward shift is 6–7 km/yr. Unusual plant diseases outbreaks caused by PPF presumed to be absent from the certain territory are regularly reported (e.g., wheat stem rust caused by *Puccinia graminis* in western Europe, see in Corredor-Moreno and Saunders, 2020), but along with migration itself, this can be ascribed to emergency of previously inactive PPF in response to favourable weather conditions. The data on PPF climate-driven migrations for Russian territory are summarized by M. Levitin (2015a, b). Wheat yellow leaf spot (*Pyrenophora tritici-repentis*) was reported as a new disease in the European south of Russia in 1985, but in 2005-2007 it was found in North-West Russia with high (70%) disease severity for some wheat cultivars. The northward migration was reported for *Fusarium graminearum*, *Ramularia collo-cygni*, and *Septoria tritici* too, regularly causing diseases with high average severity. Despite of multiple PPF migrations reported, currently we have no effective tools for exact prediction of PPF migration trajectories and success of invasion under changing climate, and careful data collection on ongoing invasions is valuable to estimate future risks.

As a conclusion, we need more data on PPF physiology and biochemistry under global change factors such as warming or CO_2 elevation to make a clear picture of future plant disease dynamics. Now it is rather difficult to delimit direct climatic effect on fungi from indirect underpinned by host plant responses. As it was noted by Váry et al. (2015), "In the bestcase scenario, the pathogen and disease-resistant host genotypes will co-evolve at such a rate that the pendulum will favour disease resistance. But there are many variables to offset this balance". Juroszek et al. (2020) on the base of multiple review papers data concluded that in most cases the increase in plant disease severity is more probable than decrease. They concede yield decline of several crops such as maize and soybean in some food-producing regions according to warming and altered precipitation, but CO_2 elevation might smooth the effect over. Regarding temperature / precipitation potential change, it is hard to say what effect on PPF, positive of wet and mild winter or negative of hot and dry vegetative season would prevail. The most relevant answer to the question may be provided by modelling studies implying all PPF life cycle stages. Unfavourable predictions of yield decline due to a number of effects not restricted to PPF only are also made in IPCC report (IPCC, 2019), but filling a range of knowledge gaps is urgent to specify PPF contribution to potential food security risks. At last, we should not forget that under environmental conditions PPF exist in equilibrium with their hosts and its upset may result in unpredictable changes of natural ecosystems (Zhan et al., 2018).

Animal and human pathogens (APF)

Fungi are known to affect a wide range of invertebrate animals and vertebrates including humans. Human pathogens causing diseases generally known as mycoses are relatively small in number (about 400) but are of great concern due to serious, often lethal harm they can make to individuals with suppressed (often due to medical interventions rather than to some disease) immunity (Boddy, 2016b). The most deleterious mammal and human pathogens belong to Ascomycota and are well adapted to inhabit hosts tissues. Such fungi are dimorphic (i.e. can switch between hyphal and yeast unicellular organization), and this morphology transition enhances virulence, facilitates infection and provides thermotolerance required by fungus existing in endotherm tissues (Thompson et al., 2011; Kabir et al., 2012). Invertebrate (predominantly insect and nematode) pathogens from Zoopagomycota and Ascomycota are regarded mainly as biocontrol agents to regulate economically important pests` populations and potential biotechnology agents due to its high proteolytic capacity (Roy et al., 2010). But "beyond the biological control" these fungi provide a trophic level undeservedly neglected in the most of biodiversity studies and this aspect is vital presuming the ongoing decline of some insect groups driven both by habitat fragmentation and destruction and climatic changes (Roy et al., 2009; Bell et al., 2019).

The data on APF responses to climate changes is rather scarce, and the most well-known example of climate effects is beyond the chapter scope considering water zoosporic *Batrachochytrium dendrobatidis* pathogenic for amphibians (Xie et al., 2016). It could be presumed that we should take account of both pathogen and host responses to changing environment to predict animal and human disease dynamics, similar to plant ones. There are some evidences of APF-caused disease distribution and incidence influenced by temperature and precipitation. Coccidioidomycosis is an endemic disease in arid regions of America caused by *Coccidioides immitis* (Eurotiomycetes, Ascomycota). Like the most APF this species has life cycle comprising saprotroph soil-inhabiting phase requiring high humidity, and pathogen invasive one needing dry conditions for air dispersal and infection. The APF sensitiveness to climate conditions was proved with monthly multivariate models (Kolivras and Comrie, 2003). The research revealed that preceding temperature and precipitation in different seasons can serve as good predictors for disease incidence and claim the significance of winter climate conditions in all-year disease dynamics.

The positive effect at least of some global change factors on endotherm vertebrate-associated APF is expected for these fungi are originally well-adapted to elevated temperature (surpassing the optima for most fungi) and CO_2 enrichment typical for their habitats (Garcia-Solache and Casadevall, 2010). Endothermy and the ability of mammals to raise their temperature further as a response to infection are presumably the main protection from APF invasion. Fungal acclimation to elevated temperatures can putatively damp the defence. The temperature elevation is suggested to stimulate mammal APF in two ways. At first, currently pathogenic species can expanse to new regions (currently tropics are the main reservoir of mammal APF). Secondly, potentially pathogenic fungi now discouraged by endothermy will be driven by selection press to thermotolerance and thus to actual pathogenicity (Garcia-Solache and Casadevall, 2010). The case can be illustrated by *Cryptococcus* (Basidiomycota) species. Some of them are notorious for cryptococcosis infection often fatal for immunocompromised persons (e.g., *C. neoformans*), but other species, such as *C. laurentii*, were rarely associated with human disease. But the incidence of opportunistic infections caused by non-*neoformans Cryptococcus* increased recently, and was particularly related to thermotolerant strains (Banerjee et al., 2013). Hypothetically basidiomycete APF will get particular prevalence for this fungal lineage is more apt to thermotolerance (Garcia-Solache and Casadevall, 2010).

CO_2 elevated concentrations deserve special attention for carbon dioxide was shown to be a signalling molecule important for disease development (Hall et al., 2010). Contrary to most signals CO_2 enters the cell by simple diffusion and does not require specific receptors to start signalling cascade. Ascomycete *Candida albicans*, an ordinary commensal resident of human body, is also a cause of widely distributed candidiasis sometimes presenting in a severe invasive form. The fungus was shown to respond to metabolic CO_2 accumulation with transition from yeast to hyphal morphology (Hall et al., 2010). Filamentation promotes adhesion and tissue invasion, thus enhancing virulence. For elevated CO_2 concentrations were shown not only facilitate pathogenicity but to suppress hosts immune response too (see references in Hall et al., 2010), at least some APF-caused diseases tentatively will be advanced.

Invertebrate APF are proved to have a potential to acclimation to future changes too at least, partly. *Metarhizium anisopliae* (Ascomycota), a generalist entomopathogenic fungus is widely studied as a biocontrol agent against pests. Its wide agricultural application is hampered by low tolerance to temperature higher than 35C, but it was possible to select thermotolerant strains by continuous culture (de Crecy et al., 2009). Such "experimental evolution" resulted in fungus` capacity to grow effectively under 37C while retaining entomopathogenicity.

Non-pest insect APF are mostly neglected in the studies of biodiversity responses to climate change (Roy et al., 2009; 2010), but *Cordyceps sinensis* pathogenic to larvae of ghost moths (Hepialidae) is an object of thorough study due to its medicinal value. This species is endemic to Tibetan plateau and is subjected to intense exploitation caused by high and inflating price for stromata and infected larvae applied as a remedy. Contrary to commercial harvesting of the most fungal species not only fruitbodies but the whole individuals of *Cordyceps* are collected and presumed that in some regions the harvesting is economically important (Winkler, 2008) the species is at great risk even without any climatic danger. Besides intensive cattle grazing destroys fungal habitats and strict specificity of *Cordyceps sinensis* limits its geographical range even more. The climate change effects considered to be essential too, but there is no unanimity in their impact on *Cordyceps* distribution. To predict future changes Shrestha and Bawa (2014) used MaxEnt modelling with 3 climate change (rate of warming) trajectories and 3 time periods (2030, 2050, and 2070). The variables include fungal locations, bioclimatic parameters, and altitude. Both increase and decrease in average elevation of proper species habitats were observed, but the main trend suggested was probable expansion of *Cordyceps* over the current state. The controversial predict is provided by Yan et al. (2017). Using species distribution modelling approach the authors predict significant range decrease and shift upward in altitude and toward the central part of the Tibetan Plateau. Following different scenarios *Cordyceps sinensis* is expected to undergo net habitat losses ranged from 4–8 till 36–39% and strongly needs protection measures (Yan et al., 2017).

Thus our knowledge on future responses of APF is based currently on analytic evidences rather than empiric ones, but it is clear that this fungal guild will be someway affected by climatic changes as well as others and further research is need to define, whether the fungi, or their hosts, or both might be endangered.

Conclusion

In short, it can be concluded that climate change biology advanced greatly during the last decades, being supplied with recently emerged high-efficiency research tools and technologies. Metabarcoding approach made direct biodiversity assessment possible which is particularly important implying hidden life style, poor morphology and microscopic size of the most of fungi. This new approach to revealing species composition *in situ* largely contributed to mycogeography. “Omics” technologies provided researchers with the information on actual functionality and contribution of different biota groups to ecosystem processes. The shift in macromycete research to really principal fungal part — substrate mycelium rather than ephemeral and only temporarily available fruitbodies is visible and appreciated. Global accumulation of open data sets allowed meta-analysis accomplishing and modelling on the base of large massive of long-term monitoring or experimental data, and global climate models emergency allowed to convert observations into predictions. But far from denying this obvious progress, it should be noted, that current research of fungal responses to climate change has much in common with "The Elephant in the Dark" parable. The exploration of the elephant by touch with each of analysts probing only a single part of the animal body surely produced quite contrasting opinions on what an elephant looks like, and the research of fungal communities` responses for only one global change factor is the same case. Recognizing of global change factors synergy and counteraction regarding fungi is required to obtain a clear picture of future community alterations. Besides, considerable gaps retain in our knowledge, and many of them regard vital for fungi biotic interactions putative shifts under environmental changes. For most symbiotic and pathogenic fungi have long evolution history shared with their hosts, their future will obviously depend on the rate of adaptation. Under climatic coadaptation scenario severe events like extinctions are less probable, but in the case of inability to "keep pace with" the disruption across the trophic levels and drastic changes in community composition and thus ecosystem functioning are more than possible. In the rare research of plant-PPF interactions in the wild under elevating temperature it was already shown that the climatic-driven shift of extinction and recolonization equilibrium lead to pathogen decline (Zhan et al., 2018). Fungal plasticity is undoubtedly high, but the question of changes rate should be addressed, for abrupt alterations are more dangerous than gradual ones providing the good chance for acclimation. Mostly models of future fungal responses operate with a number of change rates (e.g., Kremer et al., 2016) demonstrating variety of putative outcomes with the main trend "the faster — the worse", but the view of global change rate itself is rather versatile too (see in Chen et al., 2017a). Besides, the supraoptimal but tolerable conditions limits for fungi are apparently recognized insufficiently.

Currently we can claim neither general negative fungal response to changing environment, nor the contrary. Fungal responses to all global change factors studied generally show no clear trend; they are context-depending and vary across regions, taxa, and experimental design too. Experimental data and model-based predictions have little overlaps, and the underestimation of biotic interactions and complexity of abiotic factors involved are frequently results in overestimation of future disaster

(West et al., 2012; Cowden et al., 2019). As it was noted by L. Boddy (2016a) "it is extremely hard to extrapolate to effects of climate change on fungi in the field, in mixed communities, and in fluctuating environments". Without falling into excessive optimism, it should be noted too that many "worst-case scenario" models of fungal responses to climate change made about 20-30 years ago were disclaimed by more recent research failed to reveal significant changes. Only those environmental changes that lead to disruption across trophic levels and endanger the existing equilibrium may result in far-reaching and unpredictable changes of natural ecosystems. The most prominent and well acknowledged danger for fungi in the changing world is climate (and non-climate) driven anthropogenic habitat fragmentation and destruction which results in biodiversity decline and even in local extinctions.

Future prospects and trends

Despite the sharp increase in study of fungal response to changing climate and introduction of novel highly efficient research tools, there are some gaps should be filled to obtain a clearer picture of fungal potential risks under changing environment. Here we outline the most pressing issues for future research:

1. More long-term experiments should be designated for the effects considerably vary at time-scale, as it was shown in N enrichment experiments.
2. More realistic conditions should be strived for both in experimental design and in modelling. This is the case both for factors studied (rarely more than 1-2) and biotic interactions commonly neglected in climate change experiments and models. Moreover, parameters and predicators and its combinations applied in experiments are often non-matching for those that are exploited in modelling thus resulting in discrepancies and impossibility of data comparative analysis. Multi-faceted approach and interdisciplinary research is a priority for future climate change biology.
3. Data sets comprising long-term monitoring on fungal phenology and occurrence data already proved to be essential for climate change predictions (Andrew et al., 2019), and the accumulation of this information should be continued, preferably with common protocol established.
4. More narrow, region or taxa oriented research is required for many effects act on regional scale and / or vary across fungal taxa. By the way, SSF are obviously understudied despite of their vital role in global C cycling. The broad-scale

community–level responses extrapolations based on single observation should be avoided (Maček et al., 2019).

5. Some global change factors (e.g., ozone and UV radiation) effects on fungi remain yet nearly completely unknown while presumed to influence fungal performance by affecting host plants or fungal morphogenesis.
6. Some essential fungal guilds such as PPF associated with non-crop plants in natural habitats or invertebrate APF associated with non-pest insects responses to global change are generally remain unstudied.
7. Fungal responses to global change factors at the physiological, biochemical and gene level, and consequently adaptive potential, are yet poorly understood.
8. More accuracy is required for discrimination between climate change and other factors affecting fungi to avoid allegations, especially in species migrations and PPF caused plant disease dynamics study.
9. Not only estimation of global change effects, but such issues as adaptation and mitigation should be covered by research.

References

A'Bear, A.D., Jones, T.H., and Boddy, L. (2014). Potential impacts of climate change on interactions among saprotrophic cord-forming fungal mycelia and grazing soil invertebrates. Fungal Ecol. 10, 34–43 DOI: 10.1016/j.funeco.2013.01.009

Agerer, R. (2001). Exploration types of ectomycorrhizae. Mycorrhiza 11, 107–114 DOI: 10.1007/s005720100108

Agrios, G.N. (2005). Plant Pathology, 5th ed. (Amsterdam: Elsevier Academic Press)

Albertson, O., Kuyper, T.W., and Gorissen, A. (2005). Taking mycocentrism seriously: mycorrhizal fungi and plant responses to elevated CO_2. New Phytol. 167, 859–868 DOI: 10.1111/j.1469-8137.2005.01458.x

Andrew, C., Büntgen, U., Egli, S., Senn-Irlet, B., Grytnes, J.-A., Heilmann-Clausen, J., Boddy, L., Bässler, C., Gange, A.C., Heegaard, E., Høiland, K., Kirk, P.M., Krisai-Greilhüber, I., Kuyper, t.W., and Kauserud, H. (2019). Open-source data reveal how collections-based fungal diversity is sensitive to global change. Appl. Plant Sci. 7, e1227, DOI: 10.1002/aps3.1227

Andrew, C., and Lilleskov, E.A. (2009). Productivity and community structure of ectomycorrhizal fungal sporocarps under increased atmospheric CO_2 and O_3. Ecol. Letters. 12, 813–822 DOI: 10.1111/j.1461-0248.2009.01334.x

Arnolds, E. (1991). Decline of ectomycorrhizal fungi in Europe. Agricult. Ecosyst. Environ. 35, 209–244.

Banerjee, P., Haider, M., Trehan, V., Mishra, B., Thakur, A., Dogra, V., and Loomba, P. (2013). *Cryptococcus laurentii* fungemia. Indian J. Med. Microbiol. 31, 75–77 DOI: 10.4103/0255-0857.108731

Barnes, A.P., Wreford, A., Butterworth, M.H., Semenov, M.A., Moran, D., Evans, N., and Fitt, B.D.L. (2010). Adaptation to increasing severity of phoma stem canker on winter oilseed rape in the UK under climate change. J. Agric. Sci. 148, 683–694 DOI: 10.1017/s002185961000064x

Bell, J.R., Botham, M.S., Henrys, P.A., Leech, D.I., Pearce-Higgins, J.W., Shortall, C.R., Brereton, T.M., Pickup, J., and Thackeray, S.J. (2019). Spatial and habitat variation in aphid, butterfly, moth and bird phenologies over the last half century. Glob Change Biol. 25, 1982–1994 DOI: 10.1111/gcb.14592

Bebber, D., Ramotowski, M., and Gurr, S. (2013). Crop pests and pathogens move polewards in a warming world. Nature Clim. Change 3, 985–988 DOI: 10.1038/nclimate1990

Berbee, M., James, T.Y., and Strullu-Derrien, C. (2017). Early diverging fungi: Diversity and impact at the dawn of terrestrial life. Annu. Rev. Microbiol. 71, 41–59 DOI: 10.1146/annurev-micro-030117-020324

Berke, S.K. (2010). Functional groups of ecosystem engineers: A proposed classification with comments on current issues. Integr. Comp. Biol. 50, 147–157 DOI: 10.1093/icb/icq077

Birnbaum, C., Hopkins, A.J.M., Fontaine, J.B., and Enright, N.J. (2019). Soil fungal responses to experimental warming and drying in a Mediterranean shrubland. Sci. Total Environ. 683, 524–536 DOI: 10.1016/j.scitotenv.2019.05.222

Blackwell, M. and Dega, F.E. (2018). Lives within lives: Hidden fungal biodiversity and the importance of conservation. Fungal Ecol. 35, 127–134 DOI: 10.1016/j.funeco.2018.05.011

Boddy, L. (2016a). Fungi, Ecosystems, and Global Change. In The Fungi, S.C. Watkinson, L. Boddy, and N.P. Money, eds. (Amsterdam: Elsevier Academic Press), pp. 361–400. DOI: 10.1016/B978-0-12-382034-1.00011-6

Boddy, L. (2016b). Interactions with Humans and Other Animals. In The Fungi, S.C. Watkinson, L. Boddy and N.P. Money, eds. (Amsterdam: Elsevier Academic Press), pp. 293–336. DOI: 10.1016/B978-0-12-382034-1.00009-8

Boland, G.J., Melzer, M.S., Hopkin, A.A., Higgins, V., and Nassuth, A. (2004). Climate change and plant diseases in Ontario. Can. J. Plant Pathol. 26, 335–350

Bret-Harte, M.S., Shaver, G.R., Zoerner, J.P., Johnstone, J.F., Wagner, J.L., Chavez, A.S., Gunkelman IV, R.F., Lippert, S.C., and Laundre, J.A. (2001). Developmental plasticity allows *Betula nana* to dominate tundra subjected to an altered environment. Ecology 82, 18–32 DOI: 10.1890/0012-9658(2001)082[0018:DPABNT]2.0.CO;2

Brundrett, M.C. (2002). Coevolution of roots and mycorrhizas of land plants. New Phytol. 154, 275–304 DOI: 10.1046/j.1469-8137.2002.00397.x

Brundrett, M.C. and Tedersoo, L. (2018). Evolutionary history of mycorrhizal symbioses and global host plant diversity. New Phytol. 220, 1108–1115 DOI: 10.1111/nph.14976

Büntgen, U., Kauserud, H., and Egli, S. (2012). Linking climate variability to mushroom productivity and phenology. Front. Ecol. Environ. 10, 14–19 DOI: 10.1890/110064

Cairney, J.W.G. (2002). *Pisolithus* — death of the pan-global super fungus. New Phytol. 153, 199–201 DOI: 10.1046/j.0028-646X.2001.00339.x

Cairney, J.W.G., and Meharg, A.A. (1999). Influences of anthropogenic pollution on mycorrhizal fungal communities. Environ. Pollut. 106, 169–189 DOI: 10.1016/s0269-7491(99)00081-0

Camenzind, T., Hempel, S., Homeier, J., Horn, S., Velescu, A., Wilcke, W., and Rillig, M.C. (2014). Nitrogen and phosphorus additions impact arbuscular mycorrhizal abundance and molecular diversity in a tropical montane forest. Glob. Change Biol. 20, 3646–3659 DOI: 10.1111/gcb.12618

Catford, J.A., Vesk, P.A., Richardson, D.M., and Pyšek, P. (2012). Quantifying levels of biological invasion: towards the objective classification of invaded and invasible ecosystems. Glob. Change Biol. 18, 44–62. DOI: 10.1111/j.1365-2486.2011.02549.x

Chen, W.-Y., Suzuki, T., and Lackner, M., eds. (2017a). Handbook of Climate Change Mitigation and Adaptation, 2nd ed. (Cham, Switzerland: Springer). DOI: 10.1007/978-3-319-14409-2

Chen, Y.-L., Xu, Z.-W., Xu, T.-L., Veresoglou, S. D., Yang, G.-W., and Chen, B.-D. (2017b). Nitrogen deposition and precipitation induced phylogenetic clustering of arbuscular mycorrhizal fungal communities. Soil Biol. Biochem. 115, 233–242 DOI: 10.1016/j.soilbio.2017.08.024

Clemmensen, K.E., Michelsen, A., Jonasson, S., and Shaver, G.R. (2006). Increased ectomycorrhizal fungal abundance after long-term fertilization and warming of two arctic tundra ecosystems. New Phytol. 171, 391–404 DOI: 10.1111/j. 1469-8137.2006.01778.x

Corredor-Moreno, P., Saunders, D.G.O. (2020). Expecting the unexpected: factors influencing the emergence of fungal and oomycete plant pathogens. New Phytol. 225, 118–125 DOI: 10.1111/nph.16007

Cotton, T.E.A. (2018). Arbuscular mycorrhizal fungal communities and global change: An uncertain future. FEMS Microbiol. Ecol. 94, fiy179 DOI: 10.1093/ femse c/fiy179

Cowden, C.C., Shefferson, R.P., and Mohan, J.E. (2019). Mycorrhizal Mediation of Plant and Ecosystem Responses to Soil Warming. In Ecosystem Consequences of Soil Warming: Microbes, Vegetation, Fauna and Soil Biogeochemistry, J.E. Mohan, ed. (Academic Press Elsevier), pp. 157–173. DOI: 10.1016/B978-0-12-813493-1.00008-9

Crowther, T.W., Thomas, S.M., Maynard, D.S., Baldrian, P., Covey, K., Frey, S.D., van Diepen, L.T.A., and Bradford, M.A. (2015). Biotic interactions mediate soil microbial feedbacks to climate change. PNAS 112, 7033–7038 DOI: 10.1073/pnas. 1502956112

de Crecy, E., Jaronski, S., Lyons, B., Lyons, T.J. and Keyhani, N.O. (2009). Directed evolution of a filamentous fungus for thermotolerance. BMC Biotechnol. 9, 74 DOI: 10.1186/1472-6750-9-74

Dean, R., Van Kan, J.A., Pretorius, Z.A., Hammond-Kosack, K.E., Di Pietro, A., Spanu, P.D., Rudd, J.J., Dickman, M., Kahmann, R., Ellis, J., Foster, G.D. (2012). The Top 10 fungal pathogens in molecular plant pathology. Mol. Plant Pathol. 13, 414–430 DOI: 10.1111/j.1364-3703.2011.00783.x.

DeAngelis, K.M., Chowdhury, P.R., Pold, G., Romero-Olivares, A., and Frey, S. (2019). Microbial responses to experimental soil warming: Five testable hypotheses. In Ecosystem Consequences of Soil Warming: Microbes, Vegetation, Fauna and Soil Biogeochemistry, J.E. Mohan, ed. (Elsevier Academic Press), pp. 141–156. DOI: 10.1016/B978-0-12-813493-1.00007-7

Deslippe, J.R., Hartmann, M., Mohn, W.W., and Simard, S.W. (2011). Long-term experimental manipulation of climate alters the ectomycorrhizal community of *Betula nana* in Arctic tundra. Glob. Change Biol. 17, 1625–1636 DOI: 10.1111/j. 1365-2486.2010.02318.x

Dickie, I.A. and St John, M.G. (2017). Second-generation Molecular Understanding of Mycorrhizas in Soil Ecosystems. In Molecular Mycorrhizal Symbiosis, F. Martin, ed. (Hoboken: Wiley Blackwell), pp. 473–491. DOI: 10.1002/9781118951446.ch26

Dighton, J. (2016). Fungi in Ecosystem Processes, 2nd ed. (Boca Raton London NY: CRC Press, Taylor & Francis Group)

Dighton, J. and White, J.F., eds. (2017). Fungal Community Its Organization and Role in the Ecosystem, 4th ed. (Boca Raton London NY: CRC Press, Taylor & Francis Group)

Doohan, F. and Zhou, B. (2018). Fungal Pathogens of Plants. In Fungi: Biology and Applications, 3rd ed, K. Kavanagh, ed. (Hoboken, USA: Wiley Blackwell), pp. 355–388

Eveling, D.W., Wilson, R.N., Gillespie, E.S., and Bataillé, A. (1990). Environmental effects on sporocarp counts over fourteen years in a forest area. Mycol. Res. 94, 998-1002.

Fehrer, J., Réblová, M., Bambasová, V., and Vohník, M. (2019). The root-symbiotic *Rhizoscyphus ericae* aggregate and *Hyaloscypha* (Leotiomycetes) are congeneric: Phylogenetic and experimental evidence. Stud. Mycol. 92, 195–225 DOI: 10.1016/j.simyco.2018.10.004

Fernandez, C.W., Langley, A., Chapman, S., McCormack, M.L., and Koide, R.T. (2016). The decomposition of ectomycorrhizal fungal necromass. Soil Biol. Biochem. 93, 38–49 DOI: 10.1016/j.soilbio.2015.10.017

Fernandez, C.W., Nguyen, N.H., Stefanski, A., Han, Y., Hobbie, S.E., Montgomery, R.A., Reich, P.B., and Kennedy, P.G. (2017). Ectomycorrhizal fungal response to warming is linked to poor host performance at the boreal-temperate ecotone. Glob. Change Biol. 23, 1598–1609. DOI: 10.1111/gcb.13510

Ferreira, V. and Voronina, E. (2016). Impact of Climate Change on Aquatic Hypho- and Terrestrial Macromycetes. In Climate Change and Microbial Ecology: Current Research and Future Trends, J. Marxsen, ed. (Norfolk, UK: Caister Academic Press), pp. 53–72

Fransson, P. (2012). Elevated CO_2 impacts ectomycorrhiza-mediated forest soil carbon flow: fungal biomass production, respiration and exudation. Fungal Ecol. 5, 85–98 DOI: 10.1016/j.funeco.2011.10.001

Frederick, Z.A., Cummings, T.F., and Johnson, D.A. (2017). Susceptibility of weedy hosts from Pacific Northwest potato production systems to crop-aggressive isolates of *Verticillium dahliae*. Plant Dis. 101, 1500–1506 DOI: 10.1094/PDIS-01-17-0055-RE

Fukasawa, Y. and Matsuoka, S. (2015). Communities of wood-inhabiting fungi in dead pine logs along a geographical gradient in Japan. Fungal Ecol. 18, 75–82 DOI: 10.1016/j.funeco.2015.09.008

Gadd, G.M., ed. (2006). Fungi in Biogeochemical Cycles (Cambridge, UK: Cambridge University Press)

Gadd, G.M., Watkinson, S.C, and Dyer, P.S., eds. (2007). Fungi in the Environment (Cambridge, UK: Cambridge University Press)

Garcia-Solache, M.A., and Casadevall, A. (2010). Hypothesis: global warming will bring new fungal diseases for mammals. mBio 1, e00061-10 DOI: 10.1128/mBio.00061-10.

Ghini, R., Hamada, E., Angelotti, F., Costa, L.B., and Bettiol, W. (2012). Research approaches, adaptation strategies, and knowledge gaps concerning the impacts of climate change on plant diseases. Trop. Plant Pathol. 37, 5–24

Ghini, R., Bettiol, W., and Hamada, E. (2011). Diseases in tropical and plantation crops as affected by climate changes: Current knowledge and perspectives. Plant Pathol. 60, 122–132 DOI: 10.1111/j.1365-3059.2010.02403.x

Ghini, R., Hamada, E., and Bettiol, W. (2008). Climate change and plant diseases. Sci. Agric. 65 (spec. iss.), 98–107 DOI: 10.1590/S0103-90162008000700015

González-Cortés, J.C., Vega-Fraga, M., Varela-Fregoso, L., Martínez-Trujillo, M., Carreón-Abud, Y., and Gavito, M.E. (2012). Arbuscular mycorrhizal fungal (AMF) communities and land use change: the conversion of temperate forests to avocado plantations and maize fields in central Mexico. Fungal Ecol. 5, 16–23 DOI: 10.1016/j.funeco.2011.09.002

Gorissen, A. and Kuyper, T.W. (2000). Fungal species-specific response of ectomycorrhizal Scots pine (*Pinus sylvestris*) to elevated [CO_2]. New Phytol. 146, 163–168 DOI: 10.1046/j.1469-8137.2000.00610.x

Gorzelak, M.A., Hambleton, S., and Massicotte, H.B. (2012). Community structure of ericoid mycorrhizas and root-associated fungi of *Vaccinium membranaceum* across an elevation gradient in the Canadian Rocky Mountains. Fungal Ecol. 5, 36–45 DOI: 10.1016/j.funeco.2011.08.008

Grelet, G., Martino, E., Tajuddin, R., and Artz, R. (2017). Ecology of Ericoid Mycorrhizal Fungi: What Insight Have We Gained with Molecular Tools and what's

Missing? In Molecular Mycorrhizal Symbiosis, F. Martin, ed. (Hoboken: Wiley Blackwell), pp. 405–419. DOI: 10.1002/9781118951446.ch22

Griffiths, R.P., Baham, J.E., Caldwell, B.A. (1994). Soil solution chemistry of ectomycorrhizal mats in forest soil. Soil Biol. Biochem. 26, 331–337 DOI: 10.1016/0038-0717(94)90282-8

Gullino, M.L., Pugliese, M., Gilardi, G., and Garibaldi, A. (2018). Effect of increased CO_2 and temperature on plant diseases: a critical appraisal of results obtained in studies carried out under controlled environment facilities. J. Plant Pathol. 100, 371–389 DOI:10.1007/s42161-018-0125-8

Hall, R.A., De Sordi, L., Maccallum, D.M., Topal, H., Eaton, R., Bloor, J.W., Robinson, G.K., Levin, L.R., Buck, J., Wang, Y., Gow, N.A., Steegborn, C., and Mühlschlegel, F.A. (2010). CO_2 acts as a signalling molecule in populations of the fungal pathogen *Candida albicans*. PLoS Pathog. 6, e1001193 DOI: 10.1371/journal.ppat.1001193

Hasselquist, N.J. and Högberg, P. (2014). Dosage and duration effects of nitrogen additions on ectomycorrhizal sporocarp production and functioning: an example from two N-limited boreal forests. Ecol. Evol. 4, 3015–3026 DOI: 10.1002/ece3.1145

Hasselquist, N.J., Metcalfe, D.B., and Högberg, P. (2012). Contrasting effects of low and high nitrogen additions on soil CO_2 flux components and ectomycorrhizal fungal sporocarp production in a boreal forest. Glob. Change Biol.18, 3596–3605DOI: 10.1111/gcb.12001

Hays, Z. and Watson, D. (2019). Fungal Ecology, Diversity and Metabolites (Waltham Abbey, UK: ED-Tech Press)

Hiscox, J. and Boddy, L. (2017). Armed and dangerous — Chemical warfare in wood decay communities. Fungal Biol. Rev. 31, 169–184 DOI: 10.1016/j.fbr.2017.07.001

Hiscox, J., O'Leary, J., and Boddy, L. (2018). Fungus wars: basidiomycete battles in wood decay. Stud. Mycol. 89, 117–124 DOI: 10.1016/j.simyco.2018.02.003

Hu, Z., Michaletz, S.T., Johnson, D.J., McDowell, N.G., Huang, Z., Zhou, X., and Xu, C. (2018). Traits drive global wood decomposition rates more than climate. Glob. Change Biol. 24, 5259–5269 DOI: 10.1111/gcb.14357

Hýsek, J., Vavera, R. and Růžek, P. (2017). Influence of temperature, precipitation, and cultivar characteristics on changes in the spectrum of pathogenic fungi in winter wheat. Int. J Biometeorol. 61, 967–975 DOI: 10.1007/s00484-016-1276-y

IPCC (Intergovernmental Panel on Climate Change) (2019). Shukla, P.R., Skea, J., Calvo Buendia, E., Masson-Delmotte, V., Pörtner, H.-O., Roberts, D.C., Zhai, P., Slade, R., Connors, S., van Diemen, R., Ferrat, M., Haughey, E., Luz, S., Neogi, S., Pathak, M., Petzold, J., Portugal Pereira, J., Vyas, P., Huntley, E., Kissick, K., Belkacemi, M., and Malley, J., eds. Climate Change and Land: an IPCC special report on climate change, desertification, land degradation, sustainable land management, food security, and greenhouse gas fluxes in terrestrial ecosystems. In press.

Jenkinson, D.S. (1990). The turnover of organic carbon and nitrogen in soil. Phil. Trans. R. Soc. B 329, 361–368 DOI: 10.1098/rstb.1990.0177

Johnson, N.C., Gehring, C., and Jansa, J., eds. (2017). Mycorrhizal Mediation of Soils: Fertility, Structure and Carbon Storage (Amsterdam: Elsevier) DOI: 10.1016/C2015-0-01928-1

Jumpponen, A. and Jones, K.L. (2014). Tallgrass prairie soil fungal communities are resilient to climate change. Fungal Ecol. 10, 44–57 DOI: 10.1016/j.funeco.2013.11.003

Juroszek, P., Racca, P., Link, S., Farhumand, J., and Kleinhenz, B. (2020). Overview on the review articles published during the past 30 years relating to the potential climate change effects on plant pathogens and crop disease risks. Plant Pathol. 69, 179–193 DOI: 10.1111/ppa.13119

Kabir, M.A., Hussain, M.A., and Ahmad, Z. (2012). *Candida albicans*: a model organism for studying fungal pathogens. ISRN Microbiol. 2012, 538694 DOI: 10.5402/2012/538694

Kokaeva, L.Yu., Berezov, Yu.I., Zhevora, S.V., Balabko, P.N., Chudinova, E.M., Voronina, E.Yu., and Elansky, S.N. (2019). Studies on the mycobiota of blighted *Solanum dulcamara* leaves. Mikol. Fitopatol. 53, 108–114 DOI: 10.1134/S0026364819020053

Kolivras, K. and Comrie, A. (2003). Modeling valley fever (coccidioidomycosis) incidence on the basis of climate conditions. Int. J. Biometeorol. 47, 87–101 DOI: 10.1007/s00484-002-0155-x

Kremer, P., Schlüter, J., Racca, P., Fuchs, H.-J., and Lang, C. (2016). Possible impact of climate change on the occurrence and the epidemic development of cercospora leaf spot disease (*Cercospora beticola* Sacc.) in sugar beets for Rhineland-

Palatinate and the southern part of Hesse. Climat. Change 137, 481–494 DOI: 10.1007/s10584-016-1697-y

Leopold, D.R. (2016). Ericoid fungal diversity: Challenges and opportunities for mycorrhizal

Research. Fungal Ecol. 24, Part B, 114–123 DOI: 10.1016/j.funeco.2016.07.004

Levitin, M. (2015a). Plant diseases in globally changing Russian climate. J. Life Sci. 9, 476–480 DOI: 10.17265/1934-7391/2015.10.004

Levitin, M.M. (2015b). Microorganisms and global climate change. Agric. Biol. 50, 641–647 DOI: 10.15389/agrobiology.2015.5.641

Lilleskov, E.A., Fahey, T.J., Horton, T.R., and Lovett, G.M. (2002a). Belowground ectomycorrhizal fungal community change over a nitrogen deposition gradient in Alaska. Ecology, 83, 104–115. DOI: 10.2307/2680124

Lilleskov, E.A., Fahey, T.J., and Lovett, G.M. (2001). Ectomycorrhizal fungal aboveground community change over an atmospheric nitrogen deposition gradient. Ecol. Appl. 11, 397–410 DOI: 10.2307/3060897

Lilleskov, E.A., Hobbie, E.A., and Fahey, T.J. (2002b). Ectomycorrhizal fungal taxa differing in response to nitrogen deposition also differ in pure culture organic nitrogen use and natural abundance of nitrogen isotopes. New Phytol. 154, 219–231DOI:10.1046/j.1469-8137.2002.00367.x

Lindahl, B.D. and Clemmensen, K.E. (2017). Fungal Ecology in Boreal Forest Ecosystems. In Molecular Mycorrhizal Symbiosis, F. Martin, ed. (Hoboken: Wiley Blackwell), pp. 387–404. DOI: 10.1002/9781118951446.ch21

Lorberau, K.E., Botnen, S.S., Mundra, S., Aas, A.B., Rozema, J., Eidesen, P.B., and Kauserud, H. (2017). Does warming by open-top chambers induce change in the root-associated fungal community of the arctic dwarf shrub *Cassiope tetragona* (Ericaceae)? Mycorrhiza. 27, 513–524 DOI: 10.1007/s00572-017-0767-y

Luck, J., Spackman, M., Freeman, A., Trębicki, P., Griffiths, W., Finlay, K., and Chakraborty S. (2011). Climate change and diseases of food crops. Plant Pathol. 60, 113–121 DOI: 10.1111/j.1365-3059.2010.02414.x

Maček, I., Clark, D.R., Šibanc, N., Moser, G., Vodnik, D., Müller, C., and Dumbrell, A.J. (2019). Impacts of long-term elevated atmospheric CO_2 concentrations on communities of arbuscular mycorrhizal fungi. Mol. Ecol. 28, 3445–3458 DOI: 10.1111/mec.15160

Mack, M.C., Schuur, E.A.G., Bret-Harte, M.S., Shaver, G.R., and Chapin, F.S. (2004). Ecosystem carbon storage in arctic tundra reduced by long-term nutrient fertilization. Nature. 431, 440–443 DOI: 10.1038/nature02887

Madgwick, J.W., West, J.S., White, R.P., Semenov, M.A., Townsend, J.A., Turner, J.A., and Fitt, B.D.L. (2011). Impacts of climate change on wheat anthesis and fusarium ear blight in the UK. Eur. J. Plant Pathol. 130, 117–131 DOI: 10.1007/s10658-010-9739-1

Malcolm, G.M., López-Gutiérrezz, J.C., Koide, R.T., and Eissenstat, D.M. (2008). Acclimation to temperature and temperature sensitivity of metabolism by ectomycorrhizal fungi. Glob. Change Biol. 14, 1-12 DOI: 10.1111/j.1365-2486.2008.01555.x

Maaroufi, N.I., Nordin, A., Palmqvist, K., Hasselquist, N.J., Forsmark, B., Rosenstock, N.P., Wallander, H., and Gundale, M.J. (2019). Anthropogenic nitrogen enrichment enhances soil carbon accumulation by impacting saprotrophs rather than ectomycorrhizal fungal activity. Glob. Change Biol. 25, 2900-2914 DOI: 10.1111/gcb.14722

Martin, F., Díez, J., Dell, B., and Delaruelle, C. (2002). Phylogeography of the ectomycorrhizal *Pisolithus* species as inferred from nuclear ribosomal DNA ITS sequences. New Phytol. 153, 345–357 DOI: 10.1046/j.0028-646X.2001.00313.x

Martin, F., Kohler, A., Murat, C., Veneault-Fourrey, C., and Hibbett, D.S. (2016). Unearthing the roots of ectomycorrhizal symbioses. Nat. Rev. Microbiol. 14, 760−773 DOI: 10.1038/nrmicro.2016.149

Martínez-García, L.B., De Deyn, G.B., Pugnaire, F.I., Kothamasi, D., and van der Heijden, M.G.A. Symbiotic soil fungi enhance ecosystem resilience to climate change. Glob Change Biol. 23, 5228–5236 DOI: 10.1111/gcb.13785

McCormick, S.P., Stanley, A.M., Stover, N.A., and Alexander, N.J. (2011). Trichothecenes: from simple to complex mycotoxins. Toxins 3, 802–814 DOI: 10.3390/toxins3070802

Melloy, P., Hollaway, G., Luck, J.O., Norton, R.O.B., Aitken, E., and Chakraborty, S. (2010). Production and fitness of *Fusarium pseudograminearum* inoculum at elevated carbon dioxide in FACE. Glob. Change Biol. 16, 3363–3373 DOI: 10.1111/j.1365-2486.2010.02178.x

Miyamoto, Y., Terashima, Y., and Nara, K. (2018). Temperature niche position and breadth of ectomycorrhizal fungi: reduced diversity under warming predicted by a nested community structure. Glob. Change Biol. 24, 5724–573 DOI: 10.1111/gcb.14446

Mohan, J.E., Cowden, C.C., Baas, P., Dawadi, A., Frankson, P.T., Helmick, K., Hughes, E., Khan, S., Lang, A., Machmuller, M., and Taylor, M. (2014). Mycorrhizal fungi mediation of terrestrial ecosystem responses to global change: mini-review. Fungal Ecol. 10, 3–19 DOI: 10.1016/j.funeco.2014.01.005

Money, N.P. (2007). The Triumph of the Fungi: A Rotten History (NY: Oxford University Press).

Moore, D., Gange, A.C., Gange, E.G., and Boddy, L. (2008). Fruit bodies: their production and development in relation to environment. In Ecology of saprotrophic basidiomycetes, L. Boddy, J.C. Frankland, and P. van West, eds. (Oxford, UK: Elsevier), pp. 79–103.

Morgado, L.N., Semenova, T.A., Welker, J.M., Walker, M.D., Smets, E., and Geml, J. (2015). Summer temperature increase has distinct effects on the ectomycorrhizal fungal communities of moist tussock and dry tundra in Arctic Alaska. Glob. Change Biol. 21, 959–972 DOI: 10.1111/gcb.12716

Morin, E., Miyauchi, S., San Clemente, H., Chen, E.C.H., Pelin, A., de la Providencia, I., Ndikumana, S., Beaudet, D., Hainaut, M., Drula, E., Kuo, A., Tang, N., Roy, S., Viala, J., Henrissat, B., Grigoriev, I.V., Corradi, N., Roux, C., and Martin, F.M. (2019). Comparative genomics of *Rhizophagus irregularis*, *R. cerebriforme*, *R. diaphanus* and *Gigaspora rosea* highlights specific genetic features in Glomeromycotina. New Phytol. 222, 1584-1598. DOI: 10.1111/nph.15687

Mucha, J., Peay, K.G., Smith, D.P., Reich, P.B., Stefański, A., and Hobbie, S.E. (2018). Effect of simulated climate warming on the ectomycorrhizal fungal community of boreal and temperate host species growing near their shared ecotonal range limits. Microb. Ecol. 75: 348-363 DOI: 10.1007/s00248-017-1044-5

Naranjo-Ortiz, M.A. and Gabaldón, T. (2019). Fungal evolution: major ecological adaptations and evolutionary transitions. Biol. Rev. 94, 1443–1476 DOI: 10.1111/brv.12510

Nicolás, C., Martin-Bertelsen, T., Floudas, D., Bentzer, J., Smits, M., Johansson, T., Troein, C., Persson, P., and Tunlid, A. (2019). The soil organic matter decomposition mechanisms in ectomycorrhizal fungi are tuned for liberating soil organic nitrogen. ISME J. 13, 977–988 DOI: /10.1038/s41396-018-0331-6

Nickel, U.T., Weikl, F., Kerner, R., Schäfer, C., Kallenbach, C., Munch, J.C., and Pritsch, K. (2017). Quantitative losses vs. qualitative stability of ectomycorrhizal community responses to 3 years of experimental summer drought in a beech-spruce forest. Glob. Change Biol. 24, 560–576 DOI: 10.1111/gcb.13957

O`Dell, T.E., Ammirati, J.F., and Schreiner, E.G. (1999). Species richness and abundance of ectomycorrhizal basidiomycete sporocarps on a moisture gradient in the *Tsuga heterophylla* zone. Can. J. Bot. 77, 1699–1711.

Panneerselvam, P., Kumar, U., Senapati, A., Parameswaran, S., Anandan, A., Kumar, A., Jahan, A., Padhy, S.R., and Nayak, A.K. (2020). Influence of elevated CO_2 on arbuscular mycorrhizal fungal community elucidated using Illumina MiSeq platform in sub-humid tropical paddy soil. Appl. Soil Ecol. 145, 103344 DOI: 10.1016/j.apsoil.2019.08.006

Peay, K.G. and Matheny P.B. (2017). The Biogeography of Ectomycorrhizal Fungi – a History of Life in the Subterranean. In Molecular Mycorrhizal Symbiosis, F. Martin, ed. (Hoboken: Wiley Blackwell) pp. 341–361. DOI: 10.1002/9781118951446.ch19

Pickles, B.J., Egger, K.N., Massicotte, H.B., and Green, D.S. (2012). Ectomycorrhizas and climate change. Fungal Ecol. 5, 73–84 DOI: 10.1016/j.funeco. 2011.08.009

Purahong, W., Wubet, T., Kahl, T., Arnstadt, T., Hoppe, B., Lentendu, G., Baber, K., Rose, T., Kellner, H., Hofrichter, M., Bauhus, J., Krüger, D., and Buscot, F. (2018). Increasing N deposition impacts neither diversity nor functions of deadwood-inhabiting fungal communities, but adaptation and functional redundancy ensure ecosystem function. Environ. Microbiol. 20, 1693–1710 DOI: 10.1111/1462-2920.1408

Read, D.J., Leake, J.R., and Perez-Moreno, J. (2004). Mycorrhizal fungi as drivers of ecosystem processes in heathland and boreal forest biomes. Can. J. Bot. 82, 1243–1263 DOI: 10.1139/b04-123

Rillig, M.C., Treseder K.K., and Allen, M.F. (2002). Global Change and Mycorrhizal Fungi. In Ecological Studies, Vol.157. Mycorrhizal Ecology, M.G.A. van der Heijden and I.R. Sanders, eds. (Berlin, Heidelberg: Springer-Verlag), pp. 135-160.

Römheld, V., and Kirkby, E. A. (2010). Research on potassium in agriculture: needs and prospects. Plant Soil 335, 155–180 DOI: 10.1007/s11104-010-0520-1

Rosenstock, N.P. (2009). Can ectomycorrhizal weathering activity respond to host nutrient demands? Fungal Biol. Rev. 23, 107–114 DOI: 10.1016/j.fbr.2009.11.003

Rosling, A., Lindahl, B.D., Taylor, A.F.S., and Finlay, R.D. (2004). Mycelial growth and substrate acidification of ectomycorrhizal fungi in response to different minerals. FEMS Microbiol. Ecol. 47, 31–37 DOI: 10.1016/S0168-6496(03)00222-8

Roy, H.E., Hails, R.S., Hesketh, H., Roy, D.B., and Pell, J.K. (2009). Beyond biological control: non-pest insects and their pathogens in a changing world. Insect Conserv. Diver. 2, 65–72 DOI: 10.1111/j.1752-4598.2009.00046.x

Roy, H.E., Vega, F.E., Chandler, D., Goettel, M.S., Pell, J.K., and Wajnberg, E., eds. (2010).

The Ecology of Fungal Entomopathogens (Dordrecht Heidelberg London NY: Springer). DOI: 10.1007/978-90-481-3966-8

Shiryaev, A.G., Moiseev, P.A., Peintner, U., Devi, N.M., Kukarskih, V.V., and Elsakov, V.V. (2019). Arctic greening caused by warming contributes to compositional changes of mycobiota at the Polar Urals. Forests, 10, 1112 DOI: 10.3390/f10121112

Smith, S.E. and Read, D.J. (2008). Mycorrhizal Symbiosis (NY, USA: Academic Press).

Soudzilovskaia, N.A., Douma, J.C., Akhmetzhanova, A.A., van Bodegom, P.M., Cornwell, W.K., Moens, E.J., Treseder, K.K., Tibbett, M., Wang, Y.-P., and Cornelissen, J.H.C. (2015). Global patterns of plant root colonization intensity by mycorrhizal fungi explained by climate and soil chemistry. Global Ecol. Biogeogr. 24, 371–382 DOI: 10.1111/geb.12272

Spatafora, J.W., Aime, M.C., Grigoriev, I.V., Martin, F., Stajich, J.E., and Blackwell, M. (2017). The fungal tree of life: from molecular systematics to genome-scale phylogenies. Microbiol. Spectr. 5, FUNK-0053-2016 DOI:10.1128/microbiolspec.FUNK-0053-2016

Shrestha, U.B., Bawa, K.S. (2014). Impact of climate change on potential distribution of Chinese caterpillar fungus (Ophiocordyceps sinensis) in Nepal Himalaya. PLoS ONE 9, e106405 DOI:10.1371/journal.pone.0106405

Staddon, P.L., Heinemeyer, A., and Fitter, A.H. (2002). Mycorrhizas and global environmental change: research at different scales. Plant Soil. 244, 253–261 DOI: 10.1023/A:1020285309675

Sturrock, R.N., Frankel, S.J., Brown, A.V., Hennon, P.E., Kliejunas, J.T., Lewis, K.J., Worrall, J.J., and Woods, A.J. (2011). Climate change and forest diseases. Plant Pathol. 60, 133-149 DOI: 10.1111/j.1365-3059.2010.02406.x

Sulman, B.N., Moore, J.A.M., Abramoff, R., Averill, C., Kivlin, S., Georgiou, K., Sridhar, B., Hartman, M.D., Wang, G., Wieder, W.R., Bradford, M.A., Luo, Y., Mayes, M.A., Morrison, E., Riley, W.J., Salazar, A., Schimel, J.P., Tang, J., and Classen, A.T. (2018). Multiple models and experiments underscore large uncertainty in soil carbon dynamics. Biogeochemistry 141, 109–123 DOI: 10.1007/s10533-018-0509-z

Talbot, J.M., Allison, S.D., and Treseder, K.K. (2008). Decomposers in disguise: mycorrhizal fungi as regulators of soil dynamics in ecosystems under global change. Funct. Ecol. 22, 955–963 DOI: 10.1111/j.1365-2435.2008.01402.x

Taylor, J.D., Helgason, T., and Öpik, M. (2017). Molecular Community Ecology of Arbuscular Mycorrhizal Fungi. In Fungal Community Its Organization and Role in the Ecosystem, 4th ed., J. Dighton and J.F. White, eds. (Boca Raton London NY: CRC Press, Taylor & Francis Group), pp. 3–25. DOI: 10.1201/9781315119496-2

Termorshuizen, A.J. (1993). The influence of nitrogen fertilizers on ectomycorrhizas and their fungal carpophores in young stands of *Pinus sylvestris.* Forest Ecol. Manag. 57 179–189.

Thompson, D.S., Carlisle, P.L., and Kadosh, D. (2011). Coevolution of morphology and virulence in *Candida* species. Eukaryot. Cell 10, 1173–1182 DOI: 0.1128/EC.05085-11

Thorn, S., Förster, B., Heibl, C., Müller, J., and Bässler, C. (2018). Influence of macroclimate and local conservation measures on taxonomic, functional, and phylogenetic diversities of saproxylic beetles and wood-inhabiting fungi. Biodivers. Conserv. 27, 3119–3135 DOI: 10.1007/s10531-018-1592-0

Töpper, J. P., Meineri, E., Olsen, S. L., Rydgren, K., Skarpaas, O., and Vandvik, V. (2018). The devil is in the detail: Nonadditive and context-dependent plant population responses to increasing temperature and precipitation. Glob. Change Biol. 24, 4657–4666 DOI: 10.1111/gcb.14336

Trappe, J.M. (2005). A.B. Frank and mycorrhizae: the challenge to evolutionary and ecologic theory. Mycorrhiza 15, 277–281 DOI: 10.1007/s00572-004-0330-5

Treseder, K.K. (2016) Model behavior of arbuscular mycorrhizal fungi: predicting soil carbon dynamics under climate change. Botany 94, 417–423 DOI: 10.1139/cjb-2015-0245

Tunlid, A., Floudas, D., Koide, R., and Rineau, F. (2017). Soil Organic Matter Decomposition Mechanisms in Ectomycorrhizal Fungi. In Molecular Mycorrhizal Symbiosis, F. Martin, ed. (Hoboken: Wiley Blackwell), pp. 257–275. DOI: 10.1002/9781118951446.ch15

van der Heijden, M., Martin, F.M., Selosse, M.-A., and Sanders, I.R. (2015). Mycorrhizal ecology and evolution: the past, the present, and the future. New Phytol. 205, 1406−1423 DOI: 10.1111/nph.13288

Váry, Z., Mullins, E., McElwain, J.C., and Doohan, F.M. (2015). The severity of wheat diseases increases when plants and pathogens are acclimatized to elevated carbon dioxide. Glob. Change Biol. 21, 2661–2669 DOI: 10.1111/gcb.12899

Velasquez, A.C., Castroverde, C.D.M., and He, S.Y. (2018). Plant–pathogen warfare under changing climate conditions. Curr. Biol. 28, R619–R634 DOI: 10.1016/j.cub.2018.03.054

Venugopal, P., Junninen, K., Linnakoski, R., Edman, M., and Kouki, J. (2016). Climate and wood quality have decayer-specific effects on fungal wood decomposition. Forest Ecol. Manag. 360, 341–351 DOI: 10.1016/j.foreco.2015.10.023

Voronina, E., and Sidorova, I. (2017). Rhizosphere, Mycorrhizosphere and Hyphosphere as Unique Niches for Soil-inhabiting Bacteria and Micromycetes. In Advances in PGPR Research, H.B. Singh, B.K. Sarma, and C. Keswani, eds. (UK: CAB International), pp. 165–186. DOI: 10.1079/9781786390325.0165

Walker, G.M. and White, N.A. (2018). Introduction to Fungal Physiology. In Fungi: Biology and Applications, 3rd ed, K. Kavanagh, ed. (Hoboken, USA: Wiley Blackwell), pp. 1–36

Watkinson, S.C. (2016). Physiology and Adaptation. In The Fungi, S.C. Watkinson, L. Boddy, and N.P. Money, eds. (Amsterdam: Elsevier Academic Press), pp. 141–188. DOI: 10.1016/B978-0-12-382034-1.00005-0

Weber, S.E., Diez, J.M., Andrews, L.V., Goulden, M.L., Aronson, E.L., and Allen, M.F. (2019). Responses of arbuscular mycorrhizal fungi to multiple coinciding global change drivers. Fungal Ecol. 40, 62–71 DOI: 10.1016/j.funeco.2018.11.008

West, J.S., Holdgate, S., Townsend, J.A., Edwards, S.G., Jennings, P., and Fitt, B.D.L. (2012). Impacts of changing climate and agronomic factors on *Fusarium* ear blight of wheat in the UK. Fungal Ecol. 5, 53–61 DOI: 10.1016/j.funeco.2011.03.003

Winkler, D. (2008). Yartsa Gunbu (*Cordyceps sinensis*) and the fungal commodification of Tibet's rural economy. Econ. Bot. 62, 291–305 DOI: 10.1007/s12231-008-9038-3

Xie, G.Y., Olson, D.H., and Blaustein, A.R. (2016). Projecting the global distribution of the emerging amphibian fungal pathogen, *Batrachochytrium dendrobatidis*, based on IPCC climate futures. PLoS ONE 11, e0160746 DOI: 10.1371/journal.pone.0160746

Yan, Y., Li, Y., Wang, W.-J., He, J.-S., Yang R.-H., Wu, H.-J., Wang, X.-L., Jiao, L., Tanga, Z., and Yao, Y.-J. (2017). Range shifts in response to climate change of *Ophiocordyceps sinensis*, a fungus endemic to the Tibetan Plateau. Biol. Conserv. 206, 143–150 DOI: 10.1016/j.biocon.2016.12.023

Zhan, J., Ericson, L., and Burdon, J.J. (2018). Climate change accelerates local disease extinction rates in a long-term wild host-pathogen association. Glob. Change Biol. 24, 3526–3536 DOI: 10.1111/gcb.14111

Chapter 6

Impact of Climate Change on Aquatic Hyphomycetes

Verónica Ferreira

University of Coimbra, MARE - Marine and Environmental Sciences Centre, Department of Life Sciences, Coimbra, Portugal

Email: veronica@ci.uc.pt

DOI: https://doi.org/10.21775/9781913652579.06

Abstract

Freshwater fungi are important components of heterotrophic food webs in woodland streams. These organisms are pioneer colonizers of submerged litter derived from the terrestrial surroundings, and through their activities, they mineralize litter carbon and nutrients and convert dead organic matter into biomass, establishing the link between basal resources and higher trophic levels. This chapter addresses direct and indirect effects of climate change on the community composition, growth, reproduction, metabolism, and decomposing activity of aquatic hyphomycetes. Evidence so far suggests that future global climate change will affect aquatic hyphomycete activity and community structure, with consequences for the functioning of woodland streams. Several proposals are offered to advance knowledge on the effects of global climate change on these fungi.

Introduction

Food webs in small woodland streams derive most of their energy, nutrients and carbon from allochthonous litter provided by the riparian vegetation (Wallace et al., 1997). In water, litter decomposition is carried out mainly by microbes and invertebrate detritivores. Aquatic hyphomycetes are a phylogenetically heterogeneous group of freshwater fungi composed mainly of the anamorphs of Ascomycetes and Basidiomycetes. They are a large component of the microbial assemblages that colonize submerged leaf litter and generally account for more than 90% of the microbial biomass and production associated with these decomposing substrates (Hieber and Gessner, 2002; Pascoal and Cássio, 2004). These fungi are responsible for the loss of up to 67% of initial litter mass by mineralizing organic carbon and by converting it into biomass (mycelium and conidia) (Hieber and Gessner, 2002; Pascoal and Cássio, 2004; Taylor and Chauvet, 2014). The accumulation of fungal biomass on litter and the softening of litter by fungal enzymatic activities also increase the nutrient quality and palatability of litter to invertebrate detritivores, which lead to further litter mass loss (Canhoto and Graça, 2008; Bärlocher and Sridhar, 2014). Thus, aquatic hyphomycetes play a fundamental role in woodland streams by mediating litter decomposition and establishing the link between dead organic matter and higher trophic levels.

Aquatic hyphomycetes are highly sensitive to alterations in environmental conditions such as warming, nutrient enrichment, and changes in forest composition (Gulis and Suberkropp, 2003a; Ferreira et al., 2006a, 2017; Gulis et al., 2006; Ferreira and Chauvet, 2011a). This suggests that environmental change in a future climate change scenario may strongly affect aquatic hyphomycete activity and community composition, which can have impacts on ecosystem functioning and food webs. Climate models predict an increase in the atmospheric carbon dioxide concentration, ozone depletion and changes in precipitation (IPCC, 2013), which can have multiple direct and indirect impacts on freshwater ecosystems and consequently on aquatic hyphomycetes (Fig. 6.1). In this chapter, we address the response of aquatic hyphomycetes to environmental changes likely to be promoted by climate change, but these changes can also result from more direct anthropogenic activities, such as agricultural activities and forest management.

Aquatic hyphomycetes and global change

Atmospheric CO_2 – air temperature – water temperature – aquatic hyphomycetes

Climate models considering a doubling in atmospheric CO_2 concentration predict that the mean global air temperature will increase by 0.3 – 4.8 °C over this century (IPCC, 2013). This increase in air temperature will be closely followed by an increase in water temperature of streams and rivers (Pilgrim et al., 1998; Morrill et al., 2005) (Fig. 6.1). Warming of cold woodland streams will likely affect aquatic hyphomycete activity and community structure, as species have thermal optima and metabolic activities are temperature dependent (Brown et al., 2004). Indeed, evidence from laboratory experiments, field surveys over natural thermal gradients and field temperature manipulations have shown that increases in temperature generally stimulate aquatic hyphomycete growth and reproduction (Chauvet and Suberkropp,

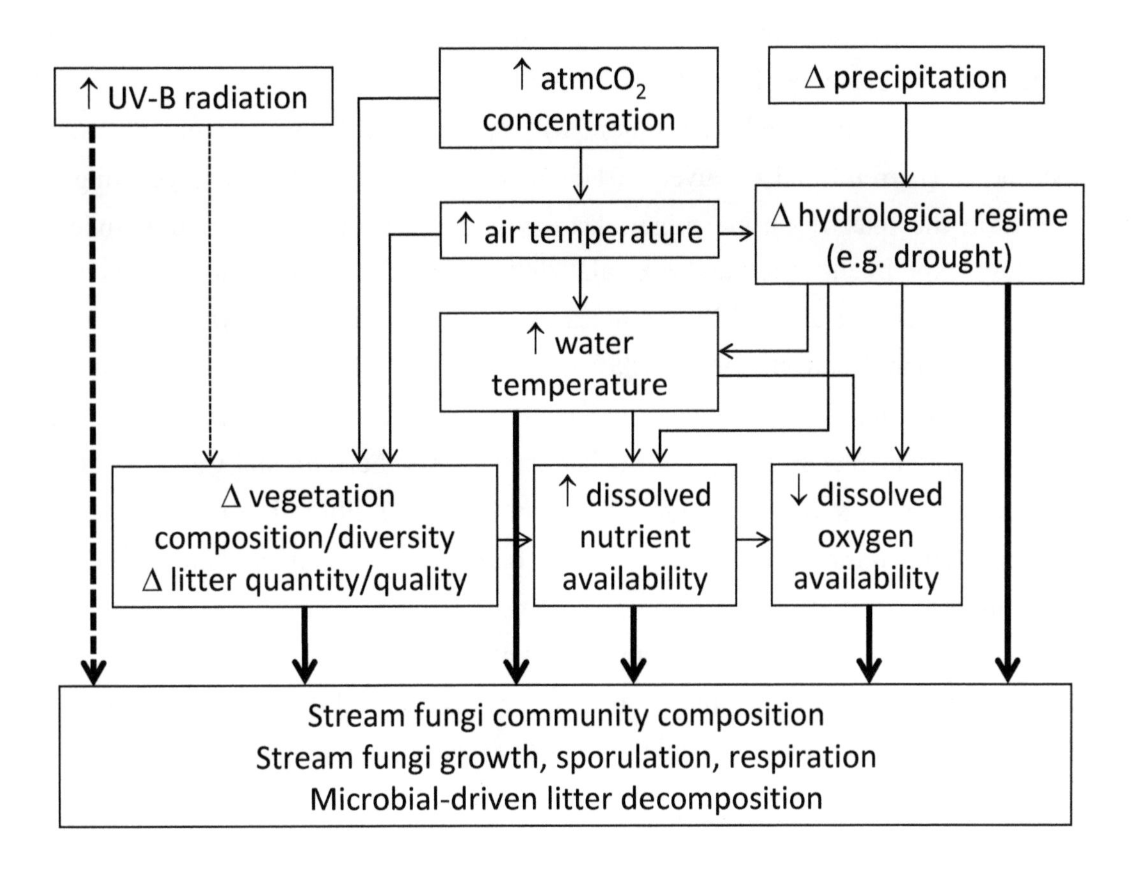

Figure 6.1 Relationships between global change factors and stream fungi. The relationships identified by the thicker lines are discussed in the text. Solid lines identify relationships of global change factors in stream fungi, whereas dashed lines shown potential relationships for which there is still no evidence.

1998; Fabre and Chauvet, 1998; Nikolcheva and Bärlocher, 2005; Bärlocher et al., 2008; Ferreira and Chauvet, 2011a, b; Fernandes et al., 2014). The exception seems to occur when the ambient water temperature is already high (> 15 °C) (Fernandes et al., 2009; Geraldes et al., 2012; Bärlocher et al., 2013), suggesting that the optimal thermal limits of fungal species may be exceeded in these cases. The temperature optima for aquatic hyphomycete species isolated from temperate regions and kept in monocultures in the laboratory have been shown to be in the range of 10 – 25 °C (Koske and Duncan, 1974; Webster et al., 1976; Graça and Ferreira, 1995; Chauvet and Suberkropp, 1998; Dang et al., 2009; Duarte et al., 2013; Ferreira et al., 2014). However, the thermal optima of species may be substantially lower (e.g. by 10 – 15 degrees) when species are in the presence of others than when they are kept in monocultures (Webster et al., 1976).

Warming stimulated respiration by aquatic hyphomycetes in laboratory experiments (Ferreira and Chauvet, 2011a, b; Ferreira et al., 2012b), suggesting an increase in the rate of carbon mineralization with increasing temperature. Indeed, laboratory experiments (Fernandes et al., 2009, 2012, 2014; Ferreira and Chauvet, 2011a, b; Ferreira et al., 2012b; Geraldes et al., 2012), field surveys (Irons et al., 1994; Fabre and Chauvet, 1998; Nikolcheva and Bärlocher, 2005; Friberg et al., 2009; Boyero et al., 2011; Ferreira and Canhoto, 2014; Martínez et al., 2014, 2015) and field experimental warming (Bärlocher et al., 2008; Ferreira and Canhoto, 2014, 2015) have reported a stimulation of microbial-driven litter decomposition with increasing temperature. An increase in microbial-driven carbon mineralization may lead to an increase in the contribution of streams to the global carbon cycle.

The effects of warming on aquatic hyphomycetes activity (including fungi-driven litter decomposition) may, however, be controlled by other factors. For instance, increases in nutrient availability may exacerbate the stimulatory effects of warming on fungal activity (Ferreira and Chauvet, 2011a; Fernandes et al., 2014), which can be of major importance if eutrophication and warming occur concomitantly (see below). Also, the stimulation of fungal activity with warming may be stronger for nutrient-poor (high C:N; e.g. oak) than for nutrient-rich (low C:N; e.g. alder, chestnut) litter species (Bärlocher et al., 2013; Gonçalves et al., 2013), which can be of major importance if future atmospheric CO_2 enrichment and warming translate into decreases in litter quality and changes in forest composition (see below). The effect of warming on fungal activity may also be stronger at lower than at warmer background water temperature (Bärlocher et al., 2013; Martínez et al., 2014; Ferreira and Canhoto, 2014, 2015), suggesting higher temperature sensitivity of fungal activity at higher latitude and elevation than in the tropics and in autumn/ winter than in spring/summer (Ferreira and Canhoto, 2014, 2015).

Warming has also been shown to lead to changes in the relative abundance of aquatic hyphomycete species (based on conidial production), which may result in changes in species dominance (Dang et al., 2009; Ferreira and Chauvet, 2011a, b; Fernandes et al., 2012; Gonçalves et al., 2013; Ferreira and Canhoto, 2015) and decreases in species richness (Rajashekhar and Kaveriappa, 2003; Bärlocher et al., 2008; Fernandes et al., 2012; Gonçalves et al., 2013; Martínez et al., 2014). Positive relationships exist between fungi species richness and fungal biomass accrual on decomposing litter and leaf consumption by detritivores (Lecerf et al., 2005; Duarte et al., 2006). Thus, a decrease in fungi richness may lead to decreases in litter decomposition (Pérez et al., 2012) that might counteract the stimulatory effect of warming. However, no interaction between fungi richness and warming on litter decomposition was found in a laboratory study (Geraldes et al., 2012). Also, if fungal species differ in nutrient concentration and enzymatic activities (Arsuffi and Suberkropp, 1988; Danger and Chauvet, 2013), changes in fungal species dominance may affect consumption by invertebrate detritivores and further litter decomposition (Arsuffi and Suberkropp, 1988; Cornut et al., 2015).

Temperature (– vegetation/hydrological regime) – nutrients – aquatic hyphomycetes
The increase in temperature will also indirectly affect aquatic communities and processes by promoting changes in other environmental variables. Warming will stimulate nutrient mineralization, which will lead to increases in dissolved nutrient availability (Murdoch et al., 2000; Rustad et al., 2001; Moss et al., 2011). Warming will also enhance evaporation, evapotranspiration, and water abstraction, which will translate into decreases in water volume and increases in the concentration of the dissolved nutrients; this will be exacerbated in cases of reduced precipitation (Murdoch et al., 2000). Additionally, warming may promote changes in the composition of native forests (e.g. increase in the representation of nitrogen-fixing species) (see below), which may lead to an increase in nitrogen availability in streams (Compton et al., 2003; Goldstein et al., 2009) (Fig. 6.1).

Aquatic hyphomycetes are able to retrieve nutrients from both the organic substrates and the water column (Suberkropp and Chauvet, 1995; Suberkropp, 1998), and therefore an increase in dissolved nutrient availability will likely stimulate fungal activity. Indeed, laboratory experiments, field surveys over moderate nutrient gradients and whole-stream nutrient manipulations have generally shown a stimulation of fungal biomass accrual, spore production, respiration and litter decomposition with increases in nutrient concentration (Suberkropp and Chauvet, 1995; Rosemond et al., 2002; Gulis and Suberkropp, 2003a, b; Niyogi et al., 2003; Ferreira et al., 2006b; Gulis et al., 2006, 2008; Rosemond et al., 2015). Reproductive activity is often the fungal variable most sensitive to increases in nutrient availability (Gulis and Suberkropp, 2003a; Ferreira et al., 2006b; Ferreira and Chauvet, 2011a). Under moderate nutrient enrichment in the field, the relationship between fungal activity and nutrient concentration is often asymptotic and nutrient saturation occurs at relatively low nutrient levels: half of the maximum rate of activity has been achieved at 16 – 303 μg NO_3-N /L and at 7 – 21 μg PO_4-P/L (Rosemond et al., 2002; Ferreira et al., 2006b; Gulis et al., 2006).

The magnitude of the stimulation of fungal activity by nutrient enrichment depends, however, on other factors. Ferreira and Chauvet (2011a) have shown a synergistic effect between nutrient enrichment and warming on fungal activity in a laboratory experiment. Also, the effect of nutrient enrichment on fungal activity is generally stronger for low quality substrates such as nutrient-poor leaves and wood (high C:N ratio), where fungal activity may be nutrient limited, than for high-quality substrates (Stelzer et al., 2003; Ferreira et al., 2006b; Gulis et al., 2006). When a given nutrient is not limiting, however, further increases in its concentration may not stimulate fungal activity (Chadwick and Huryn, 2003; Abelho and Graça, 2006; Baldy et al., 2007). Also, if the nutrient enrichment is accompanied by increases in sedimentation, increases in the concentration of toxic compounds (e.g. pesticides) or decreases in dissolved oxygen concentration, fungal activity may not be stimulated or may even be inhibited at highly polluted sites (Pascoal and Cássio, 2004; Lecerf et al., 2006; Woodward et al., 2012).

Nutrient enrichment may also affect the structure of aquatic hyphomycete communities (based on conidial production) both in the water column (Gulis and Suberkropp, 2003a, 2004) and associated with submerged substrates (Gulis and Suberkropp, 2003a; Ferreira et al., 2006b; Artigas et al., 2008; Ferreira and Chauvet, 2011a) by changing species relative abundance and/or species dominance. As discussed above, these changes in community structure might lead to changes in fungal decomposing activity (Duarte et al., 2006; Pérez et al., 2012; Cornut et al., 2015).

Water temperature/hydrological regime (– nutrients) – dissolved oxygen – aquatic hyphomycetes

The increase in water temperature will also lead to a decrease in the concentration of dissolved oxygen, both directly by decreasing oxygen solubility and indirectly via increased nutrient availability that promotes microbial respiration (Murdoch et al., 2000). Additionally, a decrease in flow may lead to a decrease in the incorporation of oxygen from the atmosphere and a reduction in mixing of the water column that may also contribute to reduce dissolved oxygen availability (Fig. 6.1). Since aquatic hyphomycetes inhabit preferentially well-oxygenated waters (Webster and Descals, 1981), decreases in oxygen availability may negatively affect these organisms. In fact, the survival of aquatic hyphomycetes species was reduced under anaerobic conditions in a laboratory experiment, but the magnitude of the effect was species specific. The origin of the strain (strains isolated from stagnant water seem to better tolerate low oxygen concentrations than strains from well aerated waters) and the duration of the exposure to anaerobic conditions (the longer the exposure the more negative are the effects) had affects (Field and Webster, 1983). Fungal species richness, biomass accrual, reproduction, and litter decomposition also decreased with decreasing oxygen saturation in another laboratory experiment (Medeiros et al., 2009; Gomes et al., 2018).

Under field conditions, fungal biomass and species richness associated with decomposing litter was positively correlated with dissolved oxygen concentration (Rajashekhar and Kaveriappa, 2003; Pascoal and Cássio, 2004). Some species, however, are able to tolerate low oxygen availability (e.g. *Flagellospora curta*), which translates into changes in community structure as oxygenation decreases, both in the laboratory and in the field (Pascoal and Cássio, 2004; Medeiros et al., 2009). These impoverished communities composed of tolerant strains, however, do not seem to be able to maintain the litter decomposition process (Pascoal and Cássio, 2004; Medeiros et al., 2009; Gomes et al., 2018). This important ecosystem function will likely be impaired if the availability of dissolved oxygen decreases.

Atmospheric CO_2/air temperature – vegetation (– nutrients) – aquatic hyphomycetes

The predicted increase in atmospheric CO_2 concentration and in air temperature over this century will likely lead to increases in forest primary production and decreases in leaf quality due to a higher plant investment in structural and secondary compounds (Rustad et al., 2001; Stiling and Cornelissen, 2007); if there is a reduction in precipitation and water stress occurs, primary production might be reduced (Wu et al., 2011). This suggests that in the future, streams may receive an increased input of poor quality litter. Also, changes in the quantity, quality and diversity of litter fall may arrive from changes in the composition of riparian forests caused by a shift in plant species distribution to higher latitudes and elevations in a future climate change scenario (Bakkenes et al., 2002; Morin et al., 2008) (Fig. 6.1).

Since aquatic hyphomycetes depend on the litter provided by the riparian forest, they can be most affected by changes in litter input. Indeed, aquatic hyphomycete biomass accrual and production at the whole-stream level were found to be higher in autumn/winter than in spring/summer due to higher organic matter standing stock in the colder months (Suberkropp, 1997). The density of conidia in transport also increased with an increase in organic matter standing stock, either seasonally or across streams (Suberkropp, 1997; Laitung et al., 2002).

Aquatic hyphomycete activity is also affected by litter quality. Fungal biomass concentration, conidial production and litter decomposition were generally lower for nutrient-poor (high C:N; e.g. oak and rhododendron leaves) than for nutrient-rich litter species (e.g. alder and maple leaves), as there probably was nutrient limitation of microbial activity in the former litter (Gessner and Chauvet, 1994; Gulis et al., 2003a, 2006; Ferreira et al., 2006b, 2012a; Gonçalves et al., 2013). In contrast, the effects of decreased litter quality induced by an enriched CO_2 atmosphere on fungal growth, activity and litter decomposition were generally small to absent (Tuchman et al., 2002, 2003; Rier et al., 2002, 2005; Ferreira et al., 2010; Kelly et al., 2010; Ferreira and Chauvet, 2011b). Although leaves grown under an enriched CO_2 atmosphere (580 – 720 ppm) generally have lower nutrient concentration and higher lignin, total phenols and condensed tannins concentration than leaves grown under an ambient CO_2 atmosphere (360 – 380 ppm) (Rier et al., 2002, 2005; Tuchman et al., 2002, 2003; Kelly et al., 2010; Ferreira and Chauvet, 2011b), there is fast leaching of secondary compounds after immersion (Rier et al., 2002, 2005; Tuchman et al., 2002). This may suggest that future changes in forest composition, and consequently on litter quality, may have stronger effects on aquatic fungi than changes in litter quality induced by increases in atmospheric CO_2 concentration.

The simultaneous effects of increased water temperature and decreased litter quality on fungal activity and community structure are difficult to anticipate, as these changes potentially have contrasting effects. Increases in water temperature seem to have stronger effects on fungal biomass, activity and litter decomposition than changes in litter quality promoted by increases in atmospheric CO_2 concentration (Ferreira and Chauvet, 2011b). In contrast, changes in litter quality due to substitution of litter species may be more important than increases in temperature on fungal activity and community structure (Gonçalves et al., 2013; Ferreira et al., 2015b).

Aquatic hyphomycete species richness seems to increase with an increase in riparian vegetation and in litter species richness (Rajashekhar and Kaveriappa, 2003; Laitung and Chauvet, 2005; Ferreira et al., 2016). Fungal community structure has also been shown to depend on the composition of the riparian vegetation (Bärlocher and Graça, 2002; Ferreira et al., 2006a) and identity/quality of the submerged litter (Canhoto and Graça, 1996; Gulis, 2001; Ferreira et al., 2006b).

Changes in forest composition with an increase in the relative abundance of nitrogen-fixing species may lead to an increase in the concentration of nitrogen in soils and in plant tissues (Rhoades et al. 2001; Goldstein et al., 2010), which may translate into an increase of the nitrogen concentration in streams (Compton et al., 2003; Goldstein et al., 2009; Shaftel et al., 2011) (Fig. 6.1). As discussed above, aquatic hyphomycete growth, reproduction and litter decomposition are generally higher in nutrient rich litter and in streams with high dissolved nutrient availability.

UV-B radiation (– vegetation) – aquatic hyphomycetes

Increases in ultraviolet-B (UV-B) radiation due to stratospheric ozone depletion have the potential to affect plants and microbes (Fig. 6.1), and consequently litter decomposition. However, the effects of increases in UV-B radiation seem to be stronger for plant UV-B-absorbing compounds than for morphologic characteristics (Searles et al., 2001; Li et al., 2012), although the information available for woody plants (including trees, the main contributor of allochthonous litter to many streams) is limited by small sample size (Li et al., 2012).

A recent systematic review showed that an increase in UV-B radiation by 30% inhibited litter decomposition in terrestrial systems, which might be attributed to an inhibition of microbial activity or a decrease in litter quality under enhanced UV-B radiation (Song et al., 2013). The effects of enhanced UV-B radiation on stream fungi were still poorly addressed. Villanueva et al. (2010) found no evidence of an effect of enhanced UV-B radiation on aquatic hyphomycete biomass, reproduction and decomposition of litter submerged in a Patagonian stream or in the laboratory.

Precipitation – hydrological regime – aquatic hyphomycetes

Future increases in temperature and changes in precipitation (IPCC, 2013) will likely affect the hydrology of many streams that will experience an increase in the frequency and intensity of extreme events (Murdoch et al., 2000) (Fig. 6.1). An increase in the frequency and duration of droughts in Mediterranean streams, for instance, will likely affect aquatic fungi. Fungal biomass and litter decomposition decreased with an increase in the duration of the dry period from 0 to 30 days, whereas the effects of the frequency of the dry period depended on the season

(Langhans and Tockner, 2006). Fungal biomass, reproduction and litter decomposition were also affected by the existence and timing of the dry period (Bruder et al., 2011). Generally, fungal activity was greater when leaves were immersed during the entire experiment than when there was a drying event. In addition, the drying event had stronger negative effects when it occurred at the beginning or middle of the incubation period than when it occurred during the last week (Bruder et al., 2011).

Conclusions

The effects of changes in water temperature, dissolved nutrient concentration, and leaf species on aquatic fungi have been addressed frequently, while changes in other factors (e.g. UV-B radiation, hydrological regime) have received less attention. Even less attention has been given to the assessment of changes in multiple factors despite the strong possibility for interactions. The effects of global change factors on aquatic fungi will depend on the background species composition and environmental characteristics and thus will likely be stream, or at least region, specific. Evidence gathered so far suggests that future global climate change will affect aquatic hyphomycete activity and community structure, which may have consequences for ecosystem functioning as these organisms mediate the transfer of carbon from dead organic matter to higher trophic levels. An increase in temperature of well-oxygenated, nutrient-enriched streams will most likely stimulate aquatic hyphomycete activity and litter decomposition.

Future trends

Future research should aim at addressing the response of aquatic hyphomycetes to changes in environmental factors at relevant spatial and temporal scales. The effects of changes in environmental factors on fungi community structure and activity often depend on background fungi community composition and environmental characteristics. Thus, the assessment of effects of global change factors on fungi should be done under realistic field conditions. For instance, whole-stream nutrient concentration increases (Rosemond et al., 2002; Gulis and Suberkropp, 2003a; Ferreira et al., 2006b) and temperature (Bärlocher et al., 2008; Ferreira and Canhoto, 2014, 2015) allow assessment of effects of nutrient enrichment and warming, respectively, on fungi at relevant spatial scales. These manipulations should be performed in distinct regions and seasons in order to accommodate for background differences in communities and climate (Gulis et al., 2008; Ferreira and Canhoto, 2014, 2015). In addition, long term experiments are needed to assess the possibility for adaptation of fungal communities to environmental change (Rosemond et al., 2015). Moreover, if we want to anticipate the response of aquatic hyphomycetes to global change, we need to assess the simultaneous change of multiple environmental factors because they will change concomitantly in the future. Any interactions only make it more difficult to anticipate the magnitude and direction of the interaction (Ferreira and Chauvet, 2011a, b; Gonçalves et al., 2013; Piggott et al., 2012, 2015; Fernandes et al., 2014; Ferreira et al., 2015b).

Although replication of ecosystem-scale experiments is logistically challenging, the implementation of coordinated distributed experiments can be a solution (Fraser et al., 2013). Coordinated distributed experiments are hypothesis-driven experiments that take place at multiple geographic locations, following a standardized research design and protocol, with data collection being synchronized and performed by multiple teams (Fraser et al., 2013). This approach increases the feasibility of the work plan and decreases the budget required as compared with a similar experimental design involving a single research team. This approach has been implemented before to assess litter decomposition along environmental gradients in streams. For instance, Boyero et al. (2011) and Seena et al. (2019) assessed the effect of temperature on alder litter decomposition and litter fungi diversity, respectively, in 19 – 22 streams along a latitudinal gradient. These projects involved research teams from about 20 countries. Woodward et al. (2012) assessed the effect of nutrient enrichment on litter decomposition in 100 European streams along a nutrient gradient (14 – 21 641 μg DIN/L; 1 – 926 μg SRP/L). This project involved 10 research teams from 9 countries.

The collaboration between stream and terrestrial ecologists can also contribute to addressing the effects of global change factors on aquatic communities and processes. Several coordinated distributed experiments exist in terrestrial systems to investigate the effect of warming, changes in precipitation, ozone depletion, and CO_2 elevation on plants (Fraser et al., 2013). The incubation of litter produced during these experiments in terrestrial systems in streams would allow the effect of global change factors on the decomposition of submerged litter mediated by changes in litter quality to be addressed. For instance, litter produced during a free-air CO_2 enrichment experiment, designed to address the effects of elevated CO_2 on trees, was used to assess the effects of changes in litter quality induced by the enriched atmosphere on microbes, invertebrates and litter decomposition in streams (Ferreira et al., 2010; Ferreira and Chauvet, 2011b). Systematic reviews of existing data on the effect of global change factors on fungal activity in streams would help summarise information, identify possible moderator variables of the effect size and identify questions needing further research (Ferreira et al., 2015a; Amani et al., 2019).

Acknowledgement

Financial support by the Portuguese Foundation for Science and Technology (FCT) (postdoctoral fellowship SFRH/BPD/76482/2011, program POPH/FSE; IF/00129/2014) is gratefully acknowledged.

References

Abelho, M., and Graça, M.A.S. (2006). Effects of nutrient enrichment on decomposition and fungal colonization of sweet chestnut leaves in an Iberian stream (Central Portugal). Hydrobiologia *560*, 239–247.

Amani, M., Graça, M.A.S., and Ferreira V. (2019). Effects of elevated atmospheric CO_2 concentration and temperature on litter decomposition in streams: a meta-analysis. Internat. Rev. Hydrobiol. 104, 14–25.

Arsuffi, T.L., and Suberkropp, K. (1988). Effects of fungal mycelia and enzymatically degraded leaves on feeding and performance of caddisfly (Trichoptera) larvae. J. N. Am. Benthol. Soc. *7*, 205–211.

Artigas, J., Romaní, A.M., and Sabater, S. (2008). Effect of nutrients on the sporulation and diversity of aquatic hyphomycetes on submerged substrata in a Mediterranean stream. Aquat. Bot. *88*, 32–28.

Bakkenes, M., Alkemade, J.R.M., Ihle, F., Leemans, R., and Latour, J.B. (2002). Assessing effects of forecasted climate change on the diversity and distribution of European higher plants for 2050. Global Change Biol. *8*, 390–407.

Baldy, V., Gobert, V., Guerold, F., Chauvet, E., Lambrigot, D., and Charcosset, J.-Y. (2007). Leaf litter breakdown budgets in streams of various trophic status: effects of dissolved inorganic nutrients on microorganisms and invertebrates. Freshwat. Biol. *52*, 1322–1335.

Bärlocher, B., and Graça, M.A.S. (2002). Exotic riparian vegetation lowers fungal diversity but not leaf decomposition in Portuguese streams. Freshwat. Biol. *47*, 1123–1135.

Bärlocher, F., Kebede, Y.K., Gonçalves, A.L., and Canhoto, C. (2013). Incubation temperature and substrate quality modulate sporulation by aquatic hyphomycetes. Microb. Ecol. *66*, 30–39.

Bärlocher, F., Seena, S., Wilson, K.P., and Williams, D.D. (2008). Raised water temperature lowers diversity of hyporheic aquatic hyphomycetes. Freshwat. Biol. *53*, 368–379.

Bärlocher, F., and Sridhar, K.R. (2014). Association of animals and fungi in leaf decomposition. In Freshwater Fungi and Fungus-like Organisms, G., Jones, K. Hyde and K.-L. Pang, eds. (Berlin: De Gruyter), pp. 413–442.

Boyero, L., Pearson, R.G., Gessner, M.O., Barmuta, L.A., Ferreira, V., Graça, M.A.S., Dudgeon, D., Boulton, A.J., Callisto, M., Chauvet, E., Helson, J.E., Bruder, A., Albariño, R.J., Yule, C.M., Arunachalam, M., Davies, J.N., Figueroa, R., Flecker, A.S., Ramírez, A., Death, R.G., Iwata, T., Mathooko, J.M., Mathuriau, C., Gonçalves, Jr. J.F., Moretti, M.S., Jingut, T., Lamothe, S., M'Erimba, C., Ratnarajah, L., Schindler, M.H., Castela, J., Buria, L.M., Cornejo, A., Villanueva, V.D., and West, D.C. (2011). A global experiment suggests climate warming will not accelerate litter decomposition in streams but may reduce carbon sequestration. Ecol. Lett. *14*, 289–294.

Brown, J.H., Gillooly, J.F., Allen, A.P., Savage, V.M., and West, G.B. (2004). Toward a metabolic theory of ecology. Ecology *85*, 1771–1789.

Bruder, A., Chauvet, E., and Gessner, M.O. (2011). Litter diversity, fungal decomposers and litter decomposition under simulated stream intermittency. Funct. Ecol. *25*, 1269–1277.

Canhoto, C, and Graça, M.A.S. (1996). Decomposition of *Eucalyptus globulus* leaves and three native leaf species (*Alnus glutinosa*, *Castanea sativa* and *Quercus faginea*) in a Portuguese low order stream. Hydrobiologia *333*, 79–85.

Canhoto, C., and Graça, M.A.S. (2008). Interactions between fungi and stream invertebrates: back to the future. In Novel Techniques and Ideas in Mycology, K.R. Sridhar, F. Bärlocher and K.D. Hyde, eds. (Hong Kong: Hong Kong University Press), pp. 305–325.

Chadwick, M.A., and Huryn, A.D. (2003). Effect of a whole-catchment N addition on stream detritus processing. J. N. Am. Benthol. Soc. *22*, 194–206.

Chauvet, E., and Suberkropp, K. (1998). Temperature and sporulation of aquatic hyphomycetes. Appl. Environ. Microb. *64*, 1522–1525.

Compton, J.E., Church, M.R., Larned, S.T., and Hogsett, W.E. (2003). Nitrogen export from forested watersheds in the Oregon coast range: the role of N_2-fixing red alder. Ecosystems *6*, 773–785.

Cornut, J., Ferreira, V., Gonçalves, A.L., Chauvet, E., and Canhoto, C. (2015). Fungal alteration of the elemental composition of leaf litter affects shredder feeding activity. Freshwat. Biol. *60*, 1755–1771.

Dang, K., Schindler, M., Chauvet, E., and Gessner, M.O. (2009). Temperature oscillation coupled with fungal community shifts can modulate warming effects on litter decomposition. Ecology *90*, 122–131.

Danger, M., and Chauvet, E. (2013). Elemental composition and degree of homeostasis of fungi: are aquatic hyphomycetes more like metazoans, bacteria or plants? Fungal Ecol. *6*, 453–457.

Duarte, D., Fernandes, I., Nogueira, M.J., Cássio, F., and Pascoal, C. (2013). Temperature alters interspecific relationships among aquatic fungi. Fungal Ecol. *6*, 187–191.

Duarte, D., Pascoal, C., Cássio, F., and Bärlocher, F. (2006). Aquatic hyphomycete diversity and identity affect leaf litter decomposition in microcosms. Oecologia *147*, 658–666.

Fabre, E., and Chauvet, E. (1998). Leaf breakdown along an altitudinal stream gradient. Arch. Hydrobiol. *141*, 167–179.

Fernandes, I., Pascoal, C., Guimarães, H., Pinto, R., Sousa, I., and Cássio, F. (2012). Higher temperature reduces the effects of litter quality on decomposition by aquatic fungi. Freshwat. Biol. *57*, 2306–2317.

Fernandes, I., Seena, S., Pascoal, C., and Cássio, F. (2014). Elevated temperature may intensify the positive effects of nutrients on microbial decomposition in streams. Freshwat. Biol. 59, 2390–2399.

Fernandes, I., Uzun, B., Pascoal, C., and Cássio, F. (2009). Responses of aquatic fungal communities on leaf litter to temperature-change events. Internat. Rev. Hydrobiol. *94*, 410–418.

Ferreira, V., and Canhoto, C. (2014). Effect of experimental and seasonal warming on litter decomposition in a temperate stream. Aquat. Sci. *76*, 155-163.

Ferreira, V., and Canhoto, C. (2015). Future increase in temperature might stimulate litter decomposition in temperate cold water streams – evidence from a stream manipulation experiment. Freshwat. Biol. *60*, 881–892.

Ferreira, V., Castagneyrol, B., Koricheva, J., Gulis, V., Chauvet, E., and Graça, M.A.S. (2015a). A meta-analysis of the effects of nutrient enrichment on litter decomposition in streams. Biol. Rev. *90*, 669–688.

Ferreira, V., Castela, J., Rosa, P., Tonin, A.M., Boyero, L., and Graça M.A.S. (2016). Aquatic hyphomycetes, benthic macroinvertebrates and leaf litter decomposition in streams naturally differing in riparian vegetation. Aquat. Ecol. *50*, 711–725.

Ferreira, V., and Chauvet, E. (2011a). Synergistic effects of water temperature and dissolved nutrients on litter decomposition and associated fungi. Global Change Biol. *17*, 551–564.

Ferreira, V., and Chauvet, E. (2011b). Future increase in temperature more than decrease in litter quality can affect microbial litter decomposition in streams. Oecologia *167*, 279–291.

Ferreira, V., Chauvet E., and Canhoto C. (2015b). Effects of experimental warming, litter species, and presence of macroinvertebrates on litter decomposition and associated decomposers in a temperate mountain stream. Can. J. Fish. Aq. Sci. 72, 206–216.

Ferreira, V., Elosegi, A., Gulis, V., Pozo, J., and Graça, M.A.S. (2006a). Eucalyptus plantations affect fungal communities associated with leaf litter decomposition in Iberian streams. Arch. Hydrobiol. *166*, 467–490.

Ferreira, V., Encalada, A.C., and Graça, M.A.S. (2012a). Effects of litter diversity on decomposition and biological colonization of submerged litter in temperate and tropical streams. Freshwat. Sci. *31*, 945–962.

Ferreira, V., Gonçalves, A.L., and Canhoto, C. (2012b). Aquatic hyphomycete strains from metal-contaminated and reference streams might respond differently to future increases in temperature. Mycologia *104*, 613–622.

Ferreira, V., Faustino, H., Raposeiro, P.M., and Gonçalves V. (2017). Replacement of native forests by conifer plantations affects fungal decomposer community structure but not litter decomposition in Atlantic island streams. Forest Ecol. Manage. *389*, 323–330.

Ferreira, V., Gonçalves, A.L., Godbold, D.L., and Canhoto, C. (2010). Effect of increased atmospheric CO_2 on the performance of an aquatic detritivore through changes in water temperature and litter quality. Global Change Biol. *16*, 3284–3296.

Ferreira, V., Gulis, V., and Graça, M.A.S. (2006b). Whole-stream nitrate addition affects litter decomposition and associated fungi but not invertebrates. Oecologia *149*, 718–729.

Ferreira, V., Gulis, V., Pascoal, C., and Graça, M.A.S. (2014). Stream pollution and fungi. In Freshwater Fungi and Fungus-like Organisms, G., Jones, K. Hyde and K.-L. Pang, eds. (Berlin: De Gruyter), pp. 389–412.

Field, J.I., and Webster, J. (1983). Anaerobic survival of aquatic fungi. Trans. Br. Mycol. Soc. *81*, 365–369.

Fraser, L.H., Herny, H.A.L., Carlyle, C.N., White, S.R., Beierkuhnlein, C., Cahill Jr., J.F., Casper, B.B., Cleland, E., Collins, S., Dukes, J.S., Knapp, A.K., Lind, E., Long, R., Luo, Y., Reich, P.B., Smith, M.D., Sternberg, M., and Turkington, R. (2013). Coordinated distributed experiments: an emerging tool for testing global hypotheses in ecology and environmental sciences. Front. Ecol. Environ. *11*, 147–155.

Friberg, N., Dybkjaer, J.B., Olafsson, J.S., Gislason, G.M., Larsen, S.E., and Lauridsen, T.L. (2009). Relationships between structure and function in streams contrasting in temperature. Freshwat. Biol. *54*, 2051–2068.

Geraldes, P., Pascoal, C., and Cássio, F. (2012). Effects of increased temperature and aquatic fungal diversity on litter decomposition. Fungal Ecol. *6*, 734–740.

Gessner, M.O., and Chauvet, E. (1994). Importance of stream microfungi in controlling breakdown rates of leaf litter. Ecology *75*, 1807–1817.

Goldstein, C.L., Williard, K.W.J., and Schoonover, J.E. (2009). Impact of an invasive exotic species on stream nitrogen levels in southern Illinois. J. Am. Water Resour. Assoc. *45*, 664–672.

Goldstein, C.L., Williard, K.W., Schoonover, J.E., Baer, S.G., Groninger, J.W., and Snyder, J.M. (2010). Soil and groundwater nitrogen response to invasion by an exotic nitrogen-fixing shrub. J. Environ. Qual. *39*, 1077–1084.

Gomes, P.P., Ferreira, V., Tonin, A.M., Medeiros, A.O., and Gonçalves, J.F. Jr. (2018). Combined effects of nutrients and dissolved oxygen on plant litter decomposition and associated fungal communities. Microb. Ecol. *75*, 854–862.

Gonçalves, A.L., Graça, M.A.S., and Canhoto, C. (2013). The effect of temperature on leaf decomposition and diversity of associated aquatic hyphomycetes depends on the substrate. Fungal Ecol. *6*, 546–553.

Graça, M.A.S., and Ferreira, R. (1995). The ability of selected aquatic and terrestrial fungi to decompose leaves in freshwater. Sydowia *47*, 167–179.

Gulis, V. (2001). Are there any substrate preferences in aquatic hyphomycetes? Mycol. Res *105*, 1088–1093.

Gulis, V., Ferreira, V., and Graça, M.A.S. (2006). Stimulation of leaf litter decomposition and associated fungi and invertebrates by moderate eutrophication: implications for stream assessment. Freshwat. Biol. *51*, 1655–1669.

Gulis, V., and Suberkropp, K. (2003a). Leaf litter decomposition and microbial activity in nutrient-enriched and unaltered reaches of a headwater stream. Freshwat. Biol. *48*, 123–134.

Gulis, V., and Suberkropp, K. (2003b). Effect of inorganic nutrients on relative contributions of fungi and bacteria to carbon flow from submerged decomposing leaf litter. Microb. Ecol. *45*, 11–19.

Gulis, V., and Suberkropp, K. (2004). Effects of whole-stream nutrient enrichment on the concentration and abundance of aquatic hyphomycete conidia in transport. Mycologia *96*, 57–65.

Gulis, V., Suberkropp, K., and Rosemond, A.D. (2008). Comparison of fungal activities on wood and leaf litter in unaltered and nutrient-enriched headwater streams. Appl. Environ. Microbiol. *74*, 1094–1101.

Hieber, M., and Gessner, M.O. (2002). Contribution of stream detritivores, fungi, and bacteria to leaf breakdown based on biomass estimates. Ecology *83*, 1026–1038.

IPCC (Intergovernmental Panel on Climate Change). (2013). Climate change 2013: The physical science basis. Working group I contribution to the IPCC fifth assessment report. Summary for policymakers.

Irons, J.G., Oswood, M.W., Stout, R.J., and Pringle, C.M. (1994). Latitudinal patterns in leaf litter breakdown, is temperature really important? Freshwat. Biol. *32*, 401–411.

Kelly, J.J., Bansal, A., Winkelman, J., Janus, L.R., Hell, S., Wencel, M., Belt, P., Kuehn, K.A., Rier, S.T., and Tuchman, N.C. (2010). Alteration of microbial communities colonizing leaf litter in a temperate woodland stream by growth of trees under conditions of elevated atmospheric CO_2. Appl. Environ. Microbiol. *76*, 4950–4959.

Koske, R.E., and Duncan, I.W. (1974). Temperature effects on growth, sporulation, and germination of some 'aquatic' hyphomycetes. Can. J. Bot. *52*, 1387–1391.

Laitung, B., and Chauvet, E. (2005). Vegetation diversity increases species richness of leaf-decaying fungal communities in woodland streams. Arch. Hydrobiol. *164*, 217–235.

Laitung, B., Pretty, J.L., Chauvet, E., and Dobson, M. (2002). Response of aquatic hyphomycete communities to enhanced stream retention in areas impacted by commercial forestry. Freshwat. Biol. *47*, 313–323.

Langhans, S.D., and Tockner, K. (2006). The role of timing, duration, and frequency of inundation in controlling leaf litter decomposition in a river-floodplain ecosystem (Tagliamento, northeastern Italy). Oecologia *147*, 501–509.

Lecerf, A., Dobson, M, Dang, C.K., and Chauvet, E. (2005). Riparian plant species loss alters trophic dynamics in detritus-based stream ecosystems. Oecologia *146*, 432–442.

Lecerf, A., Usseglio-Polatera, P., Charcosset, J.-Y., Bracht, B., and Chauvet, E. (2006). Assessment of functional integrity of eutrophic streams using litter breakdown and benthic macroinvertebrates. Arch. Hydrobiol. *165*, 105–126.

Li, F.-R., Peng, S.-L., Chen, B.-M., and Hou, Y-P. (2012). A meta-analysis of the responses of woody and herbaceous plants to elevated ultraviolet-B radiation. Acta Oecol. *36*, 1–9.

Martínez, A., Larrañaga, A., Pérez, J., Descals, E., and Pozo, J. (2014). Temperature affects leaf litter decomposition in low-order forest streams: field and microcosm approaches. FEMS Microbiol. Ecol. *87*, 257–267.

Martínez, A., Monroy, S., Pérez, J., Larrañaga, A., Basaguren, A., Molinero, J., and Pozo, J. (2015). In-stream litter decomposition along an altitudinal gradient: does substrate quality matter? Hydrobiologia *766*: 17-28. DOI: 10.1007/s10750-015-2432-9

Medeiros, A.O., Pascoal, C, and Graça, M.A.S. (2009). Diversity and activity of aquatic fungi under low oxygen conditions. Freshwat. Biol. *54*, 142–149.

Morin, X., Viner, D., and Chuine, I. (2008). Tree species range shifts at a continental scale: new predictive insights from a process-based model. J. Ecol. *96*, 784–794.

Morrill, J.C., Bales, R.C., and Conklin, M.H. (2005). Estimating stream temperature from air temperature, implications for future water quality. J. Environ. Eng. *131*, 139–146.

Moss, B., Kosten, S., Meerhoff, M., Battarbee, R.W., Jeppesen, E., Mazzeo, N., Havens, K., Lacerot, G., Liu, Z., De Meester, L., Paerl, H., and Scheffer, M. (2011). Allied attack: climate change and eutrophication. Inland Waters *1*, 101–105.

Murdoch, P.S., Baron, J.S., and Miller, T.L. (2000). Potential effects of climate change on surface-water quality in North America. J. Am. Water Res. Ass. *36*, 347–366.

Nikolcheva, L.G., and Bärlocher, F. (2005). Seasonal and substrate preferences of fungi colonizing leaves in streams: traditional versus molecular evidence. Environ. Microbiol. *7*, 270–280.

Niyogi, D.K., Simon, K.S., and Townsend, C.R. (2003). Breakdown of tussock grass in streams along a gradient of agricultural development in New Zealand. Freshwat. Biol. *48*, 1698–1708.

Pascoal, C., and Cássio, F. (2004). Contribution of fungi and bacteria to leaf litter decomposition in a polluted river. Appl. Environ. Microbiol. *70*, 5266–5273.

Pérez, J., Descals, E., and Pozo, J. (2012). Aquatic hyphomycete communities associated with decomposing alder leaf litter in reference headwater streams of the Basque Country (northern Spain). Microb. Ecol. *64*, 279–290.

Piggott, J.J., Lange, K., Townsend, C.R., and Matthaei, C.D. (2012). Multiple stressors in agricultural streams: a mesocosm study of interactions among raised water temperature, sediment addition and nutrient enrichment. PLoS One 7, e49873.

Piggott, J.J., Niyogi, D.K., Townsend, C.R., and Matthaei, C.D. (2015). Multiple stressors and stream ecosystem functioning: climate warming and agricultural stressors interact to affect processing of organic matter. J. Appl. Ecol. *52*, 1126–1134. DOI: 10.1111/1365-2664.12480

Pilgrim, J.M., Fang, X., and Stefan, H. (1998). Stream temperature correlations with air temperatures in Minnesota: implications for climate warming. J. Am. Water Res. Ass. *34*, 1109–1121.

Rajashekhar, M., and Kaveriappa, K.M. (2003). Diversity of aquatic hyphomycetes in the aquatic ecosystems of the Western Ghats of India. Hydrobiologia *501*, 167–177.

Rhoades, C., Oskarsson, H., Binkley, D., and Stottlemyer, B. (2001). Alder (*Alnus crispa*) effects on soils in ecosystems of the Agashashok river valley, northwest Alaska. Ecosience *8*, 89–95.

Rier, S.T., Tuchman, N.C., and Wetzel, R.G. (2005). Chemical changes to leaf litter grown under elevated CO_2 and the implications for microbial utilization in a stream ecosystem. Can. J. Fish. Aq. Sci. *62*, 185–194.

Rier, S.T., Tuchman, N.C., Wetzel, R.G., and Teeri, J.A. (2002). Elevated-CO_2-induced changes in the chemistry of quaking aspen (*Populus tremuloides* Michaux) leaf litter: subsequent mass loss and microbial responses in a stream ecosystem. J. N. Am. Benthol. Soc. *21*, 16–27.

Rosemond, A.D., Benstead, J.P., Bumpers, P.M., Gulis, V., Kominoski, J.S., Manning, D.W., Suberkropp, K., and Wallace, J.B. (2015). Experimental nutrient additions accelerate terrestrial carbon loss from stream ecosystems. Science *347*, 1142–1145.

Rosemond, A.D., Pringle, C.M., Ramírez, A, Paul, M.J., and Meyer, J.L. (2002). Landscape variation in phosphorus concentration and effects on detritus-based tropical streams. Limnol. Oceanogr. *47*, 278–289.

Rustad, L.E., Campbell, J.L., Marion, G.M., Norby, R.J., Mitchell, M.J., Hartley, A.E., Cornelissen, J.H.C., Gurevitch, J., GCTE-NEWS. (2001). A meta-analysis of the response of soil respiration, net nitrogen mineralization, and aboveground plant growth to experimental ecosystem warming. Oecologia *126*, 543–562.

Searles, P.S., Flint, S.D., and Caldwell, M.M. (2001). A meta-analysis of plant field studies simulating stratospheric ozone depletion. Oecologia *127*, 1–10.

Seena, S., Bärlocher, F., Sobral, O., Gessner, M.O., Dudgeon, D., McKie, B.G., Chauvet, E., Boyero, L., Ferreira, V., Frainer, A., Bruder, A., Matthaei, C.D., Fenoglio, S., Sridhar, K.R., Albariño, R.J., Douglas, M., Encalada, A.C., Garcia, E., Ghate, S.D., Giling, D.P., Gonçalves, V., Iwata, T., Landeira-Dabarca, A., McMaster, D., Medeiros, A.O., Naggea, J., Pozo, J., Raposeiro, P.M., Swan, C.M., Tenkiano, N.S.D., Yule, C.M., and Graça, M.A.S. (2019). Biodiversity of leaf litter fungi in streams along a latitudinal gradient. Sci. Total Environ. *661*, 306–315.

Shaftel, R.S., King, R.S., and Back, J.A. (2011). Alder cover drives nitrogen availability in Kenai lowland headwater streams, Alaska. Biogeochemistry *107*, 135–148.

Song, X., Peng, C., Jiang, H., Zhu, Q., and Wang, W. (2013). Direct and indirect effects of UV-B exposure on litter decomposition: a meta-analysis. PLoS ONE *8*, e68858.

Stelzer, R.S., Heffernan, J., and Likens, G.E. (2003). The influence of dissolved nutrients and particulate organic matter quality on microbial respiration and biomass in a forest stream. Freshwat. Biol. *48*, 1925–1937.

Stiling, P., and Cornelissen, T. (2007). How does elevated carbon dioxide (CO_2) affect plant-herbivore interactions? A field experiment and meta-analysis of CO_2-mediated changes on plant chemistry and herbivore performance. Global Change Biol. *13*, 1823–1842.

Suberkropp, K. (1997). Annual production of leaf-decaying fungi in a woodland stream. Freshwat. Biol. *38*, 169–178.

Suberkropp, K. (1998). Effect of dissolved nutrients on two aquatic hyphomycetes growing on leaf litter. Mycol. Res. *102*, 998–1002.

Suberkropp, K. and Chauvet, E. (1995). Regulation of leaf breakdown by fungi in streams: influences of water chemistry. Ecology *76*, 1433–1445.

Taylor, B.R., and Chauvet, E. (2014). Relative influence of shredders and fungi on leaf litter decomposition along a river altitudinal gradient. Hydrobiologia *721*, 239–250.

Tuchman, N.C., Wahtera, K.A., Wetzel, R.G., and Teeri, J.A. (2003). Elevated atmospheric CO_2 alters leaf litter quality for stream ecosystems: an in situ leaf decomposition study. Hydrobiologia *495*, 203–211.

Tuchman, N.C., Wetzel, R.G., Rier, S.T., Wahtera, K.A., and Teeri, J.A. (2002). Elevated atmospheric CO_2 lowers leaf litter nutritional quality for stream ecosystem food webs. Global Change Biol. *8*, 163–170.

Villanueva, V.D., Albariño, R., and Graça, M.A.S. (2010). Natural UVB does not affect decomposition by aquatic hyphomycetes. Internat. Rev. Hydrobiol. *95*, 1–11.

Wallace, J.B., Eggert, S.L., Meyer, J.L., and Webster, J.R. (1997). Multiple trophic levels of a forest stream linked to terrestrial litter inputs. Science *277*, 102–104.

Webster, J., and Descals, E. (1981). Morphology, distribution, and ecology of conidial fungi. In Biology of Conidial Fungi, G.T. Cole and B. Kendrick, eds. (New York and London: Academic Press), pp. 295–355.

Webster, J., Moran, S.T., and Davey, R.A. (1976). Growth and sporulation of *Tricladium chaetocladium* and *Lunulosporula curvula* in relations to temperature. Trans. Br. Mycol. Soc. *67*, 491–495.

Woodward, G., Gessner, M.O., Giller, P.S., Gulis, V., Hladyz, S., Lecerf, A., Malmqvist, B., McKie, B.G., Tiegs, S.D., Cariss, H., Dobson, M., Elosegi, A.,

Ferreira, V., Graça, M.A.S., Fleituch, T., Lacoursière, J., Nistorescu, M., Pozo, J., Risnoveanu, G., Schindler, M., Vadineanu, A., Vought, L.B.-M., and Chauvet, E. (2012). Continental-scale effects of nutrient pollution on stream ecosystem functioning. Science *336*, 1438–1440.

Wu, Z., Dijkstra, P., Koch, G.W., Peñuelas, J., and Hungate, B.A. (2011). Responses of terrestrial ecosystems to temperature and precipitation change: a meta-analysis of experimental manipulation. Global Change Biol. *17*, 927–942.

Chapter 7

Aquatic Viruses and Climate Change

Rui Zhang[1], Markus G. Weinbauer[2] and Peter Peduzzi[3,*]

[1] Institute of Marine Microbes and Ecospheres; and State Key Laboratory of Marine Environmental Science, Xiamen University (Xiang'an), Xiamen, Fujian, China; [2] Microbial Ecology and Biogeochemistry Group, Laboratoirè d'Oceanographie de Villefranche, Universitè Pierre et Marie Curie-Paris 6, Villefranche-sur-Mer, France, and CNRS, Laboratoire d'Ocèanographie de Villefranche, Villefranche-sur-Mer, France; [3] University of Vienna, Faculty of Life Sciences, Functional and Evolutionary Ecology, Limnology Unit, Vienna, Austria; * corresponding author

Email: ruizhang@xmu.edu.cn, zhangray@gmail.com, wein@obs-vlfr.fr, peter.peduzzi@univie.ac.at

DOI: https://doi.org/10.21775/9781913652579.07

Abstract

The viral component in aquatic systems clearly needs to be incorporated into future ocean and inland water climate models. Viruses have the potential to influence carbon and nutrient cycling in aquatic ecosystems significantly. Changing climate likely has both direct and indirect influence on virus-mediated processes, among them an impact on food webs, biogeochemical cycles and on the overall metabolic performance of whole ecosystems. Here we synthesise current knowledge on potential climate-related consequences for viral assemblages, virus-host interactions and virus functions, and in turn, viral processes contributing to climate change. There is a need to increase the accuracy of predictions of climate change impacts on virus-driven processes, particularly of those linked to biological production and biogeochemical cycles. Comprehension of the relationships between microbial/viral processes and global phenomena is essential to predict the influence on as well as the response of the biosphere to global change.

Introduction

Whereas most human interventions on ecosystems are immediate, the impact on climate change acts over long periods of time, leading to slight but continuous alterations. As an example, the predicted 2.6–4.8°C warming, under the conditions of Representative Concentration Pathways 8.5 (IPCC, 2013), for this century is expected to be responsible for species movements and extinctions, as well as for changes in the composition of communities, thus likely altering ecosystem functioning (Mooney et al., 2009). Since global change will probably impact all ecosystem components, it is expected that even the smallest life-forms such as Bacteria, Archaea, eukaryotic microorganisms and viruses of aquatic systems will play important roles as agents and recipients of global climate change (Danovaro et al., 2011; Hutchins and Fu, 2017).

Beside direct effects by changes in water temperature, others may occur through altered oceanic circulation along with changing habitats, biogeochemical cycles and food webs. Particularly estuaries and continental shelf regions seem to be affected by changed river runoff due to reduced or enhanced precipitation and flooding. This will result in changes in salinity and nutrients, and in the strength and seasonality of circulation patterns (Danovaro et al., 2011; Cozzi et al., 2012). Changing ocean currents, fluctuations in the depth of the surface mixed layer and stratification might influence light and nutrient availability and ultimately CO_2-dynamics and the carbon cycle (Bauer et al., 2013; Finke et al., 2017). Moreover, warming will increase the extent of the oxygen minimum zones (OMZ), which are a globally important sink for nitrogen and a source of CO_2. Another global player is the reduction of seawater pH due to enhanced CO_2 levels in the atmosphere (ocean acidification, OA). OMZ changes and OA will also influence virus-host interactions.

For inland waters and freshwater systems a whole array of human-mediated threats makes them the most affected ecosystems. Beside pollution and nutrient loads, overexploitation, flow modification, habitat destruction or degradation, climate change is designated as a significant factor for ongoing alterations of these systems (Dudgeon et al., 2006; Zweimüller et al., 2008). There is also evidence that climate warming will affect pelagic carbon metabolism and sediment delivery in lakes and carbon sequestration in streams and rivers (Battin et al., 2009; Boyero et al., 2011; Kritzberg et al., 2014).

Global climate change-related impacts are most striking in the polar regions, where temperature and other factors are changing at more than twice the global average (Hoegh-Guldberg and Bruno, 2010). Due to the major contribution of microbes to ecosystem processes, much attention has been paid to the impact of rapid climate change in Arctic and Antarctic microbial communities (e.g., the European Project on Ocean Acidification, EPOCA). Declines or shifts in these microbial ecosystems will no doubt have implications for entire food webs and biogeochemical fluxes, and polar microbiota can be viewed both as sentinels and amplifiers of global change (Vincent, 2010).

The global relevance of aquatic viruses has been evident for many years. Viruses are the smallest and most numerous biological entities in aquatic ecosystems, the majority being bacteriophages (Wommack and Colwell, 2000; Weinbauer, 2004; Suttle, 2007). They are about 10 to 15-fold more abundant than their microbial hosts, and equivalent to the carbon in ca. 75 million blue whales (ca. 10% of prokaryotic carbon by weight; Suttle, 2005). The typically high frequency of virus infection in aquatic ecosystems is known to be a major cause of prokaryotic host mortality. Host cell lysis in the marine environment was calculated to be a source of up to 10^9 tons of carbon each day, released from the biological pool (Suttle, 2007; Brussaard et al., 2008). If the main control of prokaryotic abundance is via protozoan grazing, most of the carbon will be channelled to higher trophic levels (microbial loop). In contrast, if viral lysis accounts for most prokaryotic losses, then carbon and nutrients are diverted away from larger organisms (Proctor and Fuhrman, 1990; Wilhelm and Suttle, 1999). The biogeochemical consequence of this "viral shunt" (the process of diverting organic carbon away from the grazing food chain via viral lysis of host cells) are changed rates of carbon accumulation in the photic zone (CO_2 release to the atmosphere vs. vertical transport to the meso-/bathypelagic zone). Viruses are also involved in shaping the composition of bacterial communities, either by reducing abundant host taxa or even by introducing new genetic information into their hosts (e.g., Zhang et al., 2007). They are the largest and most diverse genetic reservoir on Earth and may boost the resilience of ecosystems by sustaining multiple species with similar or identical biochemical pathways (Thingstad, 2000; Brussard et al., 2008; Jacquet et al., 2010; Brum et al., 2015; Guidi et al., 2016; Gregory et al., 2019).

Elucidating the microbial mediation of carbon-cycle feedbacks to climate change appears to be fundamental for understanding ecosystem responses. This approach would provide a mechanistic basis for carbon-climate modelling (Zhou et al., 2011). Our substantial lack of knowledge explains why the viral compartment is missing from most climate change models. This chapter tackles the growing evidence that aquatic viruses interact actively with climate alterations and are key biotic components that can influence the microbial feedback on climate change (compare also Danovaro et al., 2011; Mojica and Brussaard, 2014).

Impact on viral abundance, distribution and dynamics

In most natural aquatic environments virus communities were found to be highly dynamic, with often pronounced changes in abundance and diversity over broad ranges of space and time. Viral abundance is largely linked to host availability. The numbers of planktonic viruses commonly range between 10^4 and 10^8 ml^{-1}, being generally higher in inland waters than in marine systems; peak values were reported for very productive estuaries and lakes (Peduzzi and Luef, 2009; Parikka et al., 2017). Exceptionally high virioplankton abundance has been documented in the alkaline-saline lake Nakuru in Kenya, East Africa (up to 7 x 10^9 ml^{-1}; Peduzzi et al., 2014). Abundance may be up to three orders of magnitude lower in deep oceanic waters (e.g. Parada et al., 2006; Magagnini et al., 2007; Li et al., 2014; Liang et al., 2014; Lara et al. 2017; Muck et al. 2014), but virus-induced relative prokaryotic mortality increases with water depth: below a depth of 1000 m almost all of the prokaryotic heterotrophic production in surface sediment is transformed into organic detritus and dissolved organic matter (DOM). This "viral shunt" releases on a global scale 0.37-0.63 Gt carbon $year^{-1}$, and is an essential source of labile organic matter for the deep-sea ecosystem (Danovaro et al., 2008). Recently, Lara et al. (2017) estimated that about 145 Gt of C, 27.6 Gt of N, and 4.6 Gt of P are released annually by the viral shunt in global tropical and subtropical oceans. Moreover, on average, 33.6% (equalling 0.605 Pg C $year^{-1}$) of the globally respired carbon from fluvial systems may pass through a viral loop (Peduzzi, 2016).

Factors affecting virus dynamics and microbial host-virus interactions in the marine environment have been summarized (Mojica and Brussaard, 2014). Typically virus numbers are linked to system productivity, as a result of the balance between production and decay. Eutrophic water bodies contain more virus particles than oligotrophic systems (Parikka et al., 2017). The trophic situation can also be very variable on seasonal scales, particularly in running waters with pronounced hydrological dynamics (Peduzzi, 2016). Virus abundance apparently varies more pronouncedly on seasonal scales in inland waters than in the marine environment (Wilhelm and Matteson; 2008).

Impact of temperature

The typical viral life cycles (lytic and lysogenic) and replication rates are closely linked with host metabolism. As temperature is a major regulatory factor of microbial growth, an increase in temperature will likely affect the interaction between viruses and their hosts (Mojica and Brussaard, 2014). With increasing prokaryote growth rate, burst size can increase while the length of the lytic cycle decreases (Proctor et al., 1993; Hadas et al., 1997), boosting virus production. Higher temperatures should also increase the contact rates between viruses and hosts (Murray and Jackson, 1992) and hence potentially infection. Although evidence across systems is scarce, increases in burst size together with increases in production have been described for some natural systems (e.g. Parada et al., 2006). Danovaro et al. (2011) suggested that the observed relationship between viral abundance and temperature from different oceanic regions could be used to infer evidence for the potential effect of rising sea-surface temperatures. With these compiled data, the strongest effect was detected in temperate-open oceans: a temperature increase of only a few degrees was accompanied by a doubling in viral abundance. Nonetheless, factors influencing virioplankton distribution are clearly more complex than predicted by temperature alone. This is underlined by the observation that, when global data were grouped together, an overall decreasing pattern of viral abundance with increasing temperature was identified. This inverse relationship suggests that latitudinal changes, which influence radiation regimes and trophic characteristics, can display cascade effects on growth rates of hosts and viral infectivity (Danovaro et al., 2011). A large data set from the Atlantic Ocean revealed that temperature was important in the Sargasso Sea, but not in the northeast Atlantic (Rowe et al., 2008). In experiments over short time scales, changes in temperature did not directly influence viral abundance (Feng et al., 2009).

In the same compilation of literature data, Danovaro et al. (2011) showed that, in cold water systems, higher viral production rates were associated with warmer temperatures, whereas the relationship for systems at tropical and mid-latitudes was less clear. They concluded that different oceanic regions would respond differently to changes in surface temperatures caused by climate change. These temperature-associated changes in viral abundance and production are likely secondary effects caused by changes in host cell communities and in virus decay rates (Nagasaki and Yamaguchi, 1998; Wells and Deming, 2006). Demory et al. (2017) demonstrated that at suboptimum temperatures, lytic cycle kinetics of a *Micromonas* virus were lengthened and viral yield was reduced. Temperature increase shortened the latent periods, increased the burst size, and affected viral infectivity (Maat et al., 2017). Higher temperature will increase viral decay by enhanced activity of extracellular proteases and nucleases (Suttle and Chen, 1992; Noble and Fuhrman, 1997; Corinaldesi et al., 2010; Wei et al., 2018). In addition, alteration of temperature may affect protein stability and biomolecule elasticity of viral capsid proteins or lipid membranes, and influence the folding and binding of proteins and nucleic acids (Mojica and Brussaard, 2014). An experimental study in two contrasting Arctic systems (Lara et al., 2013) revealed that heterotrophic bacterial and viral abundance, bacterial production and grazing by protists increased at higher experimental temperatures. It remains to be elucidated whether predictable seasonal changes in hosts and viruses are applicable for climate-related long-term and large-scale changes (Danovaro et al., 2011).

The molecular mechanisms and the environmental factors behind the lysogenic decision are still not well understood (Long et al., 2008). The present climate-related changes in environmental conditions (temperature, circulation of water masses, nutrient availability, alterations in system productivity, coastal mixing with freshwater, etc.) may significantly influence viral life strategies. Lysogeny is thought to be favoured under low nutrient concentrations and high virus-to-host cell ratios (Wilson and Mann, 1997). Thus, if nutrient availability becomes limited due to increased vertical stratification caused by rising sea surface temperatures (Sarmiento et al., 2004), a substantial shift to the lysogenic life cycle may be the consequence. Conversely coastal and estuarine environments may shift to increase favouring of the lytic life-style with large-scale implications on heterotrophic and autotrophic organisms (Danovaro et al., 2011). However, Vaqué et al. (2019) found rather the inverse trend related to temperature. Knowles et al. (2016) suggested that lysogeny is also favoured at high bacterial production rates (Piggyback-the-Winner model). As temperature increases will influence growth, there could be also a change in viral life strategies. The direct and indirect effects of temperature on viruses and their interaction with hosts remain to be investigated in more detail.

Effect of changing salinity

Freshening of the oceanic environments due to climate change has been discussed particularly for poleward regions, and freshwater increases of the Arctic Ocean are suggested as being linked to respective changes of the North Atlantic as well (Peterson et al. 2002; Peltier et al., 2006). Changes in salinity can also influence viral lifestyles (Williamson and Paul, 2006), but experimental studies reveal no consistent pattern regarding the lytic-lysogenic-decision (Danovaro et al., 2011 and references therein), whereas Bettarel et al. (2011) found a switch from lytic to lysogenic life styles at high salinity. Recently, viral lysogeny and lysis shifts related to spring-neap tidal cycle were observed in a macrotidal subtropical estuary, which may be caused by freshwater-seawater mixing dynamics (Chen et al., 2019). Either induction or increased phage latent periods have been found under elevated sodium chloride concentrations. There is evidence that salinity can be of importance for the burst size of infected cells. In waters with high salinity, large burst sizes (up to 200) have been described (Guixa-Boixareu et al., 1996). Furthermore, in freshwater environments burst sizes are typically higher than in marine systems (Peduzzi and Luef, 2009). Mixing of these two environments may result in shifts of host growth rates, DOM availability and burst sizes. An interesting experimental study regarding freshwater and seawater mixing (e.g. as in estuaries) demonstrated that production rates of freshwater viruses sharply declined after seawater addition, followed by a rapid (within 48 h) recovery of the viral populations. Conversely, marine viruses were not significantly affected by mixing with freshwater (Cissoko et al., 2008). However, Marine et al. (2013) showed that freshwater prokaryote and virus communities can adapt to a controlled increase in salinity through changes in their structure and interactions. Another study has shown the possibility for freshwater viruses to cross into the marine environment and replicate normally (Sano et al. 2004). These studies and others (Bonilla-Findji et al., 2009; Wei et al. 2019) indicate that viruses can rapidly respond to major shifts in the abundance and community composition of bacterial hosts, which suffer from osmotic shock.

In a meta-analysis Danovaro et al. (2011) showed that systems with lower salinity display higher viral abundance. However, these results could reflect the more specific situation of the sampled areas rather than a general trend because this data set was largely from few estuarine systems. It may also be linked to the different nutrient availability and trophic state of the estuarine environments rather than to a direct effect of salinity on the viral assemblages. In addition, Parikka et al. (2017) showed in an across-systems study that the ratio of viruses to prokaryote abundance was inversely related to salinity.

Changes in salinity and pH can also influence the extent of virus adsorption to particles (Harvey and Ryan, 2004; Mojica and Brussaard, 2014), thus affecting virus distribution and proliferation. Further, enhanced river runoff to the sea will likely increase the particle concentration in coastal regions. In a riverine environment, for example, the quality, size and age of particles and aggregates, and the exposure time of viruses to aggregates, apparently are the key factors regulating viral abundance (Kernegger et al., 2009; Peduzzi, 2016). Importantly, particles (mainly with organic constituents) appear to play a role as viral scavengers or reservoirs rather than viral factories (Weinbauer et al. 2009). Moreover, adsorption of viruses to suspended particles can stimulate growth of the free-living prokaryotic community, e.g. by reducing viral lysis (Peduzzi and Luef, 2008; Weinbauer et al., 2009). In case salinity and particle load play a significant role in community composition, then the predicted sea-level rise, saltwater and freshwater mixing and temperature changes associated with the present global climate change could influence the success of the dominant viral and host taxa.

In those locations where climate change will increase freshwater input and nutrient concentrations, lytic life strategies could become more important through higher burst sizes, increasing infection rates due to the addition of freshwater viruses, and increased growth rates of prokaryotes and phytoplankton (Danovaro et al. 2011). An accelerated prokaryote-virus production cycle will on one hand release more components (e.g. enzymes) that are known as highly active for viral decay (Wommack and Colwell, 2000). On the other hand, increasing amounts of substrate for prokaryotes from the 'viral shunt' will result in larger burst sizes and potentially enhanced rates of adaptation and evolution. Important information could be drawn from additional experimental, proteomic and genomic data analyses in systems with gradients of salinity and from studies on the lifestyle of viruses and their host interactions. The speed and mechanisms of change, which are currently poorly understood, will be the main controlling factors in the shifts along the freshwater-marine continuum (Danovaro et al., 2011).

Ocean acidification

The uptake of anthropogenic CO_2 in ocean waters is reported to alter the pH (Riebesell et al., 2009). Nonetheless, predictions of the effect of ocean pH changes on the viral compartment are difficult. While the direct effects on marine viruses remain uncertain, it is likely that the most significant changes will be caused by the impact on their host organisms, for example on calcifying protists (Danovaro et al., 2011). Another example is the observed sharp drop in nitrification under increased CO_2 concentrations (Hutchins et al., 2009), suggesting that viruses from ammonia-oxidising bacteria and archaea might be affected. Furthermore, decreased pH can affect a broad range of physiological processes in microorganisms, e.g. those based on a proton gradient across membranes. Moreover, a significant fraction of near-surface prokaryotes was found to carry the pigment proteorhodopsin (which can act as a light-driven proton pump), being quite sensitive to even small decreases in pH (Fuhrman et al., 2008). In the so far largest CO_2 manipulation experiment (European Project on Ocean Acidification, EPOCA), complex impacts on different groups of plankton and on biogeochemical cycles were found (Riebesell et al., 2013, and special volume of Biogeosciences).

The survival, infectivity and adsorption of some bacteriophages were reported to be sensitive to pH (Harvey et al., 2014) and several studies have demonstrated an influence on the proliferation dynamics of phage-host systems (e.g., Larsen et al., 2008; Carreira et al., 2012; Traving et al., 2014; Chen et al., 2015). Highfield et al. (2017) showed that the diversity of EhV was much lower in the high-pCO_2 treatment enclosure that did not show inhibition of *E. huxleyi* growth. In the EPOCA experiment, a significant influence of ocean acidification on the relationship between viral and bacterial abundance was found (Brussaard et al., 2013). Ocean acidification did not affect lytic viral production in mesocosms experiments from the Mediterranean Sea and a Norwegian fjord, but lysogeny was stimulated and either linked to phytoplankton production (Tsiola et al., 2017) or bacterial production (Vaqué et al., 2017). In summary, profound effects related to pH-changes on microbial communities and their viruses have been detected, but the topic remains largely unexplored.

Potential impact of climate change on viruses and carbon cycling

Knowledge on the interactive role of viruses in global CO_2 fluxes is largely in its infancy. Four main CO_2 sequestration processes in the ocean are known, the physical or solubility pump, the carbonate, the biological pump and the microbial carbon pump (Legendre et al., 2015). Conversion of CO_2 into biomass and the subsequent sinking of organic and inorganic components of the plankton in the ocean interior are known as the biological carbon pump. The microbial carbon pump is the conversion of organic matter into forms with low bioavailability (Jiao et al., 2010; Legendre et al., 2015). The effect of rising atmospheric CO_2 on ocean acidification and carbonation is also receiving current attention. Factors affecting these processes are relevant to understand the functioning of the largest ecosystem on Earth, and to better comprehend the global carbon cycle and its implications on climate (Riebesell et al., 2009). It is still a largely open question how and to what extent the 'viral shunt' influences the efficiency of the biological pump, the microbial carbon pump and the sequestration of carbon in the ocean interior. After viral infection, a varying but substantial fraction of the released organic material can be more or less efficiently utilized by uninfected heterotrophic prokaryotes, thus entering again the food web. The net effect of the viral shunt is that it detours bacterial production away from being consumed by protists and ultimately converts organic matter into dissolved inorganic nutrients, including respired CO_2 (Wilhelm and Suttle, 1999; Suttle, 2007). Viral particles also make a contribution to the pools and fluxes of DOC, DON and DOP in the global oceans (Suttle, 2005; Jover et al., 2014; Zhang, et al., 2014).

In global carbon cycling models viruses are still practically ignored, making it difficult to come up with comprehensive conclusions. One scenario is that viral lysis increases the efficiency of the biological pump by enriching the proportion of carbon in the sinking particulate material (Suttle, 2007). At the same time, a viral-mediated control on the biological pump will potentially be impacted by climate change in that climate-induced warming of surface oceans could increase the mortality of the pelagic prokaryotes; this would cause an increased release of labile dissolved organic matter (DOM) that can enhance the metabolism and respiration of uninfected cells. Furthermore, a meta-analysis of data from the literature suggested that higher temperatures are associated with exponentially higher decay rates of virioplankton, based on higher levels of exoenzymatic activities (Danovaro et al., 2011). Danovaro et al. (2011) present a conceptual scheme of the impact on the biological carbon pump and of potential feedback mechanisms. In that concept, viral lysis, together with rising CO_2 concentration, cause changes in phytoplankton composition and production. The result would be a modification of both the photosynthetically produced particulate organic (POC) and inorganic (PIC, by calcifying photoautotrophs) carbon. It would also alter the particle export, in particular the relative ratio of PIC to POC in this material. Thus, the flux of CO_2 between the ocean surface and the atmosphere would be affected as a possible feedback effect on atmospheric CO_2 concentrations. The metabolism of heterotrophic prokaryotic cells could also be enhanced by rising sea-surface temperatures, thus increasing carbon consumption and respiration rates. Two potential scenarios regarding the viral-mediated control on the biological pump were suggested (Danovaro et al. 2011). On one hand, by altering the pathways of carbon fluxes in the sea - when converting living organic matter into dead particulate organic matter and DOM via cell lysis - the viral shunt could have a negative effect on the biological pump. DOM from cell lysis will be retained to a greater extent in surface waters, much of it converted to DIC through respiration and photodegradation. On the other hand, the viral shunt could favour carbon sequestration by increasing the nutrient availability for primary producers. This could increase the amount of living POC in surface waters, thus enhancing carbon export. In addition, other studies highlighted the role of viral lysis for the release of dense and refractory colloidal aggregates (Mari et al., 2005), or for the formation and size increase of organic aggregates (Peduzzi and Weinbauer, 1993), suggesting that viral mortality of host cells can favour carbon transport to the deep sea (viral shuttle; (Sullivan et al., 2017; Yamada et al., 2018). As viruses could also produce positively buoyant aggregates (Weinbauer, 2004), the role of aggregates produced by viruses can be described as viral elevator (Weinbauer et al., 2009). The influence of viral lysis on the stability and formation of aggregates may either increase the retention time of particles in the euphotic zone or increase carbon export to the ocean interior (Weinbauer et al., 2009). However, it is still unclear, whether viruses short-circuit or prime the biological pump (Brussaard et al., 2008). Changes in the amount of carbon release into the deep will no doubt influence also the functioning of deep-sea ecosystems (Smith et al., 2008). Recently, it has been shown that viruses play a more significant role for carbon fixation than previously thought; this influence is mediated by auxiliary metabolic genes (AMGs) (Puxty et al., 2016, 2018). This suggests a closer link of viral infection to the carbon cycle. In a recent review, it has been argued that metabolic reprogramming of infected cells and viral lysis alter nutrient cycling and carbon export in the ocean; however, the net impacts remain uncertain (Zimmerman et al., 2020).

Regarding inland waters the current perception is that the contribution of river networks to fluvial net heterotrophy and CO_2 outgassing has been underestimated in the past (Battin, 2008; Raymond et al., 2013; Acuña et al., this book). Almost nothing, however, is known about the significance of the viral shunt in flowing inland waters (Peduzzi and Luef, 2008; Peduzzi, 2016). Most fluvial systems harbour and transport high amounts of organic carbon (dissolved and particulate). Although viral lysis may release only a small fraction to this large organic carbon pool, it could still contribute significantly to the readily utilizable carbon (labile and semi-labile fraction) and its remineralisation in fluvial systems. This should be particularly relevant in these systems, which are characterised by a high proportion of allochthonous (terrestrial) aged and recalcitrant carbon (Peduzzi, 2016). Pollard and Ducklow (2011) reported for the Australian Bremer River that terrestrial DOC was partly returned to the atmosphere as CO_2 through bacterial respiration, assisted by bacteriophage lysis of their hosts. This short-circuits the microbial loop. Since streams and rivers are also subjected to climate change-related temperature shifts (Zweimüller et al., 2008), it is likely that the role of viruses in determining the proportion of organic material in horizontal transport and in processing and remineralisation in fluvial waters will be impacted as well.

Conclusions and outlook

The growing knowledge on virus ecology of marine and inland water environments has increased the incorporation of virus-related processes into aquatic food web models. Nonetheless, in climate change models the role of viruses is difficult to assess due to the rudimentary studies and data. Viruses can be anticipated as the ultimate nanoscale drivers/regulators of life in the effect they can have on organisms. Accordingly, aquatic viral ecology will contribute importantly, as the effects of climate and anthropogenic forcing in aquatic food webs are resolved step by step (Brussaard et al., 2008; Zimmerman et al., 2020). Changes in water temperature no doubt affect microbial growth, respiratory rates and carbon assimilation. This makes it extremely important to understand the effect of global warming on microbial communities and their central role in the carbon cycle. Temperature increases are also likely to impact the interactions between viruses and their host cells. Clearly, an improved understanding of viral responses to present climate change would enhance our chances to predict and adapt to potential consequences of such changes (compare Danovaro et al., 2011). Virus infection of microbial cells has the potential for cascading effects in food webs (Peduzzi et al., 2014). As ocean temperatures rise, precipitation patterns change and freshening of surface oceans occur, simultaneous and intermingled cascading effects may be the result. Global shifts in viral lifestyles in response to changing climate could have dramatic effects on global biogeochemical cycles via cascading effects.

One substantial drawback is the lack of long-term data sets that could be used to identify the relationships between climatic conditions and microbial processes. In a synthesis of the current knowledge and based on relevant case studies, and on meta-analyses of literature data, Danovaro et al. (2011) presented a catalogue of potential consequences and scenarios; some are outlined below together with additional important points:

1. There is some degree of consensus that the effect of temperature on virus related processes will be significant. Effects might be different at different latitudes and oceanic regions (at high latitudes rising temperatures may promote the viral compartment and depress it at the tropics). Increased vertical stratification due to climate change can lead to large-scale nutrient limitations, altering viral life strategies (lytic vs. lysogenic cycle).
2. Changes in salinity and the freshwater-marine continuum will probably impact the abundance, proliferation and life strategies of viruses. They may also influence the geographical success of dominant host taxa and subsequently the success of their viral counterparts. Freshening at the poles may increase the input and spread of freshwater groups of hosts and viruses into marine waters, facilitating enhanced crossing over of marine and freshwater taxa. Apparently, climate change-related impacts are particularly severe in polar regions and the microbiota there might be sentinels of global change.
3. It is currently unclear whether the viral shunt will result in negative or positive effects on the efficiency of the biological pump, and consequently, on the feedback of marine systems on climate. Viruses have the potential to interact with the climate through their contribution to the DOM-pool and to the biogenic particles.
4. The effect of ocean acidification on marine viruses will be largely via the effects of pH on the host organisms. Since some key metabolic processes of microorganisms are sensitive to changes in pH of the medium, this may

profoundly influence the overall functioning of microbial-mediated processes and virus-host interactions.

5. The role of inland waters, in particular streams and rivers, will be largely re-evaluated through the current perception that the contribution of river networks to fluvial net heterotrophy and CO_2 outgassing has been underestimated in the past. Increased virus activity will probably contribute significantly to the readily utilizable carbon (labile and semi-labile fraction) and its remineralisation. In fluvial waters, climate change-related temperature shifts will influence the role of viruses in determining the proportion of organic material in horizontal transport and in processing and remineralisation.

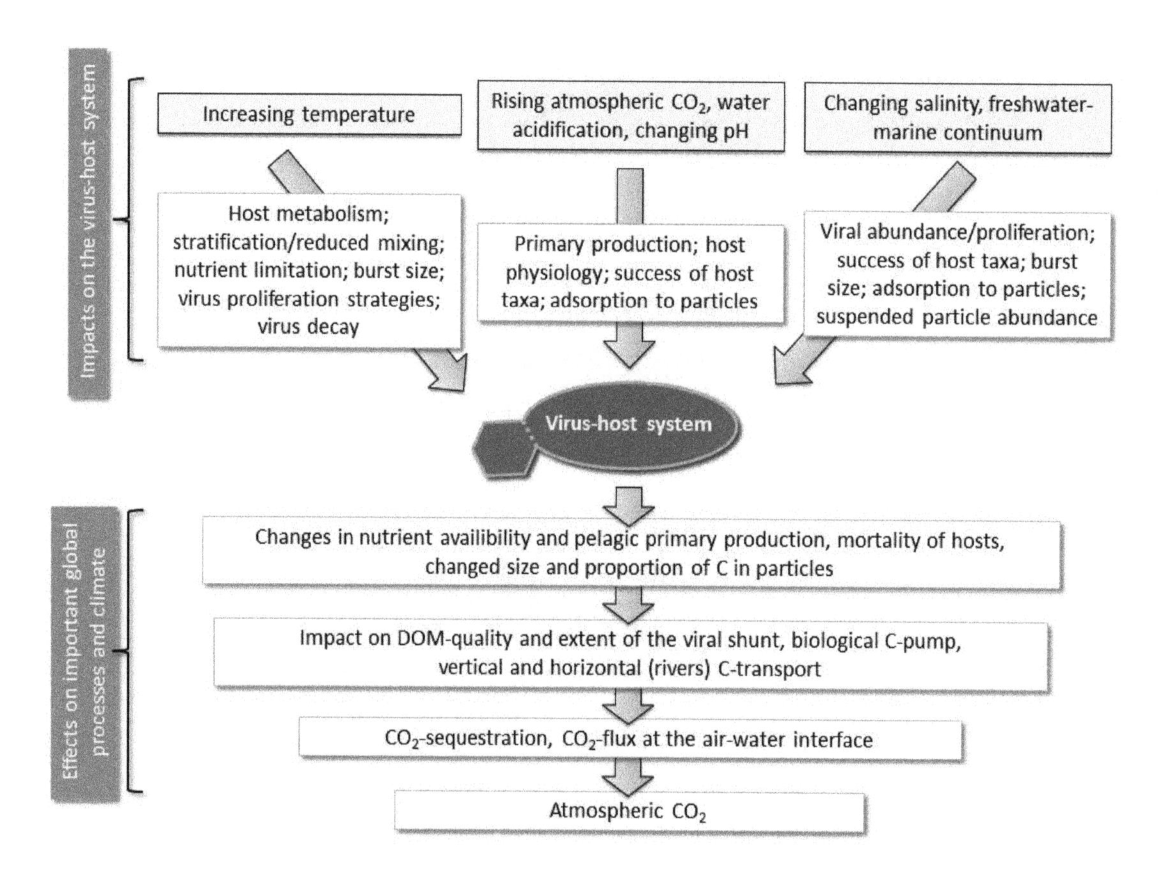

Figure 7.1 Schematic representation of some potentially important impacts of climate change on the virus-host system and potential effects and feedbacks of this biological compartment on some global processes and pools; for details see text.

6. Enhanced river runoff to the sea and mixing processes will likely increase the particle concentration at least in coastal regions; such particles are important factors regulating viral abundance. Aquatic aggregates apparently play a role as viral scavengers or reservoirs rather than viral factories, and adsorption of viruses to suspended particles may stimulate growth of the free-living prokaryotic community, e.g. by reducing viral lysis.

From all these considerations it can be concluded that aquatic viruses will be significantly influenced by climate change and that viruses, in turn, can influence processes contributing to climate change. In Fig. 7.1 some potentially important impacts, effects and relations of virus-host systems, facing a global climate change, are outlined. Under the present scenario of climate change, however, it remains unclear whether viruses will ultimately destabilise or even stabilise the dynamics of the living components in ecosystems and the related biogeochemical cycles. Due to the dearth of data it is currently unpredictable whether virus-related processes will amplify or smooth the impact of climate change on aquatic systems (Danovaro et al., 2011). These limitations inhibit incorporating virus-related processes into current climate models on the necessary spatial and temporal scales at which climate change scenarios respond and interact with this important compartment of the biosphere. This calls for focused research priorities and long-term data sets to enhance our knowledge about the role of aquatic viruses in the present climate change scenario.

References

Acuña, V., Freixa, A., Marcé, R., and Timoner, X. (this book). Ecosystem Metabolism in River Networks and Climate Change. In Climate Change and Microbial Ecology, Current Research and Future Trends, 2nd edition, J. Marxsen, ed. (Norfolk, UK: Caister Academic Press).

Battin, T.J., Kaplan, L.A., Findlay, S., Hopkinson, C.S., Marti, E., Packman, A.I., et al. (2008). Biophysical controls on organic carbon fluxes in fluvial networks. Nature Geosci. *1*, 95–100.

Battin, T.J., Luyssaert, S., Kaplan, L.A., Aufdenkampe, A.K., Richter, A., and Tranvik, L.J. (2009). The boundless carbon cycle. Nature Geosci. *2*, 598–600.

Bauer, J.E., Cai, W., Raymond, P.A., Bianchi, T.S., Hopkinson, C.S., and Regnier, P.A. (2013). The changing carbon cycle of the coastal ocean. Nature *504*, 61–70.

Bettarel, Y., Bouvier, T., Bouvier, C., Carre, C., Desnues, A., Domaizon, I., et al. (2011). Ecological traits of planktonic viruses and prokaryotes along a full-salinity gradient. FEMS Microbiol. Ecol. *76*, 360–372.

Bonilla-Findji, O., Rochelle-Newall, E., Weinbauer, M.G., Pizay, M.D., Kerros, M.E., and Gattuso, J.P. (2009). Effect of seawater–freshwater cross-transplantations on viral dynamics and bacterial diversity and production. Aquat. Microb. Ecol. *54*, 1–11.

Boyero, L., Pearson, R.G., Gessner, M.O., Barmuta, L.A., Ferreira, V., Graça, M.A.S., et al. (2011). A global experiment suggests climate warming will not accelerate litter decomposition in streams but might reduce carbon sequestration. Ecol. Lett. *14*, 289–294.

Brum, J.R., Ignacio-Espinoza, J.C., Roux, S., Doulcier, G., Acinas, S.G., Alberti, A., et al. (2015). Patterns and ecological drivers of ocean viral communities. Science. *348*, 1261498.

Brussaard, C.P.D, Wilhelm, S.W., Thingstad, F., Weinbauer, M.G., Bratbak, G., Heldal, M., et al. (2008). Global-scale processes with a nanoscale drive: the role of marine viruses. ISME J. *2*, 575–578.

Brussaard, C.P.D., Noordeloos, A.A.M., Witte, H., Collenteur, M.C.J., Schulz, K., Ludwig, A., et al. (2013). Arctic microbial community dynamics influenced by elevated CO_2 levels. Biogeosciences. *10*, 719–731.

Carreira, C., Heldal, M., and Bratbak, G. (2013). Effect of increased pCO_2 on phytoplankton–virus interactions. Biogeochemistry *114*, 391–397.

Chen, S., Gao, K., and Beardall, J. (2015). Viral attack exacerbates the susceptibility of a bloom-forming alga to ocean acidification. Global. Change Biol. *21*, 629–636.

Chen, X., Wei, W., Wang, J., Li, H., Sun, J., Ma, R., et al. (2019). Tide driven microbial dynamics through virus-host interactions in the estuarine ecosystem. Water Res. *160*, 118–129.

Cissoko, M., Desnues, A., Bouvy, M., Sime-Ngando, T., Verling, E., and Bettarel, Y. (2008). Effects of freshwater and seawater mixing on virio- and bacterioplankton in a tropical estuary. Freshw. Biol. 53, 1154–1162.

Corinaldesi, C., Dell'Anno, A., Magagnini, M., and Danovaro, R. (2010). Viral decay and viral production rates in continental-shelf and deep-sea sediments of the Mediterranean Sea. FEMS Microbiol. Ecol. *72*, 208–218.

Cozzi, S., Falconi, C., Comici, C., Čermelj, B., Kovac, N., Turk, V., et al. (2012). Recent evolution of river discharge in the Gulf of Trieste and their potential response to climate changes and anthropogenic pressure. Estuar. Coast. Shelf Sci. *115*, 14–24.

Danovaro, R., Corinaldesi, C., Dell'anno, A., Fuhrman, J.A., Middelburg, J.J., Noble, R.T., et al. (2011). Marine viruses and global climate change. FEMS Microbiol. Rev. *35*, 993–1034.

Danovaro, R., Dell'Anno, A., Corinaldesi, C., Magagnini, M., Noble, R., and Tamburini, C. (2008). Major viral impact on the functioning of benthic deep-sea ecosystems. Nature *454*, 1084–1087.

Demory, D., Arsenieff, L., Simon, N., Six, C., Rigaut-Jalabert, F., Marie, D., et al. (2017). Temperature is a key factor in *Micromonas*–virus interactions. ISME J. *11*, 601–612.

Dudgeon, D., Arthington, A.H., Gessner, M.O., Kawabata, Z., Knowler, D.J., Leveque, C., et al. (2006). Freshwater biodiversity: importance, threats, status and conservation challenges. Biol. Rev. *81*, 163–182.

Feng, Y., Hare, C.E., Leblanc, K., Rose, J.M., Zhang, Y., DiTullio, G.R., et al. (2009). Effects of increased pCO_2 and temperature on the North Atlantic spring bloom. I. The phytoplankton community and biogeochemical response. Mar. Ecol. Prog. Ser. *388*, 13–25.

Finke, J.F., Hunt, B.P.V., Winter, C., Carmack, E.C., and Suttle, C.A. (2017). Nutrients and other environmental factors influence virus abundances across oxic and hypoxic marine environments. Viruses *9*, 152.

Fuhrman, J.A., Schwalbach, M.S., and Stingl, U. (2008). Proteorhodopsins: an array of physiological roles? Nat. Rev. Microbiol. *6*, 488–494.

Gregory, A.C., Zayed, A.A., Conceicao-Neto, N., Temperton, B., Bolduc, B., Alberti, A., et al. (2019). Marine DNA viral macro- and microdiversity from pole to pole. Cell. *177*, 1109–1123.

Guidi, L., Chaffron, S., Bittner, L., Eveillard, D., Larhlimi, A., Roux, S., et al. (2016). Plankton networks driving carbon export in the oligotrophic ocean. Nature. *532*, 465–470.

Guixa-Boixareu, N., Calderón-Paz, J.I., Heldal, M., Bratbak, G., and Pedrós-Alió, C. (1996). Viral lysis and bacterivory as prokaryotic loss factors along a salinity gradient. Aquat. Microb. Ecol. *11*, 215–227.

Hadas, H., Einav, M., Fishov, l., and Zaritsky, A. (1997). Bacteriophage T4 development depends on the physiology of its host *Escherichia coli*. Microbiol. *143*, 179–185.

Harvey, R.W., and Ryan, J.N. (2004). Use of PRD1 bacteriophage in groundwater viral transport, inactivation, and attachment studies. FEMS Microbiol. Ecol. *49*, 3–16.

Highfield, A., Joint, I., Gilbert, J.A., Crawfurd, K.J., and Schroeder, D.C. (2017). Change in *Emiliania huxleyi* virus assemblage diversity but not in host genetic composition during an ocean acidification mesocosm experiment. Viruses *9*, 41.

Hoegh-Guldberg, O., and Bruno, J.F. (2010). The impact of climate change on the world's marine ecosystems. Science. *328*, 1523–1528.

Hutchins, D., and Fu, F. (2017). Microorganisms and ocean global change. Nat. Microbiol. *2*, 17058.

Hutchins, D.A., Mulholland, M.R., and Fu, F. (2009). Nutrient cycles and marine microbes in a CO_2-enriched ocean. Oceanography. *22*, 128–145.

IPCC (International Panel on Climate Change) (2013). Climate Change 2013: The Physical Science Basis (Cambridge: Cambridge University Press).

Jacquet, S., Miki, T., Noble, R., Peduzzi, P., and Wilhelm, S. (2010). Viruses in aquatic ecosystems: important advancements of the last 20 years and prospects for the future in the field of microbial oceanography and limnology. Adv. Oceanogr. Limnol. *1*, 97–141.

Jiao, N., Herndl, G.J., Hansell, D.A., Benner, R., Kattner, G., Wilhelm, S.W., et al. (2010). Microbial production of recalcitrant dissolved organic matter: long-term carbon storage in the global ocean. Nat. Rev. Microbiol. *8*, 593–599.

Jover, L.F., Effler, T.C., Buchan, A., Wilhelm, and S.W. Weitz, J.S. (2014). The elemental composition of virus particles: implications for marine biogeochemical cycles. Nat. Rev. Microbiol. *12*, 519–528.

Kernegger, L., Zweimuller, I., and Peduzzi, P. (2009). Effects of suspended matter quality and virus abundance on microbial parameters: experimental evidence from a large European river. Aquat. Microb. Ecol. *57*, 161–173.

Knowles, B., Silveira, C.B., Bailey, B.A., Barott, K., Cantu, V.A., Cobián-Güemes, A.G., et al. (2016). Lytic to temperate switching of viral communities. Nature *531*, 466–470.

Kritzberg, E.S., Granéli, W., Björk, J., Brönmark, C., Hallgren, P., Nicolle, A., et al. (2014). Warming and browning of lakes: consequences for pelagic carbon metabolism and sediment delivery. Freshw. Biol. *59*, 325–336.

Lara, E., Arrieta, J.M., Garcia-Zarandona, I., Boras, J.A., Duarte, C.M., Agustí, S., et al. (2013). Experimental evaluation of the warming effect on viral, bacterial and protistan communities in two contrasting Arctic systems. Aquat. Microb. Ecol. *70*, 17–32.

Lara, E., Vaque, D., Sa, E.L., Boras, J.A., Gomes, A., Borrull, E., et al. 2017. Unveiling the role and life strategies of viruses from the surface to the dark ocean. Sci. Adv. *3*, e1602565.

Larsen, J.B., Larsen, A., Bratbak, G., and Sandaa, R.A. (2008). Phylogenetic analysis of members of the *Phycodnaviridae* virus family, using amplified fragments of the major capsid protein gene. Appl. Environ. Microbiol. *74*, 3048–3057.

Legendre, L., Rivkin, R.B., Weinbauer, M., Guidi, L., and Uitz, J. (2015). The microbial carbon pump concept: Potential biogeochemical significance in the globally changing ocean. Prog. Oceanogr. *134*, 432–450.

Li, Y., Luo, T., Sun, J., Cai, L., Liang, Y., Jiao, N., et al. (2014). Lytic viral infection of bacterioplankton in deep waters of the western Pacific Ocean. Biogeosciences *11*, 2531–2542.

Liang, Y., Li, L., Luo, T., Zhang, Y., Zhang, R., and Jiao, N. (2014). Horizontal and vertical distribution of marine virioplankton: a basin scale investigation based on a global cruise. PLoS One *9*, e111634.

Long, A., McDaniel, L.D., Mobberley, J., and Paul, J.H. (2008). Comparison of lysogeny (prophage induction) in heterotrophic bacterial and *Synechococcus* populations in the Gulf of Mexico and Mississippi River plume. ISME J. *2*, 132–144.

Maat, D.S., Biggs, T., Evans, C., and van Bleijswijk, J.D.L. (2017). Characterization and temperature dependence of arctic *Micromonas polaris*. Viruses *9*, 134.

Magagnini, M., Corinaldesi, C., Monticelli, L.S., Domenico, E.D., and Danovaro, R. (2007). Viral abundance and distribution in mesopelagic and bathypelagic waters of the Mediterranean Sea. Deep Sea. Res. Pt I. *54*, 1209–1220.

Mari, X., Rassoulzadegan, F., Brussaard, C.P.D., and Wassmann, P. (2005). Dynamics of transparent exopolymeric particles (TEP) production by *Phaeocystis globosa* under N- or P-limitation: a controlling factor of the retention/export balance. Harmful Algae *4*, 895–914.

Marine, C., Thierry, B., Olivier, P., Emma, R.N., Corinne, B., Martin, A., et al. (2013). Freshwater prokaryote and virus communities can adapt to a controlled increase in salinity through changes in their structure and interactions. Estuar. Coast. Shelf S. *133*, 58–66.

Mojica, K.D., and Brussaard, C.P. (2014). Factors affecting virus dynamics and microbial host-virus interactions in marine environments. FEMS Microbiol. Ecol. *89*, 495–515.

Mooney, H., Larigauderie, A., Cesario, M., Elmquist, T., Hoegh-Guldberg, O., Lavorel, S., et al. (2009). Biodiversity, climate change, and ecosystem services. Curr. Opin. Environ. Sustain. *1*, 46–54.

Muck, S., Griessler, T., Köstner, N., Klimiuk, A., Winter, C., and Herndl, G.J. (2014). Fracture zones in the Mid Atlantic Ridge lead to alterations in prokaryotic and viral parameters in deep-water masses. Front. Microbiol. *5*, 264.

Murray, A.G., and Jackson, G.A. (1992). Viral dynamics: a model of the effects of size shape, motion and abundance of single-celled planktonic organisms and other particles. Mar. Ecol. Prog. Ser. *89*, 103–116.

Nagasaki, K., and Yamaguchi, M. (1998). Effect of temperature on the algicidal activity and stability of HaV (*Heterosigma akashiwo* Virus). Aquat. Microb. Ecol. *15*, 211–216.

Noble, R.T., and Fuhrman, J.A. (1997). Virus decay and its causes in coastal waters. Appl. Environ. Microbiol. *63*, 77–83.

Parada, V., Herndl, G.J., and G.Weinbauer, M. (2006). Viral burst size of heterotrophic prokaryotes in aquatic systems. J. Mar. Biol. Assoc. UK. *86*, 613–621.

Parikka, K.J., Le Romancer, M., Wauters, N., and Jacquet, S. (2017). Deciphering the virus-to-prokaryote ratio (VPR): insights into virus-host relationships in a variety of ecosystems. Biol. Rev. Camb. Philos. Soc. *92*, 1081–1100.

Peduzzi, P. (2016). Virus ecology of fluvial systems: a blank spot on the map?. Biol. Rev. Camb. Philos. Soc. *91*, 937–949.

Peduzzi, P., Gruber, M., Gruber, M., and Schagerl, M. (2014). The virus's tooth: cyanophages affect an African flamingo population in a bottom-up cascade. ISME J. *8*, 1346–1351.

Peduzzi, P., and Luef, B. (2009). Viruses. In G.E. Likens (ed.): Encyclopedia of Inland Waters (Oxford: Elsevier).

Peduzzi, P., and Luef, B. (2008). Viruses, bacteria and suspended particles in a backwater and main channel site of the Danube (Austria). Aquat. Sci. *70*, 186–194.

Peduzzi, P., and Weinbauer, M.G. (1993). Effect of concentrating the virus-rich 2-200 nm size fraction of seawater on the formation of algal flocs (marine snow). Limnol. Oceanogr. *38*, 1562–1565.

Peltier, W.R., Vettoretti, G., and Stastna, M. (2006). Atlantic meridional overturning and climate response to Arctic Ocean freshening. Geophy. Res. Lett. *33*, L06713.

Peterson, B.J., Holmes, R.M., McClelland, J.W., Vorosmarty, C.J., Lammers, R.B., Shiklomanov, A.I., et al. (2002). Increasing river discharge to the Arctic Ocean. Science. *298*, 2171–2173.

Pollard, P.C., and Ducklow, H. (2011). Ultrahigh bacterial production in a eutrophic subtropical Australian river: Does viral lysis short-circuit the microbial loop? Limnol. Oceanogr. *56*, 1115–1129.

Proctor, L.M., and Fuhrman, J.A. (1990). Viral mortality of marine bacteria and cyanobacteria. Nature. *343*, 60–62.

Proctor, L.M., Okubo, A., and Fuhrman, J.A. (1993). Calibrating estimates of phage-induced mortality in marine bacteria: ultrastructural studies of marine bacteriophage development from one-step growth experiments. Microb. Ecol. *25*, 161–182.

Puxty, R.J., Evans, D.J., Millard, A.D., and Scanlan, D.J. (2018). Energy limitation of cyanophage development: implications for marine carbon cycling. ISME J. *12*, 1273–1286.

Puxty, R.J., Millard, A.D., Evans, D.J., and Scanlan, D.J. (2016). Viruses inhibit CO_2 fixation in the most abundant phototrophs on earth. Curr. Biol. *26*, 1585–1589.

Raymond, P.A., Hartmann, J., Lauerwald, R., Sobek, S., McDonald, C., Hoover, M., et al. (2013). Global carbon dioxide emission from inland waters. Nature. *503*, 355–359.

Riebesell, U., Gattuso, J.P., Thingstad, T.F., and Middelburg, J.J. (2013). Arctic ocean acidification pelagic ecosystem and biogeochemical responses during a mesocosm study. Biogeosciences. *10*, 5619–5626.

Riebesell, U., Kortzinger, A., and Oschlies, A. (2009). Sensitivities of marine carbon fluxes to ocean change. Proc. Natl. Acad. Sci. USA. *106*, 20602–20609.

Rowe, J.M., Saxton, M.A., Cottrell, M.T., DeBruyn, J.M., Berg, G.M., Kirchman, D.L., et al. (2008). Constraints on viral production in the Sargasso Sea and North Atlantic. Aquat. Microb. Ecol. *52*, 233–244.

Sano, E., Carlson, S., Wegley, L., and Rohwer, F. (2004). Movement of viruses between biomes. Appl. Environ. Microbiol. *70*, 5842–5846.

Sarmiento, J.L., Slater, R., Barber, R., Bopp, L., Doney, S.C., Hirst, A.C., et al. (2004). Response of ocean ecosystems to climate warming. Global Biogeochem. Cycles *18*, GB3003.

Smith, C.R., De Leo, F.C., Bernardino, A.F., Sweetman, A.K., and Arbizu, P.M. (2008). Abyssal food limitation, ecosystem structure and climate change. Trends Ecol. Evol. *23*, 518–528.

Sullivan, M.B., Weitz, J.S., and Wilhelm, S. (2017). Viral ecology comes of age. Environ. Microbiol. Rep. *9*, 33–35.

Suttle, C.A. (2005). Viruses in the sea. Nature. *437*, 356–361.

Suttle, C.A. (2007). Marine viruses-major players in the global ecosystem. Nat. Rev. Microbiol. *5*, 801–812.

Suttle, C.A., and Chen, F. (1992). Mechanisms and rates of decay of marine viruses in seawatert. Appl. Environ. Microbiol. *58*, 3721–3729.

Thingstad, T.F. (2000). Elements of a theory for the mechanisms controlling abundance, diversity, and biogeochemical role of lytic bacterial viruses in aquatic systems. Limnol. Oceanogr. *45*, 1320–1328.

Traving, S.J., Clokie, M.R.J., and Middelboe, M. (2014). Increased acidification has a profound effect on the interactions between the cyanobacterium *Synechococcus* sp. WH7803 and its viruses. FEMS Microbiol. Ecol. *87*, 133–141.

Tsiola, A., Tsagaraki, T.M., Giannakourou, A., Nikolioudakis, N., Yücel, N., Herut, B., et al. (2017). Bacterial growth and mortality after deposition of Saharan dust and mixed aerosols in the Eastern Mediterranean Sea: a mesocosm experiment. Front. Mar. Sci. *3*, 281.

Vaqué, D., Lara, E., Arrieta, J.M., Holding, J., Sà, E.L., Hendriks, I.E., et al. (2019). Warming and CO_2 enhance Arctic heterotrophic microbial activity. Front. Microbiol. *10*, 494.

Vaqué, D., Boras, J.A., Torrent-Llagostera, F., Agusti, S., Arrieta, J.M., Lara, E., et al. (2017). Viruses and protists induced-mortality of prokaryotes around the Antarctic Peninsula during the austral summer. Front. Microbiol. *8*, 241.

Vincent, W.F. (2010). Microbial ecosystem responses to rapid climate change in the Arctic. ISME J. *4*, 1087–1090.

Wei, W., Wang, N., Cai, L., Zhang, C., Jiao, N., and Zhang, R. (2019). Impacts of freshwater and seawater mixing on the production and decay of virioplankton in a subtropical estuary. Microb. Ecol. *78*, 843–854.

Wei, W., Zhang, R., Peng, L., Liang, Y., and Jiao, N. (2018). Effects of temperature and photosynthetically active radiation on virioplankton decay in the western Pacific Ocean. Sci. Rep. *8*, 1525.

Weinbauer, M.G. (2004). Ecology of prokaryotic viruses. FEMS Microbiol. Rev. *28*, 127–181.

Weinbauer, M.G., Bettarel, Y., Cattaneo, R., Luef, B., Maier, C., Motegi, C., et al. (2009). Viral ecology of organic and inorganic particles in aquatic systems: avenues for further research. Aquat. Microb. Ecol. *57*, 321–341.

Wells, L.E., and Deming, J.W. (2006). Effects of temperature, salinity and clay particles on inactivation and decay of cold-active marine Bacteriophage 9A. Aquat. Microb. Ecol. *45*, 31–39.

Wilhelm, S.W., and Matteson, A.R. (2008). Freshwater and marine virioplankton: a brief overview of commonalities and differences. Freshw. Biol. *53*, 1076–1089.

Wilhelm, S.W., and Suttle, C.A. (1999). Viruses and nutrient cycles in the sea: Viruses play critical roles in the structure and function of aquatic food webs. BioScience. *49*, 781–788.

Williamson, S.J., and Paul, J.H. (2006). Environmental factors that influence the transition from lysogenic to lytic existence in the phiHSIC/Listonella pelagia marine phage-host system. Microb. Ecol. *52*, 217–225.

Wilson, W.H., and Mann, N.H. (1997). Lysogenic and lytic production in marine microbial communities. Aquat. Microb. Ecol. *13*, 95–100.

Wommack, K.E., and Colwell, R.R. (2000). Virioplankton: Viruses in aquatic ecosystems. Microbiol. Molec. Biol. Rev. *64*, 69–114.

Yamada, Y., Tomaru, Y., Fukuda, H., and Nagata, T. (2018). Aggregate formation during the viral lysis of a marine diatom. Front. Mar. Sci. *5*, 167.

Zhang, R., Weinbauer, M.G., and Qian, P.Y. (2007). Viruses and flagellates sustain apparent richness and reduce biomass accumulation of bacterioplankton in coastal marine waters. Environ. Microbiol. *9*, 3008–3018.

Zhang, R., Wei, W., and Cai, L. (2014). The fate and biogeochemical cycling of viral elements. Nat. Rev. Microbiol. *12*, 850–851.

Zhou, J., Xue, K., Xie, J., Deng, Y., Wu, L., Cheng, X., et al. (2011). Microbial mediation of carbon-cycle feedbacks to climate warming. Nature Clim. Change. *2*, 106–110.

Zimmerman, A.E., Howard-Varona, C., Needham, D.M., John, S.G., Worden, A.Z., Sullivan, M.B., et al. (2020). Metabolic and biogeochemical consequences of viral infection in aquatic ecosystems. Nat. Rev. Microbiol. *18*, 21–34.

Zweimüller, I., Zessner, M., and Hein, T. (2008). Effects of climate change on nitrate loads in a large river: the Austrian Danube as example. Hydrol. Pro. *22*, 1022–1036.

Chapter 8

Microbes in Aquatic Biofilms under the Effect of Changing Climate

Anna M. Romaní[1*], Stéphanie Boulêtreau[2], Verónica Díaz Villanueva[3], Frédéric Garabetian[4], Jürgen Marxsen[5,6], Helge Norf[7,8], Elisabeth Pohlon[5] and Markus Weitere[7]

[1] Institute of Aquatic Ecology, University of Girona, Campus de Montilivi, 17071 Girona, Spain; [2] Université de Toulouse, UPS, INP, EcoLab (Laboratoire Ecologie Fonctionnelle et Environnement), 118 route de Narbonne, F-31062 Toulouse, France; [3] Laboratorio de Limnología, INIBIOMA, Universidad Nacional del Comahue-CONICET, Quintral 1250, 8400, Argentina; [4] Université de Bordeaux, UMR 5805 EPOC, Station Marine d'Arcachon, 2 rue Jolyet, F-33120 Arcachon, France; [5] Institut für Allgemeine und Spezielle Zoologie, Tierökologie, Justus-Liebig-Universität Gießen, Heinrich-Buff-Ring 26, 35392 Gießen, Germany; [6] Limnologische Flussstation des Max-Planck-Instituts für Limnologie, Schlitz, Germany; [7] Department of River Ecology, Helmholtz-Centre for Environmental Research – UFZ, Brückstr. 3a, 39114 Magdeburg, Germany; [8] Department of Aquatic Ecosystem Analyses, Helmholtz-Centre for Environmental Research – UFZ, Brückstr. 3a, 39114 Magdeburg, Germany; * corresponding author

Email: anna.romani@udg.edu, stephanie.bouletreau@univ-tlse3.fr, diazv@comahue-conicet.gob.ar, frederic.garabetian@u-bordeaux.fr, jmar@zo.jlug.de, helge.norf@ufz.de, elisabeth.pohlon@allzool.bio.uni-giessen.de, markus.weitere@ufz.de

DOI: https://doi.org/10.21775/9781913652579.08

Abstract

The effects of climate change on aquatic biofilm structure and function is difficult to predict mainly due to biofilms being complex and dynamic assemblages of microorganisms. We review observed patterns of the effects of warming and desiccation on biofilms. Commonly observed effects of warming on biofilms include changes in the autotrophic community composition and extracellular polymeric substances, stimulation of the heterotrophic community, and changes in the microbes, protozoa and small metazoans densities and composition. The magnitude of the temperature effects depends on the biofilm successional stage, resources availability, community composition and interactions within communities including top-down effects. Temperature also affects biofilm functioning by direct control of biological activities and by selecting adapted taxa, which provide feedback on activities. Biofilm photosynthesis, respiration, denitrification and extracellular enzyme activities show differential sensitivity to temperature. Results suggest a significant effect of temperature on the nitrogen cycling and a link between the specific community composition and the biofilm temperature sensitivity. On the other hand, desiccation may produce more permanent changes on the biofilm microbial community composition than on extracellular enzyme activities, the effects also depending on species specific sensitivity and biofilm structure (such as the content of extracellular polymeric substances). At the ecosystem level, both factors (warming and desiccation) may coincide in time, but few studies have looked at the drought-temperature interactions on aquatic biofilms. Future trends might include multi-stress and short- and long-term experimental approaches. Measurements of carbon and nitrogen budgets are needed to quantify the effects of biofilm metabolism on ecosystem nutrient cycling and, at the same time, to improve biofilm models.

Introduction

Aquatic biofilms are assemblages of microorganisms (archaea, bacteria, cyanobacteria, algae, fungi, protozoa and small metazoans) attached to a surface and usually embedded in an extracellular matrix of polymeric substances. In aquatic ecosystems, biofilms are present in various compartments (e.g. benthic, hyporheic, aquifer sediments, plant material; Lock, 1993) and play a key role in the uptake or retention of inorganic and organic nutrients (Fischer *et al.*, 2002; Romaní *et al.*, 2004; Teissier *et al.*, 2007). In freshwater ecosystems, biofilms are metabolic hotspots and major sites for organic matter degradation.

Aquatic biofilms have been defined as interfaces that integrate a variety of responses to environmental changes and chemical stressors. Their short generation time results in short-term functional and structural responses, making them useful as "early warning systems" of disturbances (Sabater *et al.*, 2007). Among those disturbances, those linked to climate change might seriously affect biofilm communities either directly or indirectly. The Intergovernmental Panel on Climate

Change predicts an increase of global warming and changing patterns of rainfall frequency and distribution. In the case of global warming, it is predicted that mean surface temperatures will increase about 2.6-4.8°C for climate change scenarios by 2100 (IPCC, 2013), and river water temperature will also increase as a consequence (Mohseni and Stefan, 1999). Temperature is one of the major factors affecting different key physiological mechanisms such as respiration, growth and feeding. Thus not only are individuals affected, but also the interaction between organisms. On the other hand, in rivers and streams, the predicted changes in the discharge pattern in temperate regions will cause increased flooding and drought frequency and intensity, and a substantial change in quantity and quality of available organic matter. Drought episodes and the drastic reduction in water availability might determine changes in the biofilm community composition and functioning (Zoppini *et al.*, 2010). Both factors, warming conditions and extreme hydrological events, might interact in a perspective of climate change, where rising temperature will be combined with changes in discharge pattern.

The present chapter aims to review the knowledge on the effect of climate change on freshwater biofilms, specifically including two effects: effects of warming and effect of drought periods. The chapter is mainly focused on river and stream biofilms but many examples are also included from other aquatic habitats. Both effects are being analyzed under the view of changes in biofilm structure and functioning. Finally, the future trends and perspective in climate change effects on biofilms are discussed, including the need for multi-stress short and long-term studies and the development of improved biofilm models.

Aquatic biofilm responses to increasing water temperature

Increases in water temperature may affect both biofilm structure (e.g. biomass of the different groups, community composition, thickness, density, content of extracellular polymeric substances) and biofilm function (e.g. primary production, respiration, extracellular enzyme activities, denitrification) (Fig. 8.1). The expected changes in biofilm structure and function are discussed in this section highlighting the relevance of interactions that challenge any unidirectional prediction for such complex microbial communities.

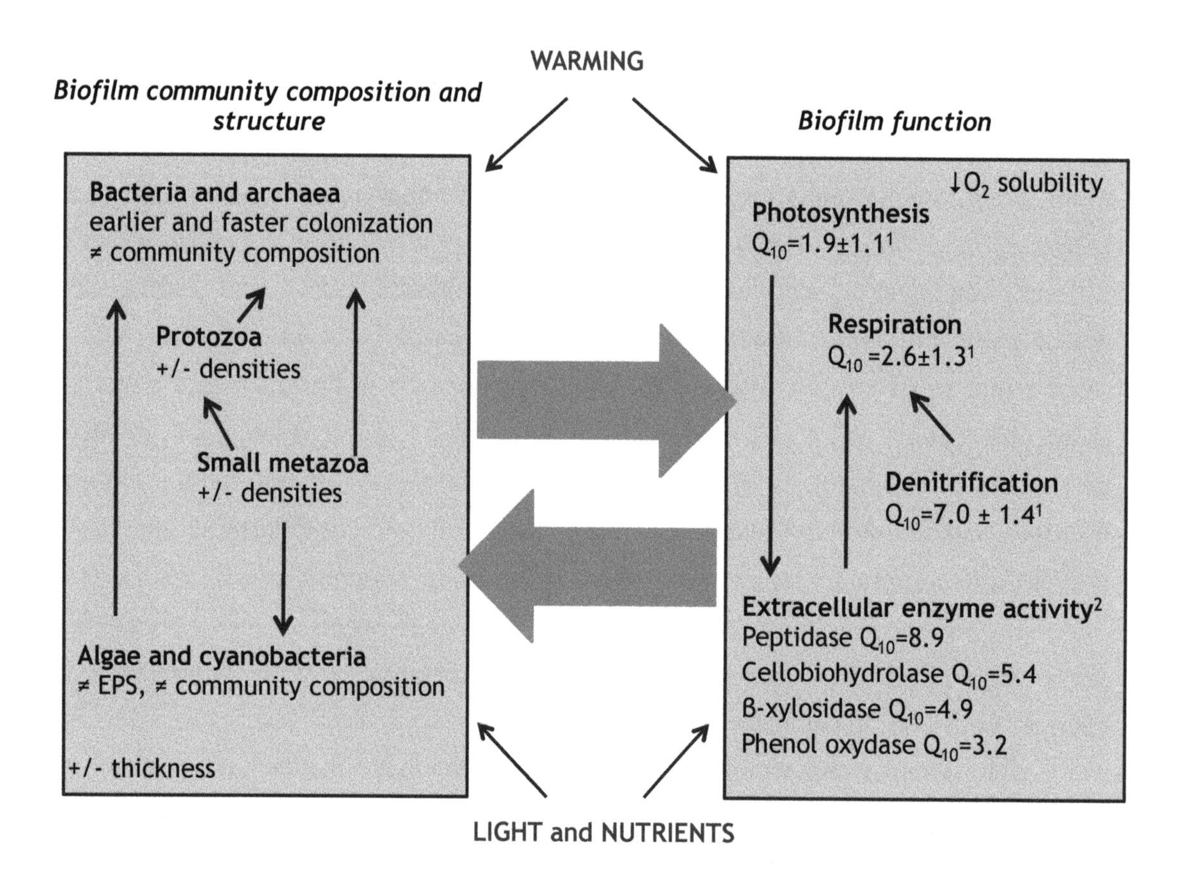

Figure 8.1 Summary of mechanisms affecting biofilm structure and functioning under warming conditions. Arrows in the biofilm structure and community composition effects indicate trophic links (grazing, organic matter resources). Superscript numbers indicate the source of the Q_{10} values obtained from freshwater biofilms: (1) Bloulêtreau *et al.* 2012, (2) calculated from the results in Ylla *et al.* 2014, for those enzymes that showed a higher sensitivity to temperature than that predicted from activation energy reported values. The "+/-"represents that the parameter (densities/thickness) may either increase or decrease depending on the biofilm successional stage and/or short/long response to temperature (see text for details). EPS = Extracellular polymeric substances.

Effect of warming on biofilm structure: the significant role of interactions

One would expect that biofilm formation and its increase in biomass through colonization will be enhanced by increasing temperatures. Research on biofilm formation on food has shown a significant effect of temperature on bacterial attachment and higher biofilm bacterial density is attained at higher temperature. In this example, the temperature effect is linked to hydrophobicity changes at the substrate surface and/or changing the regulation of specific bacterial genes (Moltz and Martin, 2005; Di Bonaventura *et al.*, 2008). In general, since temperature accelerates organismal growth within their thermal tolerance window, increased biomass would be expected. In the stream, comparing biofilm colonization sequences, bacteria colonized the substratum at higher temperature faster and earlier, whereas no differences were found in total densities at the mature biofilm (Romaní, 2010). In contrast, in that study, algal biomass was more influenced by light and nutrients than by temperature. However, a field experiment showed that density of prokaryotes and chlorophyll-*a* was higher in the biofilm developed at +3°C than in the control site (Ylla *et al.*, 2014). These field results indicate that biofilm responses to warming may not be easy to predict.

Microbial organisms in biofilms are tightly packed. Supported by the high organismal densities, structures and processes within biofilms are to a large extent controlled by interactions between species, particularly symbiotic, competitive and trophic interactions such as grazing and utilization of exudates (Romaní and Sabater, 1999; Arndt *et al.*, 2003; Hansen *et al.*, 2007). Ecological interactions in general depend on specific ecological rates such as feeding and growth rates, most of which are either directly or indirectly coupled to temperature, and thus are potentially sensitive to environmental warming (e.g. Montagnes and Weisse, 2000; Brown *et al.*, 2004; Rall *et al.*, 2012). Warming alters the strength of trophic interactions as feeding and consumption rates increase with temperature within a species-specific optimal range due to increased metabolic demand (reviewed in Clarke, 2006). Such a general pattern also applies to biofilms, where both resource uptake as well as grazing rate increase with warming (e.g. Kathol *et al.*, 2009; Díaz Villanueva *et al.*, 2011).

Indirect effects of warming on interactions within biofilms can be caused by warming effects on community structures, which are coupled to changes in interactions and ecosystem functioning (Petchey *et al.*, 1999; Fox and Morin, 2001). Whereas physiological processes and thermal tolerances are the basis for warming-related shifts in communities, the outcome of warming effects on the composition of attached microbial communities is complex. The range of coexistence changes with temperature even in simple microbial food webs (Nomdedeu *et al.*, 2012). Complex microbial communities including biofilms distinctly shift with relatively small temperature changes (e.g. Jiang and Morin, 2004; Norf and Weitere, 2010), demonstrating that the outcome of such warming-related shifts in biofilm communities is very difficult, if not impossible, to predict (cf. Abbot *et al.*, 2014).

Nevertheless, some common mechanisms and patterns of warming-related shifts could be identified:

(i) Warming may affect the *autotrophic community composition* and polymeric substances content of aquatic biofilms. Marine and freshwater biofilms change toward a cyanobacterial community (Russell *et al.*, 2013; Romaní *et al.*, 2014). Changes in temperature can determine the composition of capsular polysaccharides and increases in production of extracellular polymeric substances (Di Pippo *et al.*, 2012). These changes in autotrophic community composition and extracellular polymeric substances may determine the increase in biofilm *three-dimensional structure*. In Díaz Villanueva (2011) laboratory experiment, though chlorophyll concentration did not vary when increasing +3°C, biofilms were thicker in high temperature treatment based on confocal laser scanning microscopy observations. Similar results were observed for biofilms affected by nuclear power plant effluents that increased the water temperature by about 5°C (Rao, 2010).

(ii) Environmental warming would *stimulate the heterotrophic community* of aquatic biofilms. There is a stronger stimulation of heterotrophic processes in relation to autotrophic processes, with the latter being often light- instead of temperature-controlled (e.g. Hancke and Glud, 2005; Boulêtreau *et al.*, 2012; see subsection below). Even if studies on warming-induced increases in heterotrophy of biofilms are rare, mechanisms are probably similar to those observed for plankton systems (e.g. Dahlgren *et al.*, 2011). Switches to heterotrophy due to warming have been observed as higher densities of ciliate and rotifer in the biofilm when temperature was increased by 2-3°C in a laboratory experiment in contrast to minor changes in autotrophic biomass (Díaz Villanueva, 2011; Romaní *et al.*, 2014).

Increase in heterotrophy may be also an indirect response to increased biofilm thickness under warming conditions. Increase of biofilm thickness causes primary producers buried and shaded in the lower layers of the biofilm to be less productive than those that are fully exposed in the upper layers, resulting in more heterotrophic communities and with more organic matter accumulation.

(iii) The *successional stage* of biofilms determines the magnitude of warming effects (Norf *et al.*, 2007; Norf and Weitere, 2010; Díaz Villanueva *et al.*, 2011; Boulêtreau *et al.*, 2014). Whereas early stages usually show enhanced production and activity with warming, late succession stages show neutral or negative responses. This is related to differential controls of communities at different stages (Jackson *et al.*, 2001; Jackson, 2003; Lyautey *et al.*, 2005). Densities within early successional stages increase with warming due to enhanced colonization and growth. Within late successional stages, individual resource availability decreases as competition increases. Under such resource-limiting conditions, warming can reduce carrying capacity due to increasing metabolic costs (Carpenter *et al.*, 1992; Rosa *et al.*, 2013).

Sensitivity of shifts in prokaryote community composition due to warming also changes through successional stage. At the very young biofilm (7 days) there were no significant effects of temperature (+2ºC) on prokaryote community composition but it had a significant effect on the mature biofilm (28-day-old), especially on those biofilms developed under dark conditions (Romaní *et al.*, 2014). Similarly, in the experiment by Díaz Villanueva *et al.* (2011), differences in the bacterial community composition due to warming (+3ºC) were detected at the 30-day-old biofilm, when the biofilms were just reaching their maturity. However, in the later experiment, differences in bacterial community composition at the old biofilm (58 days) were mainly caused by differences in nutrient availability. In contrast, changes in bacterial community composition were also reported in 6-day-old biofilms underlining the relevance of ecosystem-level replication in such experiments (Boulêtreau *et al.*, 2014).

(iv) The strength of the warming effect on communities and on interactions within communities strongly depends on *resources availability* (e.g. Norf and Weitere, 2010; Díaz Villanueva *et al.*, 2011; Degerman *et al.*, 2013). Limited resource availability in the ecosystems seems to buffer warming effects due to resource instead of temperature limitation. This can also vary seasonally within single sites. Whereas ciliates within river biofilms showed strong response towards warming in winter (temperature limitation), they showed no responses during summer under assumed resource limitation (Norf and Weitere, 2010; Marcus *et al.*, 2014). At the community level, the response of algal growth to temperature may be modulated by nutrient concentration. In natural biofilms, nutrient enrichment increases algal biomass and net primary production more in summer than in winter (Guasch *et al.*, 1995). However, Díaz Villanueva *et al.* (2011) found that at eutrophic conditions, warming promoted faster biofilm colonization while biofilm carrying capacity was not affected and only depended on nutrient concentration. This suggests that if nutrients are a limiting factor, increases in water temperature might not lead to an increment in algal growth rate (Raven and Geider, 1988). Similarly, in a specific experiment dealing with biofilm-associated ammonia oxidizing bacteria, a differential effect of temperature on community composition was shown depending on ammonium concentrations (Avrahami *et al.*, 2011).

(v) Apart from resources (bottom-up), interactions within complex communities including *top-down* effects can stabilize community composition if subjected to warming within physiological tolerable temperature ranges. For example, parallel increases of growth and grazing rates with warming at moderate temperature ranges were shown to stabilize community densities of certain trophic guilds (Viergutz *et al.*, 2007). Grazing by ciliates was probably the reason why Díaz Villanueva *et al.* (2011) did not detect translation of warming-induced increases of bacterial growth into biomass within complex biofilms. A control of lower trophic levels by rotifers was shown, where rotifers were controlling the development of ciliates and prokaryotes (Romaní *et al.*, 2014). Only the density of the highest trophic level, i.e. the rotifers themselves, was distinctly enhanced by temperature. A significant effect of top-down biofilm control was also highlighted for marine biofilms and linked to changes in cyanobacteria community composition (Russell *et al.* 2013).

Top-down effects can change depending on the specific temperature considered, i.e. being within or outside the thermal tolerance window for each organism. At high ambient temperatures, even small temperature changes can result in strong effects on the balance of interactions strengths and, correspondingly, community density and compositions. For example, Viergutz *et al.* (2007) demonstrated that planktonic heterotrophic protists, which are grazing controlled by filter-feeding macrofauna, benefit strongly from extreme temperatures due to a decoupling of grazing (decreases at high temperatures) and growth rate (increases at high temperatures), whereas both rates match at moderate temperatures. Similar mechanisms are assumed for macrograzer- controlled biofilm communities. Furthermore, several species exceed their thermal optima during summer warming, leading to decreases in their abundances and a gain of species traits best adapted to warm conditions. This results in pronounced shifts in the composition of both taxonomic and functional groups within biofilms at extreme temperatures. Examples are the increasing importance of cyanobacteria among the autotrophic community with summer warming (Paerl and Huisman, 2009). As cyanobacteria differ in their characteristics from eukaryotic algae (e.g. with respect to food quality; von Elert *et al.*, 2003), such warming-induced shifts have the potential to alter trophic interactions. Summer warming can reduce the relative importance of suspension feeding peritrich ciliates within biofilm grazer communities (Norf and Weitere, 2010), altering the proportion of surface versus plankton feeders and consequently resource acquisition within biofilms. Studies on hyporheic biofilms with experimental warming reported restructuring of the entire biofilm food web as a consequence of altered primary production (Andrushchyshyn *et al.*, 2009).

The combination of these described mechanisms of warming-related shifts of biofilm structure and community composition will determine the final response. The structural changes are, at the same time, sensitive to functional changes (see next subsection and summary of mechanisms in Fig. 8.1). The strength of warming effects on biofilm structure and community composition is highly context-specific, and might range from no detectable effects to large effects within small temperature ranges.

Effect of warming on biofilm metabolism and function

The effect of warming on aquatic biofilm metabolism and function can be documented at different levels of biological organization. Temperature controls enzymatic and physiological activities, as well as population dynamics and community composition (as discussed above). The temperature dependence of biological processes is described in Fig. 8.2. Temperature therefore affects ecosystem functioning by direct control of biological activities and by selecting adapted taxa which provide feedback on activities. In addition, temperature affects solubility of dissolved elements in water such as oxygen, resulting in a cascade effect on any biological processes. Direct and indirect controls by temperature play a role in multispecies and multifunction aquatic biofilms.

Primary production has long been recognized as one of the ecosystem function of aquatic biofilms developing under light conditions (e.g. Wetzel, 1963; Lock, 1993). If neither light nor nutrients are limiting, the effect of temperature on primary production is expected to be positive, since photosynthetic enzymatic activity is enhanced by temperature (Davison, 1991; Necchi, 2006). However, beyond a certain temperature (optimum, which is species specific), algal biomass decreases. In Boulêtreau *et al.* (2012), experimental estimates of respiration and gross primary production sensitivity of stream biofilms to temperature were E_A=0.67 and 0.47 eV, respectively (E_A = activation energy). The recorded values were close to values

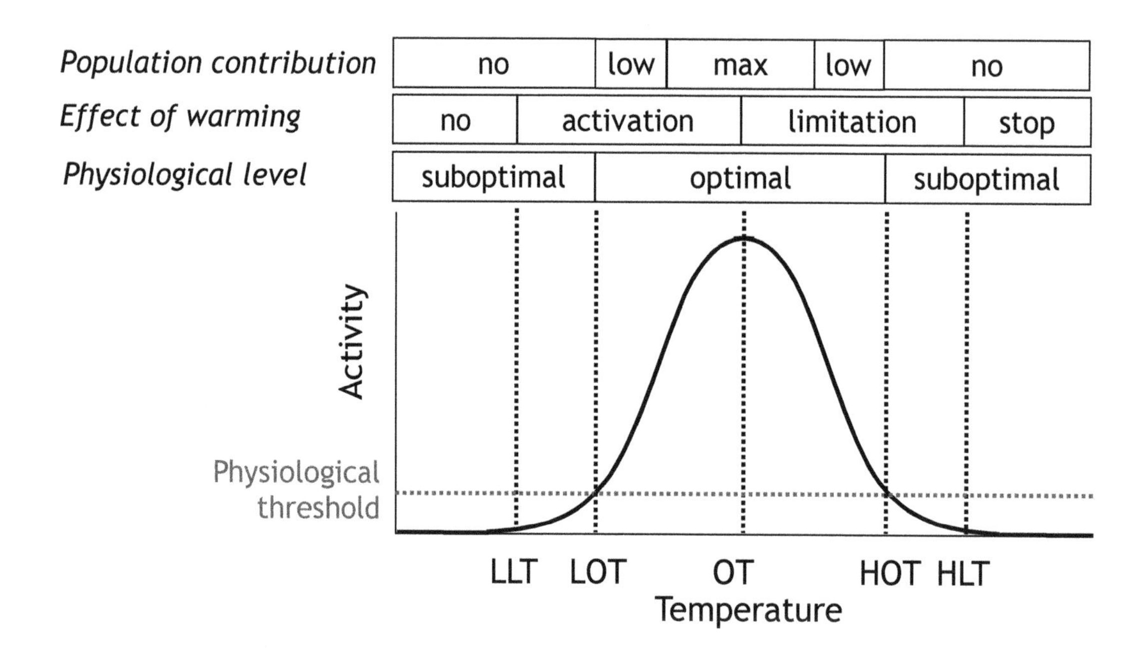

Figure 8.2 The temperature dependence of biological processes is not monotonous although it can be described by a single pattern. The theoretical curve presenting the intensity of any biological activity as a function of temperature defines optimal and suboptimal ranges of temperature limited by the low and high optimum temperature thresholds (LOT and HOT, respectively). Within this optimal range of temperature, a temperature increase may cause either activation of the considered activity (when temperature is lower than the optimal temperature (OT)) or its limitation (when temperature is higher than OT). In the activation range by temperature, and whilst the model was in debate (see Gillooly *et al.*, 2006; Clarke, 2006), the Boltzmann-Arrhenius equation predicts the exponential effect of temperature on any metabolic rate. A fundamental factor of this equation is the activation energy term E_A; it is given by the slope of the plot of the logarithm of mass-corrected metabolism rate R_T to 1/k T, where T is absolute temperature (K) and k is Boltzmann's constant (eV K^{-1}). The traditional Q_{10} factor which is limited to a 10°C-range of variation is defined by the equation

$$\frac{R_T}{R_{T_{ref}}} = \left[Q_{10}\right]^{\frac{T-T_{ref}}{10}}$$

where T_{ref} is the reference temperature and $R_{T_{ref}}$ is the metabolic rate at the reference temperature. Both E_A and Q_{10} are used to express the temperature sensitivity of different enzymes, populations or communities of interest. Outside the optimum range of temperature, namely, no activity is expected in the lower and/or upper ranges of lethal temperature (LLT and HLT). The contribution of one population of microorganism to the considered metabolism will follow the same pattern with a more or less wide range of optimal activity according to its eury- or stenotherm status, respectively.

reported for microphytobenthos and for stream biofilms (Hancke and Glud, 2004; Acuña *et al.*, 2008) and close to values of the temperature dependencies of their relative enzymatic complexes: ATP synthesis in respiratory complexes ($E_R \approx 0.65$ eV, e.g. Gillooly *et al.*, 2001) and the Rubisco carboxylation in the chloroplasts ($E_P \approx 0.32$ eV, e.g. Allen *et al.*, 2005). Consistently, it was reported that biofilm communities originating from 4 Icelandic streams across a thermal gradient exhibited similar E_A and Q_{10} values (Perkins *et al.*, 2012); the authors concluded that, whatever the ecosystem, short-term temperature dependence of respiration could be modelled using a single value.

The higher sensitivity of respiration than photosynthesis to temperature measured in river biofilms (Hancke and Glud, 2004; Uehlinger, 2006; Veraart *et al.*, 2011; Boulêtreau *et al.*, 2012) is consistent with previous measurements in other contexts and models (e.g. Pomeroy and Deibel, 1986; Allen *et al.*, 2005; Yvon-Durocher *et al.*, 2010). Necchi (2006) demonstrated that though temperature increases photosynthesis reactions, it also increases respiration and photorespiration, which conduces to a negative net photosynthesis. In the case of antagonistic or synergistic functions, this, in turn, can affect the functional budget of the biofilm. As an example, this difference in temperature sensitivity between heterotrophic self-detachment and primary production processes was explained by modelling to drive river biofilm biomass dynamics (Boulêtreau *et al.*, 2006). Given that bacterial colonization of the substratum was found to be more affected by warming than algae colonization (Romaní, 2010), major shift from autotrophy to heterotrophy and associated functional implications are to be expected from moderate warming.

Denitrification is a function of growing interest in river biofilms since it is considered as the main soluble nitrogen removal process (Peterson *et al.*, 2011; Kalscheur *et al.*, 2012; Lyautey *et al.*, 2013). Estimated values of temperature dependence of denitrification enzyme activity ($E_A = 1.42 \pm 0.24$ eV and $Q_{10} = 7.0 \pm 1.4$) were in the upper range of values reported in soils and sediments, and suggested that potential denitrification in river biofilms was more strongly temperature-dependent than photosynthesis- and/or respiration-dependent. This sensitivity varied noticeably (Boulêtreau *et al.*, 2012), unlike respiration, in relation with the number and variety of the enzymes presently known to catalyze the four reaction steps necessary to reduce NO_3^- to N_2 (Philippot and Hallin, 2005). Interestingly, quantifying the number of copies of genes encoding for proteic subunits of these enzymes permits to address the denitrifier community structure of a microbial assemblage. Moderate warming (ca. +2.5°C) proved to induce changes in the denitrifier community structure of river biofilms. These changes were linked to changes in the total bacterial community structure and likely modulated biofilm denitrifying activity (Boulêtreau *et al.*, 2014). The strong temperature dependence of denitrification observed in pond sediments was explained by a disproportionate boost of denitrification due to altered oxygen dynamics linked to change in the balance between photosynthesis and respiration and to lowered oxygen solubility (Veraart *et al.*, 2011). In the biofilm, thanks to limited diffusion processes, marked oxygen gradients and physico-chemical heterogeneity are expected to occur (Wimpenny *et al.*, 2000; Steward, 2003) favoring the settlement of a denitrifier community (Lyautey *et al.*, 2013). Such a warming induced denitrification boost is thus very likely to occur in the assembled biofilm community. The importance of the coupling between N mineralization, N uptake and denitrification processes for biofilm function demonstrated by Teissier *et al.* (2007) suggests that both warming induced changes in biochemical balance are very likely to occur in the biofilm community. Driving algae, total bacterial and denitrifier community structure, kinetics of denitrifying enzymes, balance between sources and sinks of nitrogen, warming have varying impacts on nitrogen budget in biofilms.

River and stream biofilms also play a key role in the decomposition of organic matter. This function is highly linked to the capacity to decompose high molecular weight compounds where extracellular enzymes are key actors. In soils, sensitivity of extracellular enzyme activity (EEA) to temperature has been demonstrated (Wallenstein *et al.*, 2009). Also, German *et al.* (2012) showed that temperature was affecting extracellular enzyme kinetics and found a greater sensitivity of soils from cooler climates indicating the adaptation of soil microorganisms. Significant increase of stream biofilm EEA have been measured in biofilms developing at 3°C higher than in the control conditions in a field experiment (Ylla *et al.*, 2014). Some activities such as the increase in leucine-aminopeptidase, cellobiohydrolase and β-xylosidase, showed a higher sensitivity to temperature than what would be expected by applying reported values of activation energy and the Arrhenius equation. From those results we approach the Q_{10} values for these activities, indicating that in the case of peptidase, it would be 8.9. This large value for the sensitivity to temperature for decomposition of organic nitrogen compounds together with the large Q_{10} value reported for denitrification suggest that nitrogen dynamics would be specially modified under warming conditions (Fig. 8.1). In the case of peptidase activity, the increased potential activity has been linked to the use of proteinaceous photosynthetic algal exudates. The other two enzymes that were specially enhanced by temperature showed Q_{10} values of 5.4 and 4.9 (cellobiohydrolase and β-xylosidase, respectively), indicating an increased use of complex polysaccharides such as cellulose and hemicellulose. Similarly, in a laboratory experiment it was shown that the degradation of recalcitrant dissolved organic matter (humic acids) by the biofilm was enhanced at increasing water temperature by 4°C, whereas the decomposition of more labile substances (cellobiose and the dipeptide leucine-proline) was fast and independent of temperature raise (Ylla *et al.*, 2012). Taken together, this indicates a significant greater capacity to use organic nitrogen compounds and to decompose complex compounds and polymers (such as cellulose and hemicellulose). This differential biofilm enzymatic potential may affect biogeochemical processes at the ecosystem level.

The functional response mechanisms described here might be at the same time linked to changes in community composition (as already highlighted for denitrification). The observed results in biofilm functioning (photosynthesis, respiration, denitrification, extracellular enzyme activities) may not be the unique possible response but this might be linked to changes in community composition and biofilm structure, as conceptually illustrated in Fig. 8.1.

Drought and desiccation effects on aquatic biofilms

Besides temperature, droughts are among the most important climate change disturbance events in stream ecosystems. They not only affect stream hydrology, but also the stream biota. Predicted shifts in the precipitation regime due to climate change are likely to alter both the magnitude and frequency of extreme hydraulic events (Jentsch *et al.*, 2007). Furthermore, the effect of increasing air temperature on the stream temperature is highly dependent on discharge (Webb *et al.*, 2008). Thus, during drying, the temperature of streams also increases (Gasith and Resh, 1999). Desiccation of streams is common in arid and semi-arid regions (Zoppini *et al.*, 2010; Romaní *et al.*, 2013). Headwaters in extended temperate regions, including Central Europe, have been observed to be exposed to more frequent and longer periods of drought, reflecting global climate change and enhanced human withdrawal of water (Marxsen *et al.*, 2010; Wagner *et al.*, 2011; Pohlon *et al.*, 2013a).

The effects of desiccation on bacterial community composition and EEA, as a basic function of the microbial food web (Marxsen and Fiebig, 1993), were investigated in biofilms of Mediterranean streams (Timoner *et al.*, 2012) and Central European streams (Marxsen *et al.*, 2010; Artigas *et al.*, 2012; Pohlon *et al.*, 2013 a, b). In contrast to Mediterranean streams, biofilms in the temperate regions are not adapted to desiccation. Short drought events already provoke distinct changes of bacterial community structure and enzymatic activities. Bacterial community composition shifts during desiccation toward composition in soil, exhibiting increasing proportions of Actinobacteria and Alphaproteobacteria, and decreasing proportions of Bacteroidetes and Betaproteobacteria (Fig. 8.3a). Simultaneously, the activities of extracellular enzymes decrease, most pronounced with aminopeptidases and less pronounced with enzymes involved in the degradation of polymeric carbohydrates, like α- and β-glucosidases or β-xylosidases (Fig. 8.4a, cf. Ylla *et al.*, 2010). After rewetting, EEA recovers within 14 days or even faster (Fig. 8.4b). The initial bacterial community composition is not attained within two weeks, which was the duration of the described rewetting experiment (Pohlon *et al.*, 2013b); Alphaproteobacteria remain the dominant bacterial group (Fig. 8.3b) (Marxsen *et al.*, 2010; Pohlon *et al.*, 2013b). Thus, the communities in temperate environments become more similar to communities in typical Mediterranean streambed sediments (Zoppini *et al.*, 2010; Marxsen *et al.*, 2010), although during desiccation those communities develop mostly in other directions. Trends similar to those in temperate streams were only rarely observed in Mediterranean rivers (River Pardiela, Portugal; Amalfitano *et al.*, 2008).

For most of the investigated enzymes, remarkable activity levels remain in temperate as well as in Mediterranean streams even after 13 weeks of desiccation (Marxsen *et al.*, 2010; Zoppini and Marxsen, 2011). This ensured that the complete ecosystem functions delivered by extracellular enzymes start recovering without delay upon rewetting. Also a fast metabolism recovery was measured in a biofilm in an intermittent Mediterranean stream, where not only EEA showed an immediate increase of activity after rewetting, but also photosynthetic activity recovered upon one hour of rewetting (Romaní and Sabater, 1997). This fast recovering was linked to the specific structure of this benthic community consisting of cyanobacteria stromatolitic-like mats (Sabater *et al.*, 2000).

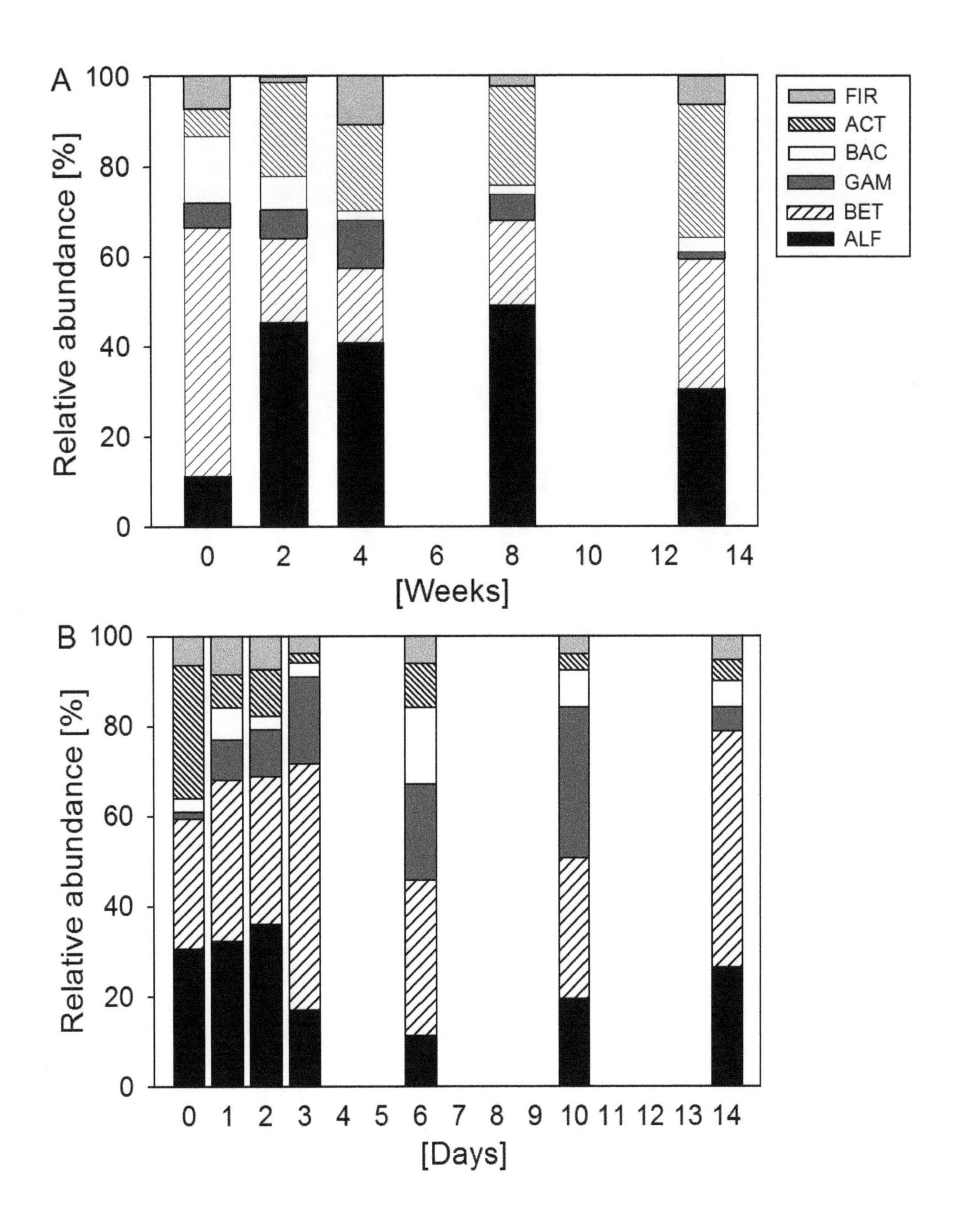

Figure 8.3 Bacterial community composition determined by CARD-FISH in wet streambed sediments (0) of the Breitenbach (Hesse, Germany) and after 2, 4, 8 and 13 weeks of artificial desiccation (A) and after 1, 2, 3, 6, 10 and 14 days of rewetting (B). Relative abundances of the investigated groups of Alphaproteobacteria (ALF), Betaproteobacteria (BET), Gammaproteobacteria (GAM), Bacteroidetes (BAC), Actinobacteria (ACT) and Firmicutes (FIR) are given. After Pohlon *et al.* (2013b).

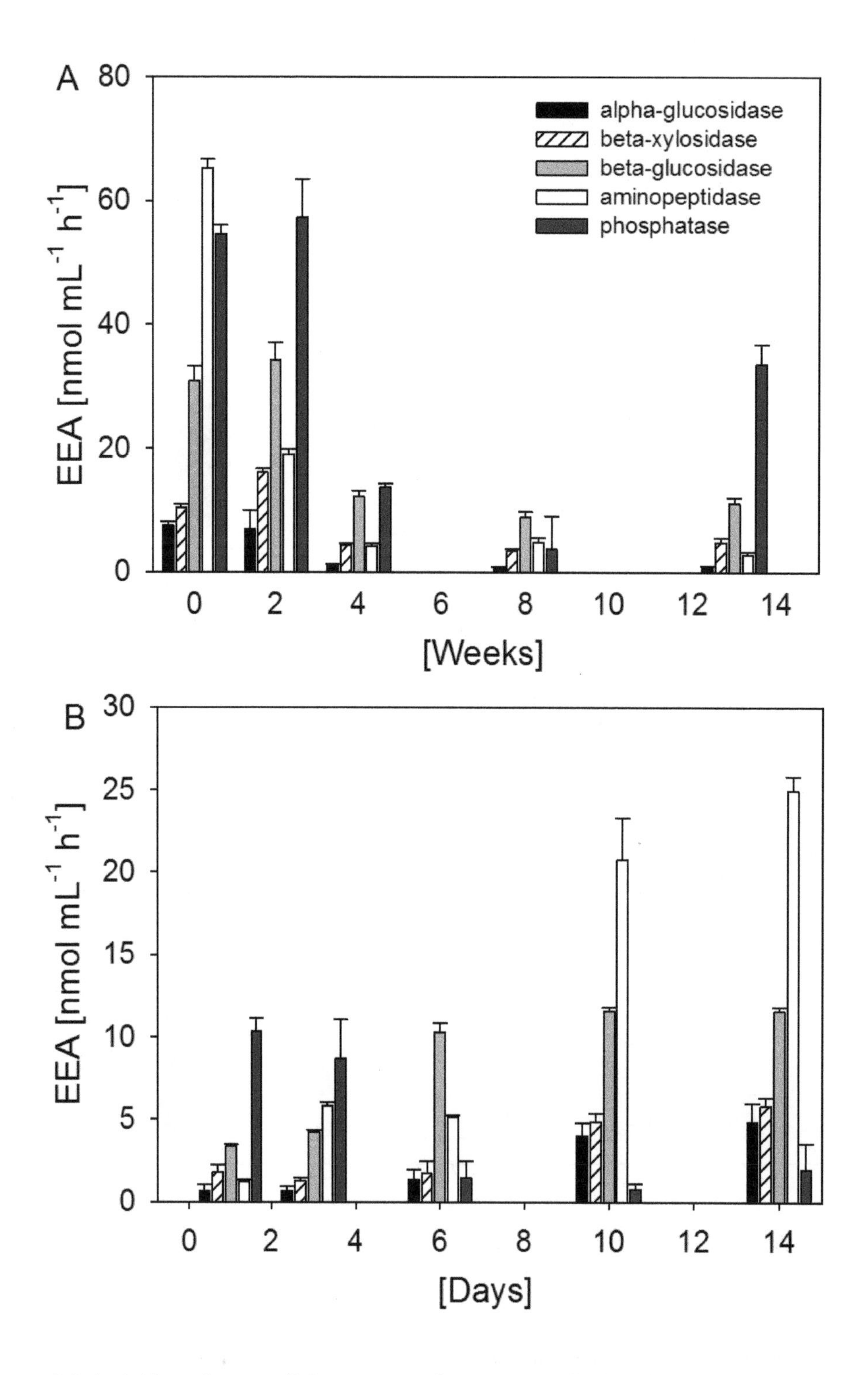

Figure 8.4 Activities of extracellular enzymes in wet streambed sediments (0) of the Breitenbach (Hesse, Germany) and after artificial desiccation over 2, 4, 8 and 13 weeks (A) and after 1, 3, 6, 10 and 14 days of rewetting (B). Mean values with standard deviations are given (n=4). After Pohlon *et al.* (2013b).

With desiccation carbon production in aquatic biofilms decreases, followed by reduction of bacterial biomass (Amalfitano *et al.*, 2008). In desiccating wetlands for example, the autotrophic biomass of biofilms decreases (Freeman *et al.*, 1994). In Mediterranean aquatic ecosystems the biomass of diatoms increases, which in turn changes the utilization of the biofilms by metazoans (Caramujo *et al.*, 2008). Fungi and Gram-positive bacteria of biofilms are assumed to be most adapted to changing moisture conditions (Harris, 1981). Also the extracellular polymeric substances (EPS) of the biofilms can delay the loss of moisture (Ophir and Gutnick, 1994). An experimentally simulated drought (by one week of flow reduction and two days of flow interruption) significantly affected biofilm structure and function by increasing the extracellular phosphatase activity, decreasing the density of live bacteria and live diatoms and decreasing photosynthetic capacity (Proia *et al.*, 2013). In this experiment, an increase of green algal biomass under drought conditions (in stagnant pools) could be responsible for the phosphorus limitation and increased phosphatase activity. However, it is generally shown that high phosphatase activity is maintained during drought periods in Mediterranean streams, suggesting phosphorus limitation (Romaní *et al.*, 2013).

Autotrophic organisms are more affected by desiccation than heterotrophic organisms, but recover faster after rewetting (Timoner *et al.*, 2012). However, some primary producers are better adapted than others to desiccation. Especially among green algae and cyanobacteria, specific structures or physiologic mechanisms occur, such as the production of EPS (see Romaní *et al.*, 2013), enabling survival during droughts. Recolonization of desiccated surfaces by biofilms occur faster in Mediterranean than in the streams of temperate regions (Artigas *et al.*, 2012), but the resettlement of algae is more dependent on the intensity of desiccation (Robson and Matthews, 2004).

Conclusions, future trends and perspectives

Taking into account the biofilm complexity, the current research indicates common mechanisms of shifts in biofilm community composition, structure and functioning due to increasing water temperature or suffering from drought episodes. Both warming conditions and desiccation affect the food web and cycling of materials through the microbial biofilm compartment. In the case of warming, significant changes in the nitrogen cycling through the biofilm may occur, as well as effects on the microbial food web interactions and organic matter use capabilities. In the case of desiccation, relevant structural and community composition changes and metabolic adaptations to the use of available materials occur. However, at the ecosystem level, both factors may coincide in time, but few studies have looked at the drought-temperature effects interaction on aquatic biofilms. In this regard, disturbance was demonstrated to strongly interact with warming (Marcus *et al.*, 2014). In that study, whereas the response of biofilm ciliate communities towards warming was negative in undisturbed biofilms (decreasing organism density), positive effects (increasing density) occurred among disturbed biofilm communities due to the resetting to earlier successional stages, which respond differently towards warming (as discussed above). This adds to an increasing number of examples showing that warming interacts with other factors, which are also altered by recent environmental changes, as has already shown for interactions with resource supply. Also, other climate change-related factors can affect community composition, and thus interactions within biofilms, such as increased shading due to cloud cover and runoff related increases in water turbidity that constrains light-related biofilm production, or increased sediment drift during extreme runoff events disturbing the biofilms. Identifying such mechanisms and identifying environmental settings under which warming alters interactions within microbial communities are major tasks for future research in order to enhance the accuracy in predicting warming effects on complex biofilm communities. Thus, to take a better figure of response mechanisms and final effects, multi-stress approaches should be applied in future experiments.

Since temperature and desiccation affect several biofilm key metabolic processes, such as photosynthetic activity, respiration, denitrification, and extracellular enzyme activities, it would be expected that this may change the role of the biofilm on biogeochemical processes in the aquatic habitat. Thus, carbon and nitrogen budgets need to be included in future experiments (including quantity and quality) in order to figure out whether metabolic changes have a direct effect on nutrient cycling at the ecosystem level. Also, most of the observed effects on biofilm metabolism are linked to community composition and structure, and vice versa. Different communities may have different sensitivity to a disturbance. Further research is needed in order to link the community composition to the observed metabolic changes.

Biofilms developing in different habitat types may show different responses to warming and desiccation. For example, benthic microbial communities developing in stream sediment or decomposing material are more resistant to drought than those developing on rocks and cobbles (Ylla *et al.*, 2010). Each biofilm is made of different consortia of bacteria, archaea, fungi, algae and their interactions, which determine a different structure and function of the whole biofilm (Wetzel, 1993; Lock, 1993; Romaní and Sabater, 1999; Frossard *et al.*, 2012). In this regard, biofilms of the light-influenced surfaces on stones and plant material are more complex than the bacteria dominated community in the sediment (Zubkov and Sleigh, 2000; Romaní *et al.*, 2004). Warming has a greater effect on the prokaryote community composition in dark-grown biofilms (such as those developing in the hyporheic zone or the bottom of cobbles and rocks) than in light-grown biofilms, suggesting that the higher complexity of biofilm structure and interactions in the light-grown biofilm buffer potential changes in prokaryote community composition (Romaní *et al.*, 2014).

Biofilms are relevant models for experimentation on global change since they are easy to handle and represent a complex community. However, further research should be performed to include theoretical background to the changes in the biomass of the different organisms colonizing the biofilm, which should also include the different steps of biofilm formation. Whereas the Boltzmann-Arrhenius equation gives a theoretical background for modeling and comparing responses of biofilm metabolism, no such equation exists for modeling biofilm biomass. The application of a theoretical model for biofilm biomass is not easy since it should include the food web interactions between all microorganisms, producing either direct or indirect, positive or negative responses. Also, experimentation should consider both short- and long-term experiments since adaptation may give to a different response, as shown by Russell *et al.* (2013). The final objective would be to have a predictive model for biofilms, including direct and indirect effects of climate change disturbances on natural assemblages and the resulting impact on their functioning.

Acknowledgements

We are grateful to Thomas Horvath, University of Koblenz-Landau, Landau, Germany, for improving the English text.

References

Abbot, K.C., Harmon, J.P., and Fabina, N.S. (2014). The challenge of predicting temperature effects on short-term predator-prey dynamics. Popul. Ecol. DOI: 10.1007/s10144-013-0426-x.

Acuña, V., Wolf, A, Uehlinger, U., and Tockner, K. (2008). Temperature dependence of stream benthic respiration in an Alpine river network under global warming. Freshwat. Biol. *53*, 2076-2088.

Allen, A.P., Gillooly, J.F., and Brown, J.H. (2005). Linking the global carbon cycle to individual metabolism. Funct. Ecol. *19*, 202-213.

Amalfitano, S., Fazi, S., Zoppini, A., Caracciolo, A.B., Grenni, P., and Puddu, A. (2008). Responses of benthic bacteria to experimental drying in sediments from Mediterranean temporary rivers. Microb. Ecol. *55*, 270-279.

Andrushchyshyn, O.P., Wilson, K,P,, and Williams, D.D. (2009). Climate change-predicted shifts in the temperature regime of shallow groundwater produce rapid responses in ciliate communities. Global Change Biol. *15*, 2518–2538.

Arndt, H., Schmidt-Denter, K., Auer, B., and Weitere, M. (2003). Protozoans and biofilms. In Fossil and Recent Biofilms, Mats and Networks, W.E. Krumbein, D.M. Paterson and G.A. Zavarzin, eds. (Kluwer Academic Publishers, Dordrecht, The Netherlands), pp. 173-189.

Artigas, J., Fund, K., Kirchen, S., Morin, S., Obst, U., Romani, A.M., Sabater, S., and Schwartz, T. (2012). Patterns of biofilm formation in two streams from different bioclimatic regions: analysis of microbial community structure and metabolism. Hydrobiologia *695*, 83-96.

Avrahami, S., Jia, Z., Neufeld, J.D., Murrell, J.C., Conrad, R., and Kuesel, K. (2011). Active Autotrophic Ammonia-Oxidizing Bacteria in Biofilm Enrichments from Simulated Creek Ecosystems at Two Ammonium Concentrations Respond to Temperature Manipulation. Appl. Environ. Microbiol. *77*, 7329-7338.

Boulêtreau, S., Garabetian, F., Sauvage, S., and Sánchez-Pérez, J.-M. (2006). Assessing the importance of a self-generated detachment process in river biofilm models. Freshwat. Biol. *51*, 901-912.

Boulêtreau, S., Salvo, E., Lyautey, E., Mastrorillo, S., and Garabetian, F. (2012). Temperature dependence of denitrification in phototrophic river biofilms. Sci. Tot. Env. *416*, 323–328.

Boulêtreau, S., Lyautey, E., Duboic, S., Compin, A., Delattre, C., Touron-Bodilis, A., Mastrorillo, S., and Garabetian, F. (2014). Warming-induced changes in denitrifier community structure modulate the ability of phototrophic river biofilms to denitrify. Sci. Tot. Environ. *466–467*, 856–863.

Brown, J.H., Gillooly, J.F., Allen, A.P., Savage, V.M., and West, G.B. (2004). Toward a metabolic theory of ecology. Ecology *85*, 1771-1789.

Caramujo, M.-J., Mendes, C.R.B., Cartaxana, P., Brotas, V., and Boavida, M.-J. (2008). Influence of drought on algal biofilms and meiofaunal assemblages of temperate reservoirs and rivers. Hydrobiologia *598*, 77-94.

Carpenter, S.R., Fisher, S.G., Grimm, N.B., and Kitchell, J.F. (1992). Global change and freshwater ecosystems. Ann. Rev. Ecol. Evol. Syst. *23*, 119–139.

Clarke, A. (2006). Temperature and the metabolic theory of ecology. Funct. Ecol. *20*, 405–412.

Dahlgren, K., Wiklund, A.K.E., and Andersson, A. (2011). The influence of autotrophy, heterotrophy and temperature on pelagic food web efficiency in a brackish water system. Aquat. Ecol. *45*, 307–323.

Davison, I.R. (1991). Environmental effects on algal photosynthesis: temperature. J. Phycol. *27*, 2–8.

Degerman, R., Dinasquet, J., Riemann, L., de Luna, S.S., and Andersson, A. (2013). Effect of resource availability on bacterial community responses to increased temperature. Aquat. Microb. Ecol. *68*, 131-142.

Di Bonaventura, G., Piccolomini, R., Paludi, D., D'Orio, V., Vergara, A., Conter, M., and Ianieri, A. (2008). Influence of temperature on biofilm formation by *Listeria monocytogenes* on various food-contact surfaces: relationship with motility and cell surface hydrophobicity. J. Appl. Microbiol. *104*, 1552–1561.

Díaz Villanueva, V., Font, J., Schwartz, T., and Romaní, A.M. (2011). Biofilm formation at warming temperature: acceleration of microbial colonization and microbial interactive effects. Biofouling *27*, 59-71.

Di Pippo, F., Ellwood, N.T.W., Guzzon, A., Siliato, L., Micheletti, E., De Philippis, R., and Albertano, P.B. (2012). Effect of light and temperature on biomass, photosynthesis and capsular polysaccharides in cultured phototrophic biofilms. J. Appl. Phycol. *24*, 211–220.

Fischer, H., Sachse, A., Steinberg, C.E.W., and Pusch, M. (2002). Differential retention of dissolved organic carbon by bacteria in river sediments. Limnol. Oceanogr. *47*, 1702-1711.

Fox, J.W., and Morin, P.J. (2001). Effects of intra- and interspecific interactions on species responses to environmental change. J. An. Ecol. *70*, 80-90.

Freeman, C., Gresswell, R., Guasch, H., Hudson, J., Lock, M.A., Reynolds, B., Sabater, F., and Sabater, S. (1994). The Role of Drought in the Impact of Climatic-Change on the Microbiota of Peatland Streams. Freshwat. Biol. *32*, 223-230.

Frossard, A., Gerull, L., Mutz, M., and Gessner, M.O. (2012). Fungal importance extends beyond litter decomposition in experimental early-successional streams. Environ. Microbiol. *14*, 2971-2983.

Gasith, A., and Resh, V.H. (1999). Streams in Mediterranean climate regions: Abiotic influences and biotic responses to predictable seasonal events. Ann. Rev. Ecol. Syst. *30*, 51-81.

German, D.P., Marcelo, K.R.B., Stone, M.M., and Allison, S.D. (2012). The Michaelis–Menten kinetics of soil extracellular enzymes in response to temperature: a cross-latitudinal study. Global Change Biol. *18*, 1468–1479.

Gillooly, J.F., Allen, A.P., Savage, V.M., Charnov, E.L., West, G.B., and Brown, J.H. (2006). Response to Clarke and Fraser: effects of temperature on metabolic rate. Funct. Ecol. *20*, 400-404.

Gillooly, J.F., Brown, J.H., West, G.B., Savage, V.M., and Charnov, E.L. (2001). Effects of size and temperature on metabolic rate. Science *293*, 2248-2251.

Guasch, H., Marti, E., and Sabater, S. (1995). Nutrient enrichment effects on biofilm metabolism in a Mediterranean stream. Freshwat. Biol. *33*, 373–383.

Hancke, K., and Glud, R.N. (2004). Temperature effects on respiration and photosynthesis in three diatom-dominated benthic communities. Aquat. Microb. Ecol. *37*, 265-281.

Hansen, S.K., Rainey, P.B., Haagensen, J.A.J., and Molin, S. (2007). Evolution of species interactions in a biofilm community. Nature *445*, 533-536.

Harris, R.F. (1981). Effect of water potential on microbial growth and activity. In Water Potential Relations in Soil Microbiology, J.F. Parr, W.R. Gardner and L.F. Elliott, eds. (American Society of Agronomy, Madison, Wisconsin, USA), pp. 23-95.

IPCC (2013) Climate Change 2013 - The Physical Science Basis. Contribution of Working Group I to the Fifth Assessment Report of the Intergovernmental Panel on Climate Change. (Stocker, T.F., D. Qin, G.-K. Plattner, M. Tignor, S.K. Allen, J. Boschung, A. Nauels, Y. Xia, V. Bex and P.M. Midgley, eds.) (Cambridge University Press, Cambridge, UK, and New York, NY, USA).

Jackson, C.R. (2003). Changes in community properties during microbial succession. Oikos *101*, 444-448.

Jackson, C.R., Churchill, P.F., and Roden, E.E. (2001). Successional changes in bacterial assemblage structure during epilithic biofilm development. Ecology *82*, 555-566.

Jentsch, A., Kreyling, J., and Beierkuhnlein, C. (2007). A new generation of climate change experiments: events, not trends. Front. Ecol. Environ. *5*, 315–324.

Jiang, L., and Morin, P.J. (2004). Temperature-dependent interactions explain unexpected responses to environmental warming in communities of competitors. J. An. Ecol. *73*, 569-576.

Kalscheur, K.N., Rojas, M., Peterson, C.G., Kelly, J.J., and Gray, K.A. (2012). Algal Exudates and Stream Organic Matter Influence the Structure and Function of Denitrifying Bacterial Communities. Microb. Ecol. *64*, 881-892.

Kathol, M., Norf, H., Arndt, H., and Weitere, M. (2009). Effects of temperature increase on the grazing of planktonic bacteria by biofilm-dwelling consumers. Aquat. Microb. Ecol. *55*, 65-79.

Lock, M.A. (1993). Attached bacteria communities in stream. In Aquatic Microbiology: an Ecological Approach, T.E. Ford, ed. (Blackwell Scientific Publications, Boston), pp. 113-138.

Lyautey, E., Jackson, C.R., Cayrou, J., Rols, J.-L., and Garabetian, F. (2005). Bacterial community succession in natural river biofilm assemblages. Microb. Ecol. *50*, 589-601.

Lyautey, E., Hallin, S., Teissier, S., Iribar, A., Compin, A., Philippot, L., and Garabetian, F. (2013). Abundance, activity and structure of denitrifier in phototrophic river biofilms (River Garonne, France). Hydrobiologia *716*, 177-187.

Marcus, H., Wey, J.K., Norf, H., and Weitere, M. (2014). Disturbance alters the response of consumer communities towards warming: A mesocosm study with biofilm-dwelling ciliates. Ecosphere *5*: art10. DOI: 10.1890/ES13-00170.1.

Marxsen, J., and Fiebig, D. (1993). Use of perfused cores for evaluating extracellular enzyme activity in stream-bed sediments. FEMS Microbiol. Ecol. *13*, 1-11.

Marxsen, J., Zoppini, A., and Wilczek, S. (2010). Microbial communities in streambed sediments recovering from desiccation. FEMS Microbiol. Ecol. *71*, 374-386.

Mohseni, O., and Stefan, H.G. (1999). Stream temperature/air temperature relationship: a physical interpretation. J. Hydrol. *218*, 128–141.

Moltz, A.G., and Martin, S.E. (2005). Formation of biofilms by *Listeria monocytogenes* under various growth conditions. J. Food Prot. *68*, 92–97.

Montagnes, D.J.S., and Weisse, T. (2000). Fluctuating temperatures affect growth and production rates. Aquat. Microb. Ecol. *21*, 97-102.

Necchi, O. (2006). Photosynthetic responses to temperature in tropical lotic macroalgae. Phycol. Res. *52*, 140–148.

Nomdedeu, M.M., Willen, C., Schieffer, A., and Arndt, H. (2012). Temperature-dependent ranges of coexistence in a model of a two-prey-one-predator microbial food web. Mar. Biol. *159*, 2423-2430.

Norf, H., Arndt, H., and Weitere, M. (2007). Impact of local temperature increase on the early development of biofilm-associated ciliate communities. Oecologia *151*, 341-350.

Norf, H., and Weitere, M. (2010). Resource quantity and seasonal background alter warming effects on communities of biofilm ciliates. FEMS Microbiol. Ecol. *74*, 361-370.

Ophir. T., and Gutnick, D.L. (1994). A Role for Exopolysaccharides in the Protection of Microorganisms from Desiccation. Appl. Environ. Microbiol. *60*, 740-745.

Paerl, H.W., and Huisman, J. (2009). Climate change: a catalyst for global expansion of harmful cyanobacterial blooms. Environ. Microbiol. Rep. *1*, 27–37.

Perkins, D.M., Yvon-Durocher, G., Demars, B.O.L., Reiss, J., Pichler, D.E., Friberg, N., Trimmer, M., and Woodward, G. (2012). Consistent temperature dependence of respiration across ecosystems contrasting in thermal history. Global Change Biol. *18*, 1300–1311.

Petchey, O.L., McPhearson, P.T., Casey, T.M., and Morin, P.J. (1999). Environmental warming alters food-web structure and ecosystem function. Nature *402*, 69-72.

Peterson, C.G., Daley, A.D., Pechauer, S.M., Kalscheur, K.N., Sullivan, M.J., Kufta, S.L., Rojas, M., Gray, K.A., and Kelly, J.J. (2011). Development of associations between microalgae and denitrifying bacteria in streams of contrasting anthropogenic influence. FEMS Microbiol. Ecol. *77*, 477-492.

Philippot, L., and Hallin, S. (2005). Finding the missing link between diversity and activity using denitrifying bacteria as a model functional community. Curr. Op. Microbiol. *8*, 234–239.

Pohlon, E., Mätzig, C., and Marxsen, J. (2013a). Desiccation affects bacterial community structure and function in temperate stream sediments. Fundam. Appl. Limnol. *182*, 123-134.

Pohlon, E., Ochoa Fandino, A., and Marxsen, J. (2013b). Bacterial community composition and extracellular enzyme activity in temperate streambed sediment during drying and rewetting. PLoS ONE *8*, e83365, 1-14. DOI: 10.1371/journal.pone.0083365.

Pomeroy, L.R., and Deibel, D. (1986) Temperature regulation of bacterial activity during the spring bloom in Newfoundland coastal waters. Science *233*, 359-361.

Proia, L., Vilches, C., Boninneau, C., Kantiani, L., Farré, M., Romaní, A.M., Sabater, S., and Guasch, H. (2013). Drought episode modulates the response of river biofilm to triclosan. Aquat. Toxicol. *127*, 36-45.

Rall, B.C., Brose, U., Hartvig, M., Kalinkat, G., Schwarzmüller, F., Vucic-Pestic, O., and Petchey, O.L. (2012). Universal temperature and body-mass scaling of feeding rates. Phil. Trans. R. Soc. B *367*, 2923-2934.

Rao, T.S. (2010). Comparative effect of temperature on biofilm formation in natural and modified marine environment. Aquat. Ecol. *44*, 463–478.

Raven, J., and Geider, R. (1988). Temperature and algal growth. New Phytol. *110*, 441–461.

Robson, B.J., and Matthews, T.G. (2004). Drought refuges affect algal recolonization in intermittent streams. Riv. Res. Appl. *20*, 753-763.

Romaní, A.M., and Sabater, S. (1997). Metabolism recovery of a stromatolitic biofilm after drought in a Mediterranean stream. Arch. Hydrobiol. *140*, 261–271.

Romaní, A.M., and Sabater, S. (1999). Effect of primary producers on the heterotrophic metabolism of a stream biofilm. Freshwat. Biol. *41*, 729-736.

Romaní, A.M. (2010) Freshwater Biofilms. In Biofouling, S. Dürr and J.C. Thomason, eds. (Wiley-Blackwell Blackwell Publishing, UK), pp. 137-153.

Romaní, A.M., Guasch, H., Munoz, I., Ruana, J., Vilalta, E., Schwartz, T., Emtiazi, F., and Sabater, S. (2004). Biofilm Structure and Function and Possible Implications for Riverine DOC Dynamics. Microb. Ecol. *47*, 316-328.

Romaní, A.M., Artigas, J., and Ylla, I. (2012). Extracellular Enzymes in Aquatic Biofilms: Microbial Interactions versus Water Quality Effects in the Use of Organic Matter. In Microbial Biofilm: Current Research and Applications, G. Lear and G. Lewis, eds. (Caister Academic Press, Norfolk, UK), pp. 153-174.

Romaní, A.M., Amalfitano, S., Artigas, J., Fazi, S., Sabater, S., Timoner, X., Ylla, I., and Zoppini, A. (2013). Microbial biofilm structure and organic matter use in Mediterranean streams. Hydrobiologia *719*, 43-58

Romaní, A.M., Borrego, C.M., Díaz-Villanueva, V., Freixa, A., Gich, F., and Ylla, I. (2014). Shifts in microbial community structure and function in light and dark grown biofilms driven by warming. Environ. Microbiol. DOI: 10.1111/1462-2920.12428.

Rosa, J., Ferreira, V., Canhoto, C., and Graça, M.A.S. (2013). Combined effects of water temperature and nutrients concentrations on periphyton respiration – implications of global change. Int. Rev. Hydrobiol. *98*, 14–23.

Russell, B.D., Connell, S.D., Findlay, H.S., Tait, K., Widdicombe, S., and Mieszkowska, N. (2013). Ocean acidification and rising temperatures may increase biofilm primary productivity but decrease grazer consumption. Phil. Trans. R. Soc. B *368*, 1471-2970.

Sabater, S., Guasch, H., and Romaní, A.M. (2000). Stromatolitic communities in Mediterranean streams: adaptations to a changing environment. Biodiv. Conserv. *9*, 379–392.

Sabater, S., Guasch, H., Ricart, M., Romaní, A.M., Vidal, G., Klünder, C., and Schmitt-Jansen, M. (2007). Monitoring the effect of chemicals on biological communities. The biofilm as an interface. Anal. Bioanalyt. Chem. *387*, 1425-1434.

Stewart, P.S. (2003). Diffusion in biofilms. J. Bacteriol. *185*, 1485-1491.

Teissier, S., Torre, M., Delmas, F., and Garabetian, F. (2007). Detailing biogeochemical N budgets in riverine epilithic biofilms. J. N. Am. Benthol. Soc. *26*, 178-190.

Timoner, X., Acuña, V., Von Schiller, D., and Sabater, S. (2012). Functional responses of stream biofilms to flow cessation, desiccation and rewetting. Freshwat. Biol. *57*, 1565-1578.

Uehlinger, U. (2006). Annual cycle and inter-annual variability of gross primary production and ecosystem respiration in a floodprone river during a 15-year period. Freshwat. Biol. *51*, 938-950.

Veraart, A.J., de Klein, J.J.M., and Scheffer, M. (2011). Warming can boost denitrification disproportionately due to altered oxygen dynamics. PLoS One *6*: e18508.

Viergutz, C.., Kathol, M., Norf, H., Arndt, H., and Weitere, M. (2007). Control of microbial communities by the macrofauna: a sensitive interaction in the context of extreme summer temperatures? Oecologia *151*, 115–124.

von Elert, E., Matrin-Creuzburg, D., and Le Goz, J.R. (2003). Absence of sterols contrains carbon transfer between cyanobacteria and a freshwater herbivore (*Daphnia galeata*). Proc. R. Soc. B *270*, 1209–1214.

Wagner, R., Marxsen, J., Zwick, P., and Cox, E.J. (2011). Central European Stream Ecosystems. The Long Term Study of the Breitenbach (Wiley-VCH, Weinheim).

Wallenstein, M.D., McMahon, S.K., and Schimel, J.P. (2009). Seasonal variation in enzyme activities and temperature sensitivities in Arctic tundra soils. Global Change Biol. *15*, 1631–1639.

Webb, B.W., Hannah, D.M., Moore, R.D., Brown, L.E., and Nobilis, F. (2008) Recent advances in stream and river temperature research. Hydrol. Proc. *22*, 902-918.

Wetzel, R.G. (1963). Primary productivity of periphyton. Nature *197*, 1026-1027.

Wetzel, R.G. (1993). Microcommunities and microgradients: linking nutrient regeneration, microbial mutualism, and high sustained aquatic primary production. Neth. J. Aquat. Ecol. *27*: 3-9.

Wimpenny, J., Manz, W., and Szewzyk, U. (2000) Heterogeneity in biofilms. Microbiol. Rev. *24*, 661-671.

Ylla, I., Sanpera-Calbet, I., Vázquez, E., Romaní, A.M., Muñoz, I., Butturini, A., and Sabater, S. (2010). Organic matter availability during pre- and post-drought periods in a Mediterranean stream. Hydrobiologia *657*, 217–232.

Ylla, I., Romaní, A.M., and Sabater, S. (2012). Labile and Recalcitrant Organic Matter Utilization by River Biofilm Under Increasing Water Temperature. Microb. Ecol. *64*, 593-604.

Ylla, I., Canhoto, C., and Romaní, A.M. (2014). Effects of warming on stream biofilm organic matter use capabilities. Microb. Ecol. DOI: 10.1007/s00248-014-0406-5.

Yvon-Durocher, G., Allen, A.P., Montoya, J.M., Trimmer, M., and Woodward, G. (2010). The temperature dependence of the carbon cycle in aquatic ecosystems. Adv. Ecol. Res. *43*, 267-313.

Zoppini, A., Amalfitano, S., Fazi, S., and Puddu, A. (2010). Dynamics of a benthic microbial community in a riverine environment subject to hydrological fluctuations (Mulargia River, Italy). Hydrobiologia *657*, 37-51.

Zoppini, A., and Marxsen, J. (2011). Importance of extracellular enzymes for biogeochemical processes in temporary river sediments during fluctuating dry–wet conditions. In Soil Enzymology, Soil Biology, G. Shukla and A. Varma, eds. (Springer, Heidelberg), pp. 103–117.

Zubkov, M.V., and Sleigh, M.A. (2000). Comparison of growth efficiencies of protozoa growing on bacteria deposited on surfaces and in suspension. J. Euc. Microbiol. *47*, 62-69.

Chapter 9

Climate Change, Microbes, and Soil Carbon Cycling

Timothy H. Keitt[1], Colin R. Addis[1], Daniel L. Mitchell[1], Andria Salas[1] and Christine V. Hawkes[1,2,*]

[1] Department of Integrative Biology, University of Texas at Austin, Austin, TX 78712, USA; The Nature Conservancy, New Haven, CT USA, Yale School of the Environment, New Haven, CT USA; [2] Department of Plant and Microbial Biology, North Carolina State University, Raleigh, NC 27965, USA; * corresponding author

Email: tkeitt@utexas.edu, craddis@utexas.edu, daniel.lewis.mitchell@gmail.com, aks2515@gmail.com, chawkes@ncsu.edu

DOI: https://doi.org/10.21775/9781913652579.09

Abstract

Microbial responses to climate change will partly control the balance of soil carbon storage and loss under future temperature and precipitation conditions. We propose four classes of response mechanisms that can allow for a more general understanding of microbial climate responses. We further explore how a subset of these mechanisms results in microbial responses to climate change using simulation modeling. Specifically, we incorporate soil moisture sensitivity into two current enzyme-driven models of soil carbon cycling and find that moisture has large effects on predictions for soil carbon and microbial pools. Empirical efforts to distinguish among response mechanisms will facilitate our ability to further develop models with improved accuracy.

Introduction

There is twice as much carbon in soils as in the atmosphere (Jenkinson et al., 1991), making belowground responses to climate change an important aspect of ecosystem responses and feedbacks to climate. Nevertheless, belowground responses to climate change remain a large source of uncertainty (Solomon et al., 2007), such that earth system models poorly predict current soil carbon pools (Todd-Brown et al., 2013). This is likely due, in part, to historical assumptions of purely abiotic controls of soil carbon cycling and the lack of a strong mechanistic framework for how soil microbes respond to environmental change and the resulting impacts on the fate of soil carbon (Chapin III et al., 2009; Ogle et al., 2010).

Soil respiration is the main pathway for the transfer of carbon from terrestrial to atmospheric pools (Schlesinger and Andrews, 2000). Soil microbes may also make a larger contribution to the building of soil organic carbon than previously thought (Kindler et al., 2006; Liang and Balser, 2008, 2011; Potthoff et al., 2008). For example, mycorrhizal fungi can be the dominant pathway through which carbon from plants enters the soil pool, with hyphal turnover representing ~60% of soil organic matter inputs and the remaining ~40% due to fine root turnover and leaf litter (Godbold et al., 2006). Furthermore, the type of mycorrhizal fungus can determine soil carbon: Averill et al. (2014) found that ecosystems dominated by plants colonized by ectomycorrhizal fungi stored 70% more soil carbon per unit nitrogen than ecosystems dominated by plants associated with arbuscular mycorrhizal fungi. Thus, the effects of climate change on the activity and physiology of the soil microbes will partly determine what proportion of annual soil carbon input is respired vs. stored in the long-term reservoir of soil organic carbon (Chapin III et al., 2002).

Shifts in microbial community composition, abundance, and function have been observed in climate change experiments manipulating temperature, precipitation, carbon dioxide, and their interactions (e.g., Castro et al., 2010; Cheng et al., 2012; Harper et al., 2005; Hawkes et al., 2011; Horz et al., 2004; Horz et al., 2005; Lindberg et al., 2002; Liu et al., 2009; Staddon et al., 2003; Zogg et al., 1997). While results appear to be site-specific, some broader patterns can be gleaned from meta-analyses. Based on 32 experimental temperature manipulations, warming increased soil respiration by 20% and net nitrogen mineralization by 46% (Rustad et al., 2001). Blankinship et al. (2011) analyzed 75 experimental climate studies and found that bacteria decreased in response to warming, fungi increased in response to altered precipitation, and total microbial biomass increased with elevated carbon dioxide (Blankinship et al., 2011); however, these groups were represented by as few as two studies, so more work is needed to confirm the generality of their responses.

The mechanisms regulating process rates in belowground ecosystems are complex, lending these systems a high potential for varied and seemingly idiosyncratic empirical behavior (May, 1976). To usefully integrate empirical data into even more complicated regional and global modeling, belowground observations must be systematized by comparison to simplified mathematical representations. There exists a long history of terrestrial ecosystem modeling, including belowground process modeling, but only recently have these addressed variation in microbial functioning (e.g., Allison, 2012; Allison et al., 2010; Orwin et al., 2011; Wang et al., 2014; Waring et al., 2013; Wieder et al., 2013). Furthermore, the role of water in regulating process rates has received comparatively little attention relative to mass-balance and nutrient-driven approaches.

In this chapter, we focus on how soil microbes respond to changes in soil moisture, which is a primary controller of soil microbial activity, but remains poorly understood relative to other factors such as temperature. We first provide an overview of what is known about microbial responses to changes in moisture availability, review microbial mechanisms that are likely to be important for improving our understanding of microbial responses, and discuss the current state of microbial and ecosystem models, In addition, as a demonstration of the potential relevance of moisture to microbial function, we integrate moisture functions into two microbial process models.

Microbial responses to altered soil moisture

The magnitude and shape of microbial functional responses to soil moisture will directly affect ecosystem responses, as well as how we model microbial process responses to climate change. Water is a primary controller of soil microbial activity (Liu et al., 2009), limiting both soil respiration and enzyme activities. However, microbial responses to moisture can be highly variable. Water availability can act as a resource that limits microbial processes (either directly or by limiting nutrient acquisition), such that microbial responses to water might be linear or at least monotonic. Yet because both too little and too much water can act as stressors for soil microbes (Davidson et al., 2012; Schimel et al., 2007; Stark and Firestone, 1995), we might also find non-linear, threshold and/or non-monotonic responses of microbial communities to water availability (e.g., Curiel-Yuste et al., 2007).

Broad consistency in microbial functional responses to moisture should allow for microbe-driven soil carbon processes to acclimatize given a change in precipitation regime, and lead to straightforward predictions based on water availability. Manzoni et al. (2011) found that the lower moisture limit for microbial activity was consistent across 15 studies, likely representing universal constraints on solute diffusion and dehydration tolerance. However, individual microbial taxa have unique physiological response curves to moisture, including both specialist and generalist strategies (Lennon et al., 2012). Therefore, we may be able to generalize the lower endpoint of microbial moisture responses, but we also expect variability in responses to increasing moisture and its upper limits. The challenge is then to understand the distribution of variability in microbial moisture responses and whether/how it influences aggregate function.

Microbial moisture responses that are specific to biomes, ecosystems, habitats, microbial functional groups, or taxa could lead to historical contingencies that modify how functions such as respiration respond to a change in precipitation regime. In this scenario, predictive models might require modifications based on local factors. For example, soil microbial responses may fundamentally differ between arid and mesic regions, because arid soils are expected to be dominated by fungi, have higher potential decomposition (particularly of recalcitrant material), and have greater decoupling of plant and microbial activities (Collins et al., 2008). Similarly, regions with a longer history of drought may have a reduced capacity to respond to water based on both climate change experiments and lab tests of soils exposed to different periods of drought (Evans and Wallenstein, 2012; Göransson et al., 2013; Meisner et al., 2013a). More complex scenarios may impede our ability to make accurate predictions for soil functional responses to climate change.

An additional key to considering microbial moisture responses is the dynamic distribution of rainfall. Rainfall punctuated by dry periods results in pulses of biological activity that can decouple plant and microbial processes (Collins et al., 2008). Pulsed rainfall distributions may become more important in the future given that predictions often include larger events and fewer days of rain (IPCC, 2007; Jentsch et al., 2007; Jiang and Yang, 2012; Leung and Gustafson, 2005). Pulsed rain events can drive transitions between alternative stable states representing high and low microbial functioning, and the duration of the pulse events may differentially affect fast and slow components of the microbial community. Transient rain events may contribute disproportionately to soil carbon cycling, particularly when microbial responses are both rapid and large, or when short-duration rain pulses allow older soil carbon pools to be accessed (Carbone et al., 2011; Collins et al., 2008; Sala and Lauenroth, 1982). In addition, microbial responses upon rewetting may outweigh apparent reductions in process rates during dry periods, particularly if extracellular enzymes are retained in dry soils (Schimel et al., 2007). Most studies of microbial pulse responses have been in single sites, however, and do not provide us with a sufficiently broad understanding of these processes.

Microbial response mechanisms

An understanding of the mechanisms underlying microbial responses is critical to generalize soil microbial contributions to soil carbon cycling and other ecosystem functions under future climate scenarios. When an environment is altered by a press disturbance such as climate change, we can observe changes in functions such as soil respiration, enzyme activity, and litter decomposition, but often we lack the information necessary to discover through what mechanisms those changes occurred. Whole-soil, aggregate functional responses result from the individual activities of a diverse community of soil microbes, meaning that different mechanisms can be operating simultaneously to create the observed function.

Here, we consider four classes of response mechanisms that are likely to be at play: physiology, community composition, feedbacks, and evolution. Traits linking individual physiology and performance to environmental conditions will lead to species sorting and compositional change over gradients (Leibold et al., 2004). Community states may also be influenced by dispersal limitation combined with landscape connectivity patterns (Ehrlen and Eriksson, 2000). In addition, positive feedbacks that reinforce alternative stable states may override sorting and immigration locally, and introduce strong history-dependence in community responses (Keitt et al., 2001; Scheffer et al., 2001). Finally, trait variation owing to evolutionary change is yet another mechanism modulating community response to environmental change and is perhaps the least understood in terms of interactions with sorting, migration and positive feedbacks (Johnson and Stinchcombe, 2007) . We discuss each of these in more detail below.

Microbial physiology

Microbes are often considered to have broad physiological capabilities and thus physiological acclimatization to environmental change may be common. Aggregate functional responses could be due to the physiological breadth of individual taxa in the community, but alternatively might represent the diversity of physiologies among taxa. Functional plasticity has been observed in microbial community responses to short-term changes in temperature and moisture (Bradford et al., 2010; Griffiths et al., 2003; Heinemeyer et al., 2006), although this is not always the case (Malcolm et al.,

2009). Dormancy is another form of plasticity that is widespread in soil microbes and can allow for avoidance of temporary periods of environmental stress such as drought (Lennon and Jones, 2011). Resuscitation of dormant taxa can result in rapid and predictable functional resilience once conditions improve (Placella et al., 2012). However, the success of dormancy strategies in the face of climate change will depend on both the persistence of dormant propagules and the nature of the new environment; for example, long-lived spores will be needed to withstand long-term drought.

Even if plasticity is common, however, we might expect that microbial responses to altered climate might be constrained by local climatic history, with larger, more variable, or less predictable responses when outside the range of historical selection pressures. Reciprocal transplants of intact soil cores between plant communities support this idea: Waldrop (2006) found little change in microbial community composition and aggregate function for soil cores transplanted from grasslands to beneath oak canopies (where environmental conditions were entirely within the range normally found in grasslands), but observed rapid changes when oak canopy cores were transplanted into grasslands (where conditions were outside the normal range). Understanding the limits of microbial physiological plasticity will provide the boundary conditions for potential microbial functional responses to altered climate.

Microbial community composition

Differences in microbial performance in altered environments can lead to shifts in community composition either via changes in the relative abundance of taxa already present or dispersal from the regional species pool. As the environment shifts, some microbial taxa will benefit more than others, resulting in changes in dominance and function (e.g., Pett-Ridge and Firestone, 2005). Dispersal will also provide new immigrant taxa, and species sorting should result in the presence of organisms best suited to the local environment (Leibold et al., 2004). Species sorting has been observed in bacterial communities (Hovatter et al., 2011; Van der Gucht et al., 2007), but not in protists (Finlay, 2002). The degree of sorting vs. mass effects can depend on dispersal, which is often assumed to be unlimited in microbes; however, recent studies support microbial dispersal limitation (Kivlin et al., 2011; Martiny et al., 2011; Öpik et al., 2009) and even a high degree of endemism (Cho and Tiedje, 2000; Talbot et al., 2014). If there are local differences in microbial species pools, empirical responses of microbes in experimental climate manipulations may be limited by available taxa. Many climate change experiments impose drought on small plots embedded in an ambient high rainfall region, which may lack a regional species pool containing drought-adapted individuals if dispersal is limited.

Positive feedbacks

When ecosystem states reinforce their own persistence, multiple stable states arise (Beisner et al., 2003). As an example, in arid rangelands, plants can reinforce their own water availability via shading and their influence on soils. The result is dramatic pattern formation that gives way abruptly to collapse under sever water limitation (Rietkerk and van de Koppel, 2008). Similar positive feedbacks could play a role in belowground microbial responses to climate change. Positive reinforcement of an existing microbial community may create resistance to climate change initially followed by a rapid shift or collapse as the degree of change grows. Invasion dynamics that could introduce change-adaptive varieties into local communities can be shutdown by positive feedbacks, sometimes referred to as Allee effects, because small founder populations cannot survive at low density (Keitt et al., 2001). Life history traits of microbes are generally not consistent with Allee effects; however an Allee effect might be observed in the case of microbial consortia requiring sufficient densities to enable cooperative functioning. Similarly, positive feedback may produce frequency dependent competitive asymmetries whereby larger established populations cannot be replaced by potentially more fit varieties invading at low densities. The net effect is a large potential for historical ecosystem and community structures to persist under change, increasing temporary resistance, but ultimately limiting resilience when an abrupt state shift occurs. Historical legacies in microbial function have been observed for drought (Evans and Wallenstein, 2012; Göransson et al., 2013; Meisner et al., 2013b), but do not always occur (Rousk et al., 2013). If positive feedbacks dominate in soil microbial communities, then historical legacies of past climate may play a strong role in how these communities change in the future and the resiliency of ecosystem functions.

Evolution

Periodic selective sweeps in microbial populations (Koch, 1974; Levin, 1981; Notley-McRobb and Ferenci, 2000) suggest the potential for rapid adaptive responses to climate change. In the lab, *Escherichia coli* adapted to altered temperature within

2000 generations (Bennett et al., 1992) and cultured soil bacteria appeared to be locally adapted to edaphic conditions at the scale of a few meters (Belotte et al., 2003). Identifying adaptation outside the lab, however, is constrained by our inability to measure fitness in complex, highly diverse communities. We are often limited to deducing adaptation through aggregate function, which results from the sum of traits in the microbial community. Comparisons of local and non-local litter decomposition provide evidence for functional specialization of microbial communities to the environment: microbial communities are generally most efficient at degrading litter from plant species growing immediately above them, termed the 'home-field advantage' (e.g., Ayres et al., 2009; Keiser et al., 2014; Strickland et al., 2009). The breadth of decomposition may also be constrained by resource history if some functional strategies are eliminated (Keiser et al., 2011). Although rapid adaptation represents a viable strategy for microbes facing climate change, existing specialization may prevent selective sweeps (Dykhuizen and Dean, 2004), which could limit responses to environmental change.

Current microbial and ecosystem models

Microbial contributions to both soil CO_2 fluxes and carbon pools in a changing climate are likely to be substantial and more explicit representation of their role could improve the accuracy of ecosystem carbon models. Nevertheless, the vast majority of ecosystem carbon cycling models do not include explicit microbial mechanisms (Chapin III et al., 2009; Ostle et al., 2009; Treseder et al., 2011) or, when microbes are included, all microbial taxa are treated as functional equivalents (Lawrence et al., 2009). This simplification arises from our historical assumption that high microbial diversity and apparently broad distributions equate to ecological redundancy (Allison and Martiny, 2008; Torsvik et al., 2002). It is clear, however, that there is high beta-diversity in both fungi (Kivlin et al., 2011; Öpik et al., 2009; Öpik et al., 2006; Öpik et al., 2010) and bacteria (Fierer et al., 2009; Lauber et al., 2009). Such differences in microbial community composition can directly influence ecosystem process rates (e.g., Fukami et al., 2010; Gulledge et al., 1997; Hawkes et al., 2011; Strickland et al., 2009).

Even simplified representations of microbial community functional groups can improve models. In an enzyme-drive biogeochemical model, for example, moving from one microbial pool (Schimel and Weintraub, 2003) to two pools of fungi and bacteria (Waring et al., 2013) significantly improves our ability to capture real patterns of carbon and nitrogen cycling by including differences in physiology between these groups. Other examples of microbial functional groups include active vs. dormant states (Blagodatsky and Richter, 1998; Hunt, 1977), generalists vs. specialists (Moorhead and Sinsabaugh, 2006), decomposers of fresh litter vs. soil organic matter (Fontaine and Barot, 2005), decomposers vs. builders of soil organic matter (Perveen et al., 2014), and ectomycorrhizal vs. saprotrophic fungi (Orwin et al., 2011).

Current state-of-the-art soil carbon models have adopted explicit representation of extracellular enzyme activities as a key factor influencing decomposition (Allison et al., 2010; Wang et al., 2013). The major innovation of these efforts is dynamically modeling enzyme pools and their influence on microbial carbon uptake through regulation of the rate of soil carbon conversion to dissolved organic carbon. For example, Wieder et al. (2013) recently modified the Community Land Model (Lawrence et al., 2011; Oleson et al., 2010) by adding microbial biomass pools and decomposition via enzyme-driven, temperature-dependent Michaelis-Menten kinetics, including aboveground, surface, and subsurface soil horizons. In doing so, they were able to explain 50% of the variation in global soil carbon pools, approximately a 20% improvement compared to traditional carbon models (Wieder et al., 2013). Other existing models differ primarily in the degree of detail and realism in representing different carbon states (e.g., Allison, 2012; Allison et al., 2010; Wang et al., 2013).

While these models explicitly include effects of temperature on process rates, they typically do not consider effects of soil moisture. The lack of soil moisture effects limits model applicability to forecasting, as it is known that soil moisture is a significant factor influencing both microbial and soil carbon dynamics (e.g., Bontti et al., 2009; Carbone et al., 2011; Curiel-Yuste et al., 2007; Inglima et al., 2009; Lellei-Kovács et al., 2011). An exception is the Dual Arrhenius Michaelis-Menten (DAMM) model of Davidson et al. (2012). The DAMM model includes both temperature and soil moisture effects to predict soil respiration. The current version of DAMM is not a dynamical state model, but it does define key rate functions that can be included in a dynamic process model.

There are other direct and indirect effects on soil microbes that might affect microbial responses to climate change and thus potentially improve model predictions. Chapin et al. (2009) point out that partitioning of carbon into roots, mycorrhizas, and exudates, as well as their effects on respiration and soil organic carbon, remain unknown for any ecosystem, despite these carbon pools having very different residence times and potential climate change responses. Mycorrhizas may increase soil organic carbon by hyphal aggregation of soil particles (Rillig and Mummey, 2006; Wilson et al., 2009) or reduce carbon pools via hyphal respiration under some climate change scenarios (Hawkes et al., 2008). Priming effects may occur in response to an increase in labile carbon exudates resulting in depletion of soil organic carbon pools, although the magnitude is soil-specific (Blagodatskaya and Kuzyakov, 2008; Paterson and Sim, 2013). Finally, trophic interactions are likely to be important, despite similar responses of soil fauna to different climate change factors in experimental settings (Blankinship et al., 2011). Although we do not address these further here, these effects likely warrant further consideration.

Integrating moisture into microbial carbon cycling models

As a first step towards integrating soil moisture into decomposition models, we have fused the DAMM model of Davidson et al. (2012) with two enzyme-based, temperature-dependent soil carbon cycling models to produce models where rates depend on both soil moisture and soil temperature (Fig. 9.1). Specifically, we focus on the models developed by Allison et al. (2010) and Wang et al. (2013), hereafter referred to as the AWB and MEND models, respectively. While we believe that soil moisture could have complex effects throughout the models, we integrate the models initially by substituting available dissolved organic carbon, as computed in the DAMM model, for the particulate dissolved organic carbon pool used in the AWB and MEND models. Additionally, we incorporate oxygen limitation by adding an additional oxygen-driven Michaelis-Menten term to the microbial biomass growth equation. The MEND model differs slightly in how temperature enters into the model, but is otherwise similar to the AWB model, except for splitting particulate organic carbon pool into three different pools and the inclusion of adsorption-desorption dynamics.

The integration of the models generates an initial set of four model-hypotheses to be compared and confronted with data: AWB (unmodified), MEND (unmodified), AWB-DAMM, and MEND-DAMM. Our initial merger of these models does not explicitly include oxygen uptake dynamics. We expect, however, that we can ignore the oxygen limitation term from the DAMM model as it will have a relatively small impact in arid ecosystems that are the focus of this initial effort. However, we

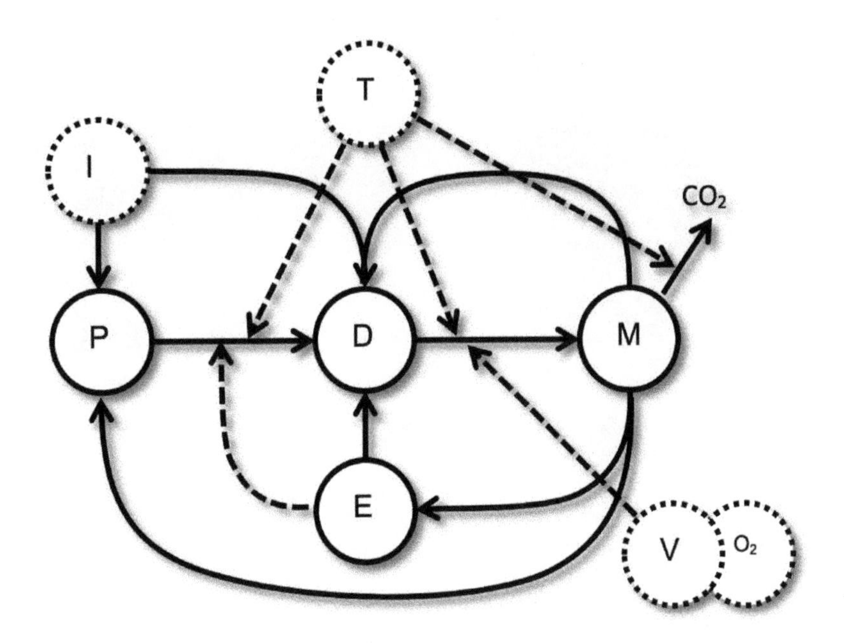

Figure 9.1 Carbon pools and fluxes in the AWB-DAMM model. Symbols are I = carbon input, P = particulate carbon, D = dissolved carbon, M = microbial biomass, E = enzymes, V = volumetric soil moisture, O_2 = oxygen and T = temperature. Dashed lines indicate effects on rates. Dashed circles represent boundary conditions. The MEND-DAMM model is analogous but with the P-pool divided into three separate pools and the E-pool divided into two separate pools.

also acknowledge that oxygen limitation may be large during extreme wetting events. In ecosystems where empirical validation suggests the inclusion of explicit oxygen dynamics, this should be introduced as a state variable.

Enzyme-driven models with sensitivity to temperature and soil moisture

Carbon exists in many forms in soils (Robertson et al., 1999). Newer models of soil carbon dynamics more finely divide pools, explicitly model biological feedbacks, and use more realistic uptake kinetics. A key factor regulating rates of carbon flux in soils is enzymatic catalysis involved in the breakdown of particulate organic carbon into biologically-available dissolved organic carbon (Schimel and Weintraub, 2003). Soil microbes produce extracellular enzymes that react with particulate carbon to produce soluble carbon, which can then be consumed. The extracellular enzymes are

Table 9.1 Symbols used in the AWB-DAMM model

Variable	Description	Units	Default value
P	Particulate organic carbon	mg cm^{-3}	111.876
D	Dissolved organic carbon	mg cm^{-3}	0.00144928
M	Microbial biomass	mg cm^{-3}	2.19159
E	Extracellular enzymes	mg cm^{-3}	0.0109579
I_p	Input rate of particulate carbon	mg cm^{-3} h^{-1}	0.0005
I_D	Input rate of dissolved carbon	mg cm^{-3} h^{-1}	0.0005
α	Soluble fraction of dead microbial matter	dimensionless	0.5
δ	Microbe death rate	h^{-1}	0.0002
v_1	Maximum rate of conversion from particulate to dissolved carbon	h^{-1}	Function of temperature
v_2	Maximum dissolved carbon uptake by microbes	h^{-1}	Function of temperature
$\acute{v}_1$	Conversion rate scaling constant	h^{-1}	10^8
$\acute{v}_2$	Uptake rate scaling constant	h^{-1}	10^8
K_1	Half-saturation of particulate carbon conversion	mg cm^{-3}	Function of temperature
K_1^I, K_1^S	Least-squares K_1 intercept and slope	mg cm^{-3}, mg cm^{-3} T^{-1}	500, 5
K_2	Half-saturation of dissolved carbon uptake by microbes	mg cm^{-3}	Function of temperature
K_2^I, K_2^S	Least-squares K_1 intercept and slope	mg cm^{-3}, mg cm^{-3} T^{-1}	0.1, 0.01
ε	Microbial carbon use efficiency	dimensionless	Function of temperature
$\varepsilon^I, \varepsilon^S$	Least-squares CUE intercept and slope	dimensionless, T^{-1}	0.63, -0.016
r_1	Rate of enzyme loss	h^{-1}	0.001
r_2	Enzyme production rate	h^{-1}	0.000005
η	Dissolved organic carbon diffusion factor	dimensionless	72.3
τ	Scaling factor for O_2 limitation	dimensionless	2.0
V	Volumetric soil moisture	dimensionless	0.24
O_2	Oxygen concentration	L O_2 L^{-1} air	Function of soil moisture
K_3	Oxygen half-saturation	L O_2 L^{-1} air	0.121

themselves a pool of dissolved organic carbon and contribute to the overall soil carbon budget.

Allison et al. (2010) developed an extracellular-enzyme-driven model of soil carbon (AWB). The model is not a full ecosystem model but captures in a simplified way the essential details of the enzyme-driven particulate-to-dissolved carbon pathway. The model includes four carbon pools: particulate organic carbon (*P*), dissolved organic carbon (*D*), microbial biomass (*M*), and extra-cellular enzymes (*E*). The AWB model is given by the following system of equations (Equations 9.1 to 9.4; symbols defined in Table 9.1):

$$\frac{dP}{dt} = I_p + (1-\alpha)\delta M - v_1\left(\frac{P}{K_1+P}\right)E \quad (9.1)$$

$$\frac{dD}{dt} = I_D + \left(\alpha\delta - v_2\frac{D}{K_2+D}\right)M + \left(v_1\frac{P}{K_1+P} + r_1\right)E \quad (9.2)$$

$$\frac{dM}{dt} = \left(\varepsilon v_2\frac{D}{K_2+D} - \delta - r_2\right)M \quad (9.3)$$

$$\frac{dE}{dt} = r_2 M - r_1 E. \quad (9.4)$$

The AWB model is sensitive to temperature, which affects both the maximum rates v_1 and v_2, the half-saturation coefficients K_1 and K_2, and carbon-use efficiency ε. The maximum rates are governed by the Arrhenius function: $v_{1,2} = \acute{v}_{1,2} e^{-5653/(T+273)}$ where T is temperature in Celsius. The numerator in the exponent is derived from the activation energy and ideal gas constant. This function captures the well-known increase in reaction rates with increasing temperature. Temperature effects on the half-saturation coefficients and carbon-use efficiency are linear regression functions constructed from empirical data: $K_1 = K_1^I + K_1^S T$, $K_2 = K_2^I + K_2^S T$ and $\varepsilon = \varepsilon^I - \varepsilon^S T$.

While the AWB model is responsive to temperature, it does not integrate soil moisture effects. Changes in soil moisture have multiple potential influences on soil carbon dynamics. First, soil moisture content is the carrier for dissolved organic carbon diffusing though the soil pore space. Hence, the rate of microbial uptake of dissolved organic carbon is dependent on available soil moisture. A second effect is the decreasing rate of gas exchange as soil pore space becomes saturated with water. Reduced gas exchange can limit aerobic respiration of acquired carbon reducing not only microbial biomass, but also production of extracellular enzymes, an energetically costly activity. The net effect is to shutdown the entire carbon feedback loop at either end of the moisture spectrum: lack of liquid transport when dry and lack of oxygen when saturated.

The effect of soil moisture on dissolved organic carbon available for microbial uptake and on gas exchange is captured in the DAMM model (Davidson et al., 2012). Although not a fully dynamical model, the DAMM model framework identifies functions to model available carbon and incorporates soil oxygen content into microbial carbon kinetics. Integration of these functions into a combined AWB-DAMM model results in the following updated equations (Equations 9.5 and 9.6) for dissolved organic carbon and microbial biomass:

$$\frac{dD}{dt} = I_D + \left(\alpha\delta - v_2 \frac{\eta D V^3}{K_2 + \eta D V^3} \frac{\tau O_2}{K_3 + O_2}\right) M + \left(v_1 \frac{P}{K_1 + P} + r_1\right) E \quad (9.5)$$

$$\frac{dM}{dt} = \left(\varepsilon v_2 \frac{\eta D V^3}{K_2 + \eta D V^3} \frac{\tau O_2}{K_3 + O_2} - \delta - r_2\right) M. \quad (9.6)$$

The term ηDV^3 models the diffusion of dissolved organic carbon to the cell surface where it is available for uptake. The parameter η is related to the diffusivity of dissolved organic carbon. In our merged model, η is scaled such that the dissolved organic carbon available for uptake is the same as in the unmodified AWB model when volumetric soil moisture is 24%, the time average of soil moistures from our study sites. The cubic dependence of available carbon on soil moisture generates a strong limitation to microbial growth at lower soil moisture content, a mechanism absent from the AWB model. Oxygen concentration is modeled purely as a function of diffusion into the soil pore space; microbial consumption is neglected. We retain the default values of soil bulk density, particle density, gas diffusion and O_2 fraction in air given in Davidson et al. (2012). With these values, the oxygen concentration is calculated as $O_2 = 0.35(0.68 - V)^{4/3}$. Similarly to the scaled factor η, the parameter τ scales the influence of O_2 such that there is no net effect on dissolved carbon concentration D when volumetric soil moisture is 24%. In the absence of this scaling, it would have been necessary to re-parameterize the AWB model. With these modifications, we have a model sensitive to both soil temperature and moisture content.

We also explore adding soil moisture sensitivity to the more elaborate enzyme-driven microbial MEND (Microbial-Enzyme-Mediated Decomposition) model proposed by Wang et al. (2013). The MEND model (Wang et al., 2013) divides the three soil carbon pools (enzymes, particulate and dissolved) of the AWB model into six separate components: (1) particulate and (2) dissolved carbon pools as in AWB, (3) an adsorbed-phase dissolved carbon pool, (4) a mineral-associated carbon pool, and two enzyme pools, (5) one acting on particulate carbon and the other (6) acting on mineral associated carbon. The MEND model gives an unprecedented level of detail in below ground carbon dynamics relative to older models that divide belowground pools into labile, recalcitrant and immobile pools without consideration of enzyme dynamics. Wang et al. (2013) demonstrate strong differences in carbon kinetics compared to these traditional models. We refer the reader to Wang et al. (2013) for additional details of the model formulation and parameterization.

We utilized the same approach applied to the AWB-DAMM model to merge the MEND model with the DAMM model. As in the AWB model (Equation 9.3), the MEND model uses Michaelis-Menten kinetics to model the rate of microbial uptake as a function of dissolved organic carbon concentration. Our modification of the MEND model (Equation 9.7) uses the dual Michaelis-Menten formulation of the DAMM model by making the substitution

$$\frac{D}{K_D+D} \Longrightarrow \frac{\eta D V^3}{K_D+\eta D V^3} \frac{\tau O_2}{K_3+O_2} \quad (9.7)$$

where again, η and τ are scaled parameters that assure a neutral influence on dissolved carbon concentration at the mean soil moisture. The half-saturation coefficient K_D is specific to MEND (see Wang et al., 2013).

Model simulations with and without soil moisture sensitivity in wet and dry sites

The models described above primarily allow us to consider physiological response mechanisms, as well as the potential for alternative stable states (positive feedbacks). We contrast outputs of the model combinations (AWB, AWB-DAMM, MEND, and MEND-DAMM) driven by soil moisture and soil temperature time series for two locations on the Edwards Plateau, TX. A steep precipitation gradient is found across the Edwards Plateau, with mean annual rainfall ranging from 90 cm in the east to 40 cm in the west, decreasing by ~10 cm every 40-50 km westward. Across this gradient, the plant communities are savanna grasslands (McMahan et al., 1984) and soils are limestone-derived clay Mollisols (Werchan et al., 1974). In order to analyze the potential differences in model outcomes based on soil moisture, we have chosen one drier and one wetter site from the west and east ends of the gradient as the basis for the simulation. Mean annual precipitation at these sites is approximately 500mm and 1000mm, respectively.

Our soil moisture and temperature time series are extracted from the National Centers for Environmental Prediction's (NCEP) reanalysis dataset (Kalnay et al., 1996). Reanalysis provides a uniform procedure for fusing all available climate data into a consistent historical record. A climate model is used to integrate data inputs and to produce gridded outputs. A stepwise update procedure is utilized to force the climate model simulation to conform to the historical climate data. As the model computes soil climate, it provides a way to obtain soil moisture and temperature time series for specific locations. Although not the same as direct measurements, the modeled reanalysis outputs nonetheless represent realistic scenarios constrained to conform to historical climate patterns.

The soil climate time series represented three years of moisture and temperature data sampled hourly beginning January 1, 2008. Moisture and temperature values are for the first 10 cm of soil below the surface. Both sites were centered at 30.44 degrees north latitude. The dry site was centered on 85.62 degrees west longitude. The moist site was centered on 77.49 degrees west longitude. To make the climate time series continuous for integration, we used cubic spline interpolation. We used the deSolve package (Soetaert et al., 2010) in R (R Development Core Team, 2011) to numerically integrate the models. Initial conditions were those given in Allison et al. (2010) and Wang et al. (2013). For each model we ran ten simulations over the climate time series, each time initializing the model with the time-average of the state variables from the previous run. This allowed the models to converge to their long-term behavior. Output of the 10th run is shown in Fig. 9.2 and summarized in Fig. 9.3.

As expected, the moist site showed greater soil moisture (Fig. 9.2A-B, 9.3A-B). Temperatures were similar between the two sites with the dry site showing greater variability. The covariance of moisture and temperature is entirely different between the two sites with the dry site showing warm-season increases in soil moisture, likely driven by summer rains, whereas the moist site exhibits the greatest soil moisture during the winter months. In addition to changes in mean values, the dry site shows greater variation, both daily and seasonally. Crucially, the dry site commonly reaches near-zero soil moisture potentially imposing a large reduction in process rates during these extreme drying events.

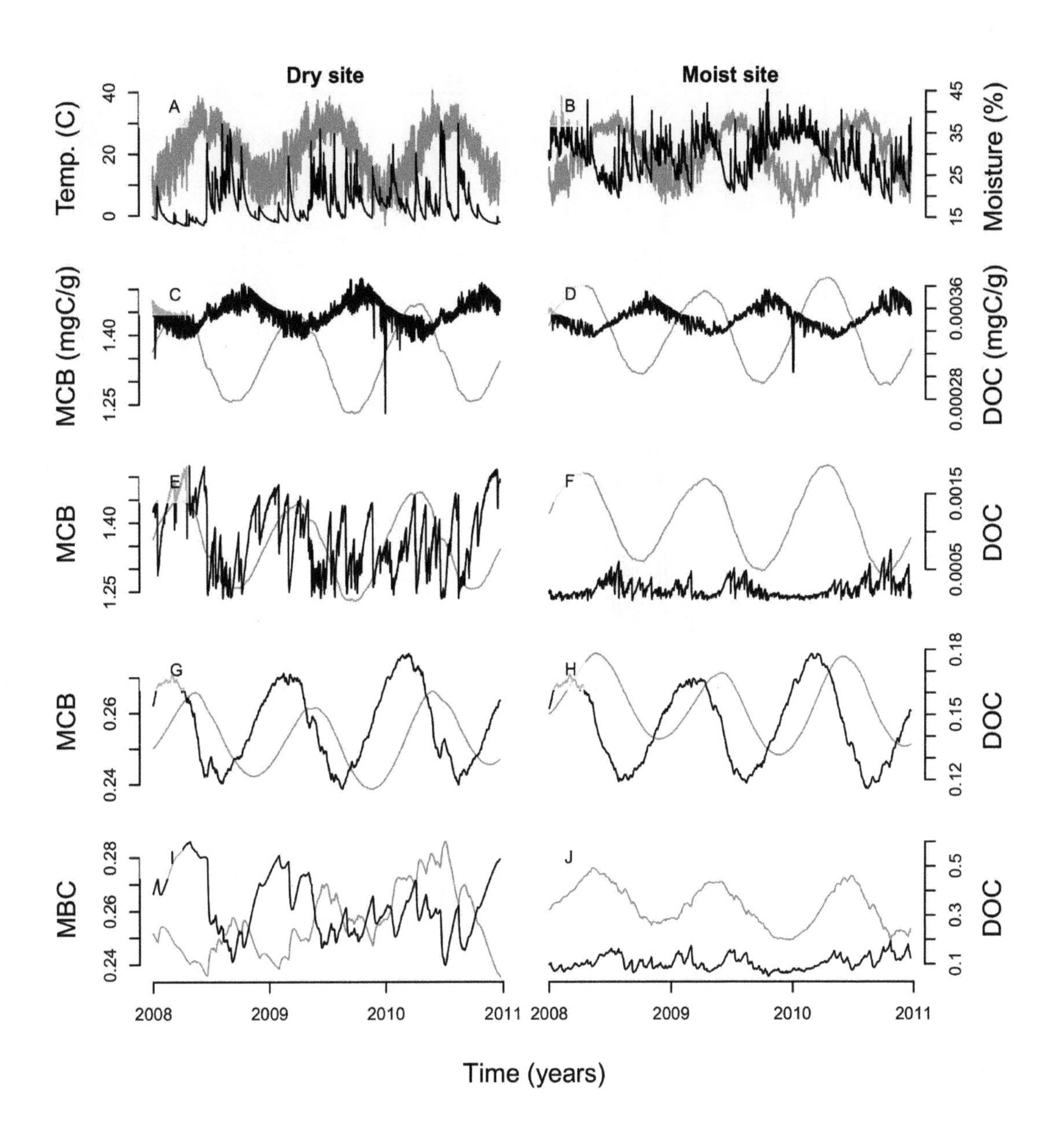

Figure 9.2 Comparison of models driven by reanalysis soil temperatures and moistures for a xeric site and mesic site. A-B: Soil moistures and temperatures for the wet and dry sites. Modeled microbial biomass (MIC) and dissolved organic carbon (DOC) for C-D: AWB model, E-F: the AWB-DAMM model, G-H: the MEND model, and I-J: the MEND-DAMM model. The time series start on January 1 and span a 3-year period.

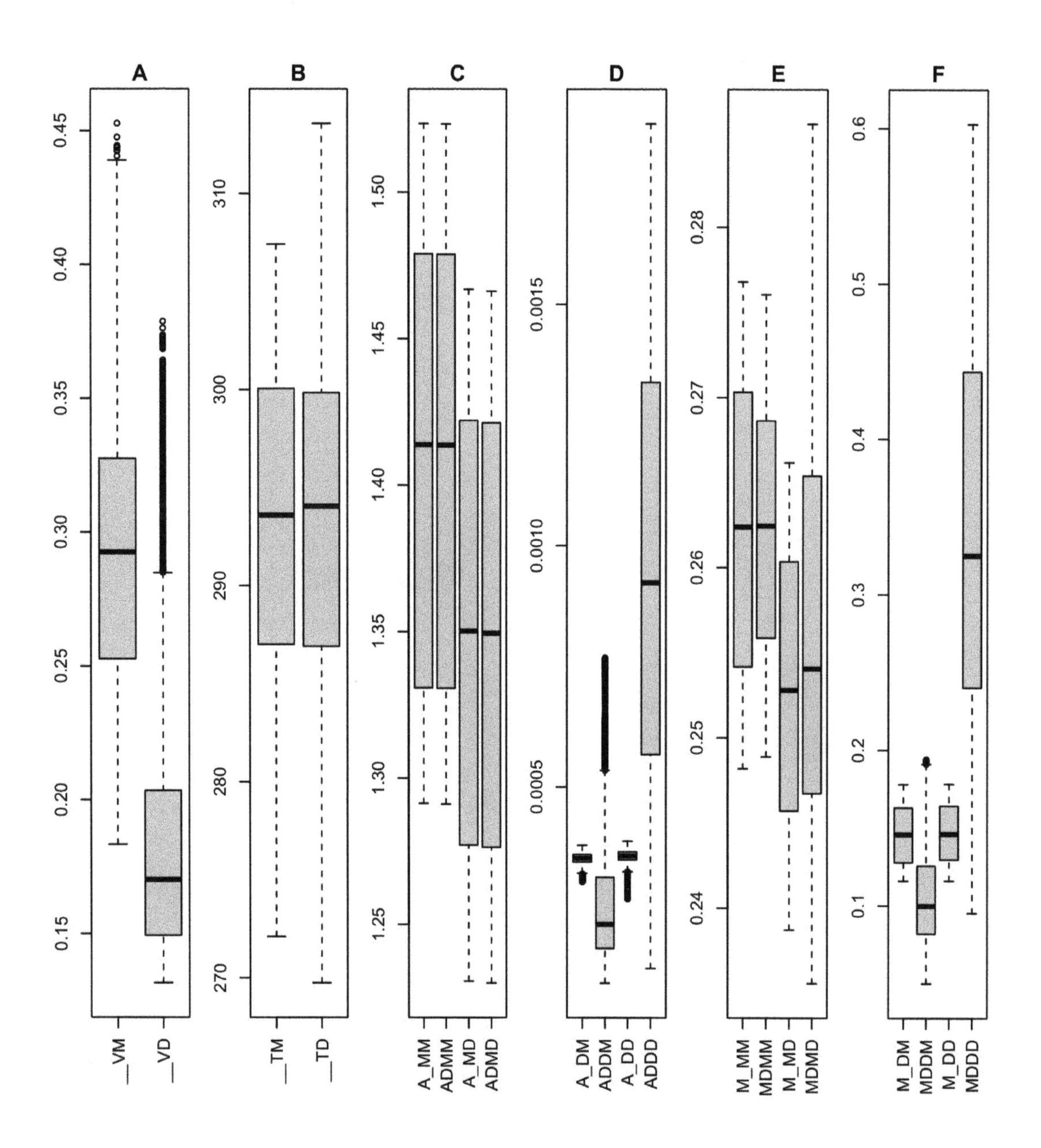

Figure 9.3 Box plot of model outputs. A: volumetric soil moisture. B: soil temperature. C: microbial biomass for AWB and AWB-DAMM. D: dissolved organic carbon for AWB and AWB-DAMM. E: microbial biomass for MEND and MEND-DAMM. F: dissolved organic carbon for MEND and MEND-DAMM. The four letter codes on the x-axis correspond to model (2 characters: A_ = AWB, AD = AWB-DAMM, M_ = MEND, MD = MEND-DAMM, __ = climate), variable (1 character: V = soil moisture, T = temperature, M = microbial carbon, D = dissolved carbon) and site (1 character: M = moist, D = dry).

All of the models (Fig. 9.2C-J) showed seasonal variation in microbial biomass driven by temperature effects. Microbial biomass peaked in late spring or early summer. Similarly, dissolved organic carbon fluctuated seasonally in the AWB and MEND models (Fig. 9.2C, D, G, H). Fluctuations in dissolved organic carbon were fully (AWB) or partially (MEND) out-of-phase with peaks in microbial biomass reflecting the uptake of dissolved organic carbon by microbes. Interestingly, dissolved organic carbon in the MEND model peaks before microbial biomass. This effect is likely the result of accelerating soil enzyme activity during the spring months and eventual drawdown of dissolved organic carbon by microbial uptake later in the season.

The most striking result is the dynamics of dissolved organic carbon in the AWB-DAMM and MEND-DAMM models (Fig. 9.2E, F, I, J; Fig. 9.3D, F). When soil moisture sensitivity is built into the model, we see strong fluctuations in dissolved carbon, largely shadowing variation in soil moisture. Interestingly, we observe a large increase in the average concentration of dissolved organic carbon in the dry site relative to the moist site for the moisture-sensitive models (Fig. 9.2E, I; Fig. 9.3D, F). There is also a marked increase in the variance of dissolved organic carbon through time, although these fluctuations have relatively minor influence on microbial biomass, which turns over more slowly and is therefore more buffered. The exception is the MEND-DAMM model in the dry site (Fig. 9.2I) where the dynamics of both microbial biomass and dissolved organic carbon appear to be strongly driven by soil moisture. Although some seasonal temperature-driven trends are apparent, finer-scale fluctuations corresponding to rapid changes in soil moisture appear to dominate. Unlike the other simulations, the MEND-DAMM output for the dry site shows strong negative covariance: increases in dissolved organic carbon are mirrored by decreases in microbial biomass indicated by a strong and rapid influence of soil moisture on microbial carbon uptake.

We also compared total respiration between the models with and without the DAMM modifications. Interestingly, respiration rates primarily tracked temperature and did not differ strongly between the models. Again the MEND-DAMM model run on the dry site showed the strongest differences in respiration rates compared to the model without the DAMM modifications. However, the differences were small and overwhelmed by the seasonal temperature-tracking dynamics.

Simulation results and microbial mechanisms

Based on our merged models, soil moisture has a strong influence on predictions for microbial enzyme-mediated ecosystem processes, including large differences in carbon pools and pool size variability between wetter and drier regions. This was particularly evident in the dynamics of dissolved organic carbon. Moisture is known to be a key factor in arid land biogeochemical cycling with pulsed rainfall patterns (Collins et al., 2008) and adding moisture as a rewetting factor to biogeochemical models for arid ecosystems can significantly improve model performance (Li et al., 2010). However, soil moisture is likely to constrain microbial processes more broadly as drought increases with climate change and, as noted by Moyano et al. (2013), current models are limited approximations and substantial work is needed to improve their predictive capacity.

Of the four proposed response mechanisms of physiology, community composition, feedbacks, and evolution, these results are primarily related to physiology. Through integration of the DAMM model with the AWB and MEND models, we have shown how dynamic moisture and temperature drivers can influence physiological process rates in a transient manner. However, the physiological plasticity built into these models may not be realistic in situations where environmental history constrains responses. For example, a history of drought can alter microbial community structure (Evans and Wallenstein, 2012), change microbial carbon use efficiency (Göransson et al., 2013), decouple microbial growth from respiration (Meisner et al., 2013a), and create feedbacks to plant communities (Meisner et al., 2013b).

The dynamic nature of soils indicated in the driving climate time series of the models presented here also has bearing on the other mechanisms. A variable environment is a mechanism that can allow for coexistence of species in communities (community composition mechanism) that would otherwise exclude each other through resource competition (Chesson, 2000). This variation is at least in part an attribute of the physiological rates incorporated into our models. Moisture and other climate variation may therefore maintain more diverse communities, and as a result more resilient communities with multiple players than can compensate for losses of past dominants when conditions change.

We did not find strong evidence of multiple stable states (positive feedback mechanism) for the ranges of climate we investigated. Nevertheless, sudden changes in dynamics could occur at extremes of soil temperature and moisture. We notice, for example, a sharp drop in dissolved organic carbon in the second cold season (around day 650) corresponding to an extreme low temperature event that influenced both the wet and dry sites. Whether this corresponds to a shift between stable states or a more simple transient effect cannot be ascertained without additional analysis. However, this observation suggests the models are capable of rapid shifts in response to extremes in climate.

Whether variability in soil climate impacts evolution is not directly addressed in the modified enzyme-driven models examined here. Similar to the community composition mechanism, a variable environment could maintain genetic polymorphisms with populations, and this genotypic variation could maintain functional phenotypic variation lending greater resilience under climate change. Evolutionary tradeoffs must certainly abound in soil microbial systems (Gudelj et al., 2010). A key tradeoff is the fast-slow spectrum characterized by maximizing rate of reproduction at the expense of survivorship versus long lifespan with lower fecundity. The observed strong variation in dissolved organic carbon – the energy resource fueling microbial population growth – under a variable soil moisture regime has bearing on the fast-slow tradeoff. Rapid changes in resources should favor the fast end of the tradeoff spectrum. However a live-fast life history provides little population buffer to prolonged or directional changes in environment and thus limit resilience to certain types of disturbance. Conversely the shorter generation times on the fast end of the spectrum could enable more rapid adaptive evolution and rescue populations impacted by major environmental changes (Gonzalez and Bell, 2013).

Conclusions and future needs

Here we demonstrate that adding moisture sensitivity to enzyme-driven models of carbon cycling changes their outcomes when parameterized for a wet and a dry ecosystem located on the same soil type. However, these outcomes have not been validated against field data and the particular moisture function that we incorporated may not produce the best fit. As ecosystem models become more and more sophisticated, we expect discovery of the tipping point at which additional realism ceases to improve the accuracy of model predictions. For example, Talbot et al. (2014) recently found high endemism in the fungal communities of North American pine forests, but this did not translate into local variation in enzymes, suggesting a high degree of functional redundancy. Distinguishing among physiological, compositional, and other microbial mechanisms will help us to generalize expectations across ecosystems. Linking of modeling activities to data collection in the field and experimental plots will promote greater understanding of critical processes and build that understanding into a new generation of advanced soil carbon models.

References

Allison, S.D. (2012). A trait-based approach for modelling microbial litter decomposition. Ecol. Lett. *15*, 1058-1070.

Allison, S.D., and Martiny, J.B.H. (2008). Resistance, resilience, and redundancy in microbial communities. Proc. Natl. Acad. Sci. USA *105*, 11512-11519.

Allison, S.D., Wallenstein, M.D., and Bradford, M.A. (2010). Soil-carbon response to warming dependent on microbial physiology. Nature Geosci. *3*, 336-340.

Averill, C., Turner, B.L., and Finzi, A.C. (2014). Mycorrhiza-mediated competition between plants and decomposers drives soil carbon storage. Nature *505*, 543-545.

Ayres, E., Steltzer, H., Simmons, B.L., Simpson, R.T., Steinweg, J.M., Wallenstein, M.D., Mellor, N., Parton, W.J., Moore, J.C., and Wall, D.H. (2009). Home-field advantage accelerates leaf litter decomposition in forests. Soil Biol. Biochem. *41*, 606-610.

Beisner, B.E., Haydon, D.T., and Cuddington, K. (2003). Alternative stable states in ecology. Front. Ecol. Environ. *1*, 376-382.

Belotte, D., Curien, J.B., Maclean, R.C., and Bell, G. (2003). An experimental test of local adaptation in soil bacteria. Evolution *57*, 27-36.

Bennett, A.F., Lenski, R.E., and Mittler, J.E. (1992). Evolutionary adaptation to temperature. I. Fitness responses of *Escherichia coli* to changes in its thermal environment. Evolution *46*, 16-30.

Blagodatskaya, E., and Kuzyakov, Y. (2008). Mechanisms of real and apparent priming effects and their dependence on soil microbial biomass and community structure: critical review. Biol. Fertility Soils *45*, 115-131.

Blagodatsky, S.A., and Richter, O. (1998). Microbial growth in soil and nitrogen turnover: a theoretical model considering the activity state of microorganisms. Soil Biol. Biochem. *30*, 1743-1755.

Blankinship, J.C., Niklaus, P.A., and Hungate, B.A. (2011). A meta-analysis of responses of soil biota to global change. Oecologia *165*, 553-565.

Bontti, E.E., Joseph P, D., Seth M, M., Mark A, G., Agnieszka, P., Michelle L, H., Stephanie, O., Ingrid C, B., William J, P., and Mark E, H. (2009). Litter decomposition in grasslands of Central North America (US Great Plains). Global Change Biol. *15*, 1356-1363.

Bradford, M.A., Watts, B.W., and Davies, C.A. (2010). Thermal adaptation of heterotrophic soil respiration in laboratory microcosms. Global Change Biol. *16*, 1576-1588.

Carbone, M., Still, C., Ambrose, A., Dawson, T., Williams, A., Boot, C., Schaeffer, S., and Schimel, J. (2011). Seasonal and episodic moisture controls on plant and microbial contributions to soil respiration. Oecologia *167*, 265-278.

Castro, H.F., Classen, A.T., Austin, E.E., Norby, R.J., and Schadt, C.W. (2010). Soil microbial community responses to multiple experimental climate change drivers. Appl. Environ. Microbiol. *76*, 999-1007.

Chapin III, F.S., Matson, P.A., and Mooney, H.A. (2002). Principles of Terrestrial Ecosystem Ecology (New York, Springer).

Chapin III, F.S., McFarland, J., McGuire, A.D., Euskirchen, E.S., Ruess, R.W., and Kielland, K. (2009). The changing global carbon cycle: linking plant–soil carbon dynamics to global consequences. J Ecol *97*, 840-850.

Cheng, L., Booker, F.L., Tu, C., Burkey, K.O., Zhou, L., Shew, H.D., Rufty, T.W., and Hu, S. (2012). Arbuscular mycorrhizal fungi increase organic carbon decomposition under elevated CO2. Science *337*, 1084-1087.

Chesson, P. (2000). Mechanisms of maintenance of species diversity. Annu. Rev. Ecol. Syst. *31*, 343-366.

Cho, J.-C., and Tiedje, J.M. (2000). Biogeography and Degree of Endemicity of Fluorescent Pseudomonas Strains in Soil. Appl. Environ. Microbiol. *66*, 5448-5456.

Collins, S.L., Sinsabaugh, R.L., Crenshaw, C., Green, L., Porras-Alfaro, A., Stursova, M., and Zeglin, L.H. (2008). Pulse dynamics and microbial processes in aridland ecosystems. J. Ecol. *96*, 413-420.

Curiel-Yuste, J., Baldocchi, D.D., Gershenson, A., Goldstein, A., Misson, L., and Wong, S. (2007). Microbial soil respiration and its dependency on carbon inputs, soil temperature and moisture. Global Change Biol. *13*, 2018-2035.

Davidson, E.A., Samanta, S., Caramori, S.S., and Savage, K. (2012). The Dual Arrhenius and Michaelis–Menten kinetics model for decomposition of soil organic matter at hourly to seasonal time scales. Global Change Biol. *18*, 371-384.

Dykhuizen, D.E., and Dean, A.M. (2004). Evolution of specialists in an experimental microcosm. Genetics *167*, 2015-2026.

Ehrlen, J., and Eriksson, O. (2000). Dispersal Limitation and Patch Occupancy in Forest Herbs. Ecology *81*, 1667-1674.

Evans, S.E., and Wallenstein, M.D. (2012). Soil microbial community response to drying and rewetting stress: does historical precipitation regime matter? Biogeochemistry *109*, 101-116.

Fierer, N., Strickland, M.S., Liptzin, D., Bradford, M.A., and Cleveland, C.C. (2009). Global patterns in belowground communities. Ecol. Lett. *12*, 1238-1249.

Finlay, B.J. (2002). Global Dispersal of Free-Living Microbial Eukaryote Species. Science *296*, 1061-1063.

Fontaine, S., and Barot, S. (2005). Size and functional diversity of microbe populations control plant persistence and long-term soil carbon accumulation. Ecol. Lett. *8*, 1075-1087.

Fukami, T., Dickie, I.A., Paula Wilkie, J., Paulus, B.C., Park, D., Roberts, A., Buchanan, P.K., and Allen, R.B. (2010). Assembly history dictates ecosystem functioning: evidence from wood decomposer communities. Ecol. Lett. *13*, 675-684.

Godbold, D., Hoosbeek, M., Lukac, M., Cotrufo, M., Janssens, I., Ceulemans, R., Polle, A., Velthorst, E., Scarascia-Mugnozza, G., De Angelis, P., *et al.* (2006). Mycorrhizal Hyphal Turnover as a Dominant Process for Carbon Input into Soil Organic Matter. Plant Soil *281*, 15-24.

Gonzalez, A., and Bell, G. (2013). Evolutionary rescue and adaptation to abrupt environmental change depends upon the history of stress. Phil. Trans. R. Soc. B *368*.

Göransson, H., Godbold, D.L., Jones, D.L., and Rousk, J. (2013). Bacterial growth and respiration responses upon rewetting dry forest soils: Impact of drought-legacy. Soil Biol. Biochem. *57*, 477-486.

Griffiths, R.I., Whiteley, A.S., O'Donnell, A.G., and Bailey, M.J. (2003). Physiological and Community Responses of Established Grassland Bacterial Populations to Water Stress. Appl. Environ. Microbiol. *69*, 6961-6968.

Gudelj, I., Weitz, J.S., Ferenci, T., Claire Horner-Devine, M., Marx, C.J., Meyer, J.R., and Forde, S.E. (2010). An integrative approach to understanding microbial diversity: from intracellular mechanisms to community structure. Ecol. Lett. *13*, 1073-1084.

Gulledge, J., Doyle, A.P., and Schimel, J.P. (1997). Different NH4+-inhibition patterns of soil CH4 consumption: A result of distinct CH4-oxidizer populations across sites? Soil Biol. Biochem. *29*, 13-21.

Harper, C.W., John, M.B., Philip, A.F., Alan, K.K., and Jonathan, D.C. (2005). Increased rainfall variability and reduced rainfall amount decreases soil CO_2 flux in a grassland ecosystem. Global Change Biol. *11*, 322-334.

Hawkes, C.V., Hartley, I.P., Ineson, P., and Fitter, A.H. (2008). Soil temperature affects carbon allocation within arbuscular mycorrhizal networks and carbon transport from plant to fungus. Global Change Biol. *14*, 1181-1190.

Hawkes, C.V., Kivlin, S.N., Rocca, J.D., Huguet, V., Thomsen, M.A., and Suttle, K.B. (2011). Fungal community responses to precipitation. Global Change Biol. *in press*.

Heinemeyer, A., Ineson, P., Ostle, N., and Fitter, A.H. (2006). Respiration of the external mycelium in the arbuscular mycorrhizal symbiosis shows strong dependence on recent photosynthates and acclimation to temperature. New Phytol. *171*, 159-170.

Horz, H.-P., Barbrook, A., Field, C.B., and Bohannan, B.J.M. (2004). Ammonia-oxidizing bacteria respond to multifactorial global change. Proc. Natl. Acad. Sci. USA *101*, 15136-15141.

Horz, H.-P., Rich, V., Avrahami, S., and Bohannan, B.J.M. (2005). Methane-oxidizing bacteria in a California upland grassland soil: diversity and response to simulated global change. Appl. Environ. Microbiol. *71*, 2642-2652.

Hovatter, S.R., Dejelo, C., Case, A.L., and Blackwood, C.B. (2011). Metacommunity organization of soil microorganisms depends on habitat defined by presence of *Lobelia siphilitica* plants. Ecology *92*, 57-65.

Hunt, H.W. (1977). A Simulation Model for Decomposition in Grasslands. Ecology *58*, 469-484.

Inglima, I., Alberti, G., Bertolini, T., Vaccari, F.P., Gioli, B., Miglietta, F., Cotrufo, M.F., and Peressotti, A. (2009). Precipitation pulses enhance respiration of Mediterranean ecosystems: the balance between organic and inorganic components of increased soil CO_2 efflux. Global Change Biol. *15*, 1289-1301.

IPCC (2007). Climate Change 2007: The Physical Basis—Summary for Policymakers.

Jenkinson, D.S., Adams, D.E., and Wild, A. (1991). Model estimates of CO2 emissions from soil in response to global warming. Nature *351*, 304-306.

Jentsch, A., Kreyling, J., and Beierkuhnlein, C. (2007). A new generation of climate-change experiments: events, not trends. Front. Ecol. Environ. *5*, 365-374.

Jiang, X., and Yang, Z.L. (2012). Projected changes of temperature and precipitation in Texas from downscaled global climate models. Clim. Res. *53*, 229-244.

Johnson, M.T.J., and Stinchcombe, J.R. (2007). An emerging synthesis between community ecology and evolutionary biology. Trends Ecol. Evol. *22*, 250-257.

Kalnay, E., Kanamitsu, M., Kistler, R., Collins, W., Deaven, D., Gandin, L., Iredell, M., Saha, S., White, G., Woollen, J., *et al.* (1996). The NCEP/NCAR 40-Year Reanalysis Project. B. Am. Meteorol. Soc. *77*, 437-471.

Keiser, A.D., Keiser, D.A., Strickland, M.S., and Bradford, M.A. (2014). Disentangling the mechanisms underlying functional differences among decomposer communities. J. Ecol. *102*, 603–609.

Keiser, A.D., Strickland, M.S., Fierer, N., and Bradford, M.A. (2011). The effect of resource history on the functioning of soil microbial communities is maintained across time. Biogeosci. Disc. *8*, 1643-1667.

Keitt, T.H., Lewis, M.A., and Holt, R.D. (2001). Allee effects, invasion pinning, and species' borders. The American Naturalist *157*, 203-216.

Kindler, R., Miltner, A., Richnow, H.-H., and Kästner, M. (2006). Fate of gram-negative bacterial biomass in soil--mineralization and contribution to SOM. Soil Biol. Biochem. *38*, 2860-2870.

Kivlin, S.N., Hawkes, C.V., and Treseder, K.K. (2011). Global diversity and distribution of arbuscular mycorrhizal fungi. Soil Biol. Biochem. *43*, 2294-2303.

Koch, A.L. (1974). The pertinence of the periodic selection phenonmenon to prokaryote evolution. Genetics *77*, 127-142.

Lauber, C.L., Hamady, M., Knight, R., and Fierer, N. (2009). Soil pH as a predictor of soil bacterial community structure at the continental scale: a pyrosequencing-based assessment. Appl. Environ. Microbiol. *75*, 5111-5120.

Lawrence, C.R., Neff, J.C., and Schimel, J.P. (2009). Does adding microbial mechanisms of decomposition improve soil organic matter models? A comparison of four models using data from a pulsed rewetting experiment. Soil Biol. Biochem. *41*, 1923-1934.

Lawrence, D.M., Oleson, K.W., Flanner, M.G., Thornton, P.E., Swenson, S.C., Lawrence, P.J., Zeng, X., Yang, Z.-L., Levis, S., Sakaguchi, K., *et al.* (2011). Parameterization improvements and functional and structural advances in Version 4 of the Community Land Model. J. Adv. Model. Earth. Syst *3*, M03001.

Leibold, M.A., Holyoak, M., Mouquet, N., Amarasekare, P., Chase, J.M., Hoopes, M.F., Holt, R.D., Shurin, J.B., Law, R., Tilman, D., *et al.* (2004). The metacommunity concept: a framework for multi-scale community ecology. Ecol. Lett. *7*, 601-613.

Lellei-Kovács, E., Kovács-Láng, E., Botta-Dukát, Z., Kalapos, T., Emmett, B., and Beier, C. (2011). Thresholds and interactive effects of soil moisture on the temperature response of soil respiration. Eur. J. Soil Biol. *47*, 247-255.

Lennon, J.T., Aanderud, Z.T., Lehmkuhl, B.K., and Schoolmaster, D.R. (2012). Mapping the niche space of soil microorganisms using taxonomy and traits. Ecology *93*, 1867-1879.

Lennon, J.T., and Jones, S.E. (2011). Microbial seed banks: the ecological and evolutionary implications of dormancy. Nat. Rev. Micro. *9*, 119-130.

Leung, L.R., and Gustafson, W.I. (2005). Potential regional climate change and implications to US air quality. Geophys. Res. Lett. *32*, L16711.

Levin, B.R. (1981). Periodic selection, infectious gene exchange, and the genetic structure of *E. coli* populations. Genetics *99*, 1-23.

Li, X., Miller, A.E., Meixner, T., Schimel, J.P., Melack, J.M., and Sickman, J.O. (2010). Adding an empirical factor to better represent the rewetting pulse mechanism in a soil biogeochemical model. Geoderma *159*, 440-451.

Liang, C., and Balser, T.C. (2008). Preferential sequestration of microbial carbon in subsoils of a glacial-landscape toposequence, Dane County, WI, USA. Geoderma *148*, 113-119.

Liang, C., and Balser, T.C. (2011). Microbial production of recalcitrant organic matter in global soils: implications for productivity and climate policy. Nat. Rev. Micro. *9*, 75.

Lindberg, N., Engtsson, J.B., and Persson, T. (2002). Effects of experimental irrigation and drought on the composition and diversity of soil fauna in a coniferous stand. J. Appl. Ecol. *39*, 924-936.

Liu, W., Zhang, Z.H.E., and Wan, S. (2009). Predominant role of water in regulating soil and microbial respiration and their responses to climate change in a semiarid grassland. Global Change Biol. *15*, 184-195.

Malcolm, G.M., López-Gutiérrez, J.C., and Koide, R.T. (2009). Little evidence for respiratory acclimation by microbial communities to short-term shifts in temperature in red pine (*Pinus resinosa*) litter. Global Change Biol. *15*, 2485-2492.

Manzoni, S., Schimel, J.P., and Porporato, A. (2011). Responses of soil microbial communities to water stress: results from a meta-analysis. Ecology *93*, 930-938.

Martiny, J.B.H., Eisen, J.A., Penn, K., Allison, S.D., and Horner-Devine, M.C. (2011). Drivers of bacterial β-diversity depend on spatial scale. Proc. Natl. Acad. Sci. USA *108*, 7850-7854.

May, R.M. (1976). Simple mathematical models with very complicated dynamics. Nature *261*, 459-467.

McMahan, C.A., Frye, R.G., and Brown, K.L. (1984). The vegetation types of Texas (Austin, Texas, USA, Texas Parks and Wildlife Department).

Meisner, A., Bååth, E., and Rousk, J. (2013a). Microbial growth responses upon rewetting soil dried for four days or one year. Soil Biol. Biochem. *66*, 188-192.

Meisner, A., De Deyn, G.B., de Boer, W., and van der Putten, W.H. (2013b). Soil biotic legacy effects of extreme weather events influence plant invasiveness. Proc. Natl. Acad. Sci. USA *110*, 9835-9838.

Moorhead, D.L., and Sinsabaugh, R.L. (2006). A Theoretical Model of Litter Decay and Microbial Interaction. Ecol. Monogr. *76*, 151-174.

Moyano, F.E., Manzoni, S., and Chenu, C. (2013). Responses of soil heterotrophic respiration to moisture availability: An exploration of processes and models. Soil Biol. Biochem. *59*, 72-85.

Notley-McRobb, L., and Ferenci, T. (2000). Experimental analysis of molecular events during mutational periodic selections in bacterial evolution. Genetics *156*, 1493-1501.

Ogle, S.M., Breidt, F.J., Easter, M., Williams, S., Killian, K., and Paustian, K. (2010). Scale and uncertainty in modeled soil organic carbon stock changes for US croplands using a process-based model. Global Change Biol. *16*, 810-822.

Oleson, K.W., Lawrence, D.M., Bonan, G.B., Flanner, M.G., Kluzek, E., Lawrence, P.J., Levis, S., Swenson, S.C., Thornton, P.E., Dai, A., *et al.* (2010). Technical Description of version 4.0 of the Community Land Model (CLM) (Boulder, CO, National Center for Atmospheric Research).

Öpik, M., Metsis, M., Daniell, T.J., Zobel, M., and Moora, M. (2009). Large-scale parallel 454 sequencing reveals host ecological group specificity of arbuscular mycorrhizal fungi in a boreonemoral forest. New Phytol. *184*, 424-437.

Öpik, M., Moora, M., Liira, J., and Zobel, M. (2006). Composition of root-colonizing arbuscular mycorrhizal fungal communities in different ecosystems around the globe. J Ecol *94*, 778-790.

Öpik, M., Vanatoa, A., Vanatoa, E., Moora, M., Davison, J., Kalwij, J.M., Reier, Ü., and Zobel, M. (2010). The online database MaarjAM reveals global and ecosystemic distribution patterns in arbuscular mycorrhizal fungi (Glomeromycota). New Phytol. *188*, 223-241.

Orwin, K.H., Kirschbaum, M.U.F., St John, M.G., and Dickie, I.A. (2011). Organic nutrient uptake by mycorrhizal fungi enhances ecosystem carbon storage: a model-based assessment. Ecol. Lett. *14*, 493-502.

Ostle, N.J., Smith, P., Fisher, R., Ian Woodward, F., Fisher, J.B., Smith, J.U., Galbraith, D., Levy, P., Meir, P., McNamara, N.P., *et al.* (2009). Integrating plant–soil interactions into global carbon cycle models. J. Ecol. *97*, 851-863.

Paterson, E., and Sim, A. (2013). Soil-specific response functions of organic matter mineralization to the availability of labile carbon. Global Change Biol. *19*, 1562-1571.

Perveen, N., Barot, S., Alvarez, G., Klumpp, K., Martin, R., Rapaport, A., Herfurth, D., Louault, F., and Fontaine, S. (2014). Priming effect and microbial diversity in ecosystem functioning and response to global change: a modeling approach using the Symphony model. Global Change Biol. *20*, 1174–1190.

Pett-Ridge, J., and Firestone, M.K. (2005). Redox Fluctuation Structures Microbial Communities in a Wet Tropical Soil. Appl. Environ. Microbiol. *71*, 6998-7007.

Placella, S.A., Brodie, E.L., and Firestone, M.K. (2012). Rainfall-induced carbon dioxide pulses result from sequential resuscitation of phylogenetically clustered microbial groups. Proc. Natl. Acad. Sci. USA *109*, 10931-10936.

Potthoff, M., Dyckmans, J., Flessa, H., Beese, F., and Joergensen, R. (2008). Decomposition of maize residues after manipulation of colonization and its contribution to the soil microbial biomass. Biol. Fertility Soils *44*, 891-895.

R Development Core Team (2011). R: A language and environment for statistical computing (Vienna, Austria, R Foundation for Statistical Computing).

Rietkerk, M., and van de Koppel, J. (2008). Regular pattern formation in real ecosystems. Trends Ecol. Evol. *23*, 169-175.

Rillig, M.C., and Mummey, D.L. (2006). Mycorrhizas and soil structure. New Phytol. *171*, 41-53.

Robertson, G.P., Coleman, D.C., Bledsoe, C.S., and Sollins, P. (1999). Standard soil methods for long-term ecological research (Oxford University Press).

Rousk, J., Smith, A.R., and Jones, D.L. (2013). Investigating the long-term legacy of drought and warming on the soil microbial community across five European shrubland ecosystems. Global Change Biol. *19*, 3872–3884.

Rustad, L., Campbell, J., Marion, G., Norby, R., Mitchell, M., Hartley, A., Cornelissen, J., Gurevitch, J., and Gcte, N. (2001). A meta-analysis of the response of soil respiration, net nitrogen mineralization, and aboveground plant growth to experimental ecosystem warming. Oecologia *126*, 543-562.

Sala, O.E., and Lauenroth, W.K. (1982). Small rainfall events - an ecological role in semi-arid regions. Oecologia *53*, 301-304.

Scheffer, M., Carpenter, S., Foley, J.A., Folke, C., and Walker, B. (2001). Catastrophic shifts in ecosystems. Nature *413*, 591-596.

Schimel, J., Balser, T.C., and Wallenstein, M. (2007). Microbial stress-response physiology and its implications for ecosystem function. Ecology *88*, 1386-1394.

Schimel, J.P., and Weintraub, M.N. (2003). The implications of exoenzyme activity on microbial carbon and nitrogen limitation in soil: a theoretical model. Soil Biol. Biochem. *35*, 549-563.

Schlesinger, W.H., and Andrews, J.A. (2000). Soil respiration and the global carbon cycle. Biogeochemistry *48*, 7-20.

Soetaert, K., Petzoldt, T., and Setzer, R.W. (2010). Solving differential equations in R: Package deSolve. Journal of Statistical Software *33*, 1-25.

Solomon, S., Qin, D., Manning, M., Chen, Z., Marquis, M., Averyt, K.B., Tignor, M., and Miller, H.L., eds. (2007). Climate Change 2007: The Physical Science Basis (Cambridge, UK, Cambridge University Press).

Staddon, P.L., Thompson, K., Jakobsen, I., Grime, J.P., Askew, A.P., and Fitter, A.H. (2003). Mycorrhizal fungal abundance is affected by long-term climatic manipulations in the field. Global Change Biol. *9*, 186-194.

Stark, J., and Firestone, M. (1995). Mechanisms for Soil Moisture Effects on Activity of Nitrifying Bacteria. Appl. Environ. Microbiol. *61*, 218-221.

Strickland, M.S., Lauber, C., Fierer, N., and Bradford, M.A. (2009). Testing the functional significance of microbial community composition. Ecology *90*, 441-451.

Talbot, J.M., Bruns, T.D., Taylor, J.W., Smith, D.P., Branco, S., Glassman, S.I., Erlandson, S., Vilgalys, R., Liao, H.-L., Smith, M.E., *et al.* (2014). Endemism and functional convergence across the North American soil mycobiome. Proc. Natl. Acad. Sci. USA.

Todd-Brown, K.E.O., Randerson, J.T., Post, W.M., Hoffman, F.M., Tarnocai, C., Schuur, E.A.G., and Allison, S.D. (2013). Causes of variation in soil carbon simulations from CMIP5 Earth system models and comparison with observations. Biogeosciences *10*, 1717-1736.

Torsvik, V., Øvreås, L., and Thingstad, T.F. (2002). Prokaryotic diversity--magnitude, dynamics, and controlling factors. Science *296*, 1064-1066.

Treseder, K., Balser, T., Bradford, M., Brodie, E., Dubinsky, E., Eviner, V., Hofmockel, K., Lennon, J., Levine, U., MacGregor, B., *et al.* (2011). Integrating microbial ecology into ecosystem models: challenges and priorities. Biogeochemistry, 1-12.

Van der Gucht, K., Cottenie, K., Muylaert, K., Vloemans, N., Cousin, S., Declerck, S., Jeppesen, E., Conde-Porcuna, J.-M., Schwenk, K., Zwart, G., *et al.* (2007). The power of species sorting: Local factors drive bacterial community composition over a wide range of spatial scales. Proc. Natl. Acad. Sci. USA *104*, 20404-20409.

Waldrop, M.P., Zak, D.R., Blackwood, C.B., Curtis, C.D., and Tilman, D. (2006). Resource availability controls fungal diversity across a plant diversity gradient. Ecol. Lett. *9*, 1127-1135.

Wang, G., Mayes, M.A., Gu, L., and Schadt, C.W. (2014). Representation of dormant and active microbial dynamics for ecosystem modeling. PLoS ONE *9*, e89252.

Wang, G., Post, W.M., and Mayes, M.A. (2013). Development of microbial-enzyme-mediated decomposition model parameters through steady-state and dynamic analyses. Ecol. Appl. *23*, 255-272.

Waring, B.G., Averill, C., and Hawkes, C.V. (2013). Differences in fungal and bacterial physiology alter soil carbon and nitrogen cycling: insights from meta-analysis and theoretical models. Ecol. Lett. *16*, 887-894.

Werchan, L.E., Lowther, A.C., and Ramsey, R.N. (1974). Soil Survey of Travis County, Texas (Washington, DC, United States Department of Agriculture Soil Conservation Service).

Wieder, W.R., Bonan, G.B., and Allison, S.D. (2013). Global soil carbon projections are improved by modelling microbial processes. Nature Clim Change *advance online publication.*

Wilson, G.W.T., Rice, C.W., Rillig, M.C., Springer, A., and Hartnett, D.C. (2009). Soil aggregation and carbon sequestration are tightly correlated with the abundance of arbuscular mycorrhizal fungi: results from long-term field experiments. Ecol. Lett. *12*, 452-461.

Zogg, G.P., Zak, D.R., Ringelberg, D.B., White, D.C., MacDonald, N.W., and Pregitzer, K.S. (1997). Compositional and functional shifts in microbial communities due to soil warming. Soil Sci. Soc. Am. J. *61*, 475-481.

Chapter 10

Environmental Change and Microbial Contributions to Carbon Cycle Feedbacks

Lei Qin[1,‡], Hojeong Kang[2,‡], Chris Freeman[3,*], Juanita Mora-Gómez[4,5] and Ming Jiang[1,6]

[1] Key Laboratory of Wetland Ecology and Environment, Northeast Institute of Geography and Agroecology, Chinese Academy of Sciences, Changchun, China; [2] School of Civil and Environmental Engineering, Yonsei University, Seoul, Korea; [3] Wolfson Carbon Capture Laboratory, School of Natural Sciences, Bangor University, Bangor, UK; [4] ISTO, CNRS UMR 7327, Université d'Orléans, BRGM, Orléans, France; [5] Le Studium Loire Valley Institute for Advanced Studies, Orléans, France; [6] Joint Key Lab of Changbaishan Wetland and Ecology, Jilin Province, Changchun, China; * corresponding author; ‡ contributed equally

Email:qinlei@iga.ac.cn, hj_kang@yonsei.ac.kr, c.freeman@bangor.ac.uk, juanita.mora.gomez78@gmail.com, jiangm@iga.ac.cn

DOI: https://doi.org/10.21775/9781913652579.10

Abstract

Microbes in soil play a key role in the global carbon cycle by metabolizing organic matter and releasing over 60 Pg of carbon per year. Since the composition and activities of microbes are strongly influenced by changes in environmental conditions such as temperature, water availability, oxygen penetration, and carbon supply, global climate change may exert climate-microbial feedbacks to accelerate or alleviate GHG emission. In the present study, we review effects of temperature rise, precipitation change, and elevated CO_2 on soil microbial composition and process rates. Furthermore, we suggest several topics that should be addressed to better understand microbial feedbacks to the future climate.

Introduction

Global environmental change caused by anthropogenic release of greenhouse gases (GHG) is considered to be one of the most serious environmental problems facing society. Many research-based scientific and engineering approaches have been conducted to address this issue - from accurate estimation of global carbon cycle to 'Carbon-Capture & Storage' techniques (see Dunn et al, this volume). Although global carbon flux through terrestrial ecosystems is estimated to be more than 120 Pg per year (Prentice, 2001), which is thus the single most significant flux from a global perspective, much less attention has been directed towards consideration of human-induced changes in carbon cycle. In particular, around the half of the flux through terrestrial ecosystems is mediated by heterotrophic metabolism of microorganisms, until now numerous studies have been conducted into microbial composition, and more recently into microbial functional and growth strategy studies, which could offer a better picture of the mechanisms regulating carbon exchange to climate change. Since global climate change includes changes in key controlling variables (e.g., temperature rise, changes in precipitation, elevated carbon dioxide-CO_2, and sea level rise) for microbial composition and process rates, climate-microbial feedbacks could amplify or dampen future climate change by increasing or decreasing GHG releases through microbial processes (Heimann and Reichstein, 2008). The present chapter aims to summarize findings in relation to this issue, looking at the effects of climate change on microbial biomass, microbial composition and processes. Furthermore, we aimed to synthesize related information to elucidate possible microbial feedbacks to GHG dynamics. Of various components of global climate change, particular attention is given to the direct effects of rising CO_2 levels, together with less direct effects such as temperature rise, and changes in drought or water-logging, as these factors can all influence microbial properties across various types of soil ecosystems.

Effects of elevated CO_2

Rising CO_2 levels can have marked indirect effects on soil microbial communities and their activity, and hence the potential for microbial feedback to climate change through its influence on plant growth and vegetation composition. Such plant-mediated indirect effects of climate change on soil microbes operate through a variety of mechanisms, with differing routes of feedback to climate change. One such mechanism concerns the indirect effects of rising atmospheric concentrations of CO_2 on soil microbes, through increased plant photosynthesis and transfer of photosynthate carbon to fine roots, mycorrhizal fungi (Johnson et al., 2005; Högberg and Read, 2006; Keel et al., 2006), and heterotrophic microbes (Zak et al., 1993;

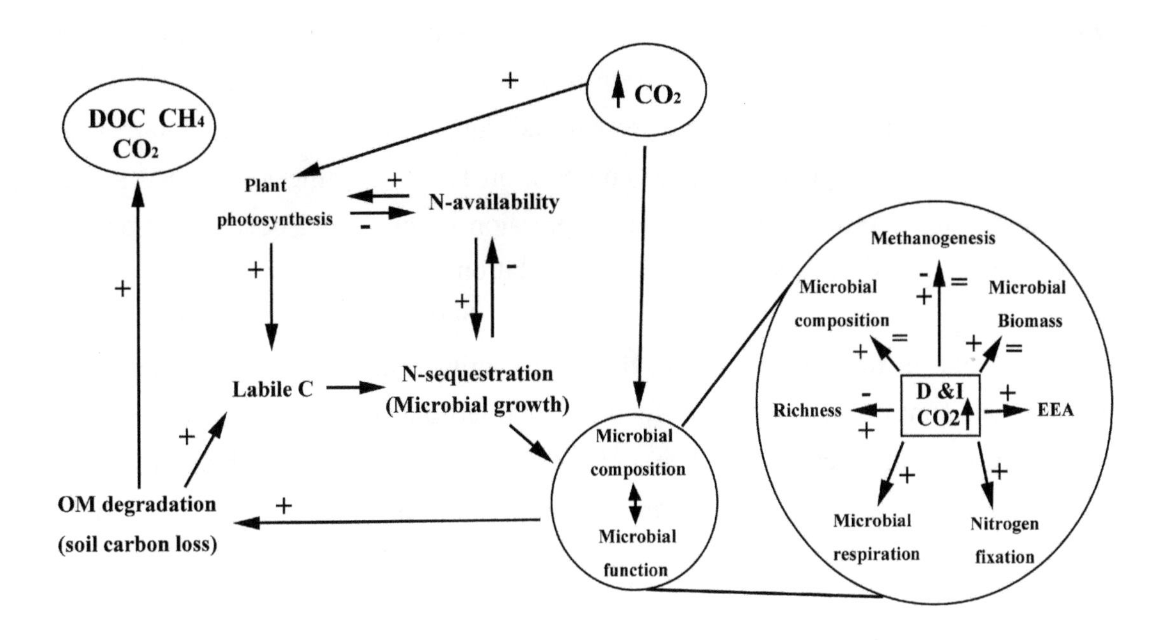

Figure 10.1 Summary of potential feedbacks and impacts of rising CO_2 levels on microbial communities. + = positive impact, - = negative impact, = = no change, EEA = extracellular enzyme activity, DOC = dissolved organic carbon, CO_2 = carbon dioxide, CH_4 = methane, D&I = direct and indirect effect of rising CO_2 atmospheric concentrations.

Bardgett et al., 2005). It is well established that elevated CO_2 increases plant photosynthesis and growth, especially under nutrient-rich conditions (Curtis and Wang, 1998) and this in turn increases the flux of carbon to roots, their symbionts and heterotrophic microbes, in form of root exudation of easily degradable sugars, organic acids and amino acids (Diaz et al., 1993; Zak et al., 1993). The consequences of increased carbon flux from roots to soil for microbial communities and carbon exchange are difficult to predict, because they vary substantially with factors such as plant identity, soil food web interactions, soil fertility and a range of other ecosystem properties (Wardle, 2002; Bardgett, 2005). But, some potential outcomes for soil microbes and carbon exchange include: (i) increases in soil carbon loss by respiration and in drainage waters as dissolved organic carbon (DOC) due to stimulation of microbial abundance and activity, and enhanced mineralization of recent and old soil organic carbon (Zak et al., 1993; Freeman et al., 2004a; Heath et al., 2005), a phenomenon known as 'priming' (Fontaine and Barot, 2005; Kuzyakov, 2006;

Dijkstra and Cheng, 2007); (ii) stimulation of microbial biomass and immobilization of soil N, thereby limiting N availability to plants, creating a negative feedback that constrains future increases in plant growth and carbon transfer to soil (Diaz et al., 1993); (iii) increased plant-microbial competition for N, leading to reduced soil N availability and microbial activity and suppression of microbial decomposition and ultimately increased ecosystem carbon accumulation (Hu et al., 2001); (iv) increased growth of mycorrhizal fungi (Klironomos et al., 1997; Staddon et al., 2004)—which receive carbon in the form of photosynthate directly from the host plant and retain this carbon, controlling its release to the soil microbial community (Högberg and Read, 2006)—and selection for beneficial fungal strains that help their host plant meet increased nutrient demands (Gamper et al., 2005), leading to a possible positive feedback on plant growth and carbon assimilation and enhanced stabilization of soil organic carbon through promotion of soil aggregation (Rillig and Mummey, 2006; Six et al., 2006); and (v) changes in root exudation are known to play a potentially important role through the promotion of methanogenesis and hence carbon loss from soil as methane (Ström et al., 2005), but the mechanisms involved remain poorly understood.

Microbial biomass might respond the feedbacks to elevated CO_2, but numerous studies have failed to find a significant response (Insam et al., 1999; Wiemken et al., 2001; Montealegre et al., 2002). For example, Wiemken et al. (2001) showed that the amounts of microbial carbon (a general marker for microbial biomass) and chitin (a marker for fungal biomass) did not respond significantly to the treatments with elevated CO_2 or nitrogen fertilizer. The results of Niklaus and Körner (1996) and Rouhier et al. (1994) suggest that responses of microbial C to elevated CO_2 are unlikely to develop in nutrient deficient ecosystems. Schortemeyer et al. (1996) reported that the size of the total heterotrophic microbial populations, in the form of microbial C in the rhizosphere of white clover or perennial ryegrass, did not change under elevated CO_2. Likewise, Schortemeyer et al. (2000) reported that microbial biomass did not increase in a natural Florida scrub ecosystem after 2 years of CO_2 enrichment. In an artificial tropical ecosystem with low nutrient availability, Insam et al. (1999) reported that microbial biomass C, ergosterol contents, and fungal hyphal lengths were not significantly altered by high CO_2 concentration, although total bacterial counts were significantly higher. Zak et al. (2000b) found that microbial biomass remained unchanged in bulk soils under elevated CO_2 after 2.5 growing seasons. Niklaus (2001) analyzed ecosystem C partitioning and soil C fluxes in grassland exposed to elevated CO_2 for 6 years. They showed that C pools increased in plants (+23%) and surface litter (+24%), but were not altered in terms of microbes and soil organisms. However, several studies conflict with such observations, and have reported an increase in soil microbial biomass under elevated CO_2 (Diaz et al., 1993; Zak et al., 1993; Dhillion et al., 1996; Pregitzer et al., 2000, Williams et al., 2000; Klamer et al., 2002). Zak et al. (1993) observed that microbial biomass C in the rhizosphere and bulk soil of *Populus grandidentata* was greater under elevated than ambient CO_2. In an acidic grassland herbaceous community, an increase of up to 80% in microbial biomass C occurred under elevated CO_2 (Diaz et al., 1993). Dhillion et al. (1996) reported that microbial biomass C was significantly higher in root region of soil from monocultures of *Bromus madritensis*, a common and sometimes dominant annual grass in Mediterranean model ecosystem plants under elevated CO_2. In tallgrass prairie exposed to elevated CO_2 for 8 years, soil microbial biomass C tended to be greater under elevated C compared to ambient treatment (Williams et al., 2000). Montealegre (2002) reported that bacterial populations increased about 1.4-fold under white clover after 3 years of CO_2 fumigation in pasture ecosystem. Liu et al. (2018) also found elevated CO_2 increased 14% microbial biomass according to 68 studies. In addition, several research groups found an increase in mycorrhizal short roots and extra-radical mycelium in response to elevated CO_2 (Ineichen et al., 1995; Lewis and Strain, 1996; Runion et al., 1997; Walker et al., 1997; Wiemken et al., 2001).

Microbial biomass essentially comprises the entire microbial community, but more effects can be seen within that community with new technique application, e.g. real-time quantitative polymerase chain reaction (qPCR), DNA-stable isotope probing (SIP) and high-throughput sequencing. There is now far greater potential for the identification of more specific processes showing responses to elevated CO_2, with potential effects on carbon cycling throughout different ecosystems. Under elevated CO_2, a differential response can be observed between microbial taxa (Xia et al., 2017), which has different functional implications for organic matter decomposition. Plant identity, soil food web interactions, soil fertility and a range of other ecosystem properties could influence the effects of elevated CO_2 on microbial community. Wang et al. (2019) used DNA-stable isotope probing (SIP) and Illumina Miseq sequencing to study bacterial community assimilating plant-derived carbon in the rhizosphere of soya bean, and found that elevated CO_2 significantly decreased the richness and diversity of the ^{13}C-assimilating bacterial community, which decreased the abundance of genera that were sensitive to labile C, but stimulated the abundance of genera metabolizing the complex C compounds. Sun et al. (2017) used PLFA analysis and found elevated CO_2 significantly increased the mean abundance of AM fungi, actinomycetes and bacteria by 11.5% and 16.7%, and 11% respectively in 10 years of elevated CO_2 open top chamber experiment in forest. Yu et al. (2018a) used Illumina MiSeq sequencing and also found that elevated CO_2 increases abundance of fungi in soybean rhizosphere, as well as it simplifies network structure by changing topological roles of operational taxonomic units and key fungal members. Yang et al. (2019) used 16S rRNA gene amplicons and a GeoChip microarray and found both taxonomic and functional gene compositions of the soil microbial community were modified by 14-years elevated CO_2 experiment, showing a significant decreasing of nitrogen fixation gene and ammonium-oxidizing gene and not significant effects on carbon fixation gene. Furthermore, Butterly et al. (2016) compared the structure and function of soil microbial communities in different soil type in dryland agriculture under elevated CO_2, founding that the N demand is the main reason for the changes of microbial community.

N is an essential element for microbial growth and consequently it is important to understand the relationship between N accumulation and availability, and microbial composition and function, with elevated CO_2. In detail, elevated CO_2 might influence N fixation by promoting plant growth and increased labile carbon in soil (Mckinley et al., 2009; Das et al., 2011), which has been found to promote diazotrophic bacteria with enhanced N_2 fixation (Billings et al., 2004; Das et al., 2011). Grover et al. (2015) also showed that high C:N ratio under elevated CO_2 favors fungi with wider C:N ratio and nutrient acquisition ability and biological nitrogen fixers. Additionally, Ochoa-Hueso et al. (2017) reported that nutrient-poor eucalypt woodland exhibited rhizosphere responses to atmospheric CO_2 enrichment that increased nitrogen availability in rhizosphere soil. Several studies have shown that elevated CO_2 can activate nitrogen fixation enzymes, which is another way to alleviate nitrogen limitation under elevated CO_2. For example, Zanetti et al. (1996) reported an increase in symbiotic nitrogen fixation activity for *T. repens* growth under enriched CO_2 atmosphere. Dakora and Drake (2000) also observed that elevated CO_2 stimulated greater N_2 fixation and nitrogenase activity in stands of the C3 sedge, *Scirpus olneyi* of the Chesapeake Bay wetland. According to the review of 77 published studies, soil

N availability could influence microbial community with elevated CO_2, as high N availability might influence C availability, C:N ratios and soil pH (Wang et al., 2018a). The N availability could also influence organic matter decomposition but differently in different soil systems. For instance, in forest ecosystem, long-term N addition has been found to inhibit organic matter decomposition mainly because inorganic N enrichment suppresses oxidative enzyme activity (Bonner et al., 2019); however, short-term N addition also could alleviate microbial nitrogen limitation and promote soil respiration in forest (Liu et al., 2019). While in peatlands, high N promotes oxidative enzyme activity and increases organic matter decomposition rates (Bragazza et al., 2006; Song et al., 2019a).

The impact of elevated CO_2 effects on soil microbes is related to extracellular enzymes. Microorganisms produce extracellular enzymes which are involved in decomposition and mineralization of organic matter. Elevated CO_2 concentration can affect extracellular enzyme activities in several ways. Dhillion et al. (1996) reported that dehydrogenase, cellulose, phosphatase, and xylanase were increased by elevated CO_2 in the root region of soil from monocultures of *Bromus madritensis*, a common and sometimes dominant annual grass in Mediterranean model ecosystem. Of the four enzymes examined, dehydrogenase and xylanase activities were significantly higher in soils under elevated CO_2 than in ambient. Moorhead and Linkins (1997) suggested that elevated CO_2 altered the soil enzyme characteristics in a tussock tundra ecosystem. They found significantly higher phosphatase activities at 680 mol/mol CO_2 on the surfaces of plant roots, mycorrhizal surfaces, and in the shallowest organic horizons soil. Although responses can vary depending on the chemical properties of organic matter, general trends of extracellular enzyme activities under elevated CO_2 conditions tend to be overall increases in enzymes or enhancement of a specific enzyme activity that is related to availability of nitrogen or phosphorus. As carbon supply to soils can increase under elevated CO_2, and as additional carbon supply is known to support additional enzyme activity (Shackle et al., 2000), microorganisms are able to release additional enzymes to acquire nitrogen or phosphate. Soil type and seasonal changes all influence the response to soil enzyme activities to elevated CO_2 (Kelley et al., 2011).

One main contribution of soil microbes to climate change is their role in soil organic matter decomposition and consequent releases of CO_2. Many studies have found microbial respiration to be significantly greater in elevated CO_2 conditions

(Rogers et al., 1992; O'Neill, 1994; Runion et al., 1994; Dhillion et al., 1996; Williams et al., 2000), which is attributed to 'priming effects' through enhanced primary production and consequent extra supply of carbon to microbes. For example, Williams et al. (2000) observed that microbial respiration was higher in tallgrass prairie exposed to elevated CO_2 for 8 years. Changes in microbial composition have also been found to promote CO_2 flux (Sun et al., 2017). Kuzyakov et al. (2019) also found elevated CO_2 accelerated soil C cycling but promotion of plant productivity and enhanced stability of soil microaggregates also contributed to soil carbon stock. Most evidence then suggests that elevated CO_2 contributes to microbial respiration and accelerated C turnover.

Another important microbial feedback between elevated CO_2 and the global carbon cycle relates to methane production. A number of studies have addressed the potential changes in trace gas emissions from wetlands exposed to elevated CO_2. For example, Drake (1992) reported CO_2 enrichment stimulated methane emissions by 80% in a salt marsh containing sedge *Scirpus olneyi*. Hutchin et al. (1995) also found a similar effect for mire peat and vegetation exposed CO_2 enrichment treatment. Allen et al. (1994) reported the similar results with a combination of increased CO_2 and temperature. Wang and Adachi (1999) also provided evidence that elevated atmospheric CO_2 concentrations could promote CH_4 production from flooded soils. Megonigal and Schlesinger (1997) who performed experiments with *Orontium aquaticum* reported CH_4 emissions increased by 136% under elevated CO_2. Liu et al. (2018) synthesized 1655 measurement from 169 published studies and showed that elevated CO_2 stimulated CH_4 fluxes by 34% from paddy and by 12% from wetlands. The increases in CH_4 emission under elevated CO_2 conditions can be explained by two mechanisms. First, elevated CO_2 often results in ample supply of carbon into soil and hence larger amounts of organic carbon are available for methanogens. Secondly, elevated CO_2 concentration might shift the relative abundance in hydrogenotrophic, acetoclastic and methylotrophic of methanogens (Guthrie, 1986; Dacey et al, 1994; Wang et al., 2018). However, some studies also found that elevated CO_2 has suppression or minor effects on CH_4 emission. For example, Saarnio et al. (1998) reported that the average release of CH_4 from *Sphagnum* samples exposed to the doubled concentration of CO_2 was significantly lower than that at ambient CO_2. Schrope et al. (1999) reported methane emissions from rice grown in a sandy soil under doubled CO_2 were 4-45 times less. Kang et al. (2001) found no significant differences for CH_4 emission on northern fen peat with *Juncus* and *Festuca* spp., although the mean value was higher under elevated CO_2 conditions. There are two possible pathways in suppression the production of CH_4. First of all, although elevated CO_2 could promote soil labile C availability via plant photosynthate, elevated CO_2 may have more effectively aerated the soil and decrease methane production by increase CH_4 oxidation (Keller et al., 2011). Secondly, the other microbes could compete C with methanogens as root oxygen could promote the terminal electron acceptors like nitrate ions, Fe (III) etc. (Laanbroek, 2010).

Effects of temperature rise

It is now 95% certain that current warming of the Earth's atmosphere is mostly a result of rising greenhouse gas concentrations by human activities (IPCC, 2013). It is generally anticipated that global warming will accelerate rates of heterotrophic microbial activity, thereby increasing the efflux of CO_2 to the atmosphere and exports of DOC by hydrologic leaching (Jenkinson et al., 1991; Davidson and Janssens, 2006; Knorr, 2013). Because rates of soil respiration are thought to be more sensitive to temperature than primary production (Jenkinson et al., 1991; Schimel et al., 1994), it is predicted that climate warming will increase the net transfer of carbon from soil to atmosphere, thereby creating a positive feedback on climate change (Jones et al., 2005). While it is well established that temperature is an important determinant of rates of organic matter decomposition, the nature of the relationship between temperature and heterotrophic microbial respiration and its potential to feedback to climate change, are far from clear (Davidson and Janssens, 2006; Trumbore, 2006).

A prime cause of this uncertainty is the inherent complexity and diversity of soil organic matter and the likelihood that the temperature-dependence of microbial decomposition of soil carbon compounds of differing chemical composition and substrate quality will vary (Davidson and Janssens, 2006). For example, there is evidence that the temperature sensitivity of litter decomposition increases as the quality of organic carbon consumed by microbes declines (Fierer et al., 2005), which is consistent with kinetic theory that indicates greater temperature sensitivity for decomposition of recalcitrant carbon pools (Knorr et al., 2005; Wallenstein et al., 2012). However, there are still many contrasting results on this subject. For example, other studies suggest that the temperature sensitivity of more recalcitrant substrates is similar (Fang et al., 2005; Conen et al., 2006) or less than (Luo et al., 2001; Melillo et al., 2002; Rey and Jarvis, 2006) that of more labile substrates, while soil C incubation data showed that temperature sensitivity of relatively more resistant organic matter is greater than that of relatively more labile substrates (Conant et al., 2008). Recently, temperature sensitivity of litter decomposition has been found to be related with microbial carbon use efficiency (CUE), high CUE inhibits respiration and low CUE favors respiration (Manzoni et al., 2012). And generally, rising soil temperatures could reduce CUE, as warming limits microbial growth by increasing the energy cost of maintaining existing biomass (Li et al., 2018a). But microbial carbon use efficiency is also influenced by soil nutrient status (Manzoni et al., 2012), for example, Sihi et al. (2016) showed that nutrient limitation can regulate carbon decomposition at higher temperature. Walker et al. (2018) and Dacal et al. (2019) likewise confirmed that short-term warming could accelerate the microbial activity temporarily, but nutrient depletion reduced microbial biomass, which constrained microbial metabolism over the ecosystem. Karhu et al. (2014) collected soils from different ecosystems along a climate gradient from the Arctic to the Amazon and showed that the strongest enhancing responses to rising temperature were observed in soils with high C:N ratios.

The uncertainty is further complicated by how reactive different microbial groups and species are to temperature change (Kandeler et al., 1998; Bardgett et al., 1999). Different microbial taxa have different pathways of C allocation, C assimilation, enzyme production, and respiration (Malik et al., 2019), and similarly, acclimation of soil microbial communities to higher temperature can vary (Kirschbaum, 2004). For example, Kim et al. (2012) have reported that the ratio of gene abundances between methanogen and bacteria decreased by 3-year warming experiment, suggesting responses of each functional group to warming environment may vary. Ali et al. (2018) found that increasing temperature stimulated cumulative respiration rates but decreased total microbial biomass, affecting fungal biomass more adversely than bacterial biomass. Furthermore, Tveit et al. (2015) also confirmed that microbial community adaption and cascade effects of substrate availability were the main reason microbial groups change with warming. Uncertain of microbial process response to climate change potentially creates unreliable model predictions of soil carbon feedbacks to climate change (Kirschbaum, 2006), to solve this issue represents a major research challenge for the future.

Wetlands are sources for CH_4 and rising temperature has been widely found to increase methanogenic population abundances and the production of CH_4 in wetlands

(Wang et al., 2018b; Chang et al., 2019; Kolton et al., 2019). Rising temperature could stimulate soil organic carbon decomposition and produce the more available substrate for methanogens (Wang et al., 2018b). A further study also showed that acetate production rate is higher than hydrogen production rate with increased soil temperature, therefore substrate asymmetric changes affect the abundance of the acetoclastic and methylotrophic of methanogens (Chang et al., 2019). These also showed that methanogens can adapt to rising temperature and substrate availability determined their activity (de Jong et al., 2018).

While overall changes in temperature are likely to have strong effects on microbial communities and decomposition processes in arctic and alpine regions, climate change-related reductions in snow cover will also be of exceptionally high importance. It has been estimated that 25% of Earth's permafrost could thaw by 2100 due to climate warming, releasing considerable amounts of otherwise protected organic matter for microbial decomposition (Anisimov et al., 1999), and creating a positive feedback on climate change (Davidson and Janssens, 2006). Also, because snow is an important insulator of soil biological processes, predicted reductions in snow cover in alpine and arctic regions will increase soil freezing, with consequences for root mortality, nutrient cycling and microbial processes of decomposition (Groffman et al., 2001; Bardgett et al., 2005). Also, some studies in subalpine forest in Colorado indicate that reduced snow cover can suppress rates of soil respiration due to a unique and highly temperature-sensitive soil microbial community that occurs beneath snow (Monson et al., 2006); such responses could have substantial consequences for winter soil microbial activity, carbon storage, and CO_2 efflux in alpine and arctic regions.

Effects of changes in precipitation

Rising GHG atmospheric levels also affects the global hydrological cycle. It is well established that both drought and flooding can have substantial direct effects on microbial physiology and the composition of the active microbial community, with important consequences for ecosystem-level carbon and nutrient flows (Schimel et al., 2007). For example, Bouskill et al. (2016) reported that drought decreased soil water content and microbial functional group with high ability for extracellular enzyme production to break down macromolecular carbon in tropical forest. Furthermore, soil microbial communities are likely to experience changes in microbial functional groups under future drought scenarios (Preece et al., 2019).

Nevertheless, effects of hydrological stressors on microbial communities and consequences for carbon exchange are likely to vary substantially across ecosystems. Increased frequency and intensity of drought in drier ecosystems, for example, may result in moisture-limiting conditions for microbial activity, creating a negative feedback on microbial decomposition and soil carbon loss (Toberman at al., 2008). This view is supported by studies in forest ecosystems that report significant reduction of phenol oxidase activity, isoenzyme diversity, soil bacterial and fungal biomass, in leaf litter decomposition during dry periods (Nardo et al., 2004; Krivtsov et al., 2006); as well as in a dry Californian annual grassland, where the addition of water increased soil phenol oxidase activity in a manipulation experiment (Henry et al., 2005). Effects of stressors on microbial communities seems to vary depending on microbial tolerance and resilience to drought; for example, drought might destabilize properties in co-occurrence networks of soil bacterial but not fungi in grassland mesocosms (de Vries et al., 2018). Zhou et al. (2018) synthesized from 65 publications about the precipitation effect on microbial communities worldwide and showed that bacteria community was influenced by precipitation, and microbial biomass had a higher response to a moderate increase than decrease in precipitation, whereas it was more sensitive to an extreme decrease than extreme increase in precipitation.

Rewetting of dry soil can also lead to a pulse of CO_2 through increased microbial mineralization of cytoplasmic solutes, and in the longer-term, by decreasing the total amount of soil organic matter that is physically protected within micro-aggregates (Fierer and Schimel, 2003). However, this response to rewetting is less pronounced in soils that are frequently exposed to natural drying and rewetting cycles (Birch, 1958; Fierer and Schimel, 2002). Barnard et al. (2013) have reported marked resilience of active bacterial community to desiccation-rewetting in annual grasslands. Likewise, Bouskill et al. (2013) noted that pre-exposure to drought increases the resistance of soil bacterial communities to extended drought.

In contrast, increased drought and drying in wetlands and peatlands will create more favorable conditions for microbial activity by lowering the water table and introducing oxygen into previously anaerobic soil. This has been shown to increase the activity of phenol oxidases (Freeman et al., 2004a; Zibilske and Bradford, 2007), which play a pivotal role in the breakdown of complex organic matter and the cycling of phenolic compounds that may interfere with extracellular enzymes (Benoit and

Starkey, 1968; Albers et al., 2004). Through what has been described as the 'enzymic latch mechanism' (Freeman et al., 2001), changes in the activity of extracellular phenol oxidases may directly affect the retention of carbon in soil through the breakdown of otherwise highly recalcitrant organic matter and by releasing extracellular hydrolase enzymes from phenolic inhibition (Freeman et al., 2001, 2004a). Because peatlands and wetlands are one of the largest stocks of terrestrial carbon (Ward et al., 2007), such enhanced breakdown of recalcitrant organic matter under drying could have major implications for the global carbon cycle (Freeman et al., 2004a).

While the increase in O_2 availability that accompanies drought promotes organic matter decomposition in wetlands and peatlands, thereby increasing CO_2 release, opposing effects occur for methanogenic pathways, in that methane emissions are reduced. Water table depth is a strong predictor of methane emissions (Roulet and Moore, 1995), and while this is generally assumed to be due to toxic effects of O_2, there is also evidence that methanogens are more sensitive to desiccation (Fetzer et al., 1993). In addition, toxic effects on methanogens of oxidized products of denitrification have been noted (Kluber and Conrad, 1998), while net methane emissions are also suppressed under drought conditions by the action of methanotrophic bacteria (King, 1992; Freeman et al., 2002). Wang et al. (2017) used GeoChip analysis revealed that deceased CH_4 production potential, rather than increased CH_4 oxidation potential, may lead to the reduction in net CH_4 emissions under drought condition. Meanwhile, increased drought also influences microbial activity through changing plant composition in peatlands. Long-term moderate drought shifted Sphagnum/herbs to high-phenolic shrubs, high content phenolic could inhibit microbial activity (Wang et al., 2015). The vascular plant expansion could influence fungal functional composition and promote soil enzyme activity through their direct effect on dissolved organic matter quality, influencing ecosystem respiration (Robroek et al., 2015; Jassey et al., 2017).

Effects of multi-environmental factors

Global climate change rarely manifests as a change in a single environmental variable but rather it occurs as multiple environmental factors, dynamics of microbial composition and function are thus determined by the balance between different interacting forces. Microbial feedback to interactive effects of different components of global climate change is now beginning to receive attention, although most of such studies focus only in two factors.

Elevated CO_2 and warming have showed different effects on carbon storage in terrestrial ecosystem (Black et al., 2017; Carrillo et al., 2018; Dietzen et al., 2019). Carrillo et al. (2018) showed that warming reduced soil C under elevated CO_2 but had no impact under ambient CO_2 in a semi-arid grassland. However, Dietzen et al. (2019) showed that accumulation of soil organic carbon under elevated CO_2 unaffected by warming and drought in temperate grassland. Elevated CO_2 and temperature have also been found to increase soil C losses in a soybean–maize ecosystem (Black et al., 2017). There are four main reasons for these carbon dynamics. Firstly, the changes in plant photosynthesis and composition changes could influence soil C through plant (McPartland et al., 2018; Cernusak et al., 2019). Second, soil moisture variability induced by temperature and elevated CO_2 could also influence soil microbial key function genes (Yu et al., 2018b). Thirdly, soil nutrient availability, as these are linked to microbial metabolism (Oldfield et al., 2018; Silva-Sanchez et al., 2019). Finally, soil carbon soil organic carbon quality also influences the microbial carbon use efficiency (Soares and Rousk, 2019). In addition, in moisture limited ecosystem, like grasslands, the precipitation also promotes the effects of warming on microbial activity and biomass (Liu et al., 2016; Li et al., 2018b). In summary, the effects of multiple climate drivers on carbon cycling depend on changes in background climate and ecosystem conditions (Song et al., 2019b).

Conclusions

The global carbon cycle is tightly coupled to soil microbial processes. The most relevant factors can be identified as; i) organic matter chemical composition, ii) the influence of root lixiviates from plants, iii) nitrogen and phosphorus availability, and iv) soil moisture. Environmental changes exert a profound influence on carbon cycling through these four pathways leading to changes in microbial community and functional gene abundance. As differences in ecosystem condition and background climate develop, in general, environmental changes modify carbon cycle feedbacks. Yet the complexity of the soil microbial community and its many roles, coupled with the plethora of ways that climate and other global changes can affect soil microbes, remain a major challenge to our understanding (Widder et al., 2016). In the future, these complexities may be addressed through collaborative approaches aimed at modelling microbial roles in carbon-cycle regulation (Wieder et al., 2013).

Future trends

Our understanding about the feedbacks between microbial regulation of carbon cycle and climate change is still in its infancy. Many challenges remain not least of which is the extreme diversity within microbial communities. But perhaps the greatest challenge will arise from appreciating how microbial diversity and function respond to climate change across entire ecosystem to planetary scales. For the future, we identify five major challenges in this research field.

Firstly, interactive effects of different components of global climate change must be considered. Although recent studies have employed 2 or even 3 factorial designs to emulate future climate, most of the studies are still focused on singular changes. For example, precipitation change in winter will affect not only water availability but also insulation of soil surface and hence modify temperature regime substantially. Likewise, CO_2 fertilization effects can strongly be limited or enhanced by nitrogen availability in many terrestrial ecosystems. Additionally, under the current global situation, soils are exposed simultaneously to both climate change and human pollution by organic and inorganic materials. Studies linking those dimensions are even more scarce and our knowledge of how they interreact, how sensitive or resilient are soil microbial communities, or how GHG emission are affected by that interaction is limited. Despite of logistical difficulties, future studies should incorporate as many environmental components as possible to better understand microbial feedback to global carbon cycle.

Secondly, soil microbial communities are extremely diverse and one of the greatest challenges concerns understanding how microbial diversity responds to climate change and the functional consequences of this for ecosystem carbon exchange, including the uptake, stabilization and release of carbon from soil as GHGs. The major hurdle here is that many microbes are uncultivable, and the function of these non-cultivable microbes is poorly understood because it is difficult to test how they respond to, or modify, their environment. However, relatively new advances in sequencing analysis (metagenomics and metatranscriptomics) and in-situ measurements of ecosystem processes, such as stable isotope probing (SIP), enable linking of changes in microbial diversity to ecosystem function, by focusing on functional genes that are important for biogeochemical processes; and through directly labelling DNA, RNA and phospholipid fatty acids (PLFA) of organisms participating in particular pathways. These tools have changed the way microbial ecologists explore the ecophysiology of microbial populations in the natural environment, because they enable study of the metabolic capabilities of uncultivable microorganisms, thus providing insights into the underlying processes regulating carbon flow in through different components of the soil microflora. Despite the promising advances in this field, there is still a lack of information on most of the microbial metabolic pathways of carbon degradation, also the enormity of the vast microbial diversity remains to be fully explored with many regions still to be surveyed.

Thirdly, as already discussed, the diversity of carbon substrates in soil is considerable, and a major challenge is to understand the complexity of this carbon and how climate change and its interaction with other environmental factors affects its availability to enzymes that catalyze its degradation.

Fourth challenge, as discussed throughout this review, soil microbes and their activities are inextricably linked to aboveground communities (including plants, herbivores, pathogens and parasites), but also to belowground organisms (bacteria, fungi, archaea, virus, protist, and invertebrates). Interactions between different biological groups can have a positive, negative or no impact on the microbial species and processes (Faust and Raes, 2012). Interactions might be affected by change in environmental conditions, for example interaction network properties can change under elevated CO_2 (Zhou et al., 2010), and might potentially influence the microbial response to climate change. Understanding the effects of climate change on carbon dynamics therefore requires explicit consideration of the feedbacks that occur between aboveground and belowground communities through carbon flow and food web structure, and their response to climate change.

Finally, future research on the microbial contribution to carbon cycle feedbacks upon climate change should include mathematical models as part of the studies on soil microbial dynamics and processes (Malik et al., 2019). Mathematical models should also seek a bridge that allows us to link microbial processes to up-scale climate change models; and, in that way, improve our capacity to predict future scenarios of global climate.

References

Ali, R.S., Poll, C., and Kandeler, E. (2018). Dynamics of soil respiration and microbial communities: Interactive controls of temperature and substrate quality. Soil Biol. Biochem. *127*, 60–70.

Albers, D., Migge, S., Schaefer, M., and Scheu, S. (2004). Decomposition of beech leaves and spruce needles in pure and mixed stands of beech and spruce. Soil Biol. Biochem. *36*, 155–164.

Allen, L.H. Jr., Albrecht, S.L., Colon, W., and Covell, S.A. (1994). Effects of carbon dioxide and temperature on methane emission of rice. Int. Rice Research Notes *19*, 43.

Anisimov, O.A., Nelson, F.E., and Pavlov, A.V. (1999). Predictive scenarios of permafrost development under conditions of global climate change in the XXI century. Earth Cryology *3*, 15–25.

Bardgett, R.D., Bowman, W.D., Kaufmann, R., and Schmidt, S.K. (2005). A temporal approach to linking aboveground and belowground ecology. Trends Ecol. Evol. *20*, 634–641.

Bardgett, R.D., Kandeler, E., Tscherko, D., Hobbs, P.J., Bezemer, T.M., Jones, T.H., and Thompson, L.J. (1999). Below-ground microbial community development in a high temperature world. Oikos *85*, 193–203.

Barnard, R.L., Osborne, C.A., and Firestone, M.K. (2013). Responses of soil bacterial and fungal communities to extreme desiccation and rewetting. ISME J. *7*, 2229–2241.

Benoit, R.E., and Starkey, R.L. (1968). Enzyme inactivation as a factor in the inhibition of decomposition of organic matter by tannins. Soil Science *105*, 203–208.

Billings, S.A., Schaeffer, S.M., and Evans, R.D. (2004). Soil microbial activity and N availability with elevated CO_2 in Mojave Desert soils. Global Biogeochem. Cycles *18*, GB1011.

Birch, H. (1958). The effect of soil drying on humus decomposition and nitrogen availability. Plant Soil *10*, 9–31.

Black, C.K., Davis, S.C., Hudiburg, T.W., Bernacchi, C.J., and DeLucia, E.H. (2017). Elevated CO_2 and temperature increase soil C losses from a soybean–maize ecosystem. Global Change Biol. *23*, 435–445.

Bonner, M.T.L., Castro, D., Schneider, A.N., Sundstrom, G., Hurry, V., Street, N.R., and Nasholm, T. (2019). Why does nitrogen addition to forest soils inhibit decomposition? Soil Biol. Biochem. *137*, 107570.

Bouskill, N.J., Lim, H.C., Borglin, S., Salve, R., Wood, T.E., Silver, W.L., and Brodie, E.L. (2013). Pre-exposure to drought increases the resistance of tropical forest soil bacterial communities to extended drought. ISME J. *7*, 384–394.

Bragazza, L., Freeman, C., Jones, T., et al. (2006). Atmospheric nitrogen deposition promotes carbon loss from peat bogs. PNAS *103*, 19386–19389.

Butterly, C.R., Phillips, L.A., Wiltshire, J.L., Franks, A.E., Armstrong, R.D., Chen, D.L., Mele, P.M., and Tang, C.X. (2016). Long–term effects of elevated CO_2 on carbon and nitrogen functional capacity of microbial communities in three contrasting soils. Soil Biol. Biochem. *97*, 157–167.

Bouskill, N.J., Wood, T.E., Baran, R., et al. (2016). Belowground response to drought in a tropical forest soil. I. Changes in microbial functional potential and metabolism. Front. Microbiol. *7*, 1–11.

Carrillo, Y., Dijkstra, F., LeCain, D., Blumenthal, D., and Pendall, E. (2018). Elevated CO_2 and warming cause interactive effects on soil carbon and shifts in carbon use by bacteria. Ecol. Lett. *21*, 1639–1648.

Conant, R.T., Drijber, R.A., Haddix, M.L., Parton, W.J., Paul, E.A., and Plante, A.F. (2008). Sensitivity of organic matter decomposition to warming varies with its quality. Global Change Biol. *14*, 868–877.

Conen, F., Leifeld, J., Seth, B., and Alewell, C. (2006). Warming mobilises young and old soil carbon equally. Biogeosciences *3*, 515–519.

Curtis, P.S., and Wang, X.Z. (1998). A meta-analysis of elevated CO_2 effects on woody plant mass, form, and physiology. Oecologia *113*, 299–313.

Cernusak, L.A., Haverd, V., Brendel, O., et al. (2019). Robust response of terrestrial plants to rising CO_2. Trends Plant Sci. *24*, 578-586.

Chang, K.Y., Riley, W.J., Bordie, E.L., et al. (2019). Methane production pathway regulated proximately by substrate availability and distally by temperature in a high latitude mire complex. J Geophys. Res-biogeo. *124*, 3057-3074.

Davidson, E.A., and Janssens, I.A. (2006). Temperature sensitivity of soil carbon decomposition and feedback to climate change. Nature *440*, 165-173.

Dietzen, C.A., Larsen, K.S., Ambus, P.L., et al. (2019). Accumulation of soil carbon under elevated CO_2 unaffected by warming and drought. Global Change Biol. *25*, 2970-2977.

de Jong, A.E.E., in't Zandt, M.H., Meisel, O.H., et al. (2018). Increases in temperature and nutrient availability positively affect methane-cycling microorganisms in Arctic thermokarst lake sediments. Environ. Microbiol. *20*, 4314-4327.

Dacey, V.W.H., Drake, B.G., and Klug, M.J. (1994). Stimulation of methane emission by carbon dioxide enrichment of marsh vegetation. Nature *370*, 47–49.

Dacal, M., Bradford, M.A., Plaza, C., Maestre, F.T., Garcia-Palacios, P. (2019). Soil microbial respiration adapts to ambient temperature in global drylands. Nat. Ecol. Evol. *3*, 232–238.

Das, S., Bhattacharyya, P., and Adhya, T.K. (2011). Impact of elevated CO_2, flooding, and temperature interaction on heterotrophic nitrogen fixation in tropical rice soils. Biol. Fert. Soils *47*, 25–30.

Dakora, F.D., and Drake, B.G. (2000). Elevated CO_2 stimulates associative N_2 fixation in a C3 plant of the Chesapeake Bay wetland. Plant Cell Environ. *23*, 943–953.

de Vries, F.T., Griffiths, R.I., Bailey, M., et al. (2018). Soil bacterial networks are less stable under drought than fungal networks. Nat. Commun. *9*, 3033.

Dhillion, S.S., Roy, J., and Abrams, M. (1996). Assessing the impact of elevated CO_2 on soil microbial activity in a Mediterranean model ecosystem. Plant Soil *187*, 333–342.

Diaz, S., Grime, J.P., Harris, J., and McPherson, E. (1993). Evidence of a feedback mechanism limiting plant-response to elevated carbon-dioxide. Nature *364*, 616–617.

Dijkstra, F.A., and Cheng, W.X. (2007). Interactions between soil and tree roots accelerate long-term soil carbon decomposition. Ecol. Lett. *10*, 1046–1053

Drake, B.G. (1992). A field study of the effects of elevated CO_2 on ecosystem processes in a Chesapeake Bay wetland. Aust. J. Bot. *40*, 579–595.

Fang, C.M., Smith, P., Moncrieff, J.B., and Smith, J.U. (2005). Similar response of labile and resistant soil organic matter pools to changes in temperature. Nature *433*, 57–59

Faust, K., and Raes, J. (2012). Microbial interactions: From networks to models. Nat. Rev. Microbiol. *10*, 538–550.

Fetzer, S., Bak, F., and Conrad, R. (1993). Sensitivity of methanogenic bacteria from paddy soil to oxygen and desiccation. FEMS Microbiol. Ecol. *12*, 107–115.

Fierer, N., and Schimel, J. (2002). Effects of drying-rewetting frequency on soil carbon and nitrogen transformations. Soil Biol. Biochem. *34*, 777–787.

Fierer, N., and Schimel, J. (2003). A proposed mechanism for the pulse in carbon dioxide production commonly observed following the rapid rewetting of a dry soil. Soil Sci. Soc. Am. J. *67*, 798–805.

Fierer, N., Craine, J.M., McLauchlan, K., and Schimel, J.P. (2005). Litter quality and the temperature sensitivity of decomposition. Ecology *86*, 320–326.

Fontaine, S., and Barot, S. (2005). Size and functional diversity of microbe populations control plant persistence and long-term soil carbon accumulation. Ecol. Lett. *8*, 1075–1087.

Freeman, C., Nevison, G.B., Kang, H., Hughes, S., Reynolds, B., and Hudson, J.A. (2002). Contrasted effects of simulated drought on the production and oxidation of methane in a mid-Wales wetland. Soil Biol. Biochem. *34*, 61–67.

Freeman, C., Ostle, N.J., and Kang, H. (2001). An enzymic 'latch' on a global carbon store. Nature *409*, 149.

Freeman, C., Ostle, N.J., Fenner, N., and Kang, H. (2004a). A regulatory role for phenol oxidase during decomposition in peatlands. Soil Biol. Biochem. *36*, 1663–1667.

Freeman, C., Fenner, N., Ostle, N.J., Kang, H., Dowrick, D.J., Reynolds, B., Lock, M.A., sleep, D., Hughes, S., and Hudson, J. (2004b). Dissolved organic carbon export from peatlands under elevated carbon dioxide levels. Nature *430*, 195–198.

Gamper, H., Hartwig, U.A., and Leuchtmann, A. (2005). Mycorrhizas improve nitrogen nutrition of *Trifolium repens* after 8 yr of selection under elevated atmospheric CO_2 partial pressure. New Phytol. *167*, 531–542.

Groffman, P.M., Driscoll, C.T., Fahey, T.J., Hardy, J.P., Fitzhugh, R.D., and Tierney, G.L. (2001). Colder soils in a warmer world: a snow manipulation study in northern hardwood forest. Biogeochemistry *56*, 135–150.

Grover, M., Maheswari, M., Desai, S., Gopinath, K.A., and Venkateswarlu, B. (2015). Elevated CO_2: Plant associated microorganisms and carbon sequestration. Appl. Soil Ecol. *95*, 73–85.

Guthrie, P.D. (1986). Biological methanogenesis and the CO_2 greenhouse effect. J. Geophy. Res. *91*, 10847–10851.

Heath, J., Ayres, E., Possell, M., Bardgett, R.D., Black, H.I.J., Grant, H., Ineson, P., and Kerstiens, G. (2005). Rising atmospheric CO_2 reduces soil carbon sequestration. Science *309*, 1711–1713.

Heimann, M., and Reichstein, M. (2008). Terrestrial ecosystem carbon dynamics and climate feedbacks. Nature *451*, 289–292.

Henry, H., Juarez, J.D., Field, C.B., and Vitousek, P.M. (2005). Interactive effects of elevated CO_2, N deposition and climate change on extracellular enzyme activity and soil density fractionation in a Californian annual grassland. Global Change Biol. *11*, 1808–1815.

Högberg, P., and Read, D.J. (2006). Towards a more plant physiological perspective on soil ecology. Trends Ecol. Evol. *21*, 548–554.

Hu, S., Chapin, F.S., Firestone, M.K., Field, C.B., and Chiariello, N.R. (2001). Nitrogen limitation of microbial decomposition in a grassland under elevated CO_2. Nature *409*, 188–191.

Hutchin, P.R., Press, M.C., Lee, J.A., and Ashenden, T.W. (1995). Elevated concentrations of CO_2 may double methane emissions from mires. Global Change Biol. *1*, 25–128.

Ineichen, K., Wiemken, V., and Wiemken, A. (1995). Shoots, roots and ectomycorrhizal formation of pine seedlings at elevated atmospheric carbon dioxide. Plant Cell Environ. *18*, 703–707.

Insam, H., Bååth, E., Berreck, M., Frostegård, A., Gerzabek, M.H., Kraft, A., Schinner, F., Schweiger, P., and Tschuggnall, G. (1999). Responses of the soil microbiota to elevated CO_2 in an artificial tropical ecosystem. J. Microbiol. Methods *36*, 45–54.

IPCC (2013). Climate Change 2013: The Physical Science Basis (Cambridge: Cambridge University Press).

Jassey, V.E.J., Reczuga, M.K., Zielinska, M., et al. (2017). Tipping point in plant–fungal interactions under severe drought causes abrupt rise in peatland ecosystem respiration. Global Change Biol. *24*, 972–986.

Jenkinson, D.S., Adams, D.E., and Wild, A. (1991). Model estimates of CO_2 emissions from soil in response to global warming. Nature *351*, 304–306.

Johnson, D., Kresk, M., Wellington, E.M.H., Stott, A.W., Cole, L., Bardgett, R.D., Read, D.J., and Leake, J.R. (2005). Soil invertebrates disrupt carbon flow through fungal networks. Science *309*, 1047.

Jones, C., McConnell, C., Coleman, K., Cox, P., Falloon, P., and Jenkinson, D. (2005). Global climate change and soil carbon stocks; predictions from two contrasting models for the turnover of organic carbon in soil. Global Change Biol. 11, 154–166.

Karhu, K., Auffret, M.D., Dungait, J.A.J. et al. (2014). Temperature sensitivity of soil respiration rates enhanced by microbial community response. Nature *513*, 81–84.

Kelley, A.M., Fay, P.A., Polley, H.W., Gill, R.A., and Jackson, R.B. (2011). Atmospheric CO_2 and soil extracellular enzyme activity:a meta-analysis and CO_2 gradient experiment. Ecosphere *2*, UNSP96.

Kandeler, E., Tscherko, D., Bardgett, R.D., Hobbs, P.J., Kampichler, C., and Jones, T.H. (1998). The response of soil microorganisms and roots to elevated CO_2 and temperature in a terrestrial model ecosystem. Plant Soil *202*, 251–262.

Kang, H.J., Freeman, C., and Ashendon, T.W. (2001). Effects of elevated CO_2 on fen peat biogeochemistry. Sci. Total Environ. *279*, 45–50.

Keel, S.G., Siegwolf, R.T.W., and Körner, C. (2006). Canopy CO_2 enrichment permits tracing the fate of recently assimilated carbon in a mature deciduous forest. New Phytologist *172*, 319–329.

Keller, J.K. (2011). Wetlands and the global carbon cycle: what might the simulated past tell us about the future? New Phytol. *198*, 789-792.

Kolton, M., Marks, A., Wilson, R.M., et al. (2019). Impact of warming on greenhouse gas production and microbial diversity in anoxic peat rom a sphagnum-dominated bog (Grand Rapids, Minnesota, United States). Front. Microbiol. *26*,1-13.

Kim, S-Y., Freeman, C., Fenner, N., and Kang, H. (2012) Functional and structural responses of bacterial and methanogen communities to 3 year-warming incubation in different depths of peat mire. Appl. Soil Ecol. *57*, 23–30.

King, G.M. (1992). Ecological aspects of methane oxidation, a key determinant of global methane dynamics. Adv. Microb. Ecol. *12*, 431–468.

Kirschbaum, M.U.F. (2004). Soil respiration under prolonged soil warming: are rate reductions caused by acclimation or substrate loss? Global Change Biol. *10*, 1870–1877.

Kirschbaum, M.U.F. (2006). The temperature dependence of organic-matter decomposition—still a topic of debate. Soil Biol. Biochem. *38*, 2510–2518.

Klamer, M., Roberts, M.S., Levine, L.H., Drake, B.G., and Garland, J.L. (2002). Influence of elevated CO_2 on the fungal community in a coastal scrub oak forest soil investigated with terminal-restriction fragment length polymorphism analysis. Appl. Environ. Microbiol. *68*, 4370–4376.

Klironomos, J.N., Rillig, M.C., Allen, M.F., Zak, D.R., Kubiske, M., and Pregitzer, K.S. (1997). Soil fungal-arthropod responses to *Populus tremuloides* grown under enriched atmospheric CO_2 under field conditions. Global Change Biol. *3*, 473–478.

Kluber, H.D., and Conrad, R. (1998). Effects of nitrate, nitrite, NO and N_2O on methanogenesis and other redox processes in anoxic rice field soil. FEMS Microbiol. Ecol. *25*, 301–318.

Knorr, W., Prentice, I.C., House, J.I., and Holland, E.A. (2005). Longterm sensitivity of soil carbon turnover to warming. Nature *433*, 298–301.

Krivtsov, V., Bezginova, T., Salmond, R., Liddell, K., Garside, A., Thompson, J., Palfreyman, J.W., Staines, H.J., Brendler, A., Griffiths, B., and Watling, R. (2006). Ecological interactions between fungi, other biota and forest litter composition in a unique Scottish woodland. Forestry *79*, 201–216

Kuzyakov, Y. (2006). Sources of CO_2 efflux from soil and review of partitioning methods. Soil Biol. Biochem. *38*, 425–448.

Kuzyakov, Y., Horwath, W.R., Dorodnikov, M., and Blagodatskaya, E. (2019). Review and synthesis of the effects of elevated atmospheric CO_2 on soil processes: No changes in pools, but increased fluxes and accelerated cycles. Soil Biol. Biochem. *128*, 66–78.

Knorr, H.K. (2013). DOC-dynamics in a small headwater catchment as driven by redox fluctuations and hydrological flow paths – are DOC exports mediated by iron reduction/oxidation cycles? Biogeosciences 10, 891-904.

Laanbroek, H.J. (2010). Methane emission from natural wetlands: interplay between emergent macrophytes and soil microbial processes. A mini-review. Ann. Bot. *105*, 141-153.

Lewis, J.D., and Strain, B.R. (1996). The role of mycorrhizas in the response of *Pinus taeda* seedlings to elevated CO_2. New Phytol. *133*, 431–443.

Liu, W.X., Allison, S.D., Xia, J.Y., Liu, L.L., and Wan, S.Q. (2016). Precipitation regime drives warming responses of microbial biomass and activity in temperate steppe soils. Biol. Fert. Soils. *52*, 469–477.

Li, J.W., Wang, G.S., Mayes, M.A., Allison, S.D., Frey, S.D., Shi, Z., Hu, X.M., Luo, Y.Q., and Melillo, J.M. (2018a). Reduced carbon use efficiency and increased microbial turnover with soil warming. Global Change Biol. *25*, 900–910.

Li, G., Kim, S., Han, S.H., Chang, H., Du, D.L., and Son, Y. (2018b). Precipitation affects soil microbial and extracellular enzymatic responses to warming. Soil Biol. Biochem. *120*, 212–221.

Liu, Y., Chen, Q.M., Wang, Z.X., Zheng, H.F., Chen, Y.M., Chen, X., Wang, L.F., Li, H.J., and Zhang, J. (2019). Nitrogen addition alleviates microbial nitrogen limitations and promotes soil respiration in a subalpine coniferous forest. Forests *10*, 1038.

Liu, S., Ji, C., Wang, C., et al. (2018). Climatic role of terrestrial ecosystem under elevated CO_2: a bottom-up greenhouse gases budget. Ecol. Lett. 21,1108-1118.

Luo, Y., Wan, S., and Hui, D. (2001). Acclimization of soil respiration to warming in tall grass prairie. Nature *413*, 622–625.

Malik, A.A., Martiny, J.B.H., Brodie, E.L., Martiny, A.C., Treseder, K.K., and Allison, S.D. (2019). Defining trait-based microbial strategies with consequences for soil carbon cycling under climate change. ISME J. *14*, 1–9.

Manzoni, S., Taylor, P., Richter, A., Porporato, A., and Agren, G.I. (2012). Environmental and stoichiometric controls on microbial carbon-use efficiency in soils. New Phytol. *196*, 79–91.

McKinley, D.C., Romero, J.C., Hungate, B.A., Drake, B.G., Megonigal, J.P. (2009). Does deep soil N availability sustain long-term ecosystem responses to elevated CO_2? Global Change Biol. *15*, 2035-2048.

McPartland, M.Y., Kane, E.S., Falkowski, M.J., et al. (2018). The response of boreal peatland community composition and NDVI to hydrologic change, warming, and elevated carbon dioxide. Global Change Biol. *25*, 93-107.

Megonigal, J.P., and Schlesinger, W.H. (1997). Enhanced CH_4 emissions from a wetland soil exposed to Elevated CO_2. Biogeochemistry *37*, 77–88.

Melillo, J.M., Steudler, P.A., Aber, J.D., Newkirk, K., Lux, H., Bowles, F.P., Catricala, C., Magill, A., Ahrens, T., and Morrisseau, S. (2002). Soil warming and carboncycle feedbacks to the climate system. Science *298*, 2173–2175.

Monson, R.K., Lipson, D.L., Burns, S.P., Turnipseed, A.A., Delany, A.C., Williams, M.W., and Schmidt, S.K. (2006). Winter soil respiration controlled by climate and microbial community composition. Nature *439*, 711–714.

Montealegre, C.M., van Kessel, C., Russelle, M.P., and Sadowsky, M.J. (2002). Changes in microbial activity and composition in a pasture ecosystem exposed to elevated atmospheric carbon dioxide. Plant Soil *243*, 197–207.

Moorhead, D.L., and Linkins., A.E. (1997). Elevated CO_2 alters belowground exoenzyme activities in tussock tundra. Plant Soil *189*, 321–329.

Nardo, C.D., Cinquegrana, A., Papa, S., Fuggi, A., and Fioretto, A. (2004). Laccase and peroxidase isoenzymes during leaf litter decomposition of *Quercus ilex* in a Mediterranean ecosystem. Soil Biol. Biochem. *36*, 1539–1544.

Niklaus, P.A., and Körner, C. (1996). Responses of soil microbiota of a late successional alpine grassland to long term CO_2 enrichment. Plant Soil *184*, 219–229.

Niklaus, P.A., Wohlfender, M., Siegwolf, R., and Körner, C. (2001). Effects of six years atmospheric CO_2 enrichment on plant, soil, and soil microbial C of a calcareous grassland. Plant Soil *233*, 189–202.

Ochoa-Hueso, R., Hughes, J., Delgado-Baquerizo, M., Drake, J.E., Tjoelker, M.G., Pineiro, J., and Power, S.A. (2017). Rhizosphere-driven increase in nitrogen and phosphorus availability under elevated atmospheric CO_2 in a mature *Eucalyptus* woodland. Plant Soil *416*, 283–295.

O'Neill, E. (1994). Responses of soil biota to elevated atmospheric carbon dioxide. Plant Soil *165*, 55–65.

Oldfield, E.E., Crowther, T.W., and Bradford, M.A. (2018). Substrate identity and amount overwhelm temperature effects on soil carbon formation. Soil Biol. Biochem. *124*, 218–226.

Preece, C., Verbruggen, E., Liu, L., Weedon, J.T., and Penuelas, J. (2019). Effects of past and current drought on the composition and diversity of soil microbial communities. Soil Biol. Biochem. *131*, 28–29.

Pregitzer, K.S., Zak, D.R., Maziasz, J., DeForest, J., Curtis, P.S., and Lussenhop, J. (2000). Interactive effects of atmospheric CO_2 and soil-N availability on fine roots of *Populus tremuloides*. Ecol. Appl. *10*, 18–13.

Prentice, I., Farquhar, G., Fasham, M., et al. (2001).The carbon cycle and atmospheric carbon dioxide. In Climate Change 2001: The Scientific Basis. Contributions of Working Group I to the Third Assessment Report of the Intergovernmental Panel on Climate Change, J.T. Houghton, Y. Ding, D.J. Griggs, M. Noguer, P.J. van der Linden, et al., eds. (Cambridge, UK: Cambridge Univ. Press), pp. 183–238.

Rey, A., and Jarvis, P. (2006). Modelling the effect of temperature on carbon mineralization rates across a network of European forest sites (FORCAST). Global Change Biol. *12*, 1894–1908.

Rillig, M.C., and Mummey, D.L. (2006). Mycorrhizas and soil structure. New Phytol. *171*, 41–53.

Robroek, B.J.M., Jassey, V.E.J., Kox, M.A.R., Berendsen, R.L., Mills, R.T.E., Cecillon, L., Puissant, J., Meima-Franke, M., Bakker, P.A.H.M., and Bodelier, P.L.E. (2015). Peatland vascular plant functional types affect methane dynamics by altering microbial community Structure. J. Ecol. 103, 925–934.

Rogers, H.H., Prior, S.A., and O'Neill, E.G. (1992). Cotton root and rhizosphere responses to free-air CO_2 enrichment. Crit. Rev. Plant Sci. *11*, 251–263.

Rouhier, H., Billes, G., El Kohen, A., Mousseau, M., and Bottner, P. (1994). Effect of elevated CO_2 on carbon and nitrogen distribution within a tree (*Castanea sativa* Mill.)-soil system. Plant Soil *162*, 281–292.

Roulet, N.T., and Moore, T.R. (1995). The effect of forestry drainage practices on the emissions of methane from northern peatlands. Can. J. Forest Res. *25*, 491–499.

Runion, G.B., Curl, E.A., Rogers, H.H., Backman, P.A., Rodriguez-Kabana, R., and Helms, B.E. (1994). Effects of free-air CO_2 enrichment on microbial on microbial populations in the rhizosphere and phyllosphere of cotton. Agric. For. Meteorol. *70*, 117–130.

Runion, G.B., Mitchell, R.J., Rogers, H.H., Prior, S.A., and Counts, T.K. (1997). Effects of nitrogen and water limitation and elevated atmospheric CO_2 on ectomycorrhiza of longleaf pine. New Phytol. *137*, 681–689.

Saarnio, S., Alm, J., Martikainen, P.J., and Silvola, J. (1998). Effects of raised CO_2 on potential CH_4 production and oxidation in, and CH_4 emission from, a boreal mire. Ecology *86*, 261–268.

Schimel, D.S., Braswell, B.H., Holland, E.A., McKeown, R., Ojima, D.S., Painter, T.H., Parton, W.J., and Townsend, A.R. (1994). Climatic, edaphic, and biotic controls over storage and turnover of carbon in soils, Global Biogeochem. Cycles *8*, 279–293.

Schimel, J.P., Balser, T.C., and Wallenstein, M. (2007). Microbial stress-response physiology and its implications for ecosystem function. Ecology *88*, 1386–1394.

Schortemeyer, M., Dijkstra, P., Johnson, D.W., and Drake, B.G. (2000). Effects of elevated atmospheric CO_2 concentration on C and N pools and rhizosphere processes in a Florida scrub oak community. Global Change Biol. *6*, 383–391.

Schortemeyer, M., Hartwig, U.A., Hendrey, G.R., and Sadowsky, M.J. (1996). Microbial community changes in the rhizospheres of white clover and perennial ryegrass exposed to free air carbon dioxide enrichment (FACE). Soil Biol. Biochem. *28*, 1717–1724.

Schrope, M.K., Chanton, J.P., Allen, L.H., and Baker, J.T. (1999). Effect of CO_2 enrichment and elevated temperature on methane emissions from rice, *Oryza sativa*. Global Change Biol. *5*, 587– 599.

Shackle, V., Freeman, C., and Reynolds, B. (2000). Carbon supply and the regulation of enzyme activity in constructed wetlands. Soil Biol. Biochem. 32, 1935–1940.

Sun, J.F., Xia, Z.W., He, T.X., et al. (2017). Ten years of elevated CO_2 affects soil greenhouse gas fluxes in an open top chamber experiment. Plant Soil *420*, 435–450.

Six, J., Frey, S.D., Thiet, R.K., and Batten, K.M. (2006). Bacterial and fungal contributions to carbon sequestration in agroecosystems. Soil Sci. Soc. Am. J. *70*, 555–569.

Silva-Sanchez, A., Soares, M., and Rousk, J. (2019). Testing the dependence of microbial growth and carbon use efficiency on nitrogen availability, pH, and organic matter quality. Soil Biol. Biochem. *134*, 25–35.

Sihi, D., Inglett, P.W., and Inglett, K.S. (2016). Carbon quality and nutrient status drive the temperature sensitivity of organic matter decomposition in subtropical peat soils. Biogeochemistry *131*, 103–119.

Soares, M., and Rousk, J. (2019). Microbial growth and carbon use efficiency in soil: Links to fungal-bacterial dominance, SOC-quality and stoichiometry. Soil Biol. Biochem. *131*, 195–205.

Song, Y.Y., Song, C.C., Ren, J.S., Zhang, X.H., and Jiang, L. (2019a). Nitrogen input increases *Deyeuxia angustifolia* litter decomposition and enzyme activities in a marshland ecosystem in Sanjiang Plain, Northeast China. Wetlands *39*, 549–557.

Song, J., Wan, S., Piao, S., et al. (2019b). A meta-analysis of 1,119 manipulative experiments on terrestrial carbon-cycling responses to global Change. Nat. Ecol. Evol. *3*, 1309–1320.

Staddon, P.L., Jakonsen, I., and Blum, H. (2004). Nitrogen input mediates the effects of free-air CO_2 enrichment on mycorrhizal fungal abundance. Global Change Biol. *10*, 1687–1688.

Ström, L., Mastepanov, M., and Christensen, T.R. (2005). Species specific effects of vascular plants on carbon turnover and methane emissions from wetlands. Biogeochemistry *75*, 65–82.

Tveit, A.T., Urich, T., Frenzel, P., and Svenning, M.M. (2015). Metabolic and trophic interactions modulate methane production by Arctic peat microbiota in response to warming. PNAS *112*, E2507–E2516.

Toberman, H., Freeman, C., Evans, C., Fenner, N., and Artz, R.R.E. (2008). Summer drought decreases soil fungal diversity and associated phenol oxidase activity in upland *Calluna* heathland soil FEMS Microbiol. Ecol. *66*, 426–436.

Trumbore, S. (2006). Carbon respired by terrestrial ecosystems—recent progress and challenges. Global Change Biol. *12*, 141–153.

Walker, T.W.N., Kaiser, C., Strasser, F., Herbold, C.W., Leblans, N.I.W., Woebken, D., Janssens, I.A., Sigurdsson, B.D., and Richter, A. (2018). Microbial temperature sensitivity and biomass change explain soil carbon loss with warming. Nat. Clim. Change *8*, 1121–1021.

Wallenstein, M.D., Haddix, M.L., Lee, D.D., Conant, R.T., and Paul, E.A. (2012). A litter-slurry technique elucidates the key role of enzyme production and microbial dynamics in temperature sensitivity of organic matter decomposition. Soil Biol. Biochem. *47*, 18–26.

Walker, R.F., Geisinger, D.R., Johnson, D.W., and Ball, J.T. (1997). Elevated atmospheric CO_2 and soil N fertility effects on growth, mycorrhizal colonization, and xylem water potential of juvenile ponderosa pine in a field soil. Plant Soil *195*, 25–36.

Wang, C., Liu, D., and Bai, E. (2018a). Decreasing soil microbial diversity is associated with decreasing microbial biomass under nitrogen addition. Soil Biol. Biochem.*120*,126-133.

Wang, C., Jin, Y.G., Ji, C., et al. (2018b). An additive effect of elevated atmospheric CO_2 and rising temperature on methane emissions related to methanogenic community in rice paddies. Agr. Ecosyst. Environ. *257*, 165–174.

Wang, H., Yu, L.F., Zhang, Z.H., Liu, W., Chen, L.T., Cao, G.M., Yue, H.W., Zhou, J.Z., Yang, Y.F., Tang, Y.H., and He, J.S. (2017). Molecular mechanisms of water table lowering and nitrogen deposition in affecting greenhouse gas emissions from a Tibetan alpine wetland. Global Change Biol. *23*, 815–829.

Wang, H.J., Richardson, C.J., and Ho, M.C. (2015). Dual controls on carbon loss during drought in peatlands. Nat. Clim. Change *5*, 584–587.

Wang, Y.H., Yu, Z.H., Li, Y.S., Wang, G.H., Tang, C., Liu, X.B., Liu, J.J., and Xie, Z.H. (2019). Elevated CO_2 alters the structure of the bacterial community assimilating plant-derived carbon in the rhizosphere of soya bean. Eur. J. Soil Sci. *70*, 1212–1220.

Wang, B., and Adachi, K. (1999). Methane production in a flooded soil in response to elevated atmospheric carbon dioxide concentrations. Biol. Fertil. Soils *29*, 218–220.

Ward, S.E., Bardgett, R.D., McNamara, N.P., Adamson, J.K., and Ostle, N.J. (2007). Long-term consequences of grazing and burning on northern peatland carbon dynamics. Ecosystems *10*, 1069–1083.

Wardle, D.A. (2002). Communities and Ecosystems: Linking the Aboveground and Belowground Components. (Princeton, USA: Princeton University Press).

Wiemken, V., Laczko, E., Ineichen, K., and Boller, T. (2001). Effects of elevated carbon dioxide and nitrogen fertilization on mycorrhizal fine roots and the soil microbial community in Beech- Spruce ecosystems on siliceous and calcareous soil. Microb. Ecol. *42*, 126–135.

Williams, M.A., Rice, C.W., and Owensby, C.E. (2000). Carbon dynamics and microbial activity in tallgrass prairie exposed to elevated CO_2 for 8 years. Plant Soil *227*, 127–137.

Widder, S., Allen, R.J., Pfeiffer, T., et al. (2016). Challenges in microbial ecology: building predictive understanding of community function and dynamics. ISME J. *10*, 2557–2568.

Wieder, W.R., Bonan, G.B., and Allison, S.D. (2013). Global soil carbon projections are improved by modelling microbial processes. Nat. Clim. Change *3,* 909–912.

Xia, W.W., Jia, Z.J., Bowatte, S., and Newton, P.C.D. (2017). Impact of elevated atmospheric CO_2 on soil bacteria community in a grazed pasture after 12-year enrichment. Geoderma *285*, 19–26.

Yang, S.H., Zheng, Q.S., Yuan, M.T., et al. (2019). Long-term elevated CO_2 shifts composition of soil microbial communities in a Californian annual grassland, reducing growth and N utilization potentials. Sci. Total Environ. *652*, 1474–1481.

Yu, Z.H., Li, Y.S., Hu, X.J., et al. (2018a). Elevated CO_2 increases the abundance but simplifies networks of soybean rhizosphere fungal community in Mollisol soils. Agr. Ecosyst. Environ. *264*, 94–98.

Yu, H., Deng, Y., He, Z.L., et al. (2018b). Elevated CO_2 and warming altered grassland microbial communities in soil top-layers. Front. Microbiol. *9*, 1790.

Zak, D.R., Pregitzer, K.S., Curtis, P.S., and Holmes, W.E. (2000b). Atmospheric CO_2 and the composition and function of soil microbial communities. Ecol. Appl. *10*, 47–59.

Zak, D.R., Pregitzer, K.S., Curtis, P.S., Teeri, J.A., Fogel, R., and Randlett, D.L. (1993). Elevated atmospheric CO_2 and feedback between carbon and nitrogen cycles. Plant Soil *151*, 105–117.

Zanetti, S., Hartwig, U.A., Lüscher, A., Hebeisen, T., Frehner, M., Fischer, B.U., Hendrey, G.R., Blum, H., and Nösberger, J. (1996). Stimulation of symbiotic N_2 fixation in *Trifolium repens* L. under elevated atmospheric pCO_2 in a grassland ecosystem. Plant Physiol. *112*, 575–583.

Zibilske, L.M., and Bradford, J.M. (2007). Oxygen effects on carbon, polyphenols, and nitrogen mineralization potential in soil. Soil Sci. Soc. Am. J. *71*, 133–139.

Zhou, Z., Wang, C., and Luo, Y. (2018). Response of soil microbial communities to altered precipitation. Global Ecol. Biogeogr. *27*, 1121–1136.

Zhou, J., Deng, Y., Luo, F., He, Z., Tu, Q., and Zhi, X. (2010). Functional molecular ecological networks. mBio. *1*, e00169–10.

Chapter 11

Climate Change and Nitrogen Turnover in Soils and Aquatic Environments

Gero Benckiser

Justus Liebig University Giessen, Giessen, Germany

Email: gero.benckiser@umwelt.uni-giessen.de

DOI: https://doi.org/10.21775/9781913652579.11

Abstract

Many more than 10,500 soil types link the bio-, geo-, aqua- with the atmosphere. After the invention of technical N_2-fixation (TNF) by Haber and Bosch (H-B) the BNF (biological N_2-fixation) input was almost doubled and the bio-, geo-, and aquaspheres are now the largest greenhouse gas flux drivers (CO_2, CH_4, NH_3, NO, N_2O) on Earth. After introduction of NH_4^+ by BNF and TNF and its conversion to NO_3^- by autotrophic nitrifying bacteria and archaea, denitrifying bacteria, archaea, and fungi together with plants as gas conduits return after reduction of the oxidized N species $NO_3^-$and NO_2^- the gaseous N species NO, N_2O and N_2 to the atmosphere.

The almost doubled N-input after H-B invention enables farmers to nearly approach Nature's high productivity, which is based on N shortage and biodiversity, by monocultures, often surpassing plant N demand and soil N buffer capacity. In consequence, increasing amounts of NO_3^- flow towards the groundwater and CO_2, CH_4, and N_2O emissions enhance the atmospheric temperature. Particularly Western Asian and Northern African regions with characteristic low and unpredictable rainfall, long dry seasons, scarce water resources, rural poverty, high dependence on limited cropping/livestock agriculture, and low levels of technological adaptations have to suffer under greenhouse gas effects. In approaching Nature's productivity achievements and being successful in food security and stabilizing ecosystem functioning, farmers and industry TNF product designers must understand (a) the mechanisms behind the individual genes holding promise of a better N absorption by an adapted germplasm and (b) how pollution costs are reducible. On both aspects scientist are doing concentrated research and on the concerned progress in aquatic and soil ecosystems nitrogen cycling this chapter is focusing.

Introduction

Marine and freshwater submerged soils ('Hydroaquents') cover about 70.7% of the 510 million km^2 of the earth's surface and terrestrial soils the soil taxonomy of the USA subdivides into many more than 10,500 aerated, occasionally and permanently submerged soil types, covering 29.3% of the earth's surface (Eurostat, 2014). Within terrestrial soils, 7.9% are associated with forests and 9.6% with agriculture. The water bodies above 'Hydroaquents' are increasingly used for aquaculture (Fig. 11.1) with predominating anaerobic and fermentation as major energy gaining processes. Anaerobic respiration adds to conserve carbon and nitrogen cycle related energy (Fig. 11.2; Benckiser, 1997; FAO, 2013). The carbon and nitrogen cycles interconnect the life functioning by being largest greenhouse gas (CO_2, CH_4, NH_3, NO, N_2O) flux drivers on earth (Delwiche, 1977; Field and Raupach, 2004; Benckiser and Schnell, 2007; Ravishankara et al., 2009; Ottow, 2011; Bottomley, 2012). The lithosphere (mantle- derived rocks, crust, and sediments) primarily stores carbon residues, the bio-, hydro- and atmosphere much less (0.05%), whereas nitrogen residues are primarily in the atmosphere (78.08% (v/v) as N_2 and in traces as NO, NO_2, N_2O and NH_3 with comparably small amounts bound in bio- and pedosphere or dissolved in water (Capone et al., 2006). Nitrate ions are water dissolved mobile, move through aquatic and terrestrial environments before they are reduced to NO, N_2O, N_2 and NH_3 and returned to the atmosphere.

During the last decades, increasingly installed wastewater treatment plants and aquaculture systems in marine and limnic environments (Fig. 11.1) improve besides well-managed soils the living standards in economically developing countries. Fishes grown inside of nets in marine, limnic environments are fed intensively with N rich food that favours activity of autotrophic nitrifying bacteria and archaea, which oxidize protein derived NH_4^+ to NO_3^- in the water body and the soil sediments below. Biofilms develop on fishpond walls of limnic aquaculture systems and nitrate used as alternative electron acceptor is a source of N_2O (Briggs and Funge-Smith, 1994; Sime-Ngando et al., 2011; FAO, 2013; Flemming and Wuertz, 2019). Protozoa and nematodes also attracted biofilm forming bacteria and archaea by N rich fish food graze on the bacterial lawns, excrete surplus NH_4^+ and thus accelerate N cycling, nitrification and denitrification. The linked food chain benefits (Fig. 11.2; Bloem et

A

B

Figure 11.1 Fishponds in the Mediterranean Sea close to the island of Malta, in which tuna are raised (photo Benckiser, 2012).

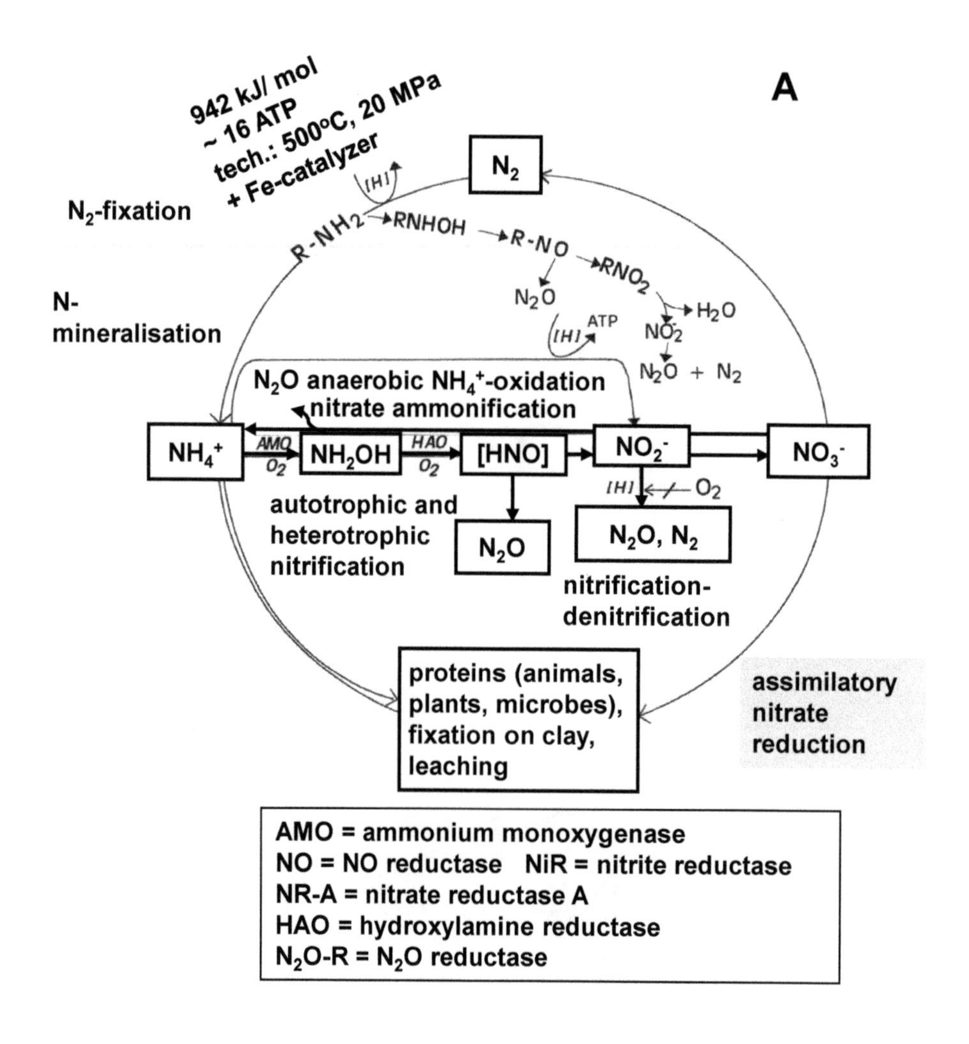
A
942 kJ/ mol
~ 16 ATP
tech.: 500°C, 20 MPa
+ Fe-catalyzer
N2
N2-fixation
R-NH2 → RNHOH → R-NO → RNO2
N2O
H2O
NO2
N2O + N2
ATP
N-mineralisation
N2O anaerobic NH4+-oxidation
nitrate ammonification
NH4+
AMO
O2
NH2OH
HAO
O2
[HNO]
NO2-
NO3-
O2
autotrophic and heterotrophic nitrification
N2O
N2O, N2
nitrification-denitrification
proteins (animals, plants, microbes), fixation on clay, leaching
assimilatory nitrate reduction
AMO = ammonium monoxygenase
NO = NO reductase NiR = nitrite reductase
NR-A = nitrate reductase A
HAO = hydroxylamine reductase
N2O-R = N2O reductase

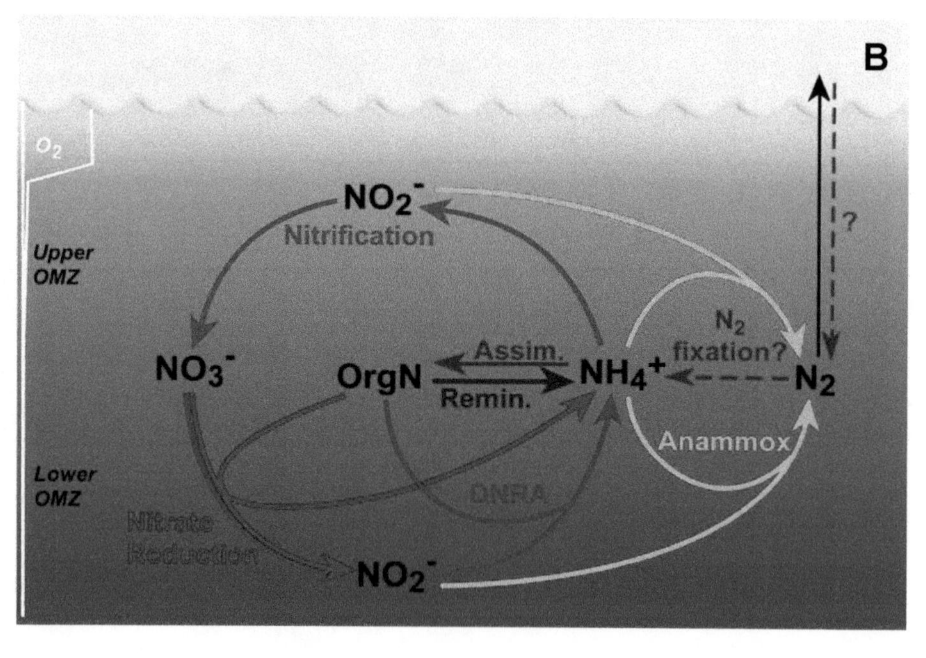
B
O2
Upper OMZ
Lower OMZ
NO2-
Nitrification
NO3-
OrgN
Assim.
Remin.
NH4+
N2 fixation?
N2
?
Anammox
DNRA
Nitrate Reduction
NO2-

Figure 11.2 Nitrogen cycle in terrestrial ecosystems (A; Benckiser, 1997) and the Peruvian oxygen minimum zone (OMZ) of the Eastern Tropical South Pacific (B; Reproduced with permission from Lam et al., 2009. Proc. Natl. Acad. Sci. U.S.A. *106*, 4752–4757). **Anaerobic ammonium oxidation** Anammox (*yellow*) is directly coupled to NO_3^-–reduction (*red*) and aerobic ammonia oxidation (first step of nitrification, *green* and source of NO_2) = predominant pathway for nitrogen loss in the Peruvian OMZ. The required NH_4^+ by anammox originates from **dissimilatory nitrate reduction to ammonium** (DNRA, *blue*) and remineralization of organic matter via nitrate reduction and probably from micro-aerobic respiration. Micro-aerobic conditions in OMZ are the occurrence of nitrification, which diminishes in importance from shelf to open ocean and lower OMZ. NH_4^+-production caused by NO_3^--reduction and DNRA occurs in the lower OMZ and offshore. **Assimilation** (Assim., *grey*). **Remineralization** (Remin., *brown*). **Nitrogen fixation,** spatially coupled to nitrogen loss near the OMZ (*grey dashes*).

al., 1988; Sturm et al., 1994; Bonkowski and Clarholm, 2016; Cowan et al., 2019; Gu et al., 2019).

The breaking of atmospheric N≡N into its N atoms to form NH_4^+ by BNF or TNF requires almost 10^3 kJ/mol to fix about 3.9 x 10^9 Tg TNF annually (Fig. 11.3). Roughly 1.0 × 10^4 and 1.7 × 10^5 Tg of nitrogen flow into plants and animals, respectively, and in the soil clay–humus–biomass fraction, seawater and sediments are intermittently stored around 0.5 × 10^6, 9.9 × 10^4 and 2.0 × 10^8 Tg nitrogen (Fig. 11.4; Delwiche, 1977). Not into biomass, N-compounds incorporated NH_4^+ and its oxidation product, the mobile nitrate ions, pollute surface and ground (drinking) waters or move into the atmosphere as gaseous N_2O (Figs 11.2- 11.6; Ravishankara et al, 2009; Reid and Greene, 2012; Kanter and Searchinger, 2018). Now the world is expecting and awaiting under the umbrella of biotechnology the development of (a) an *inter alia* germplasm improvement and a monoculture systems adapted TNF addition, (b) a reemphasizing of legumes and symbiotic N_2-fixation, (c) site-(region-) specific, on mycorrhiza, N_2 fixation support based inter cropping plant growth schemes, (d) a better understanding of nutrient transport traits including exudation, (e) a marriage of plant N demand and food security combining strategies, (f) a stronger considering of N in global climate orientated agri- and aquaculture as well as animal stocking management, and (g) environmental pollution concerned industrial arrangements (Vance, 2001; Ravishankara et al, 2009; Steffen et al., 2011).

Oxygenases and reductases liberate sun energized, in glycosylic, aromatic, proteinous compounds, NH_4^+ or H_2O captured electrons and transfer them together with H^+ on CO_2 to form biomass or on O_2, NO_3^- and various other oxides to generate energy (Figs 11.2, 11.4, 11.5; Benckiser, 1997; Gralnick and Newman, 2007; Lam et

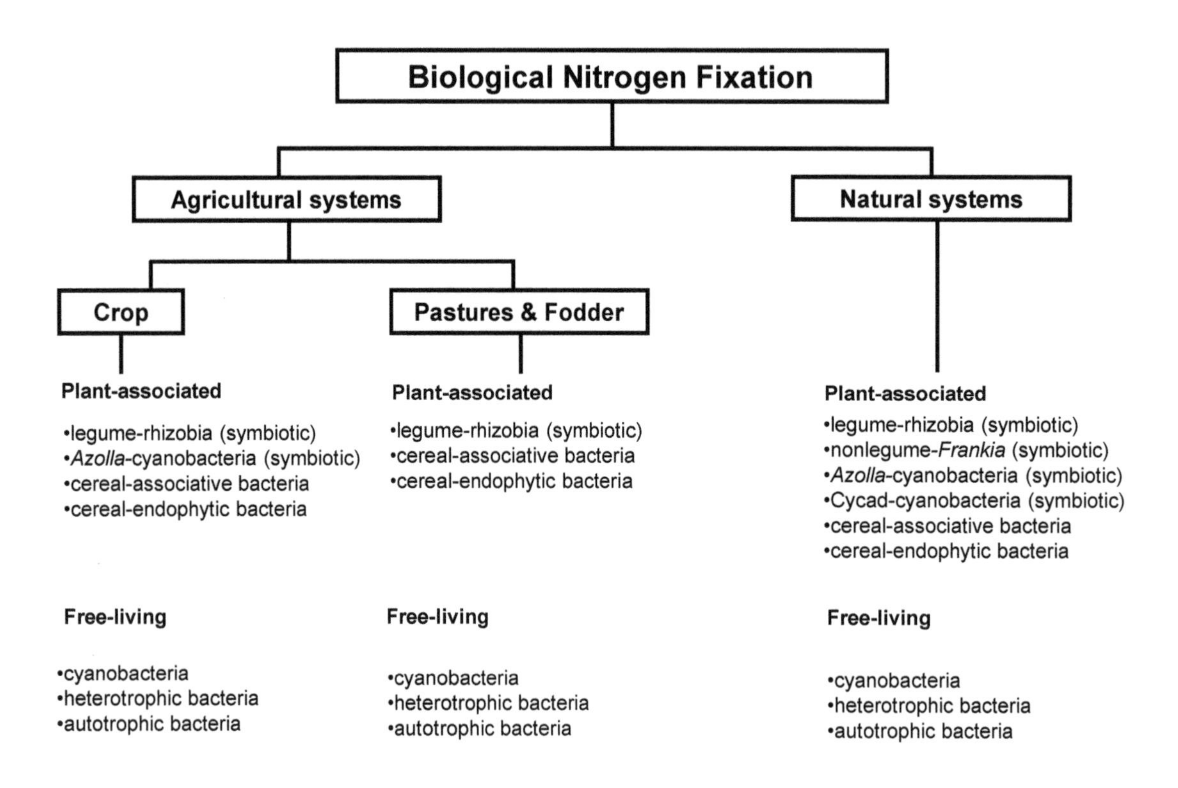

Figure 11.3 Biological nitrogen fixation in agricultural and natural systems (from Herridge et al., 2008).

al., 2009; Fuchs et al., 2011). To supply themselves adequately with nutrients plants translocate huge amounts of photosynthates into their rhizospheres in order to attract microbes by enslaving rhizobia symbionts (Benckiser, 2007). With a fine-tuned information exchange between above and belowground plant parts, high biomass productivity is possible. This is achievable via plant produced nitrification inhibitors, controlling microbial multiplication and nitrate, N_2O formation, and a TNF, BNF based N nutrition to which spot wise distributed animal excrements and a wide C/N ratio litter fall contribute (typical plant material C/N ratios: barley, 85:1; rye, 82:1; wheat, 80:1; oat straw 70:1; corn stalks 60:1; pea straw, 29:1; rye cover crop, 26:1; mature alfalfa, 25:1; ideal microbe diet, 24:1; clover, 21:1; manure, 17:1; young alfalfa, 15:1; hairy vetch, 11:1; root residue, humus, 10:1; soil microbes, 3–10:1) (https://www.nrcs.usda.gov/wps/PA.../download?cid...ext=pdf; Delwiche, 1977; Benckiser, 1997; Gralnick and Newman, 2007; Ma et al, 2008; Majumdar, 2008; Lam et al., 2009; Meng et al., 2009; Abasi et al, 2011; Fuchs et al., 2011; Carvalhais et al.,

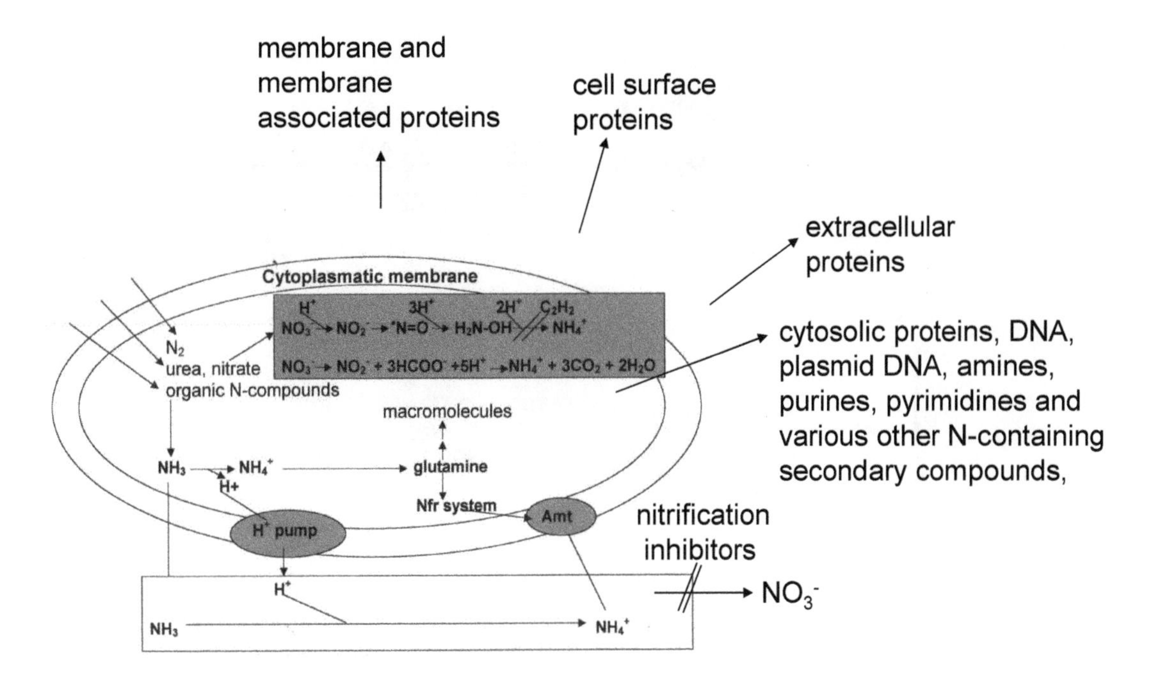

Figure 11.4 Uptake of biologically and technically fixed N_2, NH_3, urea, nitrate and organic N-compounds into plant cells and the exchange of surplus N with the surrounding environment. The about 3.1% N in biomass is N loss source. With plant excreted nitrification inhibitors Nature tries to minimize N losses (adapted from Kennedy et al.,1997; Simon, 2002; Horchani et al., 2011; Benckiser et al., 2013; Nacry et al., 2013).

2011; Brückner and Mösch, 2012; Gleason et al., 2012; Göpel and Görke, 2012; Baikova et al., 2013; Subbaro et al., 2017; Kanter and Searchinger, 2018).

Monoculturing farmers try to compensate N shortage with TFN fertilization to approach nature's high productivity based on N shortage and biodiversity, because they experience that satisfying plant N demand leads to higher productivities. Scientists invest hard work to understand the mechanisms behind the individual genes holding promise of a more effective N utilization and to improve technical N_2 fixation (Chen et al., 2018; Andersen et al., 2019; Flemming and Wuertz, 2019; Benckiser 2021). Plants, through their root thickness limited in entering small, nutrient rich soil pores, translocate specifically composed photosynthesis products into their rhizospheres for selecting out of about 10^{30} microbes, produced on Earth per year, the right ones as symbionts (Benckiser and Schnell, 2007; Reid and Greene, 2012; Nacry

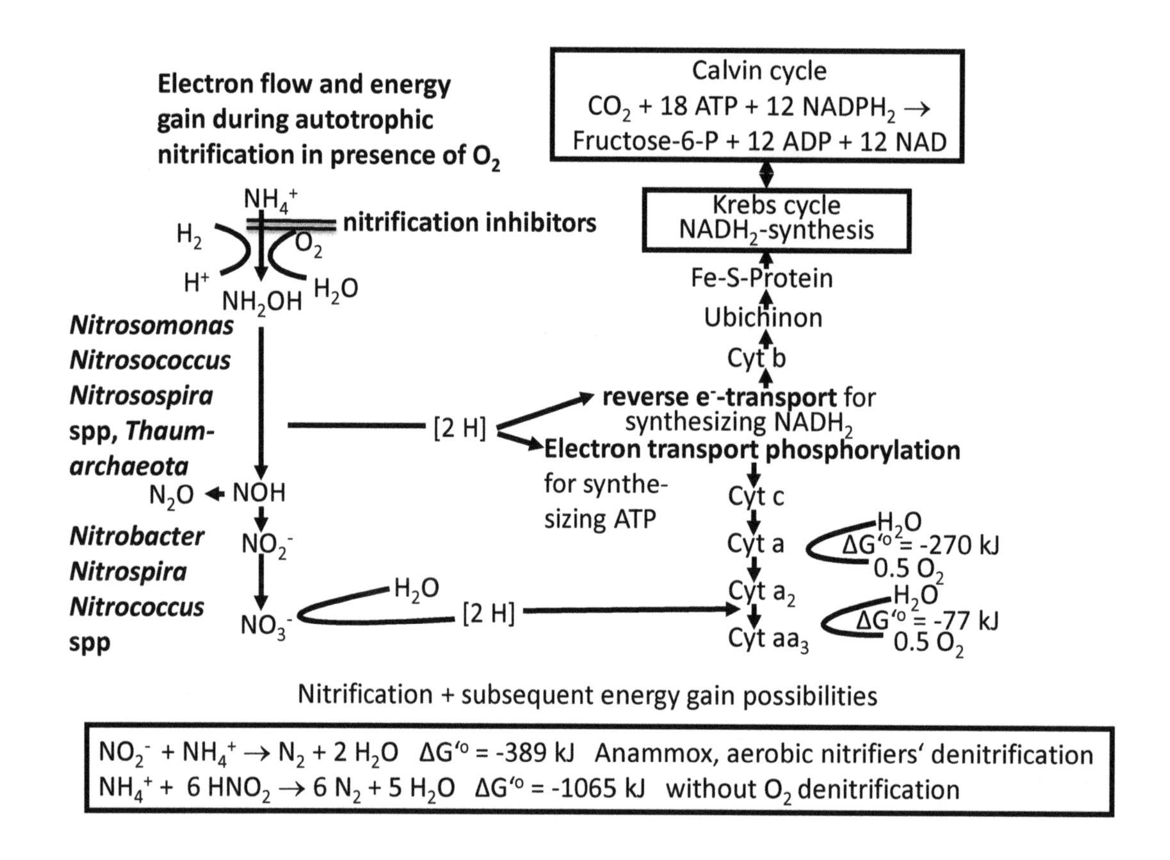

Figure 11.5 The nitrification process and subsequent denitrification possibilities to gain energy anaerobically. With nitrification inhibitor (NI) stabilized nitrogen-fertilizers, ammonium oxidation in the field can be controlled.

et al., 2013). With modern and available techniques as ribosomal 16S-RNA (16S-rRNA) sequencing, with stable isotopes combined fluorescence-in-situ-hybridisation (FISH), modern imaging technologies as Raman spectroscopy, NanSIMS (nano-secondary ion mass spectroscopy), and many other analytic methods we are able visualizing and quantifying single cell activities and ongoing transactions at the soil atom level, deciphering step by step the importance of quorum sensing, the role of viruses in controlling microbes, the capabilities of microbes, plants and animals for making nutrients available, and understanding N cycling details better (Sturm et al., 1994; Vance, 2001; Vogel and Babel, 2007; Montzka et al., 2011; Sutton et al., 2011; de Long, 2012; Reid and Greene, 2012; Santos et al., 2012; Nacry et al., 2013; Asano and Wagai, 2014; Hartmann et al., 2014; Daims et al., 2015; Vincent et al., 2017; Kitzinger et al, 2019; Vereecken et al., 2019; Benckiser, 2017, 2019, 2021). With N over-fertilization, which guarantees food security, simultaneously the climate relevant gases CO_2, N_2O, and CH_4 are produced in huge amounts, although we need to reduce these emissions significantly. To understand, how these gaseous emissions are released and can be controlled, not only the American Academy of Microbiology recommends concentrating research on CO_2-assimilation, methane formation, N_2-fixation, nitrification and denitrification; the challenges to improve technical N_2 fixation in an industrializing world will be met only when the mechanisms behind the individual genes for nitrogen absorption are fully understood and incorporated into an adapted germplasm (Vance, 2001; Montzka et al., 2011; Sutton, 2011; Selvaraju, 2013; Benckiser, 2021).

Following the American Academy of Microbiology recommendation, this chapter summarizes our present knowledge on nitrogen cycling. It discusses perspectives how landscapes, soils, surface, ground- and waste waters, and an atmosphere less polluted by reactive nitrogen species as NH_3, NO_3^-, NO_2^-, NO, NO_2, N_2O from agri- and aquaculture could be approached. Only in the European Union (EU) nitrogen pollution causes costs estimated between € 70 billion € (US$ 100 billion) and 320 billion € per year by increasing atmospheric temperature that hits presently particularly the Western Asia and Northern Africa (WANA) regions (Sutton, 2011; Selvaraju, 2013).

Nitrogen fluxes between atmosphere, terrestrial and aquatic environments

In the pre-industrial era, only BNF replenished the soil N pool. Based on BNF ($N_2 + 6\ e^- + 8H^+ \rightarrow 2\ NH_4^+$; Fig. 11.2) related N shortage nature evoluted high biodiversity and a high productivity. Since the invention of technical N_2 fixation (TNF) by Haber and Bosch in the beginning of the last century (Fig. 11.6), considered as the most important invention of the twentieth century, scientist try to improve its N fixing efficiency under reduced energy consumption and agriculturists to increase yields (Smil, 2001; Montoya et al., 2004; Arrigo, 2005; Arashiba et al., 2011; Glibert et al., 2014; Manabe et al., 2017; Chen et al., 2018; Bezdek and Chirik, 2019; Gruber, 2019; Horel et al., 2019; Benckiser, 2021). TNF enables manipulating N inputs and losses by leaching, denitrification and ammonia volatilization in the manifold types of agricultural systems, which had developed as monocultures for around 3000 years. For eliminating the gap between BNF and crop N demand. in monocultures, e.g., 120

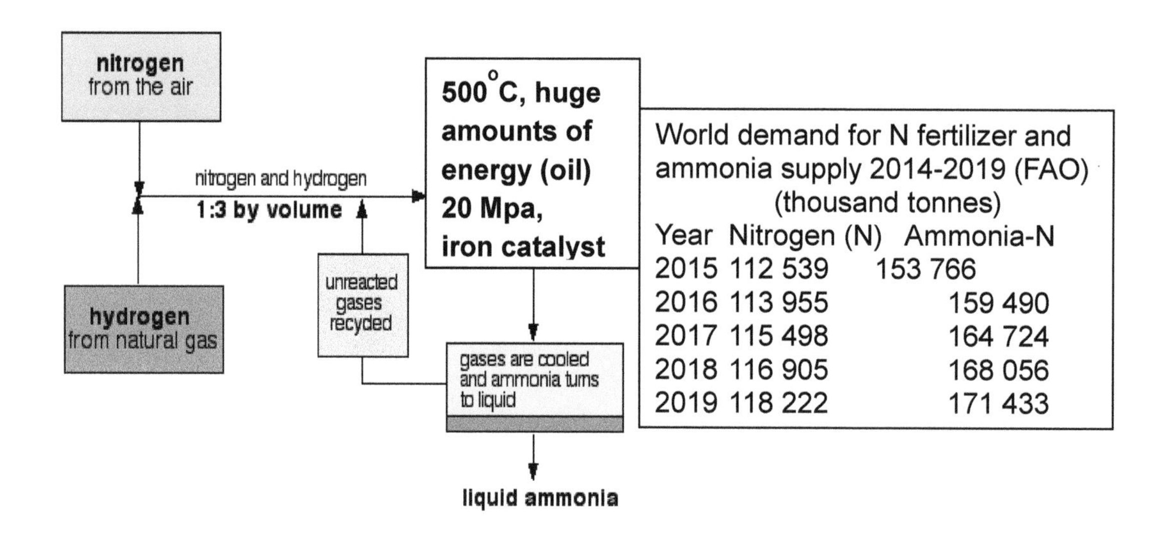

Figure 11.6 Technical N≡N fixation (TNF), invented by Haber and Bosch, and the World demand for N fertilizer and ammonia supply. At least in the industrialized world TNF almost doubled the biological N≡N fixation (BNF) input and the US geographer Smil (2001) considers TNF as the most important invention of the twentieth century, in spite of its tremendous, worldwide debated climate relevant N pollution potential.

kg technically produced N per hectare and year are given as urea, NH_4^+ or NO_3^- in addition to N inputs received by lightning and BNF, which may amount worldwide to about 5 and 60 Tg N/year (range 40-100 Tg N/year), respectively (Tables 11.1, 11.2). These BNF input values are apparently smaller than earlier published estimates of 100-290 Tg N/year (Vitousek and Howarth, 1991; Vitousek et al., 2013). Although the technically produced, reactive N species urea, NH_4^+ and NO_3^- need high amounts of fossil fuel, monocultures often receive for N demand covering soil buffer capacity surpassing fertilizer amounts that enable to produce roughly 3000 million tonnes of crop residues worldwide (Karlen et al., 2006). Marine organisms live in an environment that annually receives 125 Tg N by BNF (range 60-200 Tg N) and are long-term adapted to such N amounts in contrast to terrestrial plants and belowground organisms which had to develop differently (Fowler et al., 2013). Plants, even in N over-fertilized soils, fed inhabiting organisms by transporting phloem into the rhizospheres *inter alia* for overcoming N shortage and to support N_2 fixing rhizobia associates, free-living N_2 fixing bacteria and archaea, oxygen-evolving photosynthetic active and anaerobic photosynthetic active bacteria. This N adaptation

Table 11.1 Estimates of annual N_2 fixation rates in agricultural systems by rhizobia in symbiosis with crop, pasture, fodder legumes, numerous genera of bacteria associated with non-leguminous species and free-living bacteria (Herridge et al., 2008). Data on land areas of the different agricultural systems are from FAOSTAT (2005), Smil (1999) and Cleveland et al. (1999).

Agent	Agricultural system	Area (Mha)	Rate of N_2 fixation (kg/ha/year)	Crop N fixed (Tg/year)	Comments on validity of global N_2 fixation estimates
Legume-rhizobia	Crop (pulse and oilseed) legumes	186	115	21	May be a robust estimate and substantially higher than the Smil (1999) estimate of 10 Tg fixed
Legume-rhizobia	Pasture and fodder legumes	110	110-227	12-25	Difficult to accurately assess because of uncertainty in legume areas and production
Azolla-cyanobacteria, cyanobacteria	Rice	150	33	5	Smil (1999) estimate of 5 Tg N/year reasonable, although primarily based on C_2H_2 reduction technique
Endophytic, associative and free-living bacteria	Sugarcane	20	25	0.5	Large variations in apparent N_2 fixation, using natural 15N abundance, makes estimations difficult and speculative
Endophytic, associative and free-living bacteria	Crop lands other than that used for legumes and rice	800	<5	<4	N_2 fixation likely to be <5 kg N/ha/year and total of < 4 Tg N/year, but not sufficient data to provide more robust values
Endophytic, associative and free-living bacteria	Extensive, tropical savannahs primarily used for grazing	1390	<10	<14	Cleveland et al. (1999) estimates of 42 Tg N/year likely to be high. Not sufficient data to provide more robust values

Table 11.2 Estimates of N_2 fixation in wetland rice fields (kg N/ha/crop) and theoretical maximum potential (value and assumptions; Roger and Ladha, 1992; Kulssooriya and Magana-Arachi, 2016).

Component	Reported range of estimates	Theoretical maximum potential and assumptions
BNF associated with rice rhizosphere	1-7 kg N/ha/crop	40 kg N/ha/crop All rhizospheric bacteria are N_2-fixers, C flow through rhizosphere is 1 t/ha/crop, and 40 mg N is fixed per g C.
BNF associate with straw	2-4 kg N/t straw	35 kg N/ha/crop 5 t of straw is applied, and 7 mg N is fixed per g of straw.
Total heterotrophic BNF	31 g N/ha/crop	60 kg N/ha/crop All C input (2 t/crop) is used by N_2-fixers.
Blue-green algae	80 g N/ha/crop	70 kg N/ha/crop Photosynthetic aquatic biomass is composed exclusively of N_2-fixing BGA (C/N = 7), and primary production is 0.5 t C/ha/crop.
Azolla	20-150 kg N/ha/crop (experimental plots); 10-50 kg N/ha/crop (field trials)	224 kg N/ha/crop One *Azolla* standing crop is 140 kg N/ha, and two *Azolla* crops are grown per rice crop, Ndfa is 80%.
Legume green manures	20-260 kg N/ha/crop	260 kg N/ha in 55 days *Sesbania rostrata* is used as green manure, 290 kg N/ha is accumulated in 50-60 d, and Ndfa is 90%.

strategy is under discussion, *inter alia* the efficiency to increase of the grain legume-rhizobium symbiosis concerned which is able to contribute to the soil nitrogen budget with about 21 Tg N/year, the forage and fodder legumes with 12-25 Tg, and non-leguminous crops with 12-20 Tg (Buresh et al. 1980; Leigh 2000; Cabello et al., 2004; Crutzen et al., 2008; FAO, 2016; Hunt et al, 2016; Guyonnet et al., 2018; Pedersen et al., 2018; Turk-Kubo et al., 2018; Wang et al., 2018; Moreira-Coello, 2019). When *Azolla* or legumes used as green manure with rice cultivation, the amounts of N_2 fixation could be 50–100 kg N/ha/crop (Ladha et al., 2000) by unveiling that in a lowland continuous double rice cropping system, both total and plant-available soil nitrogen pools could be maintained on a long-term basis, regardless of whether urea or green manure was the source of nitrogen. After 27 crops, the cumulative positive N balance was estimated at 1244, 348, 646 and 1039 kg N/ha in control, urea, *Sesbania*, and *Azolla* treatments, respectively. By considering all N inputs and outputs, except N losses, the calculated soil total N increases in the *Sesbania* treatment were 344 kg and in the *Azolla* treatment 541 kg (Fig. 11.3, Tables 11.1, 11.2). As reported BNF can play a vital role in rice-rice systems for soil N replenishing, but out of food security and income reasons it is still common to fertilize rice-rice production systems with TNF nitrogen in non-irrigated as well as flooded soils (Barraquio et al., 1983; App et al., 1984; Roger and Ladha, 1992; Sturm et al., 1994; Ladha et al., 2000; Kulasooriya and Mangana-Aracchi, 2016).

Protein formation and high/low-affinity N transport systems (HATS/LATS)

The availability of N in terrestrial and aquatic ecosystems depends on BNF and TNF, atmospheric N deposition, organismal N uptake, excrement distributions, on protozoan and nematode grazing on bacterial biofilms and ammonium excreting, viral shunts N inputs, and on food web activities (Benckiser, 2019, 2021). Such N inputs and conversion activities do apparently not satisfy the N demand of high yielding plant varieties in monocultures. However, TNF invention enables farmers to achieve better food security by applying arbitrary amounts of fertilizers. Out of lacking knowledge about plant demands, farmers often over-fertilize fields with nitrogen, although from 1991 to 2007 field N overload reductions from 133 to 105 kg/ha or by around 20% have been reported (Sturm et al., 1994, Lindblom et al., 2017; Mahal et al., 2019; Umweltbericht, GEO-66, 2019). Alone the enzyme ribulose-1,5-bisphosphate-carboxylase/-oxygenase (RuBisCO) involved in the photosynthesis process requires 20–40% of the N that is *inter alia* as NH_4^+ entrapped in soil clay or loosed during N cycling (Fig. 11.2; Scherer, 1988; Vincent et al., 1993; Osaki et al., 1993; Filleur et al., 2005; Miller et al., 2007; Krouk et al., 2010; Nacry et al., 2013). Thus, it appears meaningful to supplement the BNF-N input with TNF (world demand for N fertilizer and ammonia supply, Fig. 11.6). Plant roots exude phloem to attract and feed bacteria and archaea, able to fix N_2, to convert NH_4^+ into NO_3^-, or to provide other services like carbon cycling. Plants have the capability to sense soil N pools and have preferential nitrogen uptake behaviour: nitrate > ammonium > amino acids (Fig. 11.4; Filleur et al., 2005; Alvarez et al., 2011; Nacry et al., 2013). By diffusion mobile NO_3^- ions reach the plant cytosol easier than NH_4^+, amino acids and all other N containing compounds which also interact with soil clay and humus. After plant uptake and protein formation, surplus NH_4^+ is excreted back to the soil that plant root associates and soil microbes can reintegrate in their specific architectures and NH_4^+ reaches with help from high/low-affinity transport systems (HATS/LATS) the plant interior (Benckiser, 1997; Ladha et al., 1998; Benckiser and Schnell, 2007; Miller et al., 2007; Lima et al., 2010; Kant et al., 2011; Li et al., 2018). Roots are capable to sense locally concentrated soil N-, NO_3^--, NH_4^+- urea-, amino acids-pools and direct internal/external signals and roots with their inherent architecture, plasticity and physiology towards nutrient-rich patches. Within hours after internal N status communication, plants have regulated by targeted key transporters their external N utilization efficiency (biomass production per unit N taken up by the plant) and N acquisition efficiency (the proportion of N taken up as compared to that available in the soil) (Zhang et al., 2007; Alvarez et al., 2012; Lynch and Brown, 2012). The possession of sensors, the capability to combine plant external and internal signals enhances N acquisition guaranteeing efficiency, needed that plants can successful compete among each other and with microbes for N, essential for health and high productivity stabilization (Hodge et al., 2000; Kojima et al., 2007; DeAngelis et al., 2008; Hartmann et al., 2014; Fischer et al., 2016; Abbasian et al., 2018; Abisado et al., 2018).

External, soil NO_3^-, NH_4^+, urea and amino acids concentration derived signals, sensed, transmitted and plant inside transported, transfer after internal circulation between organismal organs signals downwards to the roots and inform the roots about the plant N demand. Nitrate, responsive for up to 10% transcriptome changes, acts as local stimulator and systemic repressor of lateral root growth and downregulates the plant's uptake of NH_4^+ (Zhang et al., 2007; Krouk et al., 2010; Alvarez et al., 2012). Particularly DUR3 genes in N starving plants respond very fast to NO_3^- and NH_4^+ uptake rate reduction and depress leaf expansion. Hormones and amino acids cycling through xylem and phloem sap integrate the N status of all organs (Cooper and Clarkson, 1989; Kanno et al., 2012; Wang et al., 2012; Nacry et al., 2013; Kiba and Krapp, 2016; Ohkubo et al., 2017). Thus N homeostasis and growth reduction/expansion of aboveground organs are ensured, and the conveyed information between shoot and roots precedes any measurable symptom of N deficiency. However, plant N acquisition tuning under a mixed N nutrition waits deciphering.

The use of urea, nitrate and ammonium fertilizers increased significantly after the TNF invention and related nitrification bursts led to fluctuating NO_3^- concentrations in the soil solution from few micros to 1–10 mmol/L (Sturm et al., 1994). At low soil NO_3^- or NH_4^+ concentrations in the 1 μmol/L range the high-affinity transport systems (HATS) scavenge nitrogen ions by following Michaelis–Menten uptake dynamics (Filleur et al., 2005). When the NO_3^- or NH_4^+ soil concentrations are high (typically >0.5–1 mmol/L), low-affinity transport systems (LATS) supersede HATS (Miller et al., 2007). Interacting HATS/LATS control mechanisms are the main controlling factors for root N intake (Kojima et al., 2007; Nacry et al., 2013). The membrane transporters HATS and LATS carry NO_3^-, NH_4^+, amino acids and urea in- and outside of cells. They belong to five protein families, namely NRT1, NRT2, AMT, AAP, LHT, Pro T2, to uncharacterized peptide transporters, and to a multigene family with at least 3–4 members (DUR3, NRT2 proteins). HATS and LATS, identified and characterized in *Arabidopsis thaliana*, are also detected in phylogenetically distant grass and cereal plants (Plett et al. 2010; Tsay et al., 2011; Wang et al., 2012; Nacry et al., 2013; Wen and Kaiser, 2018). The large NRT/PTR gene family comprises low affinity NO_3^- transporters, nitrite, peptide and carboxylic acid transporters, and NRT2 proteins transporters. NRT1, NRT2 transporters (NO_3^-), high-affinity urea transporters (DUR3), AAP transporters (amino acids) and high/low-affinity transporters of the AMT family (NH_4^+) and such equipped plant species can manage root in- and efflux of internal and external NH_4^+ and amino acids across the electrochemical potential gradient of membranes in time and concentration manner. HATS display a much higher flexibility in transporting amino acids across the high plasmalemma concentration gradient than LATS (e.g., 1–

10 mmol/L inside the cell against 0.1–10 μmol/L outside; Kojima et al., 2007; Lesuffleur et al., 2007; Neuhäuser et al., 2007; Rentsch et al., 2007; Lanquar et al., 2009; Barbier-Brygoo et al., 2011; Lehmann et al., 2015; Wen and Kaiser, 2018). Nitrogen-starving plants activate a local autoregulatory mechanism of AMT1 transporters, an allosteric feedback inhibition, but with respect to AMT1 transporters, it is also speculated that the amino acid efflux from roots occurs simply through leakage instead of active transport (Näsholm et al., 2009).

Root N in- and effluxes at N satiety rely on two major modulating mechanisms. One is based on the nitrogen status of the whole plant, the other on the long distance molecule signal behaviour of NO_3^-, NH_4^+, phytohormones, microRNAs and cytokines, whereby the role of amino acids in repressing root NO_3^- and NH_4^+ uptake systems is partially still hypothetic and awaits elucidation at the molecular level (Ho et al., 2009; Zhao et al., 2011; Nacry et al., 2013; Kiba et al., 2016). With manure spreading, antibiotics as sulfadiazine are distributed. Sulfadiazine may influence plant nitric oxide production and bacterial nitrate reductases as was found with *Medicago truncatula*. Such an interfering with key processes can manipulate the complex nitrogen cycling processes in the root rhizosphere complex and is needed to be further debated (Ollivier et al., 2010; Horchani et al., 2011; Nacry et al., 2013).

Nitrification in aquatic and terrestrial environments

Nitrifying communities grow autotrophically with CO_2 and NH_3 and due to this capability even sandstone pores are inhabited, not only soils, freshwaters or marine environments (Koops et al., 2003; Phoenix et al., 2012; de Vries and Bobbink, 2017). In 9-fold higher amounts than in the pre-industrial era, N is deposited since 1960 and nitrifying communities have to oxidize increasing amounts of the reactive N species NH_3 to NO_2^- and NO_3^- (Fig. 11.6; Benckiser, 2021). With the increased N inputs microbial and plant community composition shifts are predicted, the more because crops take up only 30–50% of the N provided by BNF, TNF, plant residues, local fecal distributions, and decaying cadavers. Thus, soil and aquatic bacteria, archaea, protozoa, nematodes, fungi and animals benefit from N not utilized by plants (Imsande and Touraine, 1994; Benckiser et al., 1997; Tilman et al., 2002; Koops et al., 2003; Jones et al., 2005; Benckiser and Schnell, 2007; Sutton, 2011; Belete et al., 2018). Not consumed N, temporarily stored in clay particles and humus, reach as mobile NO_3^- and toxic NO_2^- ions the groundwater, stress as NO and N_2O the atmosphere, cause pollution costs, and NO_3^-, NO_2^-, NO, and N_2O serve at strongly oxygen reduced and anaerobic conditions denitrifying bacteria, archaea, fungi as electron (e^-) acceptors (Benckiser, 1997; Sutton et al., 2011). Oxidized N species using denitrifying bacteria can generate nearly 90% of the energy at O_2 respiration gainable. Lytic viral replication and surplus NH_4^+ excreting protozoa and nematodes support ecosystems not only nutritionally, but they also contribute to microbial population regulation, disease/parasite transmission, habitat engineering, nutrient cycling and climate shaping (Hayatsu Tago and Saito, 2008; Marusenko et al., 2013; Bonkowski and Clarholm, 2016; Benckiser, 2019; Hammerschlag et al., 2019; Benckiser, 2021). Protozoan and nematode biofilm grazing resembles that of herbivores stimulating grass budding. Such indirectly stimulated nitrifying and denitrifying microbes use electrons (e^-) and protons (H^+) stored in NH_4^+ or in organic substances to oxidize NH_4^+ to nitrite and nitrate and to reduce NO_3^-, NO_2^-, NO, N_2O to N_2 (Sherr et al., 1983; Bamforth, 1988; Bloem et al., 1988; Crab et al., 2007; Craig, 2010; Sutton, 2011; Bonkowski and Clarholm, 2016; Benckiser, 2019; Shaw et al., 2019). Ammonium oxidizing bacteria (AOB) also use e^- and H^+, stored in NH_4^+, to convert in the Calvin Cycle atmospheric CO_2 into biomass; ammonium oxidizing archaea (Thaumarchaeota) use e^- and H^+ stored in NH_4^+ to form via the hydroxyproprionate-hydroxybutyrate cycle biomass with HCO_3^- (Fig. 11.5; *Nitrobacter* Winogradsky, 1892, GBIF Secretariat, 2017, GBIF Backbone Taxonomy, Checklist dataset https://doi.org/10.15468/39omei accessed via GBIF.org on 2019-07-26; Berg et al., 2007; Lu et al., 2018).

Meanwhile we know that chemolithoautotrophic nitrifying bacteria and archaea are ubiquitously present in nature and conserve energy by oxidizing NH_4^+ to NO_3^- in two steps (Figs 11.2, 11.5). Nitrifying bacteria (NOB), belonging to the "nitroso" group (*Nitrosomonas*, *Nitrosospira*, *Nitrosovibrio*, *Nitrosolobus*, *Nitrosococcus* species), oxidize NH_4^+ to NO_2^- and NOB, belonging to the "nitro" group (*Nitrobacter*, *Nitrospira*, *Nitrospina*, *Nitrococcus* species), catalyse in a second oxidation step NO_2^- to NO_3^- (Koops et al., 2003; Sahrawat, 2008; Daims et al., 2015; Khanichaidecha et al., 2018; Koch et al., 2019; Zhang et al., 2019). To the families *Crenarchaeota* and *Thaumarchaeota* belonging nitrifying archaea (NOA) conserve energy with NH_3 as electron donator, whereas heterotrophically growing nitrifying and denitrifying fungi generate energy by using e^- and H^+ stored in -C-NH_2 (Burth et al., 1982; Umma and Sandhya, 1997; Leininger et al., 2006; Berg et al., 2007; Boyle-Yarwood et al., 2008; Schleper, 2010; Boyd et al., 2011; Hatzenpichler, 2012; Stahl and de la Torre, 2012; Taylor et al., 2012; Zhu et al., 2015; Lu et al., 2018; Ahlgren et al., 2019; Kong et al., 2019; Vandekerckhove et al., 2019; Zhang et al., 2019).

Across ecosystems the rather uniformly distributed nitrification process plays a key role in N cycling (Fig. 11.2) since the oxygenation of the Earth and because the majority of bacteria and archaea still resists cultivation attempts we still have to rely on tools such as comparative 16S rRNA sequence analyses and fluorescence-in-situ-hybridization (FISH) combined with rRNA-targeted probes to understand nitrifying community functioning (Amann et al., 1995; Jansson and Hofmockel, 2018; Kitzinger et al., 2019; Martiny, 2019; Meng et al., 2019; Wu et al., 2019). At least we know that the ammonia monooxygenase (AMO) of NOB and the (*amo*)-like gene sequences of ammonia-oxidizing archaea (AOA) are ubiquitously distributed, and NOA often outnumber NOB (Hatzenpichler, 2012). The AOA diversity and physiology distinctiveness and the AOA role in carbon and nitrogen cycling and greenhouse gas (N_2O) emission is debated and very likely we must revise our understanding of nitrification fundamentally as comparing studies with ^{18}O labelled AOA and AOB in the marine environment indicate (Lam et al., 2009; He et al., 2018). In regions of low dissolved O_2 concentrations (<10 μmol/L) ammonia-oxidizing archaea (AOA) and bacteria (AOB) are key contributors to ammonia oxidation and to dissimilatory nitrate reduction to ammonium (Strous et al., 1997; Lam et al., 2009; Hatzenpichler, 2012). In the oxygen minimum zone (OMZ) of oceans anaerobic ammonium oxidation (anammox) can substantially satisfy the NO_2^- requirement by nitrate reduction (67% or more) and aerobic ammonia oxidation (33% or less), whereby the oxygen ratio in NO_2^- seems to be the same by AOA and AOB (one atom originates from H_2O and one from O_2).

At water saturation nitrifying AOB and AOA can switch to NO_3^-, NO_2^-, NO and N_2O respiration as denitrifying bacteria, archaea, and heterotrophic fungi, although there are differences between AOA and AOB concerning NO_2^-, NO_3^-, and N_2O formation (Burth et al., 1982; Cabello et al., 2004; Doi et al., 2009). Jung et al. (2019) studied by using two AOA pure cultures (*Nitrosocosmicus oleophilus* and *Nitrosotenuis chungbukensis*) and an AOB strain (*Nitrosomonas europaea*) the effect of decreasing pH on N_2O formation by nitrifiers in an ammonia-oxidizing archaeon at low pH and found that acidification boosted nitrite transformation into N_2O in all strains, but the incorporation rate was different for each ammonia oxidizer. AOA are able to denitrify nitrite to N_2O enzymatically at low pH and from nitrite both nitrogen atoms originate during N_2O production by *N. oleophilus*. Also swine wastes, where ammonia oxidizers are generally outnumbered by oxygen consuming heterotrophs and thus only found at low concentration levels, demonstrate that AOB and AOA are flexible in switching between NH_4^+ oxidation to nitrate and denitrification and can adapt and function as the disappearance of up to 20 mg NH_3/L swine waste by ammonia oxidization is indicating (ST-Arnaud et al., 1991). In acid, coniferous forest soils with limited NOB nitrification AOA and nitrifying fungi are majorly responsible for NO_3^- formation by oxidizing R-NH_2 directly to NO_3^- at maximum rates of 8% (0.08 to 0.012 μg N/g/day; Killham, 1990; Nugroho et al., 2007; Zhu et al., 2014; Pajares and Bohannan, 2016; Jung et al., 2019). In less acid soils autotrophic nitrification reach NH_3-N oxidation rates to NO_3^- between 0.96 to 2.64 μg NH_3-N/g/day, a grassland soil (pH 6.2) may convert 84% of the NH_4^+ to NO_3^-, an arable soil of pH 4.5 and 6.5 between 85 and 91%, and a forest soil (pH 4.4) around 29% (Killham, 1990; Barraclough and Puri, 1995).

The high variation in NH_4^+ to NO_3^- oxidation suggests that in such different environments as soils, aquatic and waste water not only the pH, also other biotope characteristics (such as O_2 availability, degradation resistance, soil texture, humus components, occurrence of nitrification inhibitors, farm and waste water management) co-determine the nitrification process (Hassett et al., 1987; Killham, 1990; Majumdar et al., 2001; Abasi et al., 2011; Subbaro et al., 2017; Fuertes-Mendizábal et al., 2019). From marine sediments (Danish, Belgian, Japanese coast, Narragansett Bay), stream and lake sediments (Patuxent River estuary, Lake Verret, Little Lake, Lac des Allemands, LA), fresh water (Calcasieu River, LA), and tropical marine and fresh water polyculture fishponds (Thailand) nitrification rates between 0 and 67 mg N/m^2/day are reported, from paddy soils 11 to 152 mg N/m^2/day (Hargreaves, 1998).

Nitrosomonas europaea or the AOA *Nitrosocosmicus oleophilus* and *Nitrosotenuis chungbukensis* gain energy by reducing the electron acceptor NO_2^- with the electron donor NH_2OH. This as nitrifiers' denitrification termed process may even end as NH_3 (nitrate ammonification; Wrage-Mönnig et al., 2001; Lam et al., 2009; Rivera-Monroy et al., 2010; Lawton et al., 2013; Yu et al., 2018; Jung et al., 2019). The N_2 introduced by BNF and TNF returns through denitrification again as N_2, admixed with varying amounts of N_2O to the atmosphere, whereby heterotrophic denitrification receives electrons by organic matter degradation, fungi from R-NH_2,

Table 11.3 Influence of the nitrification inhibitors dicyandiamide (DCD) and 3,4-dimethylpyrazole phosphate (DMPP) control on the CO_2, CH_4, and N_2O emissions and the global warming potential caused by the DMPP treated summer barley (1997), corn (1998) and winter wheat (1999) cropped plots in comparison to control plots, which never received DMPP. For the data base see Weiske et al. (2001).

	Σ g CO_2-C/ha/day	Σ g CH_4-C/ha/day	Σ g N_2O-N/ha/day	CO_2-equivalents[1)]
Control	5,840,594	-130	828	6,094,553 (100%)
DCD	5,456,339	-130	614	5,643,788 (93%)
DMPP	4,185,470	-166	425	4,313,752 (70%)

[1)] = CO_2 + (CH_4 x 21) + (N_2O x 310)

nitrification, nitrifiers' denitrification or anammox from NH_2OH, and plants, farmers and waste water managers control N-use-efficiencies with nitrification inhibitors and get asides more detailed insights into the interacting N_2 fixation-nitrification-denitrification complex, into the ecophysiology of the involved organisms, into biodiversity compositions (Figs 11.2-11.5, Tables 11.1-11.3; Seitzinger, 1988; Killham, 1990; Leininger et al, 2006; Klotz and Stein, 2008; Sahrawat, 2008; Sime-Ngando and Niquil, 2011; Granger and Wankel, 2016; He et al., 2018; Yin et al., 2018; Kitzinger et al., 2019; Lepère et al., 2019; Niu et al., 2019; Wu et al., 2019).

Wetlands constructed, composed out of soil, flooded with waste water and aerated by using the aerating power of e.g. reed plants, having as rice plants the ecological advantage to survive under flooded conditions by having developed an aerenchym tissue for transporting oxygen into the rhizosphere to sustain root respiration (Benckiser et al., 1984; Fey et al., 1999). In the aerated root zone nitrifying bacteria and archaea can oxidize NH_4^+ to NO_3^- that diffuses in rhizosphere distant soil sections where plants can take up the by diffusion distributed NO_3^- and denitrifying and anammox bacteria and archaea can convert at prevailing O_2 shortage NO_3^- to N_2 (N_2O, NH_3; Benckiser et al., 1996; Fey et al., 1999; Mander et al., 2008; Kim et al., 2010; Bai et al., 2014). Nitrate ions are present since TNF invention at increasing amounts in nearly all environments and concerning the climate influencing nitrification-denitrification complexity is debated and the NO_3^- conversion rates is by denitrification are measured in biotopes as the epilimnetic sediment of lake 227

(Experimental Lakes Area, Western Ontario, Canada) with an average rate of about 15 mg $N/m^2/d$ and in the euphotic zone below the epilimnetic sediment, known as efficient sink of NO_3^--N, where inflowing compounds as chlorothalonil can strongly impact the NO_3^- converting potential (Chan and Campbell, 1980; Lindau et al., 1988; Crutzen et al., 2008; Chen et al, 2018; Su et al., 2019).

Heterotrophic denitrification and anammox

The denitrification process, already in the early Precambrian developed, was discovered in its importance as terminal electron acceptor for cellular bioenergetics only a century ago (Zumft, 1997). Chemolithotrophic, phototrophic, diazotrophic, and organotrophic bacteria, archaea, and fungi perform denitrification; an example is *Paracoccus denitrificans* that apparently needs arginine-95 for recruiting superoxide to the active site of the FerB flavoenzyme (Burth et al., 1982; Benckiser et al., 1996; Lloyd et al., 1987; Zumft, 1997; Fey et al., 1999; Sreenivas and Sharma, 2005; Bai et al., 2014; Zeng et al., 2014, Conthe et al., 2019; Hidalgo-García et al., 2019; Nadeau et al., 2019; Sedláček and Kučera, 2019). Heterotrophic denitrification is based on about 50 genes. They enable gaining energy in a range from aerated to anaerobic conditions by achieving under anaerobic conditions nearly 90% of the energy per mol glucose, possible to reach by O_2 respiration. Beyond the denitrification apparatus, the genes encode and activate periplasmic (Nap) or membrane-bound nitrate reductases (Nar). *Nap/nar, nor* - nitric oxide reductases (NirK/*cd*$_1$Nir) are responsible for the reduction of NO_2^- to NO and the nitrous oxide reductase NosZ, the product of the *nosZ* gene, a periplasmic, homodimeric metalloprotein of 130 kDa that contains two copper centers, Cu_A and Cu_Z, in each monomer for the reduction of N_2O to N_2 (Zumft, 1997; Benckiser, 1997; Cabello et al., 2004; Jones et al., 2013; Shah, 2018; Toyofuku and Yoon, 2018; Zhang et al., 2019). Nitrate and deriving N-oxides act as electron acceptors and signal carriers for distinct N-oxide metabolizing enzymes. They are intimately connected to fundamental cellular processes such as primary and secondary transport, translocation of proteins, biogenesis of cytochrome c, anaerobic gene regulation, metalloprotein assembling, the biosynthesis of the cofactors molybdopterin and heme D1 or the transcription factors of the greater FNR family (Zumft, 1997; Jones et al., 2013; Hidalgo-García et al., 2019). The enzymatic transformation of NO_3^- to gaseous intermediates depends on the Fe, Cu and Mo redox chemistry, on the concentration of O_2 in soil water and the diffusional O_2 resupply into the root and soil zone, the plant-soil microflora, fauna O_2 consumption, NO_3^-

uptake, and the presence of nitrification/denitrification inhibitors (NI, DI; Benckiser, 1997; Benckiser and Schnell, 2007; Hawkes et al., 2007; Crutzen et al., 2008; Enwall et al., 2010; Braker and Conrad, 2011; Pastorelli et al., 2011; Bardon et al., 2016; Hidalgo-García et al., 2019).

The carbon cycle dependent heterotrophic denitrification differs from that of denitrifying nitrifier and anammox assemblages, which use NH_4^+ or NH_2OH derived electrons to reduce N-oxides instead of electrons stored in organic matter. Nitrifier denitrification and anammox were first detected in waste water and oceans (Strous et al., 1997; Lam et al., 2009; Ward et al., 2009; Wrage-Mönnig et al., 2018; Cao et al., 2019). Electrons derived from NH_4^+ or NH_2OH and energized by the sun during the photosynthesis process are channeled by nitrifying/denitrifying bacteria, archaea and fungi from the e^--acceptor NO_3^- to NO_2^-, NO and N_2O and finally to N_2 to gain ATP (Figs 11.2, 11.5). Denitrification, which is found among halophilic hyperthermophilic archaea, N_2 fixing nitrifying bacteria and numerous other bacterial species, fungi, and even in mitochondria, returns the N_2 fixed by BNF and TNF to the atmosphere and is thus fundamental in nitrogen cycling. Among this process the nitrate reductase genes *narG* and *napA* and the nitrite reductase genes *nirK* and *nirS* play key roles (Elliott and Gilmour, 1971; Burth et al., 1982; Weger and Tupin, 1989; Benckiser, 1997, 2007; Bremner and Blackmer, 2007; Bru et al., 2007; Flores-Mireles et al., 2007; Smith et al., 2007; Itakura et al., 2008; Li et al., 2008; Barth et al., 2009; Demaneche et al., 2009; Palmer et al., 2009; van Alst, 2009; Wertz et al., 2009; Fischer et al., 2010; Simpson et al., 2010; Alvarez et al., 2011; Ishii et al., 2011; Nishimura et al., 2011; Pena et al., 2012; Svenningsen et al., 2012; Shoun et al., 2012; Jung et al., 2013; Dungel et al., 2017; Hidalgo-García et al., 2019; Jang et al., 2019).

NO_3^-- and NO_2^--reducing microbes are involved in regulation of organic matter availability, in ammonium, nitrite and nitrate dependent primary and secondary biomass production, and co-determine the N_2O/N_2 ratio in permanently flooded paddy, water saturated terrestrial soils after rainfall, in water bodies and sediments of rivers, lakes and oceans and in waste water treatment plants (Benckiser et al., 1996; Miyahara et al., 2010; Lam et al., 2009; Cuhel, 2010; Ishi et al., 2011; Ji et al., 2015; Dai et al., 2019). In acidic fen material (pH 4.7 to 5.2) or low pH, forest soils the pH limits denitrification and particularly co-shapes N_2O emissions, found between 1 and 105 nmol N_2O/h/g soil dry weight (Benckiser, 1997; Palmer et al., 2010; Fungo et al., 2019). The denitrifying soil resident *Pseudomonas mandelii* exhibits in the presence of the N_2O reductase inhibitor C_2H_2 at pH 6, 7 and 8 *nir*S gene (nitrite reductase) and *cnor*B gene (nitric oxide reductase) activities, which are 539- and 6,190-fold, respectively, higher than at pH 5 (Saleh-Lakha et al., 2009). Along a nitrate concentration gradient at three sites in the Colne estuary, UK, the numbers of the functional denitrification (DN) and dissimilatory nitrate reduction to ammonium (DNRA) genes (*nar*G, *nap*A, *nir*S, and *nrf*A) decreased in dependence of pH and nitrate concentration (Dong et al., 2009). In the Arabian Sea, the world's oceans largest and most intense oxygen minimum zone (OMZ), AN (anammox), DN, and DNRA are responsible for 87–99% oceanic N_2 production, *Magnetospirillum gryphiswaldense*, typically found in OMZ, to ~35% (Dong et al., 2009; Ward et al, 2009; Palmer et al., 2010; Dörsch et al., 2012; Li et al., 2012; Bertagnolli and Stewart, 2018; Quian et al., 2019). Worldwide-extrapolated annual denitrification rates for land and sea areas, mostly based on denitrification studies with NO_3^- as e^- acceptor, are 1.2 x 10^2 and 4.0 x 10^1 Tg N, respectively, but meanwhile we know that N_2O and N_2 producing processes are of diverse nature (Delwiche, 1977). The α-Proteobacterium *Ochrobactrum anthropi* YD50.2 (type strain ATCC 49188), e.g., possesses a nir and nor operon, a functional copper containing NO_2- reductase that enables to denitrify under anaerobic conditions up to 40 mmol/L NO_2^- to dinitrogen; but how such bacteria organize the toleration of the toxic electron acceptor NO_2^- is little understood (Collman et al., 2008; Kuznetsova et al., 2008; Doi et al., 2009; van der Berg et al., 2017). Nitrate respiring *Pseudomonas aeruginosa, Ralstonia eutropha* and *Escherichia coli* strains do not have a copper containing NO_2^- reductase available, but tolerate under non-denitrifying aerobic conditions NO_2^- levels up to 100 mmol/L. Novel denitrifying bacterial strains as *O. anthropi* YD50.2 with historically evolved NO_2^- tolerance may protect others from acidic NO_2^-, but unfortunately species as *O. anthropi* are not widely disseminated among heterotrophic denitrifiers (Doi et al., 2009).

An environment that favours denitrification is the alimentary canal of the earthworm families Lumbricidae and Aporrectodeae, less of *Octolasion* and *Octochaetus* (Wüst et al., 2009, 2011). Earthworms ingest large amounts of carbohydrates which provide electrons and protons, favouring the *nar*G, *nap*A, *nir* K, *nir*S, *nrf*A, and *nos*Z encoded conversion of NO_3^- and NO_2^- to N_2O and N_2 (Berks et al., 1997; Zumft, 1997; Depkat-Jakob et al., 2010). The nosZ sequences detected in alimentary canals of *L. rubellus* and *O. multiporus* resemble those of not yet cultured

Gram-positive and negative soil bacteria of the genera *Actinobacteria, Bradyrhizobium, Azospirillum, Rhodopseudomonas, Rhodospirillum, Pseudomonas, Oligotropha,* and *Sinorhizobium.* H_2 and organic acids (e.g., succinate, acetate, formate) produced in the earthworm gut indicate that in the alimentary canal of earthworms all redox processes including denitrification concomitantly occur. The reduction equivalents provided by formate dehydrogenase primes conditions for a complete nitrate reduction to N_2 (Benckiser, 1997, 2017; Benckiser and Schnell, 2007; Wüst et al., 2009, 2011; Donaldson et al., 2016). Mobile, digging earthworms distribute denitrifying microbes and fungi, probably as spores, in aerated soils, and thus contribute to the overall N_2O and N_2 release (Lubbers et al., 2013; Benckiser, 2017).

Denitrifying fungi prefer amides (proteins) as electron donators and formate in combination with oxygen regulates the N_2O release. But the related NIR and P450nor activity has been proven only with a few soil and waste water fungal isolates. Thus, it is widely unknown how fungal denitrification really performs (Burth et al., 1982; Ma et al., 2008; Shoun et al., 2012). It is speculated, that fungi contribute to the nitrous oxide emission from soils by around 65% and that denitrifying fungi apparently use nicotinamide adenine dinucleotide derived electrons to reduce NO to N_2O (Ma et al., 2008; Bergaust et al., 2012).

Plants control NO_3^- reduction and N_2O emission by excreting nitrification inhibitors (NI) (Subbarao et al., 2017). Ubiquitous denitrifying prokaryotes as *Pseudomonas aeruginosa* take up the NI. Such NIs interfere with the numerous genes, controlled through cell-to-cell signals, and increase in *P. aeruginosa* together with the quorum-sensing signal 2-heptyl-3-hydroxy-4-quinolone the nitrite reductase activity by inhibiting nitric oxide reductase, whereby iron chelation seems primarily regulating denitrification (Toyofuku et al., 2008).

Compared with the pre-industrial period, industrial N outputs over-fertilize terrestrial biotopes, narrow the soil C_{H2O}/NO_3^- ratio and the related denitrification activity, responsible for more than two-thirds of the global N_2O emissions. The presently available denitrification control possibilities are alone not sufficiently constructive. Thus N_2O emission reduction concepts are developed, including the plant N_2O conduit effect, under the headings (i) soil chemistry and microbiology management, (ii) crop plant engineering concerning intercropping, crop rotation and biological N_2 fixation, (iii) agriculture adapted to plant N demand by orientating towards the preindustrial period, and (iv) a less fossil fuel consuming TNF production (Bakken et al., 2012; Sanford et al., 2012; Thompson et al., 2012; Lenhart et al., 2019). and the N_2O/N_2 ratio

Nitrous oxide effluxes from plants

Regarding N_2O gas exchange between terrestrial soils, waste water treatment plants, and free living and gut hosted microbial communities with the atmosphere, plant conduit and virus replication side effects are little quantified by the soil cover method and generally neglected (Mosier et al., 1998; Chen et al., 1999 and 2008; Benckiser, 2019 and 2021; Lenhart et al., 2019). From N hot spots in N over-fertilized soils, it is the general opinion that N_2O enters the atmosphere by diffusion. Yet, shoots of canola (*Brassica napus* L.; Chang et al., 1998), *Lolium perenne* L. (Chen et al., 1999), and herbicide-treated soybean (Lifeng et al., 2000) enclosed in chambers revealed 20% higher N_2O concentration in comparison with unplanted controls in the *Brassica napus* L. variants. Nitrous oxide beech (*Fagus sylvatica*) leaves emission measurements indicated plant-mediated N_2O releases (Pihlatie et al., 2005). The global, plant integrated nitrogen cycle depend nutritionally on bacterial or technical N_2 fixation, NH_4^+ conversion into NO_3^-, and plant internal biochemical transformations (Figs 11.4, 11.7). Rice and reed plants, equipped with an aerenchym tissue for transporting O_2 from the shoots to the roots, make conceivable that such a spongy tissue transports soil N_2O to the atmosphere and that plants may function as gas conduits (Benckiser et al., 1984; Cheng et al., 2007). After testing with 32 plant species N_2O transport with the transpiration stream with focus on barley (*Hordeum vulgare* L.) Lenhart et al. (2019) concluded that vegetation acts as N_2O chimney and should find more consideration in greenhouse gas emission models. N_2O flux from the leave surface of 32 plant species, measured by gas chromatography and stable isotope mass spectrometry ($d^{15}N$, $d^{18}O$, $d^{15}N$ suspended particle species) and plant species, belonging to 22 different plant families, varied between 0.4 and 173 ng N_2O and 0.2 and 5.3 mg CO_2/g leaf dry weight/h. Although the N_2O transport from the soil pores to the atmosphere through plant conduits is not conclusively resolved, the available data indicate, that this may be an important part of the greenhouse gas flux (Fig. 11.7; Smart and Bloom, 2001; Lenhart et al., 2019).

From ^{15}N signatures and 2-phenyl-4,4,5-tetramethylimidazole-1-oxyl3-oxide (PTIO) measurements can be deduced that tobacco (*Nicotiana tabacum*) wild-type leaves release NO (Planchet et al., 2006; Takahashi and Morikawa, 2014). The signal

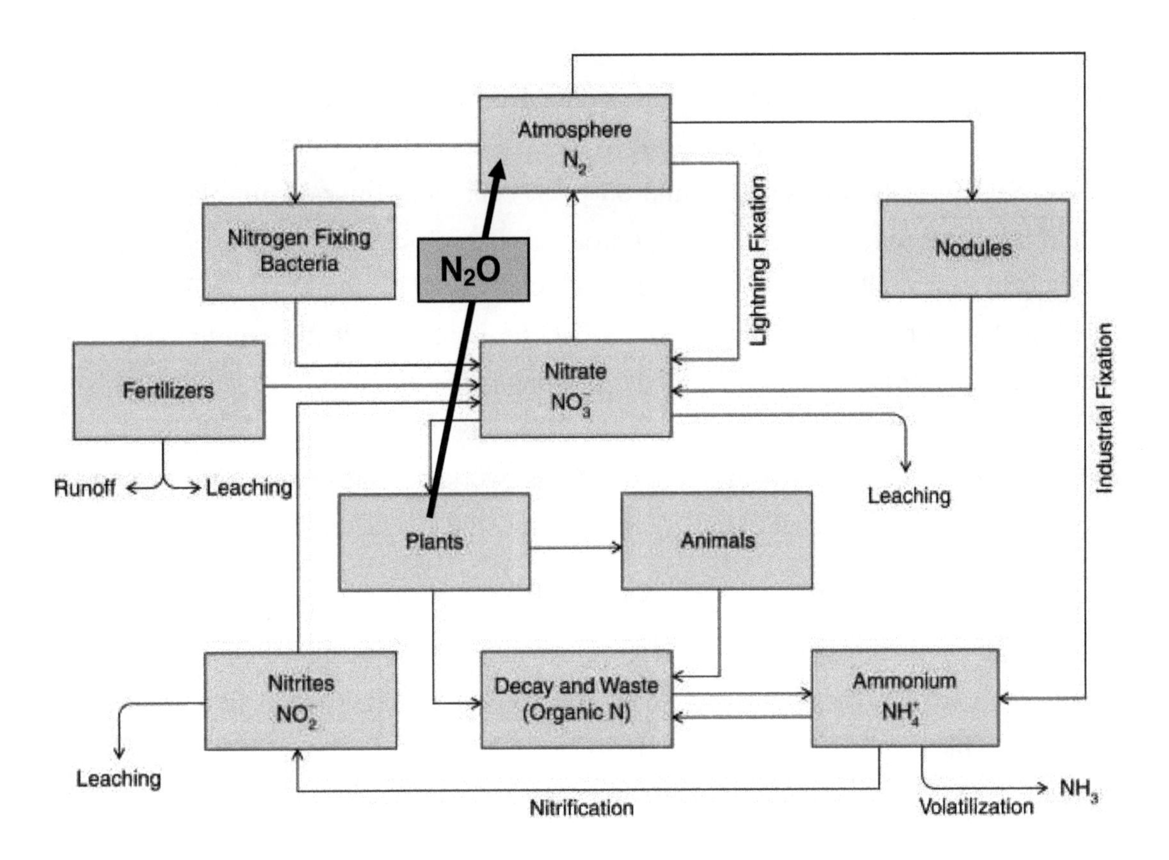

Figure 11.7 Plants and the schematic global nitrogen cycle. The most abundant elemental dinitrogen (N_2) gas in the atmosphere converts in soils and aquatic systems to ammonium (NH_4^+), organic nitrogen containing compounds, mobile nitrate (NO_3^-) ions, toxic nitrite (NO_2^-), NO and N_2O. Plant gas conduits channel NO and N_2O from soils towards the atmosphere.

compound NO is very likely produced from entering NO_3^- and the exogenous plant growth regulator NO_2 (PTIO + NO $\rightarrow$ NO_2 + PTI $\rightarrow$ $N_2O_3 + H_2O \rightarrow 2\ NO_2^- + 2H^+$). The vegetative and reproductive plant growth regulator NO_2 may contribute to the plant conduit idea as well as the NO and N_2O release concerned mitochondrial electron transport, the nitrate reductase (NR). NO_2 reaches the phyllosphere and there residing microorganisms may convert NH_4^+ into oxidized N species and eventually reduce NO_2 and N_2O (Chang et al., 1998; Chen et al., 1999; Meyer et al., 2005; Nacry et al., 2013; Bruhn et al., 2014; Bringel and Couée, 2015; Nie et al., 2015; Machacova et al., 2016; Subbarao et al., 2017; Wei et al., 2017; Aballos et al., 2018; Hu et al., 2019; Lenhart et al., 2019). NO_2^- can be rate limiting during NO_2^- reduction to NO, because the NR produces larger amounts of NO than required for signaling. More meaningful research is needed to understand better the plant conduit role in the NO and N_2O release (Meyer et al., 2005).

N_2O formation in soils, in and on plant cells, and fluxes from plant surfaces should be understood for assessing NO production changes and better methods as the diaminofluorescin-chemoluminescence approach by DAF-2 and DAF-2DA should be developed to carry out NO measurements, NO production reflecting more exact. The activity of the nitrous oxide system (NOS) in plants is estimated to 1% in comparison to a 2 to 3 orders of magnitude higher NO flux from soils. In relation, more information is needed how the signal NO is working for repurposing the plant internal NO signal (Planchet et al., 2006; Wright and Nemhauser, 2019).

Nitrification inhibitors

The globally growing plants exude 400-800 Tg phloem into their rhizospheres that mainly consists out of sugars (90 – 95%) and attracts wide C/N ratio litter degrading microbes which narrow the C/N ratio to > 15 N and increase the plant nutrient absorbability (Farrar and Jones, 2000; Bais et al., 2006; Kehr, 2006; Nacry et al., 2013; Kaiser et al., 2015; Gorka et al., 2019). The exudates, excreted by plants, contain nitrification inhibiting (NI) compounds (Fig. 11.5) with which the conversion of NH_3 into reactive oxidized N species (NO_3^-, NO_x, N_2O, NH_3 or NO_3^-) is controlled and the N_2O/N_2 ratio balanced (Subbarao et al, 2017; Domeignoz-Horta et al., 2018; Canarini et al., 2019; Verhoeven et al., 2019; Zistl-Schlingmann et al., 2019). Plants evolutionary adapted to N shortage, try controlling the by nitrifying/denitrifying activities co-determined N availability to that the soil fauna spot-wise distribute N-rich excrements. NI help controlling the competing between soil microflora, -fauna, and plants for N (Taroncher-Oldenburg et al., 2003; Benckiser and Schnell, 2007; Braker et al., 2012; Bier et al., 2015; Pajares and Bohannan, 2016; Muhammed et al., 2018; Oldfield et al., 2019; microbial genome atlas Rodriguez et al., 2018; Rütting et al., 2018; Cowan et al., 2019; Gu et al., 2019). Water saturation forces bacteria, archaea, and fungi to respire with oxidized N species, the electron acceptors NO_3^- NO_2-, NO, and N_2O for gaining energy. A luxury C disposability is supporting in this context and if the availability of NO_3^-, NO_2-, NO, and N_2O is low at a luxury C disposability denitrifying bacteria, archaea and fungi must act economically and reduce N_2O widely to N_2. In reverse, a high NO_3^- availability at low C usability, often observable in N fertilized, high yielding monoculture fields, denitrifying bacteria, archaea and fungi less reduced N_2O is released in higher amounts (Benckiser, 1997). For a better N captivity plants try to control the urease and ammonium monooxygenase (AMO), the nitrification activities, with for example flavonoids (3-methoxy furano-2', 3', 7, 8-flavone, $C_{18}H_{12}O_4$) and try protecting the NO_3^- availability by inhibiting (minimizing) denitrification, *inter alia* with procyanidin (Majumdar, 2008; Bardon et al., 2016; Dignac et al., 2017; Kafarski and Talma, 2018). As plants farmers act and employ industrially designed urease and nitrification/denitrification inhibitors (UI, NI, DI) to achieve better plant N use efficiencies, satisfying yields, and low greenhouse gas emission rates (Zerulla et al., 2001; Benckiser, 2017, 2019).

For preventing food insecurities with monocultures farmers have two options, (a) differently as nature is demonstrating to achieve a high biomass productivity with N shortage and a sustained high biodiversity farmers compensate N shortage by TNF fertilization or employ (b) industrially designed UI, NI, DI stabilized TNF fertilizers to satisfy plant N demand and minimizing N losses. In use are the UI n-butylthiophosphoric triamide (NBTP) and the NIs 2-chloro-6-(trichloromethyl)-pyridine (nitrapyrin), etridiazole (5-ethoxy-3-tri-chloromethyl -1,2,4-thiadiazole; Dwell), 3-mercapto-1,2,4-triazole, 4-amino-1,2,4-triazole, 3-methyl pyrazole-1-carboxamide, 2-ethinylpyridine, 3,4-dimethylpyrazolephosphate (DMPP) and dicyandiamide (DCD). Most marketed are nitrapyrin, DCD or 3,4-DMPP stabilized N-fertilizers under the product names N-serve, Alzon, and Entec, respectively, but increasingly tested and applied are combined UI, NI stabilized N fertilizers as Ensin (DCD, 1,2,4 triazole mixture), Agrotain (DCD 81,2%, NBPT 6.5%), Dempsa (3,4-dimethylpyrazole succinic mixture) or Piadin (1H-1,2,4 triazole-3-methylpyrazole mixture) (Mc Carty, 1999; Li et al., 2008; Roth et al., 2009; Sima et al., 2013; Huérfano et al., 2016; Wu et al., 2017; Tindaon and Benckiser, 2019). The NI DMPP, applied at a recommended concentration of 0.2–0.5 μg 3,4-DMPP/g dry soil indiscriminately binds onto the complex of membrane-bound proteins, inclusively the AMO, whereas the also broadly marketed NI DCD blocks the electron transport in the AMO cytochromes and must be applied at a 10 times higher concentration than DMPP for being as effective (Li et al., 2008; Tindaon and Benckiser, 2019). Advantages of DMPP and DCD are that they resemble nature's humus components and reduce N_2O emission by 26-49% (Table 11.3), whereby dehydrogenase activity (DHA), occurring in all organisms, is only impaired, when the field recommended application rate is significantly surpassed (Tindaon and Benckiser, 2019). In spite of the UI, NI, and DI benefits, an European Commission (EC) cost/benefit analysis hesitates to recommend the application of NI stabilized N-fertilizers, because they are relatively expensive, are under warm climate field conditions little effectively, the crop yield improvements are insufficiently tested and not conclusively documented, and in postharvest soils NH_4^+ conserving by NI may enhance the emission of N_2O (Ali et al., 2008; Mahmood et al., 2011; PICCMAT, 2011; Scheer et al., 2017). In spite of all such critics UI, NI, DI stabilized fertilizers are as pre-exercised by plants a valuable tool for controlling N cycling, securing high monoculture yields, and help approaching plant adapted N fertilizations, C/N ratio fine-tuning to reduce N_2O emissions significantly (Table 11.2; Zumft, 1997; Benckiser, 2019; Tindaon and Benckiser, 2019).

Nitrogen flows and future N_2O emission perspectives

For sustaining a satisfying grassland productivity the British Grassland Society recommends to apply in addition to the BNF-N inputs annually 120 kg TNF-N. Such recommendation related the global use of synthetic nitrogen fertilizers increased particularly from approximately 12 million tonnes N in the early 1960s to 82 million tonnes at the turn of the millennium and to 112 million tonnes in 2011 (FAO, 2016). From the 171,433 thousand tonnes of spread ammonia in 2019 (Fig. 11.6) add an intensified livestock/biogas production and from the N surplus crops take up only 30–50% and the plant demand surpassing spread N fertilizer amounts also surpass the soil N buffer capacity and NO_3^-, typically less than 10 mg/L in the groundwater enriches often to more than 50 mg nitrate/L and N recommendations started to be controversially debated because the annual N fertilizer surplus inputs of 50-80 kg N/ ha/year stress soil, water, air quality, biodiversity, and human health (Vitousek and Horwarth, 1991; Imsande and Touraine, 1994; Bacon, 1995; Jones et al., 2005; Sutton et al., 2011; Belete et al., 2018; FAO, 2016, 2019). Thus, a major global challenge is to increase N fertilizer uptake efficiency and to reduce the emissions of N_2O by concomitantly guaranteeing food security (Craswell and Godwin, 1984; White and Brown, 2010; Sutton et al., 2011; Erisman et al., 2013; FAO, 2016, 2019). Natural N shortage provokes an adapted biodiversity, which in TNF fertilized monocultures must adapt to high N availabilities, particularly in regions with an intensive animal manure and digestates accumulating livestock/biogas production (Byrnes and Bumb, 1998; FAO, 2016; Bourke et al., 2019).

Besides an industrial livestock management the industrialization of the world and a concentration of people in cities, producing sewage sludge in huge amounts which must be either spread on a limited area of agricultural soils *inter alia* used as vineyards and suggested in the 1960 decade was to ameliorate agriculture in regions of low level food security as in sub-Saharan Africa, or to deposit sewage sludge in dumps, or burning it (Pretty et al., 2011; FAO, 2019). The wastewater, produced by an urbanizing, in cities concentrating population, increased in its amount by 68% during the last decades and must be purified with respect to climate relevance. For recycling the produced and spread TNF, the huge amounts of sewage sludge and animal manures which particularly in regions with animal overstocking stress the environment, the UN Department of Economic and Social Affairs requests to develop soil, ground- and drinking water protecting strategies with solutions latest 2050. Alone the third most important anthropogenic greenhouse gas N_2O increased in the atmosphere from 1750 to 2011 by 20% or to 324 ppb and in terms of CO_2-eq contributes N_2O to the global total greenhouse gas emissions by 8% (https:// www.cnbc.com/.../two-thirds-of-global-population-will-liv...; Goulding et al., 2008; Pretty et al., 2011; Reay et al., 2012).

In finding fertilization concepts orientated on the plant N demand as well as on N_2O emissions and in realizing an environmentally integrated waste water treatment farmers and waste managers have it not easy in adjusting their measures to the environmental needs. N cycling concerned the nitrite reductase controlling genes nirK and nirS, the dissimilatory nitrate reduction to ammonium (DNRA), the anaerobic to $NO3^-$ reduction coupled ammonia oxidation (Anammox) and the N_2O reductase controlling nosZ genes, responsible for the N_2O reduction to N_2, must be managed in a way that N_2 is major end product. Concepts beyond N-cycling and for achieving an acceptable compromise between all microbial, plants, animal and human participants are organic and precision farming. Also waste water treatment plants must continuously adapt on increasing inflowing organic/inorganic loads. That farmers and waste managers can guarantee food and water security they should know how roots and of which plants sense soil N depots. For being efficient in using UI, NI, and DI for controlling the N_2O release farmers should be detailed informed how the soil nitrification-denitrification complex works and how modern breeding technology improves plant N uptake. Also for implementing organic and /or precision farming to preserve soil productivity in monocultures the ideas must intensively discussed with farmers for enabling them stress the climate and groundwater less or better not (Zumft, 1997; Braker and Conrad, 2011; Gattinger et al., 2012; Leifeld et al., 2013; Skinner et al., 2014; Santín et al., 2017; Maktabifard et al., 2018; Benckiser, 2019; Khanna and Kaur, 2019).

Numerous N_2O measurements and dose response evaluations show that N responds directly to UI and NI applications in fertilized soils (Tindaon and Benckiser, 2019).Thus UI and NI can be included in N cycling, nitrification/denitrification management, because nitrification/denitrification are key processes and major atmospheric N_2O concentration drivers (Davidson, 2009; Butterbach-Bahl et al., 2011). Environmental factors, such as oxygen depletion, soil acidity, temperature, frost-thaw cycles, carbon, nitrate, copper availability, plant N uptake, management practices (N fertilizer type, application rate, technique, timing, organic manuring, irrigation) are N_2O emission influencers. Although atmospheric N_2O measurements suggest higher N_2O values of 2–2.5 or even 3–5%, the IPCC still assumed in the year 2006 that 1% of the N applied as mineral and organic fertilizer and released crop residues is emitted as N_2O (Crutzen et al., 2008; Davidson, 2009). N_2O emission measurements carried out in 104 mineral N-fertilized and unfertilized (control) field plots unveiled that the N_2O release from fertilized plots increased by factors between 1.1 and 15.3 and on this data basis a maximum N surplus threshold of 0 to 50 kg N/ha was calculated for N_2O-efficient cropping systems in a world of high and growing food, feed, fuel, and fibre demands (Eichner, 1990; Kaiser and Ruser, 2000; Ruser et al., 2001; van Groeningen et al., 2010; Pfab et al., 2012). Since the N fertilisation praxis still differs from the recommended N surplus threshold of 0 to 50 kg N/ha, an accurate prediction of crop N fertilizer demands is crucial. Climate integrating crop rotations with a right cultivar and N fertilizer form selection can realistically planned not before application intensity, irrigation, and crop protection measures, cover crops, crop residue incorporation, and animal stocking intensities are fully understood also side effect concerned (Benckiser and Ottow, 1982; Eichner, 1990; Raun and Johnson, 1999; Wiesler, 1998; Wiesler et al., 2001; Monteny et al, 2006; Twining et al, 2007; Baggs, 2011; Van den Heuvel et al., 2011; Pfab et al., 2012; Thomson et al., 2012; Sullivan et al., 2013; Seiz et al., 2014; Benckiser, 2019; Tindao and Benckiser, 2019). Is the NH_4^+ oxidation to NO_3^- during the cropping season co-shaped by UI, NI, and DI farmers, waste managers have to consider that in postharvest soils the inhibition slows and the emission of N_2O may increase again, but when the crop residues are removed after harvest as shown for a broccoli, lettuce, cauliflower rotation the emission of N_2O may decrease over a 15 months period from 18 kg to less than 5 kg N_2O-N/ha (Baggs et al., 2000; Seiz et al., 2014; Scheer et al., 2017; Köbke et al., 2018). By using all presently available possibilities skilled farmers and waste water treatment plant managers have already tools for improving their greenhouse gas management and can approach nature high biomass productivity on basis of N shortage and a high biodiversity by plant demand adapted N fertilization and biodiversity enhancement in agriculture production systems (Smith et al., 2008, 2019).

References

Abalos, D., van Groenigen, J.W., and De Deyn, G.B. (2018). What plant functional traits can reduce nitrous oxide emissions from intensively managed grasslands? Global Change Biol. 24, e248-e258. DOI: 10.1111/gcb.13827

Abasi, M.K., Hina, M., and Tahir, M.M. (2011). Effect of *Azadirachta indica* (Neem), sodium thiosulphate and calcium chloride on changes in nitrogen transformation and inhibition of nitrification in soil incubated under laboratory conditions. Chemosphere 82, 1629-1635. DOI: 10.1016/j.chemosphere.2010.11.044

Abbasian, F., Ghafar-Zadeh, E., and Magierowski, S. (2018). Microbiological sensing technologies: A review. Bioengineering 5, 20. DOI: 10.3390/bioengineering 5010020

Abisado, R.G., Benomar, S., Klaus, J.R., Dandekar, A.A., and Chandler, J.R., (2018) Bacterial quorum sensing and microbial community interactions. mBio 9, e02331-17. DOI: 10.1128/mBio.02331-17

Ahlgren, N.A., Fuchsman, C.A., Rocap, G., and Fuhrman, J.A. (2019). Discovery of several novel, widespread, and ecologically distinct marine Thaumarchaeota viruses that encode amoC nitrification genes. ISME J. 13, 618-631.

Ali, R., Iqbal, J., Tahir, G.R., Mahmood, T., and Pak, J. (2008). Effect of 3,5-dimetylpyrazole and nitrapyrin on nitrification under high soil temperature. Pakistani J. Bot. 40: 1053-1062.

Alvarez, L., Bricio, C., Gomez, M.J., and Berenguer, J. (2011). Lateral transfer of the denitrification pathway genes among *Thermus thermophilus* strains. Appl. Environ. Microbiol. 77, 1352-1358. DOI: 10.1128/AEM.02048-10

Alvarez, J.M., Vidal, E.A., and Gutiérrez, R.A. (2012). Integration of local and systemic signaling pathways for plant N responses. Curr. Opin. Plant Biol. 15, 185-191. DOI: 10.1016/j.pbi.2012.03.009

Amann, R.I., Ludwig, W., and Schleifer, K.H. (1995). Phylogenetic identification and in situ detection of individual microbial cells without cultivation. Microbiol. Rev. 59, 143-169.

Andersen, S.Z., Colic, V., Yang, S., Schwalbe, J.A., Nielander, A.C., McEnaney, J.M., Enemark-Rasmussen, K., Baker, J.G., Singh, A.R., Rohr, B.A., Statt, M.J., Blair, S.J., Mezzavilla, S., Kibsgaard, J., Vesborg, P.C.K., Cargnello, M., Bent, S.F., Jaramillo, T.F., Stephens, I.E.L., Norskov, J.K., and Chorkendorff, I. (2019). A rigorous electrochemical ammonia synthesis protocol with quantitative isotope measurements. Nature 570, 504-507. DOI: 10.1038/s41586-019-1260-x

Angel, R., Panhölzl, C., Gabriel, R., Herbold, C.W., Wanek, W., Richter, A., Eichorst, S.A., and Woebken, D. (2018). Application of stable-isotope labelling techniques for the detection of active diazotrophs. Environ. Microbiol. 20, 44-61. DOI: 10.1111/1462-2920.13954

App, A.A., Santiago, T., Daez, C., Menguito, C., Ventura, W., Tirol, A., Po, J., Watanabe, I., DeDatta, S.K., and Roger, P. (1984). Estimation of the nitrogen balance for irrigated rice and the contribution of phototrophs. Field Crops Res. 19, 17-27. DOI: 10.1016/0378-4290(84)90003-0

Arashiba, K., Miake, Y., and Nishibayashi, Y. (2011). A molybdenum complex bearing PNP-Type pincer ligands leads to the catalytic reduction of dinitrogen into ammonia. Nature Chem. 3, 120-125. DOI: 10.1038/nchem.906

Arrigo, K.R. (2005). Marine microorganisms and global nutrient cycles. Nature 437, 349-355. DOI: 10.1038/nature04159

Asano, M., and Wagai, R. (2014). Evidence of aggregate hierarchy at micro- to submicron scales in an allophanic Andisol. Geoderma 216, 62-74. DOI: 10.1016/ j.geoderma.2013.10.005

Bacon, P.E., ed. (1995). Nitrogen Fertilization in the Environment (New York: Marcel Dekker Inc,)

Baggs, E.M. (2011). Soil microbial sources of nitrous oxide: recent advances in knowledge, emerging challenges and future direction. Curr. Opin. Environ. Sustain. 3, 321-327. DOI: 10.1016/j.cosust.2011.08.011

Bai, Y., Liang, J., Liu, R., Hu, C., and Qu, J. (2014). Metagenomic analysis reveals microbial diversity and function in the rhizosphere soil of a constructed wetland. Environ. Technol. 35, 2521-2527. DOI: 10.1080/09593330.2014.911361

Baikova, L.G., Pesina, T.I., Kurkjian, S.R., Tang, Zh., Kireenko, M.F., Tikhonova, L.V., and Pukh, V.P. (2013). On the method for determining the true strength of inorganic glasses. Tech. Phys. 58, 1447-1452. DOI: 10.1134/S1063784213100034

Bais, H.P., Weir, T.L., Perry, L.G., Gilroy, S., and Vivanco, J.M. (2006). The role of root exudates in rhizosphere interactions with plants and other organisms. Ann. Rev. Plant Biol. 57, 233-266. DOI: 10.1146/annurev.arplant.57.032905.105159

Bakken, L.R., Bergaust, L., Liu, B., and Frostegard, A. (2012). Regulation of denitrification at the cellular level: a clue to the understanding of N_2O emissions from soils. Phil. Trans. R. Soc. B 367, 1226-1234.

Bamforth, S.S. (1988). Interactions between protozoa and other organisms. Agri. Ecosy. Environm. 24, 229-234. DOI: 10.1016/0167-8809(88)90068-0

Barbier-Brygoo, H., De Angeli, A., Filleur, S., Frachisse, J.M., Gambale, F., Thomine S., and Wege, S. (2011). Anion channels/transporters in plants: from molecular bases to regulatory networks. Annu. Rev. Plant Biol. 62, 25-51. DOI: 10.1146/annurev-arplant-042110-103741

Bardon, C., Poly, F., Piola, F., Pancton, M., Comte, G., Meiffren, G., and Haichar Fel, Z. (2016). Mechanism of biological denitrification inhibition (BDI): procyanidins induce an allosteric transition of the membrane-bound NO_3-reductase through membrane alteration. FEMS Microbiol. Ecol. 92, fiw034. DOI: 10.1093/femsec/fiw034.

Barraclough, D., and PurI, G. (1995). The use of 15N pool dilution and enrichment to separate the heterotrophic and autotrophic pathways of nitrification. Soil Biol. Biochem.27, 17-22. DOI: 10.1016/0038-0717(94)00141-M

Barraquio, W.L., Ladha, J.K., & Watanabe, I. (1983). Isolation and identification of N_2-fixing *Pseudomonas* associated with wetland rice. Can. J. Microbiol. 29, 867-873. DOI: 10.1139/m83-141

Barth, K.R., Isabella, V.M., and Clark, V.L. (2009). Biochemical and genomic analysis of the denitrification pathway within the genus *Neisseria*. Microbiology 155, 4093-4103.

Belete, F., Dechassa, N., Molla, A., and Tana, T. (2018). Effect of split application of different *N* rates on productivity and nitrogen use efficiency of bread wheat (*Triticum aestivum* L.). Agric. Food Secur. 7, 92. DOI: 10.1186/s40066-018-0242-9

Benckiser, G., and Ottow, J.C.G. (1982). Metabolism of the plasticizer di-n-butylphthalate by *Pseudomonas pseudoalcaligenes* under anaerobic conditions, with nitrate as the only electron acceptor Appl. Environ. Microbiol. 44, 576-578.

Benckiser, G., Santiago, S., Neue, H.U., Watanabe, I., and Ottow, J.C.G. (1984). Effect of fertilization on exudation, dehydrogenase activity, iron-reducing populations and Fe^{++} formation in the rhizosphere of rice *(Oryza sativa* L.) in relation to iron toxicity. Plant Soil 79, 305-316.

Benckiser; G., Eilts, R., Linn, A., Lorch, H.-J., Sümer, E., Weiske, A., and Wenzhöfer, F. (1996). N_2O emissions from different cropping systems and from aerated, nitrifying and denitrifying tanks of a municipal waste water treatment plant. Biol. Fertil. Soils 23, 257-265.

Benckiser, G., ed. (1997). Fauna in Soil Ecosystems (New York, USA: Marcel Dekker).

Benckiser, G. (2007). Growth, denitrification and nitrate ammonification of the rhizobial strain TNAU 14 in presence and absence of C_2H_4 and C_2H_2. Ann. Microbiol. 57, 509-514. DOI: 10.1007/BF03175347

Benckiser, G., and Schnell, S. (2007). Biodiversity in Agricultural Production Systems (Boca Raton, USA: Taylor and Francis).

Benckiser, G. (2017). Eusocial ant nest management, template for land development. WSEAS Trans. Environ. Dev. 13, 204-215.

Benckiser. G., (2019). Plastics, Micro- and Nanomaterials, and Virus-Soil Microbe-Plant Interactions in the Environment. In Plant Nanobionics, Vol. 1 Advances in the Understanding of Nanomaterials Research and Application, R. Prasad, ed. (Cham, Switzerland: Springer Nature AG), pp. 83-101. DOI: 10.1007/978-3-030-12496-0

Benckiser, G. (2021). Editorial and the soil virome and its nitrogen dependent nutrient recycling role – an estimate for a hectare grassland. In Soil and Recycling Management in the Anthropocene Era, G. Benckiser, ed. (Cham, Switzerland: Springer), in press.

Berg, I.A., Kockelkorn, D., Buckel, W., and Fuchs, G. (2007). A 3-hydroxypropionate/ 4-hydroxybutyrate autotrophic carbon dioxide assimilation pathway in Archaea. Science 318, 1782-1786. DOI: 10.1126/science.1149976

Berks, B.C., Ferguson, S.J., Moir, J.W.B., and Richardson, D.J. (1995). Enzymes and associated electron transport systems that catalyse the respiratory reduction of nitrogen oxides and oxyanions. Biochim. Biophys. Acta Bioenerg. 1232, 97-173. DOI: 10.1016/0005-2728(95)00092-5

Bertagnolli, A.D., and Stewart, F.J. (2018). Microbial niches in marine oxygen minimum zones. Nature Rev. Microbiol. 16, 723-729. DOI: 10.1038/ s41579-018-0087-z

Bezdek, M.J., and Chirik, P.J. (2019), A fresh approach to synthesizing ammonia from air and water. Nature 568, 464-466. DOI: 10.1038/d41586-019-01213-7

Bier, R.L., Bernhardt, E.S., Boot, C.M., Graham, E., Hall, E.K., Lennon, J.T., Nemergut, D.R., Osborne, B.B., Ruiz-González, C., and Schimel, J.P. (2015). Linking microbial community structure and microbial processes: an empirical and conceptual overview. FEMS Microbiol. Ecol. 91, fiv113. DOI: 10.1093/femsec/fiv113

Bloem, J., Starink, L.M., Bar-Glissen, M.-J.B., and Cappenberg, T.E. (1988). Protozoan grazing, bacterial activity, and mineralization in two-stage continuous cultures. Appl. Environ. Microbiol. 54, 3113-3121. DOI: 10.1128/AEM.54.12.3113-3121.1988

Bonkowski, M., and Clarholm, M. (2016). Stimulation of plant growth through interactions of bacteria and protozoa: Testing the auxiliary microbial loop hypothesis. Acta Protozool. 55, 237-247.

Bourke, S.A., Iwanyshyn, M., Kohn, J., and Hendry, M.J. (2019). Sources and fate of nitrate in groundwater at agricultural operations overlying glacial sediments. Hydrol. Earth Syst. Sci. 23, 1355-1373. DOI: 10.5194/hess-23-1355-2019

Boyd, E.S., Hamilton, T.L., and Peters, J.W. (2011). An alternative path for the evolution of biological nitrogen fixation. Front. Microbiol. 2, 205. DOI: 10.3389/fmicb.2011.00205

Boyle-Yarwood, S.A., Bottomley, P.J., and Myrold, D.D. (2008). Community composition of ammonia oxidizing bacteria and archaea in soils under stands of red alder and Douglas-fir in Oregon. Environ. Microbiol. 10, 2956-2965. DOI: 10.1111/j.1462-2920.2008.01600.x

Braker, G., and Conrad, R. (2011). Diversity, structure, and size of N_2O-producing microbial communities in soils—what matters for their functioning? Adv. Appl. Microbiol. 75, 33-70. DOI: 10.1016/B978-0-12-387046-9.00002-5

Bremner, J.M., and Blackmer, A.M. (1978). Nitrous oxide: emission from soils during nitrification of fertilizer nitrogen. Science 199, 295-296.

Bringel, F., and Couée, I. (2015). Pivotal roles of phyllosphere microorganisms at the interface between plant functioning and atmospheric trace gas dynamics. Front. Microbiol. 6, 486. DOI: 10.3389/fmicb.2015.00486

Bru, D., Sarr, A., and Philippot, L. (2007). Relative abundances of proteobacterial membrane-bound and periplasmic nitrate reductases in selected environments. Appl. Environ. Microbiol. 73: 5971-5974.

Bruhn D, Albert KR, Mikkelsen TN, Ambus P (2014) UV-induced N_2O emission from plants Atmospheric Environment 99, 206-214. DOI: 10.1016/j.atmosenv. 2014.09.077

Brückner, S., and Mösch, H.-U. (2012). Choosing the right lifestyle: adhesion and development in *Saccharomyces cerevisiae*. FEMS Microbiol. Rev.36, 25-58. DOI: 10.1111/j.1574-6976.2011.00275.x

Buresh, R.H., Casselman, M.E., and Patrick, W.H. (1980). Nitrogen fixation in flooded soil systems. A review. Adv. Agronomy 33, 149-192. DOI: 10.1016/S0065-2113(08)60166-2

Burth, I., Benckiser, G., and Ottow, J.C.G. (1982). N_2O-Freisetzung aus Nitrit (Denitrifikation) durch ubiquitäre Pilze unter aeroben Bedingungen. Naturwissenschaften 69, 598-599.

Butterbach-Bahl, K., Nemitz, E., Zaehle, S., Billen, G., Boeckx, P., Erisman, J.W., Garnier, J., Upstill-Goddard, R., Kreuzer, M., Oenema, O., Reis, S., Schaap, M, Simpson, D., de Vries, W., Winiwarter, W., and Sutton, M.A. (2011). Nitrogen as a Threat to European Greenhouse Balance. In The European Nitrogen Assessment: Sources, Effects, and Policy Perspectives, M.A. Sutton, C.M. Howard, J.W. Erisman, G. Billen, A. Bleeker, P. van Grennfelt, H. Grinsve, and B. Grizzetti, eds. (Cambridge, New York: Cambridge University Press), pp. 434-462.

Byrnes, B., and Bumb, B.L. (1998). Population growth, food production and nutrient requirements. J. Crop Prod. 1, 1-27.

Cabello. P., Roldan, and Moreno-Vivian, C. (2004). Nitrate reduction and the nitrogen cycle in archaea. Microbiol. 150, 3527-3546. DOI: 10.1099/mic.0.27303-0

Canarini, A., Kaiser, C., Merchant, A., Richter, A., and Wanek, W. (2019). Root exudation of primary metabolites: mechanisms and their roles in plant responses to environmental stimuli. Front. Plant Sci. 10, 157. DOI: 10.3389/fpls.2019.00157

Cantarella, H., Otto, R., Soares, J.R., and de BritoSilva, A.G. (2018). Agronomic efficiency of NBPT as a urease inhibitor: A review. J. Adv. Res. 13, 19-27. DOI: 10.1016/j.jare.2018.05.008

Cao, S., Du, R., Zhang, H., and Peng, Y. (2019). Understanding the granulation of partial denitrification sludge for nitrite production. Chemosphere 236, 124389. DOI: 10.1016/j.chemosphere.2019.124389

Capone, D.G., Popa, R., Flood, B., & Nealson, K.H. (2006). Follow the nitrogen. Science 312, 708-709. DOI: 10.1126/science.1111863

Cardini, U., Bednarz, V.N., van Hoytemal, N., Rovere, A., Naumann, M.S., Al-Rshaidat, M.M.D., and Wild, C. (2016). Budget of primary production and dinitrogen fixation in a highly seasonal Red Sea coral reef. Ecosystems 19, 771-785. DOI: 10.1007/s10021-016-9966-1

Carvalhais, L.C., Dennis, P.G., Fedoseyenko, D., Hajirezael, M.R., Boriss, R., and von Wiren, N. (2011). Root exudation of sugars, amino acids, and organic acids by maize as affected by nitrogen, phosphorus, potassium, and iron deficiency. J. Plant Nutr. Soil Sci. 174, 3-11. DOI: 10.1002/jpln.201000085

Chan, Y.K., and Campbell, N.E.R. (1980). Denitrification in Lake 227 during summer stratification. Can. J. Fish. Aquat. Sci. 37, 506-512. DOI: 10.1139/f80-065

Chang, C., Janzen, H.H., Nakonechny, E.M., and Cho, C.M. (1998). Nitrous oxide emission through plants. Soil Sci. Soc. Am. J. 62, 35-38. DOI: 10.2136/sssaj1998.03615995006200010005x

Chen, D., Li, Y., Grace, P., and Mosier, A.R. (2008). N_2O emissions from agricultural lands: A synthesis of simulation approaches. Plant Soil 309, 169-189. DOI: 10.1007/s11104-008-9634-0

Chen, J.G., Crooks, R.M., Seefeldt, L.C., Bren, K.L., Bullock, R.M., Darensbourg, Y., Holland, P.L., Hoffman, B., Janik, M.J., Jones, A.K., Kanatzidis, M.G., King, P., Lancaster, M., Lymar, S.V., Pfromm, P., Schneider, W.F., and Schrock, R.R. (2018). Beyond fossil fuel–driven nitrogen transformations. Science 360, 6391, DOI: 75910.1126/science.aar6611

Chen, X., Boeckx, P., Shen, S., and Van Cleemput, O. (1999). Emission of N_2O from rye grass (*Lolium perenne* L.). Biol. Fertil. Soils 28, 393-396. DOI: 10.1007/s003740050510

Cheng, X., Peng, R., Chen, J., Luo, Y., Zhang, Q., An, S., Chen, J., and Li, B. (2007). CH_4 and N_2O emissions from *Spartina alterniflora* and *Phragmites australis* in experimental mesocosms. Chemosphere 68, 420-427. DOI: 10.1016/j.chemosphere.2007.01.004

Collman, J.P., Yang, Y., Dey, A., Decreau, R.A., Ghosh, S., Ohta, T., and Solomon, E.I. (2008). A functional nitric oxide reductase model. Proc. Natl. Acad. Sci. U.S.A. 105, 15660-15665. DOI: 10.1073/pnas.0808606105

Conthe, M., Lycus, P., Arntzen, M., da Silva, A.R., Frostegård, Å., Bakken, L.R., Kleerebezem, R., and van Loosdrecht, M.C.M. (2019). Denitrification as an N_2O sink. Water Res. 151, 381-387. DOI: 10.1016/j.watres.2018.11.087

Cooper, H.D., and Clarkson, D.T. (1989). Cycling of amino-nitrogen and other nutrients between shoots and roots in cereals - A possible mechanism integrating shoot and root in the regulation of nutrient uptake. J. Exp. Bot. 40, 753-762. DOI: 10.1093/jxb/40.7.753

Cowan, N., Levy, P., Moring, A., Simmons, I., Bache, C., Stephens, A., Marinheiro, J., Brichet, J., Song, L., Pickard, A., McNeill, C., McDonald, R., Maire, J., Loubet, B., Voylokov, P., Sutton, M., and Skiba, U. (2019). Nitrogen use efficiency and N_2O and NH_3 losses attributed to three fertiliser types applied to an intensively managed silage crop Biogeosciences Discuss., in review. DOI: 10.5194/bg-2019-90

Crab, R., Avnimelech, Y., Defoirdt, T., Bossier, P., and Verstraete, W. (2007). Nitrogen removal techniques in aquaculture for a sustainable production. Aquaculture 270, 1-14. DOI: 10.1016/j.aquaculture.2007.05.006

Craig, T.P. (2010). The resource regulation hypothesis and positive feedback loops in plant–herbivore interactions. Popul. Ecol. *52*, 461-473. DOI: 10.1007/s10144-010-0210-0

Craswell, E.T., and Godwin, D.C. (1984). The efficiency of nitrogen fertilizers applied to cereals in different climates. Adv. Plant Nutr. 1, 1-55.

Crutzen, P.J., Mosier, A.R., Smith, K.A., and Winiwarter, W. (2008). N_2O release from agro-biofuel production negates global warming reduction by replacing fossil fuels. Atmos. Chem. Phys. 8, 389-395. DOI: 10.5194/acp-8-389-2008

Cuhel, J., Simek, M., Laughlin, R.J., Bru, D., Cheneby, D., Watson, C.J., and Philippot, L. (2010). Insights into the effect of soil pH on N_2O and N_2 emissions and denitrifier community size and activity. Appl. Environ. Microbiol. 76, 1870-1878. DOI: 10.1128/AEM.02484-09

Dai, Z., Li, Y., Zhang, X., Wu, J., Luo, Y., Kuzyakov, Y., Brookes, P.C,, and Xu, J. (2019). Easily mineralizable carbon in manure–based biochar added to a soil influences N_2O emissions and microbial–N cycling genes. 30, 406-416. DOI: 10.1002/ldr.3230

Daims, H., Lebedeva, E.V., Pjevac, P., Han, P., Herbold, C., Albertsen, M., Jehmlich, N., Palatinszky, M., Vierheilig, J., Bulaev, A., Kirkegaard, R.H., von Bergen, M., Rattei, T., Bendinger, B., Nielsen, P.H., and Wagner, M. (2015). Complete nitrification by *Nitrospira* bacteria. Nature 528, 504-509. DOI: 10.1038/nature16461

Davidson, E.A. (2009). The contribution of manure and fertilizer nitrogen to atmospheric nitrous oxide since 1860. Nat. Geosci. 2, 659-662.

DeAngelis, K.M., Lindow, S.E., and Firestone, M.K. (2008). Bacterial quorum sensing and nitrogen cycling in rhizosphere soil. FEMS Microbiol. Ecol. 66, 197-207. DOI: 10.1111/j.1574-6941.2008.00550.x

Delwiche, C.C. (1977). Energy relations in the global nitrogen cycle. Ambio 6, 106-111.

Demaneche, S., Philippot, L., David, M.M., Navarro, E., Vogel, T.M., and Simonet, P. (2009). Characterization of denitrification gene clusters of soil bacteria via a metagenomic approach. Appl. Environ. Microbiol. 75, 534-537.

Depkat-Jakob, P.S., Hilgarth, M., Horn, M.A., and Drake, H.L. (2010). Effect of earthworm feeding guilds on ingested dissimilatory nitrate reducers and denitrifiers in the alimentary canal of the earthworm. Appl. Environ. Microbiol. 76, 6205-6214. DOI: 10.1128/AEM.01373-10

DeVries, W.M.F., and Bobbink, R. (2017). Nitrogen deposition impacts on biodiversity in terrestrial ecosystems: Mechanisms and perspectives for restoration. Biol. Conserv. 212, 387-389. DOI: 10.1016/j.biocon.2017.01.017

Dignac, M.-F., Derrien, D., Barré, P., Barot, S., Cécillon, L., Chenu, C., Chevallier, T., Freschet, G.T., Garnier, P., Guenet, B., Hedde, M., Klumpp, K., Lashermes, G., Maron, P.-A., Nunan, N., Roumet, C., and Basile-Doelsch (2017). Increasing soil carbon storage: mechanisms, effects of agricultural practices and proxies. A review. Agron. Sustain. Dev. 37, 14. DOI: 10.1007/s13593-017-0421-2

Doi, Y., Takaya, N., and Takizawa, N. (2009). Novel denitrifying bacterium *Ochrobactrum anthropi* YD50.2 tolerates high levels of reactive nitrogen oxides. Appl. Environ. Microbiol. 75, 5186-5194. DOI: 10.1128/AEM.00604-09

Dolgin, E. (2019). The secret social lives of viruses. Nature 570, 290-292. DOI: 10.1038/d41586-019-01880-6

Domeignoz−Horta, L.A., Philippot, L,. Peyrard, C., Bru, D, Breuil, M.−C., Bizouard, F., Justes, E., Mary, B., Léonard, J., and Spor, A. (2018). Peaks of in situ N_2O emissions are influenced by N_2O−producing and reducing microbial communities across arable soils. Glob. Chang. Biol. 24, 360-370. DOI: 10.1111/gcb.13853

Donaldson, G.P., Lee, S.M., and Mazmanian, S.K. (2016). Gut biogeography of the bacterial microbiota. Nat. Rev. Microbiol. 14, 20-32. DOI: 10.1038/nrmicro3552

Dong, L.F., Smith, C.J., Papaspyrou, S., Sokratis, Stott, A., Osborn, A.M., and Nedwell, D.B. (2009). Changes in benthic denitrification, nitrate ammonification, and anammox process rates and nitrate and nitrite reductase gene abundances along an estuarine nutrient gradient (the Colne Estuary, United Kingdom). Appl. Environ. Microbiol. 75, 3171-3179.

Dörsch, P., Braker, G., and Bakken, L.R. (2012). Community-specific pH response of denitrification: 660 experiments with cells extracted from organic soils. FEMS Microbiol. Ecol. 79, 530-541. DOI: 10.1111/j.1574-6941.2011.01233.x

Dungel, P., Penzenstadler, C., Ashmwe, M., Dumitrescu, S., Stoegerer, T., Redl, H., Bahrami, S., and Kozlov, A.V. (2017). Impact of mitochondrial nitrite reductase on hemodynamics and myocardial contractility. Sci. Rep. 7, 12092. DOI: 10.1038/s41598-017-11531-3

Eichner, M.J. (1990). Nitrous oxide emissions from fertilized soils: summary of available data. J. Environ. Qual. 19, 272-280.

Elliott, L.F., and Gilmour, C.M. (1971). Growth of *Pseudomonas stutzeri* with nitrate and oxygen as terminal electron acceptors. Soil Biol. Biochem. 3, 331-335. DOI: 10.1016/0038-0717(71)90043-5

Enwall, K., Throbäck, I.N., Stenberg, M., Söderström, M., and Hallin, S. (2010). Soil resources influence spatial patterns of denitrifying communities at scales compatible with land management. Appl. Environ. Microbiol. 76, 2243-2250.

Erisman, J.W., Galloway, J.N., Seitzinger, S., Bleeker, A., Dise, N.B., Petrescu, A.M.R., Leach, A.M., and de Vries, W. (2013). Consequences of human modification of the global nitrogen cycle. Philos. Trans. R. Soc. Lond. B Biol. Sci. 368, 20130116. DOI: 10.1098/rstb.2013.0116

Fischer, M., Alderson, J., van Keulen, G., White, J., and Sawers, R.G. (2010). The obligate aerobe *Streptomyces coelicolor* A3(2) synthesizes three active respiratory nitrate reductases. Microbiol. 156, 3166-3179.

FAO (2016). World fertilizer trends and outlook to 2019; Summary report. ISBN 978-92-5-108692-6 (www.fao.org/publications)

FAO (2019). The State of the World's Biodiversity for Food and Agriculture, J. Bélanger and D. Pilling, eds, (Rome: FAO Commission on Genetic Resources for Food and Agriculture Assessments). http://www.fao.org/3/CA3129EN/CA3129EN.pdf, http://www.fao.org/statistics/databases/en/

Farrar, J.F., and Jones, D.L. (2000). The control of carbon acquisition by roots. New Phytol. 147, 43-53. DOI: 10.1046/j.1469-8137.2000.00688.x

Fey, A., Benckiser, G., and Ottow, J.C.G. (1999). Emissions of nitrous oxide from a constructed wetland using a groundfilter and macrophytes in waste-water purification of a dairy farm. Biol. Fertil. Soils 29, 354-359

Field, C.B., and Raupach, M.R. (2004). The Global Carbon Cycle. Scope Series 62 (Washington, DC: Island Press).

Filleur, S., Walch-Liu, P., Gan, Y., and Forde, B.G. (2005). Nitrate and glutamate sensing by plant roots. Biochem. Soc. Trans. 33, 283-286. DOI: 10.1042/BST0330283

Fischer, M., Triggs, G.J., and Krauss, T.F. (2016). Optical sensing of microbial life on surfaces. Appl. Environ. Microbiol. 82, 1362-1371. DOI: 10.1128/AEM.03001-15

Flemming, H.-C., and Wuertz, S. (2019). Bacteria and archaea on Earth and their abundance in biofilms. Nat. Rev. Microbiol. 17, 247-260. DOI: 10.1038/s41579-019-0158-9ature

Flores-Mireles, A.L., Winans, S.C., and Holguin, G. (2007). Molecular characterization of diazotrophic and denitrifying Bacteria associated with mangrove roots. Appl. Environ. Microbiol. 73, 7308-7321. DOI: 10.1128/AEM.01892-06

Fowler, D., Coyle, M., Skiba1, U., Sutton, M.A., Cape, J.N., Reis, S., Sheppard, L.J., Jenkins, A., Grizzetti, B., Galloway, J.N., Vitousek, P., Leach, A., Bouwman, A.F., Butterbach-Bahl, K., Dentener, F., Stevenson, D., Amann, M., and Voss, M. (2013), The global nitrogen cycle in the twenty first century. Phil. Trans. R. Soc. B 368, 20130164. DOI: 10.1098/rstb.2013.0164

Freney, J.R., Denmead, O.T., Watanabe, I., and Craswell, E.T. (1981). Ammonia and nitrous oxide losses following applications of ammonium sulfate to flooded rice. Aust. J. Agric. Res 32, 37-45.

Freney, J.R., Trevitt, A.C.F., De Datta, S.K., Obcemea W.N., and Real, J.G. (1990). The interdependence of ammonia volatilization and denitrification as nitrogen loss processes in flooded rice fields in the Philippines. Biol Fertil Soils 9, 31-36.

Fuchs, G., Boll, M., and Heider, J. (2011). Microbial degradation of aromatic compounds - from one strategy to four. Nat. Rev. Microbiol. 9, 803-816. DOI: 10.1038/nrmicro2652

Fuertes-Mendizábal, X.F., Fernández-Diez, K., Estavillo-Carmen, J.M., and Menéndez, G.-M.S. (2016). The new nitrification inhibitor 3,4-dimethylpyrazole succinic (DMPSA) as an alternative to DMPP for reducing N_2O emissions from wheat crops under humid Mediterranean conditions. Eur. J. Agronomy 80, 78-87. DOI: 10.1016/j.eja.2016.07.001

Fuertes-Mendizábal, T., Huérfano, X., Vega-Mas, I., Torralbo, F., Menéndez, S., Ippolito, I.A., Kammann, C., Wrage-Mönnig, N., Cayuela, M.L., Borchard, N., Spokas, K., Novak, J., González-Moro, M.B., González-Murua, C., and Estavillo, J.M. (2019). Biochar reduces the efficiency of nitrification inhibitor 3,4-dimethylpyrazole phosphate (DMPP) mitigating N_2O emissions. Scient. Rep. 9, 2346. DOI: 10.1038/s41598-019-38697-2

Funge-Smith, S.J., and Briggs, M.R.P. (1998). Nutrient budgets in intensive shrimp ponds: implications for sustainability Aquaculture 164, 117-133. DOI: 10.1016/S0044-8486(98)00181

Fungo, B., Chen, Z., Butterbach-Bahl, K., Lehmannn, J., Saiz, G., Braojos, V., Kolar, A., Rittl, T.F., Tenywa, M., Kalbitz, K., Neufeldt, H., and Dannenmann, M. (2019). Nitrogen turnover and N_2O/N_2 ratio of three contrasting tropical soils amended with biochar, Geoderma 348, 12-20. DOI: 10.1016/j.geoderma.2019.04.007

Gattinger, A., Muller, A., Haeni, M., Skinner, C., Fliessbach, A., Buchmann, N., Mäder, P., Stolze, M., Smith, P., El-Hage Scialabba, N., and Niggli, U. (2012). Enhanced top soil carbon stocks under organic farming. Proc. Natl. Acad. Sci. U.S.A. 109, 18226-18231. DOI: 10.1073/pnas.1209429109

Gleason, F.H., Crawford, J.W., Neuhauser, S., Henderson, L.E., and Lilje, O. (2012). Resource seeking strategies of zoosporic true fungi in heterogeneous soil habitats at the microscale level. Soil Biol. Biochem. 45, 79-88. DOI: 10.1016/j.soilbio.2011.10.011

Glibert, P.M., Maranger, R., Sobota, D.J., and Bouwman, L. (2014). The Haber Bosch–harmful algal bloom (HB–HAB) link. Environ. Res. Lett. 9, 105001. DOI: 10.1088/1748-9326/9/10/105001

Göpel, Y., Görke, B. (2012). Rewiring two-component signal transduction with small RNAs. Curr. Opin. Microbiol. 15, 132-139. DOI: 10.1016/j.mib.2011.12.001

Gorka, S, Dietrich, M., Mayerhofer, W., Gabriel, R., Wiesenbauer, J., Martin, V., Zheng, Q., Imai, B., Prommer, J., Weidinger, M., Schweiger, P., Eichorst, S.A., Wagner, M., Richter, A., Schintlmeister, A., Woebken, D., and Kaiser, C. (2019). Rapid transfer of plant photosynthates to soil bacteria via ectomycorrhizal hyphae and its interaction with nitrogen availability. Front. Microbiol. 10, 168. DOI: 10.3389/fmicb.2019.00168

Goulding, K., Jarvis, S., and Whitmore, A. (2008). Optimizing nutrient management for farm systems. Phil. Trans. R. Soc. B, 667-680.

Gralnick, J.A., and Newman, D.K. (2007). Extracellular respiration. Mol. Microbiol. 65, 1-11. DOI: 10.1111/j.1365-2958.2007.05778.x

Granger, J., and Wankel, S.D. (2016). Isotopic overprinting of nitrification on denitrification as a ubiquitous and unifying feature of environmental nitrogen cycling. Proc. Natl. Acad. Sci. U.S.A. 113, E6391-E6400. DOI: 10.1073/pnas.1601383113

Granli, T., and Bøckman, O.Ch. (1994). Nitrous oxide from agriculture. Nor. J. Agric. Sci. Suppl. 12, 7-124.

Gruber, N. (2019). Consistent patterns of nitrogen fixation identified in the ocean. Nature 566, 191-193. DOI: 10.1038/d41586-019-00498-y

Gu, B., Lam, S.K., Reis, S., van Grinsven, H., Ju, X.,Yan, X., Zhou, F., Liu, H., Cai, Z., Galloway, J.N., Howard, C., Sutton, M.A., and Chen, D. (2019). Toward a generic analytical framework for sustainable nitrogen management: application for China. Environ. Sci. Technol. 53, 1109-1118. DOI: 10.1021/acs.est.8b06370

Guyonnet, J.P., Cantarel, A.A.M., Simon, L., and el Zahar Haichar, F. (2018). Root exudation rate as functional trait involved in plant nutrient–use strategy classification. Ecol. Evol. 8, 8573-8581. DOI: 10.1002/ece3.4383

Hammerschlag, N. (2019). Quantifying shark predation effects on prey: dietary data limitations and study approaches. Endang. Species Res. 38, 147-151. DOI: 10.3354/esr00950

Hansen, B., Thorling, L., Schullehner, J., Termansen, M., and Dalgaard, T. (2017). Groundwater nitrate response to sustainable nitrogen management Sci. Rep. 7, 8566. DOI: 10.1038/s41598-017-07147-2

Hargreaves, J.A. (1998). Nitrogen biochemistry of aquaculture ponds. 166, 181-212. DOI: 10.1016/S0044-8486(98)00298-1

Hartmann, A., Rothballer, M., Hense, B.A., and Schröder, P. (2014). Bacterial quorum sensing compounds are important modulators of microbe-plant interactions. Front. Plant Sci. 5, 131. DOI: 10.3389/fpls.2014.00131

Hassett, D.J., Bisesi, M.S., and Hartenstein, R. (1987). Bactericidal action of humic acids. Soil Biol. Biochem. 19, 111-113.

Hatzenpichler, R. (2012). Diversity, physiology, and niche differentiation of ammonia-oxidizing Archaea. Appl. Environ. Microbiol. 78, 7501-7510. DOI: 10.1128/AEM.01960-12

Hayatsu, M., Tago, K, and Saito, M. (2008). Various players in the nitrogen cycle: Diversity and functions of the microorganisms involved in nitrification and denitrification. Soil Sci. Plant Nutr. 54, 33-45. DOI: 10.1111/j. 1747-0765.2007.00195.x

He, H., Zhen, Y., Mi, T., Fu, L., and Yu, Z. (2018). Ammonia-Oxidizing archaea and bacteria differentially contribute to ammonia oxidation in sediments from adjacent waters of Rushan Bay, China. Front. Microbiol. 9, 116. DOI: 10.3389/fmicb. 2018.00116

Hidalgo-García, A., Torres, M.J., Salas, A., Bedmar, E.J., Girard, L., and Delgado, M.J. (2019). *Rhizobium etli* produces nitrous oxide by coupling the assimilatory and denitrification pathways. Front. Microbiol. 10, 980. DOI: 10.3389/fmicb.2019.00980

Ho, C.H., Lin, S.H., Hu, H.C., and Tsay, Y.F. (2009). CHL1 functions as a nitrate sensor in plants. Cell 138, 1184-1194. DOI: 10.1016/j.cell.2009.07.004

Hodge, A., Robinson, D., and Fitter, A. (2000). Are microorganisms more effective than plants at competing for nitrogen? Trends Plant Sci. 5, 304-308. DOI. 10.1016/ S1360-1385 (00)01656-3

Horchani, F., Prévot, M., Boscari, A., Evangelisti, E., Meilhoc, E., Bruand, C., Raymond, P., Boncompagni, E., Aschi-Smiti, S., Puppo, A., and Brouquisse, R. (2011). Both plant and bacterial nitrate reductases contribute to nitric oxide production in *Medicago truncatula* nitrogen-fixing nodules. Plant Phys. 155, 1023-1036. DOI: 10.1104/pp.110.166140

Horel, Á., Gelybó, G., Potyó, I., Pokovai, K., and Bakacsi, Z. (2019). Soil nutrient dynamics and nitrogen fixation rate changes over plant growth in Temperate Soil. Agronomy 9, 179. DOI: 10.3390/agronomy 9040179

Hu, Z., Wessels, H.J.C.T., van Alen, T., Jetten, M.S.M., and Kartal, B. (2019). Nitric oxide-dependent anaerobic ammonium oxidation. Nature Comm. 10, 1244. DOI: 10.1038/s41467-019-09268-w

Hunt, B.P.V., Bonnet, S., Berthelot, H., Conroy, B.J., Foster, R.A., and Pagano, M. (2016). Contribution and pathways of diazotroph-derived nitrogen to zooplankton during the VAHINE mesocosm experiment in the oligotrophic New Caledonia lagoon. Biogeosci. 13, 3131-3145. DOI: 10.5194/bg-13-3131-201

Imsande, J., and Touraine, B. (1994). N demand and the regulation of nitrate uptake. Plant Physiol. 105, 3-7. DOI: 10.1104/pp.105.1.3

IPCC (2006). Agriculture, forestry and other land use. In: 2006 IPCC Guidelines for National Greenhouse Gas Inventories, Vol. 4, S. Egglestone et al., eds. (Kanagawa, Japan: IGES Hayama).

IPCC (2007). Climate Change 2007: Synthesis Report.

Ishii, S., Ashida, N., Otsuka, S., and Senoo, K. (2011). Isolation of oligotrophic denitrifiers carrying previously uncharacterized functional gene sequences. Appl. Environ. Microbiol. 77, 338-342. DOI: 10.1128/AEM.02189-10

Ishi, S., Ikeda, S., Minammisawa, K., and Sendoo, K. (2011). Nitrogen cycling in rice paddy environments: past achievements and future challenges. Microbes Environ. 26, 282-292. DOI: 10.1264/jsme2.ME11293

Itakura, M., Tabata, K., Eda, S., Mitsui, H., Murakami, K., Yasuda Kiriko, J., and Minamisawa, K. (2008). Generation of *Bradyrhizobium japonicum* mutants with increased N_2O reductase activity by selection after introduction of a mutated *dnaQ* Gene. Appl. Environ. Microbiol. 74, 7258-7264.

Jang, J., Sakai, Y., Senoo, K., and Ishii, S. (2019) Potentially mobile denitrification genes ientified in *Azospirillum sp.* Strain TSH58. Appl. Environ. Microbiol. 9, e02474-18. DOI: 10.1128/AEM.02474-18

Jansson, J.K., and Hofmockel, K.S. (2018). The soil microbiome - from metagenomics to metaphenomics. Environ. Microbiol. 43, 162-168. DOI: 10.1016/j.mib.2018 01.013

Ji, B., Yang, K., Zhu, L., Jiang, Y., Wang, H., Zhou, J., and Zhang, H. (2015). Aerobic denitrification: A review of important advances of the last 30 years. Biotechn. Bioprocess Eng. 20, 643-651. DOI: 10.1007/s12257-015-0009-0

Jones, C.M., Graf, D.R.H., Bru, D., Philippot, L., and Hallin, S. (2013). The unaccounted yet abundant nitrous oxide-reducing microbial community: a potential nitrous oxide sink. ISME J. 7, 417-426. DOI: 10.1038/ismej.2012.125

Jones, D.L., Healey, J.R., Willett, V.B., Farrar, J.F., and Hodge, A. (2005). Dissolved organic nitrogen uptake by plants—an important N uptake pathway? Soil Biol. Biochem. 37, 413-423. DOI: 10.1016/j.soilbio.2004.08.008

Jung, J., Choi, S., Jung, H., Scow, K.M., and Park, W. (2013). Primers for amplification of nitrous oxide reductase genes associated with Firmicutes and Bacteroidetes in organic-compound-rich soils. Microbiol. 159, 307-315.

Jung, M.-Y., Gwak, J.H., Rohe, L., Giesemann, A., Kim, J.-G., Well, R., Madsen, E.L., Herbold, C.W., Wagner. M., and Rhee, S.-K. (2019). Indications for enzymatic denitrification to N_2O at low pH in an ammonia-oxidizing archaeon. ISME J. 13, 2633-2638. DOI: 10.1038/s41396-019-0460-6

Kafarski, P., and Talma, M. (2018). Recent advances in design of new urease inhibitors: A review. J. Adv. Res. 13, 101-112. DOI: 10.1016/j.jare.2018.01.007

Khanna, A., and Kaur, S. (2019). Evolution of internet of things (IoT) and its significant impact in the field of precision agriculture. Comput. Electron. Agric. 157, 218-231. DOI: 10.1016/j.compag.2018.12.039

Kaiser, C., Kilburn, M.R., Clode, P.L., Fuchslueger, L., Koranda, M., Cliff, J.B., Solaiman, Z.M., and Murphy, D.V. (2015). Exploring the transfer of recent plant photosynthates to soil microbes: mycorrhizal pathway vs direct root exudation. New Phytol. 205, 1537-1551. DOI: 10.1111/nph.13138

Kaiser, E.A., and Ruser, R. (2000). Nitrous oxide emissions from arable soils in Germany – An evaluation of six long-term field experiments. J. Plant Nutr. Soil Sci. 163, 249-260.

Karlen, D.L., Lal, R., Follett, R.F., Kimble, J.M., Hatfield, J.L., Miranowski, J.M., Cambardella, C.A., Manale, A., Anex, R.P., and Rice, C.W. (2006). Crop residues: the rest of the story. Environm. Sci. Technol. 43, 8011-8015. DOI: 10.1021/es9011004

Kanno, Y., Hanada, A., Chiba, Y., Ichikawa, T., Nakazawa, M., Matsui, M., Koshiba, T., Kamiya, Y., and Seo, M. (2012). Identification of an abscisic acid transporter by functional screening using the receptor complex as a sensor. Proc. Natl. Acad. Sci. U.S.A. 109, 9653-9658. DOI: 10.1073/pnas.1203567109

Kant, S., Peng, M., and Rothstein, S.J. (2011). Genetic regulation by NLA and MicroRNA827 for maintaining nitrate-dependent phosphate homeostasis in *Arabidopsis.* PLoS Genetics 7, e1002021. DOI: 10.1371/journal.pgen.1002021

Kanter, D., and Searchinger, T. (2018). A technology-forcing approach to reduce nitrogen pollution. Nat. Sustain. 1, 544-552. DOI: 10.1038/s41893-018-0143-8

Keerthisinghe, D.G., Freney, J.R., and Mosier, A.R. (1993). Effect of wax-coated calcium carbide and nitrapyrin on nitrogen loss and methane emission from dry-seeded flooded rice. Biol. Fertil. Soils 16, 71-75.

Kehr, J. (2006). Phloem sap proteins: their identities and potential roles in the interaction between plants and phloem-feeding insects. J. Exp. Bot, 57, 767-774. DOI: 10.1093/jxb/erj087

Khanitchaidecha, W., Nakaruk, A., Ratananikom, K., Eamrat, R., and Kazama, F. (2018). Heterotrophic nitrification and aerobic denitrification using pure-culture bacteria for wastewater treatment. J. Water Reuse Desal. 9, 10-17. DOI: 10.2166/wrd.2018.064

Kiba, T., and Krapp, A. (2016). Plant nitrogen acquisition under low availability: Regulation of uptake and root architecture. Plant Cell Physiol. 57, 707-714. DOI: 10.1093/pcp/pcw052

Killham, K. (1990). Nitrification in coniferous forest soils. Plant Soil 128, 31-44,

Kim, S.-W., Miyahara, M., Fushinobu, S., Wakagi, T., and Shoun, H, (2010). Nitrous oxide emission from nitrifying activated sludge dependent on denitrification by ammonia-oxidizing bacteria. Bioresour. Technol. 101, 3958-3963. DOI: 10.1016/j.biortech.2010.01.030

Kitzinger, K., Padilla, C.C., Marchant, H.K., Hach, P.F., Herbold, C.W., Kidane, A.T., Könneke, M., Littmann, S., Mooshammer, M., Niggemann, J., Petrov, S., Richter, A., Stewart, F.J., Wagner, M., Kuypers, M.M.M., and Bristow, L.A. (2019). Cyanate and urea are substrates for nitrification by Thaumarchaeota in the marine environment. Nat. Microbiol. 4, 234-243. DOI: 10.1038/s41564-018-0316-2

Klotz, M.G., and Stein, L.Y. (2008). Nitrifier genomics and evolution of the nitrogen cycle. FEMS Microbiol. Lett. 278, 146-156. DOI: 10.1111/j.1574-6968.2007.00970.x

Koch, H., van Kessel, M.A.H.J., and Lücker, S. (2019). Complete nitrification: insights into the ecophysiology of comammox *Nitrospira*. Appl. Microbiol. Biotechnol. 103, 177-189. DOI: 10.1007/s00253-018-9486-3

Köbke, S., Senbayram, M., Pfeiffer, B., Nacke, H., and Dittert, K. (2018). Post-harvest N_2O and CO_2 emissions related to plant residue incorporation of oilseed rape and barley straw depend on soil NO_3^- content. Soil Tillage Res. 179, 105-113. DOI: 10.1016/j.still.2018.01.013

Kojima, S., Bohner, A., Gassert, B., Yuan, L., and von Wiren, N. (2007). AtDUR3 represents the major transporter for high-affinity urea transport across the plasma membrane of nitrogen deficient *Arabidopsis* roots. Plant J. 52, 30-40. DOI: 10.1111/j.1365-313X.2007.03223.x

Kong, Y., Ling, N., Xue, C., Chen, H., Ruan, Y., Guo, J., Zhu, C., Wang, M., Shen, Q., and Guo, S. (2019). Long−term fertilization regimes change soil nitrification potential by impacting active autotrophic ammonia oxidizers and nitrite oxidizers as assessed by DNA stable isotope probing. Environ. Microbiol. 21, 1224-1240. DOI: 10.1111/1462-2920.14553

Koops, H.P., Purkhold, U., Pommerening-Roser, A., Timmermann, G., and Wagner, M. (2003). The Lithoautotrophic Ammonia Oxidizers. In The Prokaryotes: An Evolving Electronic Resource for the Microbiological Community, M. Dworkin et al., eds. (New York, NY, USA: Springer Verlag), pp. 2302-2309.

Krouk, G., Crawford, N.M., Coruzzi, G.M., and Tsay, Y.F. (2010a). Nitrate signaling: adaptation to fluctuating environments. Curr. Opin. Plant Biol. 13, 266-273. DOI: 10.1016/j.pbi.2009.12.003

Kulasooriya, S.A., and Magana-Arachchi, D.N. (2016). Nitrogen fixing cyanobacteria: their diversity, ecology and utilisation with special reference to rice cultivation. J. Natn. Sci. Foundation Sri Lanka 44,111-128. DOI: 10.4038/jnsfsr.v44i2.7992

Kuznetsova, S., Zauner, G., Aartsma, T., Engelkamp, H., Hatzakis, N., Rowan, A.E., Nolte, R.J.M., Christianen, P.C.M., and Canters, G. (2008). The enzyme mechanism of nitrite reductase studied at single-molecule level. Proc. Natl. Acad. Sci. U.S.A. 105, 3250-3255. DOI: 10.1073/pnas.0707736105

Ladha, J.K., Dawe, D., Ventura, T.S., Singh, U., Ventura, W., and Watanabe, I. (2000). Long-Term effects of urea and green manure on rice yields and nitrogen balance. Soil Sci. Soc. Am. J. 64, 1993-2001. DOI: 10.2136/ sssaj2000.6461993x

Lam, P., Lavika, G., Jensen, M.M., van de Vossenberg, J., Schmid, M., Woebkena, D., Gutierrez, D., Amann, R., Jetten, M.M., and Kuypers, M.M.M. (2009). Revising the nitrogen cycle in the Peruvian oxygen minimum zone. Proc. Natl. Acad. Sci. U.S.A. 106, 4752-4757. DOI: 10.1073/pnas.0812444106

Lanquar, V., Loqué, D., Hörmann, F., Yuan, L., Bohner, A., Engelsberger, W.R., Lalonde, S., Schulze, W.X., von Wirén, N., and Frommer, W.B. (2009). Feedback inhibition of ammonium uptake by a phospho-dependent allosteric mechanism in *Arabidopsis*. Plant Cell 21, 3610-3622. DOI: 10.1105/tpc.109.068593

Lawton, T.J., Bowen, K.E., Sayavedra-Soto, L.A., Arp, D.J., and Rosenzweig, A.C. (2013). Characterization of a nitrite reductase Involved in nitrifier denitrification. J. Biol. Chem. 288, 25575-25583. DOI: 10.1074/jbc.M113.484543

Lehmann, J,, Kuzyakov, Y., Pan, G.-X., and Ok, Y.S. (2015). Biochars and the plant-soil interface. Plant Soil 395, 1-5. DOI: 10.1007/s11104-015-2658-3

Leifeld, J., Angers, D.A., Chenu, C., Fuhrer, J., Kätterer, T., and Powlson, D.S. (2013). Organic farming gives no climate change benefit through soil carbon sequestration. Proc. Natl. Acad. Sci. U.S.A. 110, E984. DOI: 10.1073/pnas.1220724110

Leigh, J.A. (2000). Nitrogen fixation in methanogens: The archaeal perspective. Curr. Iss. Mol. Biol. 2, 125-131.

Leininger, S., Urich, T., Schloter, M., Schwark, L., Qi, J., Nicol, G.W., and Prosser, J.I, (2006). Archaea predominant among ammonia-oxidizing prokaryotes in soils. Nat. Lett. 442, 806-809.

Lenhart, K., Behrendt, T., Greiner, S., Steinkamp, J., Well, R., Giesemann, A., and Keppler, F. (2019). Nitrous oxide effluxes from plants as a potentially important source to the atmosphere. New Phythol. 221, 1398-1408. DOI: 10.1111/nph.15455

Lepère, C., Domaizon, I., Humbert, J.-F., Jardillier, L., Hugoni, M., and Debroas, D. (2019). Diversity, spatial distribution and activity of fungi in freshwater ecosystems. Peer J. 7, e6247. DOI: 10.7717/peerj.6247

Lesuffleur, F., Paynel, F., Bataillé, M.-P., Le Deunff, E., and Cliquet, J.-B. (2007). Root amino acid exudation: measurement of high efflux rates of glycine and serine from six different plant species. Plant Soil 294, 235-246. DOI: 10.1007/s11104-007-9249-x

Li, H., Liang, X., Chen, Y., Lian, Y., Tian, G., and Ni, W. (2008). Effect of nitrification inhibitor DMPP on nitrogen leaching, nitrifying organisms, and enzyme activities in a rice-oilseed rape cropping system. J. Environ. Sci. 20, 149-155. DOI: 10.1016/S1001-0742(08)60023-6

Li, H.-B., Zhang, L.-P., and Chen, S.-F. (2008). *Halomonas korlensis* sp. nov., a moderately halophilic, denitrifying bacterium isolated from saline and alkaline soil. Int. J. System. Evolution. Microbiol. 58, 2582-2588. DOI: 10.1099/ijs.0.65711-0

Li, T., Zhang, W., Yin, J., Chadwick, D.R., Norse, D., Lu, Y., Liu, X., Chen, X., Zhang, F., Powlson, D.S., and Dou, Z. (2018). Enhanced-efficiency fertilizers are not a panacea for resolving the nitrogen problem. Glob. Chang. Biol. 24, e511-e521. DOI: 10.1111/gcb.13918

Li, Y., Katzmann, E., Borg, S., and Schüler, D. (2012). The periplasmic nitrate reductase Nap is required for anaerobic growth and involved in redox control of magnetite biomineralization in *Magnetospirillum gryphiswaldense*. J. Bacteriol. 194, 4847-4856. DOI: 10.1128/JB.00903-12

Lifeng, Z., Boeckx, P., Guanxiong, C., and van Cleemput, O. (2000). Nitrous oxide emission from herbicide-treated soybean. Biol. Fertil. Soils 32, 173-176. DOI: 10.1007/s003740000

Lima, J.E., Kojima, S., Takahashi, H., and von Wirén, N. (2010). Ammonium triggers lateral root branching in *Arabidopsis* in an ammonium transporter 1;3-dependent manner. Plant Cell 22, 3621-3633. DOI: 10.1105/tpc.110.076216

Lindau, C., De Laune, R.D., and Jones, G.L. (1988). Fate of added nitrate and ammonium-nitrogen entering a Louisiana Gulf Coast Swamp Forest. J. Water Pollut. Control Fed. 60, 386-390.

Lindau, C.W., DeLaune, R.D., Patrick Jr., W.H., and Bollich, P.K. (1990). Fertilizer effects on dinitrogen, nitrous oxide, and methane emissions from lowland rice. Soil Sci. Soc. Am. 54, 1789-1794.

Lindblom, J., Lundström, C., Ljung, M., and Jonsson, A. (2017). A review of combine sensors for precision farming. Precis. Agric 18: 309-331. DOI: 10.1023/A: 1013823603735

Lloyd, D., Boddy, L., and Davies, K.J.P. (1987). Persistence of bacterial denitrification capacity under aerobic conditions: The rule rather than the exception. FEMS Microbiol. Ecol. 3, 185-190. DOI: 10.1111/j.1574-6968.1987.tb02354.x

Lu, X., Nicol, G.W., and Neufeld, J.D. (2018). Differential responses of soil ammonia-oxidizing archaea and bacteria to temperature and depth under two different land uses. Soil Biol. Biochem. 120, 272-282. DOI: 10.1016/j.soilbio.2018.02.017

Lubbers, I.M., López González, E., Hummelink, E.W.J., and Van Groenigen, J.W. (2013). Earthworms can increase nitrous oxide emissions from managed grassland: A field study. Agric. Ecosyst. Environ. 174, 40-48. DOI: 10.1016/j.agee.2013.05.001

Lynch, J.P., and Brown, K.M. (2012). New roots for agriculture: exploiting the root phenome. Philos. Trans. R. Soc. Lond. B Biol. Sci. 367(1595), 1598-1604. DOI: 10.1098/rstb.2011.0243

Ma, W.K., Farrell, R.E., and Siciliano, S.D. (2008). Soil formate regulates the fungal nitrous oxide emission pathway. Appl. Environ. Microbiol. 74, 6690-6696. DOI: 10.1128/AEM.00797-08

Machacova, K., Bäck, J.,Vanhatalo, A., Halmeenmäki, E., Kolari, P., Mammarella, I., Pumpanen, J., Acosta, M., Urban, O., and Pihlatie, M. (2016). *Pinus sylvestris* as a missing source of nitrous oxide and methane in boreal forest Sci. Rep. 6, 23410. DOI: 10.1038/srep23410

Mahal, N.K., Osterholz, W.R., Miguez, F.E., Poffenbarger, H.J., Sawyer, J.E., Olk, D.C., Archontoulis, S.V., and Castellano, M.J. (2019). Nitrogen fertilizer suppresses mineralization of soil organic matter in maize agroecosystems. Front. Ecol. Evol. 13, 59. DOI: 10.3389/fevo.2019.00059

Mahmood, T., Ali, R., Latif, Z., and Ishaque, W. (2011). Dicyandiamide increases the fertilizer N loss from an alkaline calcareous soil treated with 15N-labelled urea under warm climate and under different crops. Biol. Fert. Soils 147, 619-631. DOI: 10.1007/s00374-011-0559-z

Majumdar, D. (2008) Unexploited botanical nitrification inhibitors prepared from *Karanja* plant. Nat. Prod. Rad, 7, 58-67.

Maktabifard, M., Zaborowska, E., and Makinia, J. (2018). Achieving energy neutrality in wastewater treatment plants through energy savings and enhancing renewable energy production. Rev. Environ. Sci. Biotechnol. 17, 655-689. DOI: 10.1007/s11157-018-9478-x

Manabe, R., Nakatsubo, H., Gondo, A., Murakami, K., Ogo, S., Tsuneki, H., Ikeda, M., Ishikawa, A., Nakai, H., and Sekine, Y. (2017). Electrocatalytic synthesis of ammonia by surface proton hopping. Chem. Sci. 8, 5434-5439. DOI: 10.1039/C7SC00840F

Mander, Ü., Lõhmus, K., Mauring, T., Nurk, K., and Augustin, J. (2008). Gaseous fluxes in the nitrogen and carbon budgets of subsurface flow constructed wetlands. Sci. Total Environ. 404, 343-353. DOI: 10.1016/j,scitotenv.2008.03.014

Martiny, A.C. (2019). High proportions of bacteria are culturable across major biomes. ISME J. 13, 2125-2128.

Marusenko, Y., Huber, D.P., and Hall, S.J. (2013). Fungi mediate nitrous oxide production but not ammonia oxidation in arid land soils of the southwestern US. Soil Biol. Biochem. 63, 24-36. DOI: 10.1016/j.soilbio.2013.03.018

Meng, S., Torto Alalibo, T., Chibucos, M.C. et al. (2009). Common processes in pathogenesis by fungal and oomycete plant pathogens, described with gene ontology terms. BMC Microbiol. 9, S7. DOI: 10.1186/1471-2180-9S1-S7

Miller, A.J., Fan, X., Orsel, M., Smith, S.J., and Wells, D.M. (2007). Nitrate transport and signalling. J. Exp. Bot. 58, 2297-2306. DOI: 10.1093/jxb/erm066

Miyahara, M., Kim, S.-W., Fushinobu, S., Takaki, K., Yamada, T., Watanabe, A., Miyauchi, K., Endo, G., Wakagi, T., and Shoun, H. (2010). Potential of aerobic denitrification by *Pseudomonas stutzeri* TR2 to reduce nitrous oxide emissions from wastewater treatment plants. Appl. Environ. Microbiol. 76, 4619-4625. DOI: 10.1128/AEM.01983-09

Monteny, G.-J., Bannink, A., and Chadwick, D. (2006). Greenhouse gas abatement strategies for animal husbandry. Agric. Ecosyst. Environ. 112, 163-170.

Montzka, S.A., Dlugokencky, E.J., and Butler, J.H. (2011). Non-CO_2 greenhouse gases and climate change. Nature 476, 43-50.

Moreira-Coello, V., Mouriño-Carballido, B., Marañón, E., Fernández-Carrera, A., Bode, Sintes, E., Zehr, J.P., Turk-Kubo, K., and Varela, M.M. (2019). Temporal variability of diazotroph community composition in the upwelling region off NW Iberia. Sci. Rep. 9, 3737. DOI: 10.1038/s41598-019-39586-4

Morikawa, K., Matsumoto, M., Shinano, T., Iyoda, M., and Tadano, T. (1993). Productivity of high-yielding- crops III. Accumulation of ribulose-1,5-bisphosphate carboxylase/oxygenase and chlorophyll in relation to productivity of high-yielding crops Soil Sci. Plant Nutr. 39, 399-408. DOI: 10.1080/00380768.1993.10419780

Mosier, A., Kroeze, C., Nevison, C., Nevison, C., Oenema, O., Seitzinger, S., and van Cleemput, O. (1998). Closing the global N_2O budget: nitrous oxide emissions through the agricultural nitrogen cycle. Nutr. Cycl. Agroecosyst. 52, 225-248. DOI: 10.1023/A:1009740530221

Muhammed, S.E., Coleman, K., Wu, L., Bell, V.A., Davies, J.A.C., Quinton, J.N., Carnell, E.J., Tomlinson, S.J., Dore, A.J., Dragosits, U., Naden, P.S., Glendining, M.J., Tipping, E., and Whitmore, A.P. (2018) Impact of two centuries of intensive agriculture on soil carbon, nitrogen and phosphorus cycling in the UK. Sci. Total Environ. 634, 1486-1504. DOI: 10.1016/j.scitotenv.2018.03.378

Nacry, P., Bouguyon, E., and Gojon, A. (2013). Nitrogen acquisition by roots: physiological and developmental mechanisms ensuring plant adaptation to a fluctuating resource. Plant Soil 370, 1-29. DOI: 10.1007/s11104-013-1645-9

Nadeau, S.A., Roco, C.A., Debenport, S.J., Anderson, T.R., Hofmeister, K.L., Walter, M.T., and Shapleigh, J.P. (2019). Metagenomic analysis reveals distinct patterns of denitrification gene abundance across soil moisture, nitrate gradients. Environ. Microbiol. 21, 1255-1266. DOI: 10.1111/1462-2920.14587

Näsholm, T., Kielland, K., and Ganeteg, U. (2009). Uptake of organic nitrogen by plants New Phytol. 182, 31-48. DOI: 10.1111/j.1469-8137.2008.02751.x

Neuhäuser, B., Dynowski, M., Mayer, M., and Ludewig, U. (2007) Regulation of NH_4^+ transport by essential cross talk between AMT monomers through the carboxyl tails. Plant Physiol. 143, 1651-1659. DOI: 10.1104/pp.106.094243

Nie, Y., Li, L., Wang, M., Tahvanainen, T., and Hashidoko, Y. (2015). Nitrous oxide emission potentials of *Burkholderia* species isolated from the leaves of a boreal peat moss *Sphagnum fuscum*. Biosci. Biotechnol. Biochem. 79, 2086-2095. DOI: 10.1080/09168451.2015.1061420

Nishimura, T., Teramoto, H., Inui, M., and Yukawa, H. (2011). Gene expression profiling of *Corynebacterium glutamicum* during anaerobic nitrate respiration: induction of the SOS response for cell survival. J. Bacteriol. 193, 1327-1333.

Niu, J., Kasuga, I., Kurisu, F., and Furumai, H. (2019). Growth competition between ammonia-oxidizing archaea and bacteria for ammonium and urea in a biological activated carbon filter used for drinking water treatment. Environ. Sci. Water Res. Technol. 5, 231-238. DOI: 10.1039/C8EW00541A

Nugroho, R.A., Roling, W.F.M., Laverman, A.M., and Verhoef, H.A. (2007). Low nitrification rates in acid Scots pine forest soils are due to pH-related factors. Microb. Ecol. 53, 89-97. DOI: 10.1007/s00248-006-9142-9

Ohkubo, Y., Tanaka, M., Tabata, R., Ogawa-Ohnishi, M., and Matsubayashi, Y. (2017). Shoot-to-root mobile polypeptides involved in systemic regulation of nitrogen acquisition. Nature Plants 3, 17029. DOI: 10.1038/nplants.2017.29

Oldfield, E.E., Bradford, M.A., and Wood, S.A. (2019). Global meta-analysis of the relationship between soil organic matter and crop yields. Soil 5, 15-32. DOI: 10.5194/soil-5-15-2019

Ollivier, J., Töwe, S., Bannert, A., Hai, B., Kastl, E.-M., Meyer, A., Su, M.X., Kleineidam, K., and Schloter, M. (2011). Nitrogen turnover in soil and global change FEMS Microbiol. Ecol. 78, 3-16. DOI: 10.1111/j.1574-6941.2011.01165.x

Ottow, J.C.G. (2011). Mikrobiologie von Böden: Biodiversität, Ökophysiologie und Metagenomik (Heidelberg: Springer-Lehrbuch). DOI: 10.1007/978/3-642-0024-5

Pajares, S., and Bohannan, B.J.M. (2015). Ecology of nitrogen fixing, nitrifying, and denitrifying microorganisms in tropical forest soils. Front. Microbiol. 7, 1045. DOI: 10.3389/fmicb.2016.01045

Palmer, K., Drake, H.L., and Horn, M.A. (2010). Association of novel and highly diverse acid- tolerant denitrifiers with N_2O fluxes of an acidic fen. Appl. Environ. Microbiol. 76, 1125-1134.

Pastorelli, R., Landi, S., Trabelsi, D., Piccolo, R., Mengoni, A., Bazzicalupo, M., and Pagliai, M. (2011). Effects of soil management on structure and activity of denitrifying bacterial communities. Appl. Soil Ecol. 49, 46-58. DOI: 10.1016/j.apsoil. 2011.07.002

Pedersen, J.N., Bombar, D., Paerl, R.W., and Riemann, L. (2018). Diazotrophs and N_2-fixation associated with particles in coastal estuarine waters. Front. Microbiol. 9, 2759. DOI: 10.3389/fmicb.2018.02759

Peña, A., Busquets, A., Gomila, M., Bosch, R., Nogales, B., García-Valdés, E., Lalucat, J., and Bennasar, A. (2012). Draft genome of *Pseudomonas stutzeri s*train ZoBell (CCUG 16156), a marine isolate and model organism for denitrification studies. J. Bacteriol. 194, 1277-1278.

Peoples, M.B., Freney, J.R., and Mosier, A.R. (1995). Minimizing Gaseous Losses of Nitrogen. In Nitrogen Fertilization in the Environment, P.E. Bacon, ed. (New York, NY: Marcel Dekker) pp. 565-602.

Pfab, H., Palmer, I., Buegger, F., Fiedler, S., Müller, T., and Ruser, R. (2012). Influence of a nitrification inhibitor and of placed N-fertilization on N_2O fluxes from a vegetable cropped loamy soil. Agric. Ecosyst. Environ. 150, 91-101. DOI: 10.1016/ j.agee.2012.01.001

PICCMAT - Policy Incentives for Climate Change Mitigation Techniques (2011). Impact assessment. Final Report Summary - ec.europa.eu/smart-regulation/impact/ia_ carried... 2011/ sec_2011_1153_en.pdf

Pihlatie, M., Ambus, P., Rinne, J., Pilegaard, K., and Vesala, T. (2005). Plant-mediated nitrous oxide emissions from beech (*Fagus sylvatica*) leaves. New Phytol. 168, 93-98. DOI: 10.1111/j.1469-8137.2005.01542.x

Pretty, J., Toulmin, C., and Williams, S. (2011). Sustainable intensification in African agriculture. Int. J. Agr. Sus. 9, 5-24. DOI: 10.3763/ijas.2010.0583

Qian, W., Ma, B., Li, X., Zhang, Q., and Peng, Y. (2019). Long-term effect of pH on denitrification: High pH benefits achieving partial-denitrification. Bioresour. Technol. 278, 444-449. DOI: 10.1016/j.biortech.2019.01.105

Raun, W.R., and Johnson, G.V. (1999). Improving nitrogen use efficiency for cereal production. Agronomy J. 91, 357-363. DOI: 10.2134/ agronj1999.00021962009100030001x

Ravishankara, A.R., Daniel, J.S., and Portmann, R.W, (2009). Nitrous oxide (N_2O): the dominant ozone-depleting substance emitted in the 21st century. Science 326, 123-125. DOI: 10.1126/science.1176985

Reay, D.S., Davidson, E.A., Smith, K.A., Smith, P., Melillo, J.M., Dentener, F., and Crutzen, P.J. (2012). Global agriculture and nitrous oxide emissions. Nat. Clim. Change 2, 410-416. DOI: 10.1038/NCLIMATE1458

Reid, A., and Greene, S.E. (2012). How microbes can help feed the world. A report from the American Academy of Microbiology. Available online: academy.asm.org

Rentsch, D., Schmidt, S., and Tegeder, M. (2007). Transporters for uptake and allocation of organic nitrogen compounds in plants. FEBS Lett. 581, 2281-2289. DOI: 10.1016/j.febslet.2007.04.013

Rivera-Monroy, V.H., Lenaker, P., Twilley, R.R., Delaune, R.D., Lindau, C.W., Nuttle, W., Robinson, E.H., Fulweiler, W., and Castañeda-Moya, E. (2010). Denitrification in coastal Louisiana: A spatial assessment and research needs. J. Sea Res. 63, 157-172. DOI: 10.1016/j.seares.2009.12.004

Rodriguez, M.F.H., Lopes, C., Zuberi, K., Montojo, J., Bader, G.D., and Morris, Q. (2018). GeneMANIA update 2018. Nucleic Acids Res. 46, W60-W64. https://doi.org/10.1093/nar/gky311

Roger, P.A., and Ladha, J.K. (1992). Biological N_2 fixation in wetland rice fields: Estimation and contribution to nitrogen balance. Plant Soil 141, 41-55. DOI: 10.1007/978-94-017-0910-1_3

Roth, C.R., Roberts, T.L., and Norman, R.J. (2009). Variable rates of Agrotain® on ammonia volatilization loss of urea applied to a DeWitt silt loam. B.R. Wells Rice Res. Studies Ark. Rice Res. Ser. 571, 240-245.

Rütting, T., Aronsson, H., and Delin, S. (2018). Efficient use of nitrogen in agriculture. Nutr. Cycl. Agroecosyst. 110, 1-5. DOI: 10.1007/s10705-017-9900-8

Ruser, R., Flessa, H., Schilling, R., Beese, F., and Munch, J.C. (2001). Effects of crop-specific field management and N fertilization on N_2O emissions from a fine-loamy soil. Nutr. Cycl. Agroecosys. 59, 177-191. DOI: 10.1023/A:101751220

Sahrawat, K.L., and Keeney, D.R. (1986). Nitrous oxide emissions from soils. Adv. Soil Sci. 4, 103-148.

Sahrawat, K.L. (2008). Factors affecting nitrification in soils. Commun. Soil. Sci. Plant Anal. 39, 1436-1446. DOI: 10.1080/00103620802004235

Saleh-Lakha, S., Shannon, K.E., Henderson Sherri, L., Goyer, C., Trevors, J.T., Zebarth, B.J., and Burton, D.L. (2009). Effect of pH and temperature on denitrification gene expression and activity in *Pseudomonas mandelii.* Appl. Environ. Microbiol. 75, 3903-3911. DOI: 10.1128/AEM.00080-09

Sanford, R.A., Wagner, D.D., Wu, Q., Chee-Sanford, J.C., Thomas, S.H., Cruz-García, C., Rodríguez, G., Massol-Deyá, A., Krishnani, K.K., Ritalahti, K.M., Nissen, S., Konstantinidis, K.T., and Löffler, F.E. (2012). Unexpected nondenitrifier nitrous oxide reductase gene diversity and abundance in soils. Proc. Natl. Acad. Sci. U.S.A. 109, 19709-19714. DOI: 10.1073/pnas.1211238109

Santín, I., Barbu, M., Pedret, C., and Vilanova, R. (2017). Control strategies for nitrous oxide emissions reduction on wastewater treatment plants operation. Water Res. 125, 466-477. DOI: 10.1016/j.watres.2017.08.056

Santos, I.R., Eyre, B., and Glud, R.N. (2012). Influence of porewater advection on denitrification in carbonate sands: Evidence from repacked sediment column experiments. Geochim. Cosmochim. Acta 96, 247-258. DOI: 10.1016/j.gca.2012.08.018

Scheer, C., Rowlings, D., Firrell, M., Deuter, P., Morris, S., Riches, D., Porter, I., and Grace, P. (2017). Nitrification inhibitors can increase post-harvest nitrous oxide emissions in an intensive vegetable production system. Sci. Rep. 7, 43677. DOI: 10.1038/srep43677

Schleper, C. (2010). Ammonia oxidation: different niches for bacteria and archaea? ISME J. 4, 1092-1094. DOI: 10.1038/ismej.2010.111

Seiz, P., Schulz, R., Heger, A., Armbruster, M., Müller, T., Wiesler, F., and Ruser, R. (2014). Einfluss von N-Düngung, Nitrifikationshemmstoff und Abfuhr der Ernterückstände auf die N_2O-Freisetzung zweier gemüsebaulich genutzter Böden. VDLUFA Schriftenreihe, Kongressband 2013 Berlin (Darmstadt: VDLUFA-Verlag).

Sedláček, V., and Kučera, I. (2019). Arginine-95 is important for recruiting superoxide to the active site of the FerB flavoenzyme of *Paracoccus denitrificans*. FEBS Lett. 593, 697-702. DOI: 10.1002/1873-3468.13359

Selvaraju, R. (2013). Implications of Climate Change for Agriculture and Food Security in the Western Asia and Northern Africa Region. In Climate Change and Food Security in West Asia and North Africa, pp. 27-51. DOI: 10.1007/978-94-007-6751-5_2

Shaw, E.A., Boot, C.M., Moore, J.C., Wall, D.H., and Barone, J.S. (2019). Long-term nitrogen addition shifts the soil nematode community to bacterivore-dominated and reduces its ecological maturity in a subalpine forest. Soil Biol. Biochem. 130, 177-184. DOI: 10.1016/j.soilbio.2018.12.007

Sherr, B.F., Sherr, E.B., and Berman, T. (1983). Grazing, growth and ammonium excretion rates of a heterotrophic microflagellate fed with four species of bacteria. Appl. Environ. Microbiol. 45, 1196-1201.

Shoun, H., Fushinobu, S., Jiang, L., Kim, S.W., and Wakagi, T. (2012). Fungal denitrification and nitric oxide reductase cytochrome P450nor. Phil. Trans. R. Soc. B 367, 1186-1194.

Sime-Ngando, T., and Niquil, N. (2011). Editorial: 'Disregarded' microbial diversity and ecological potentials in aquatic systems: A new paradigm shift ahead. Hydrobiologia 659, 1-4. DOI: 10.1007/s10750-010-0511-5

Šima, T., Krupička, J. and Nozdrovický, L. (2013). Effect of nitrification inhibitors on fertiliser particle size distribution of the DASA® 26/13 and Ensin® fertilisers. Agronomy Res. 11, 111-116.

Sime-Ngando, T., Lefevre, E., and Gleason, F.H. (2011). Hidden diversity among aquatic heterotrophic flagellates: ecological potentials of zoosporic fungi. Hydrobiologia 659, 5-22. DOI: 10.1007/s10750-010-0230-y

Simpson, P.J.L., Richardson, D.J., and Codd, R. (2010). The periplasmic nitrate reductase in *Shewanella*: the resolution, distribution and functional implications of two NAP isoforms, NapEDABC and NapDAGHB. Microbiol. 156, 302-312.

Skinner, C., Gattinger, A., Muller, A., Mäder, P., Flieβbach, A., Stolze, M., Ruser, R., and Niggli, U. (2014). Greenhouse gas fluxes from agricultural soils under organic and non-organic management — A global meta-analysis. Sci. Total Environ. 468-469, 553-563. DOI: 10.1016/j.scitotenv.2013.08.098

Smart, D.R., and Bloom, A.J. (2001). Wheat leaves emit nitrous oxide during nitrate assimilation. Proc. Natl. Acad. Sci. U.S.A. 98, 7875-7878. DOI: 10.1073/pnas.131572798

Smil, V. (2001). Enriching the Earth: Fritz Haber, Carl Bosch, and the Transformation of World Food Production (Cambridge, MA and London: MIT Press).

Smith, C.J., Nedwell, D.B., Dong, L.F., and Osborn, A.M. (2007). Diversity and abundance of nitrate reductase genes (*nar*G and *nap*A), nitrite reductase genes (*nir*S and *nrf*A), and their transcripts in estuarine sediments. Appl. Environ. Microbiol. 73, 3612-3622. DOI: 10.1128/AEM.02894-06

Smith, P., Martino, D., Cai, Z., Gwary, D., Janzen, H., Kumar, P., McCarl, B., Ogle, S., O'Mara, F., Rice, C., Scholes, B., Sirotenko, O., Howden, M., McAllister, T., Pan, G., Romanenkov, V., Schneider, U., Towprayoon, S., Wattenbach, M., and Smith, J.

(2008). Greenhouse gas mitigation in agriculture. Philos. Trans. R. Soc. Lond. B. Biol. Sci. 363, 789-813. DOI: 10.1098/rstb.2007.2184

Smith, P., Adams, J., Beerling, D.J., Beringer, T., Calvin, K.V., Fuss, S., Griscom, B., Hagemann, N., Kammann, C., Kraxner, F., Minx, J.C., Popp, A., Renforth, P., Vicente, J.L., and Keesstra, S. (2019). Impacts of land-based greenhouse gas removal options on ecosystem services and the United Nations sustainable development goals. Ann. Rev. Environ. Resources 44, 255-286. DOI: 10.1146/annurev-environ-101718-033129

Sreenivas, B., and Sharma, S.D. (2005). An update on stable isotope research in understanding the Precambrian atmospheric oxygenation. Indian J. Geochem. 20, 103-120.

Stahl, D.A., and de la Torre, J.R. (2012). Physiology and diversity of ammonia-oxidizing archaea. Ann. Rev. Microbiol. 66, 83-101. DOI: 10.1146/annurev-micro-092611-150128

Steffen, W., Persson, A., Deutsch, L., Zalasiewicz, J., Williams, M., Richardson, K., Crumley, C., Crutzen, P., Folke, C., Gordon, L., Molina, M., Veerabhadran, R., Rockstrom, J., Scheffer, M., Schellnhuber, H.J., and Svedin, U. (2011). The anthropocene: from global change to planetary stewardship. Ambio 40, 739-761. DOI: 10.1007/s13280-011-0185-x

Strous, M., van Gerven, E., Zheng, P., Kuenen, J.G., and Jetten, M.S.M. (1997). Ammonium removal from concentrated waste streams with the anaerobic ammonium oxidation (Anammox) process in different reactor configurations. Water Res. 131, 1955-1962. DOI: 10.1016/S0043-1354(97)00055-9

Sturm, H., Buchner, A., and Zerulla, W. (1994). Gezielter Düngen (Frankfurt: Verlags-Union Agrar,).

Su, X., Chen, Y., Wang, Y., Yang, X., and He, Q. (2019). Impacts of chlorothalonil on denitrification and N_2O emission in riparian sediments: Microbial metabolism mechanism. Water Res. 148, 188-197.

Subbarao, G.V., Arango, J., Masahiro, K., Hooper, A.M., Yoshihashi, T., Ando, Y., Nakahara, K., Deshpande, S., Ortiz-Monasterio, I., Ishitani, M., Peters, M., Chirinda, N., Wollenberg, L., Lata, J.C., Gerard, B., Tobita, S., Rao, I.M., Braun, H.J., Kommerell, V., Tohme, J., and Iwanaga, M. (2017). Genetic mitigation strategies to tackle agricultural GHG emissions: The case for biological nitrification inhibition technology. Plant Sci. 262, 165-168. DOI: 10.1016/j.plantsci.2017.05.004

Sullivan, M., Gates, A., Appia-Ayme, C., Rowley, G., and Richardson, D.J. (2013). Copper control of bacterial nitrous oxide emission and its impact on vitamin B12-dependent metabolism. Proc. Natl. Acad. Sci. U.S.A. 110, 19926-19931. DOI: 10.1073/pnas.1314529110

Sutton, M.A., Howard, C.M., Erisman, J.W., Billen, G., Bleeker, A., Grennfelt, P., van Grinsven, H., and Grizzetti, B., eds. (2011). The European Nitrogen Assessment: Sources, Effects and Policy Perspectives (Cambridge, UK, New York: Cambridge University Press), pp. 1-6.

Svenningsen, N.B., Heisterkamp, I.M., Sigby-Clausen, M., Larsen, L.H., Nielsen, L.P., Stief, P., and Schramm, A. (2012). Shell biofilm nitrification and gut denitrification contribute to emission of nitrous oxide by the invasive freshwater mussel *Dreissena polymorpha* (Zebra mussel). Appl. Environ. Microbiol. 78, 4505-4509.

Taroncher-Oldenburg, G., Griner, E.M., Francis, C.A., and Ward, B.B. (2003). Oligonucleotide microarray for the study of functional gene diversity in the nitrogen cycle in the environment. Appl. Environ. Microbiol. 69, 1159-1171 DOI: 10.1128/AEM.69.2.1159-1171.2003

Taylor, A.E., Zeglin, L.H., Wanzek, T.A., Myrold, D.D., and Bottomley, P.J. (2012). Dynamics of ammonia-oxidizing archaea and bacteria populations and contributions to soil nitrification potentials. ISME J. 6, 2024-2032. DOI: 10.1038/ismej.2012.51

Thomson, A.J., Giannopoulos, G., Pretty, J., Baggs, E.M., and Richardson, D.J. (2012). Biological sources and sinks of nitrous oxide and strategies to mitigate emissions. Phil. Trans. R. Soc. B 367, 1157-1168. DOI: 10.1098/rstb.2011.0415

Tilman, D., Cassman, K.G., Matson, P.A., Naylor, R., and Polasky (2002). Agricultural sustainability and intensive production practices. Nature 418, 671-677.

Tindaon, F., and Benckiser, G. (2019). Evaluation of Side Effects of Nitrification-inhibiting Agrochemicals in Soils. In Plant Growth Promoting Rhizobacteria (PGPR): Prospects for Sustainable Agriculture, R.Z. Sayyed et al., eds. (Singapore: Springer Nature). DOI: 10.1007/978-981-13-6790-8_6

Toyofuku, M., Nomura, N., Kuno, E., Tashiro, Y., Nakajima, T., and Uchiyama, H. (2008). Influence of the *Pseudomonas* quinolone signal on denitrification in *Pseudomonas aeruginosa.* J. Bacteriol. 190, 7947-7956 DOI: 10.1128/JB.00968-08

Turk-Kubo, K.A., Connell, P., Caron, D., Hogan, M.E., Farnelid, H.M., and Zehr, J. (2018). *In Situ* diazotroph population dynamics under different resource ratios in the

North Pacific subtropical gyre. Front. Microbiol. 9, 1616. DOI: 10.3389/fmicb.2018.01616

Twining, B.S., Mylon, S.E., and Benoit, G. (2007). Potential role of copper availability in nitrous oxide accumulation in a temperate lake. Limnol. Oceanogr. 52, 1354-1366.

Uma, B., and Sandhya, S. (1997). Pyridine degradation and heterocyclic nitrification by *Bacillus coagulans*. Can. J. Microbiol. 43, 595-598. DOI: 10.1139/m97-085

Van Alst, N.E., Wellington, M., Clark, V.L., Haidaris, C.G., and Iglewski. B.H. (2009). Nitrite reductase NirS is required for type III secretion system expression and virulence in the human monocyte cell line THP-1 by *Pseudomonas aeruginosa.* Infec. Immun. 77, 4446-4454.

van den Berg, E.M., Rombouts, J.L., Kuenen, J.G., Kleerebezem, R., and van Loosdrecht, M.C.M. (2017). Role of nitrite in the competition between denitrification and DNRA in a chemostat enrichment culture. AMB Express. 7, 91. DOI: 10.1186/s13568-017-0398-x

Van den Heuvel, R.N., Bakker, S.E., Jetten, M.S.M., and Hefting, M.M. (2011). Decreased N_2O reduction by low soil pH causes high N_2O emissions in a riparian ecosystem Geobiology 9, 294-300. DOI: 10.1111/j.1472-4669.2011.00276.x

Van Groenigen, J.W., Velthof, G.L., Oenema, O., and van Groenigen, K.J. (2010). Towards an agronomic assessment of N_2O emissions: a case study for arable crops. Eur. J. Soil Sci. 61, 903-913. DOI: 10.1111/j.1365-2389.2009.01217.x

Vance, C.P. (2001). Symbiotic nitrogen fixation and phosphorus acquisition, plant nutrition in a world of declining renewable resources. Plant Physiol. 127, 390-397. DOI: 10.1104/pp.010331

Vandekerckhove, T.G.L., Kerckhof, F.-M., De Mulder, C., Vlaeminck, S.E., and Boon, N. (2019). Determining stoichiometry and kinetics of two thermophilic nitrifying communities as a crucial step in the development of thermophilic nitrogen removal. Water Res. 156, 34-45. DOI: 10.1016/j.watres.2019.03.008

Vereecken, H., Pachepsky, Y., Bogena, H., and Montzka, C. (2019). Upscaling Issues in Ecohydrological Observations. In Ecohydrology Book Series (ECOH) Vol. 2, X. Li and H. Vereecken, eds. (Springer Nature), pp. 435-454. DOI: 10.1007/978-3-662-48297-1_14

Verhoeven, E., Barthel, M., Yu, L., Celi, L., Said-Pullicino, D., Sleutel, S., Lewicka-Szczebak, D., Six, J., and Decock, C. (2019). Early season N_2O emissions under variable water management in rice systems: source-partitioning emissions using isotope ratios along a depth profile. Biogeosciences 16, 383-408. DOI: 10.5194/bg-16-383-2019

Vincent, C., Rowland, D., Na, C.-I., and Schaffer, B. (2017). A high-throughput method to quantify root hair area in digital images taken in situ. Plant Soil 412, 61-80. DOI: 10.1007/s11104-016-3016-9

Vitousek, P.M., and Horwarth, R.W. (1991). Nitrogen limitation on land and in the sea: how can it occur? Biochemistry 13, 87-115.

Vitousek, P.M., Menge, D.N.L., Reed, S., and Cleveland, C. (2013). Biological nitrogen fixation: Rates, patterns and ecological controls in terrestrial ecosystems. Phil. Trans. R. Soc. B 368, 20130119. DOI: 10.1098/rstb.2013.0119

Wang, C., Zheng, M.M., Hu, A.Y., Zhu, C.Q., and Shen, R.F. (2018). Diazotroph abundance and community composition in an acidic soil in response to aluminium-tolerant and aluminium-sensitive maize (*Zea mays* L.) cultivars under two nitrogen fertilizer forms. Plant Soil 424, 463-478. DOI: 10.1007/s11104-017-3550-0

Wang, W.-H., Köhler, B., Cao, F.-Q., Liu, G.-W., Gong, Y.-Y., Sheng, S., Song, Q.-C., Cheng, X.-Y., Garnett, T., Okamoto, M., Qin, R., Mueller-Roeber, B., Tester, M., and Liu, L.-H. (2012). Rice DUR3 mediates high-affinity urea transport and plays an effective role in improvement of urea acquisition and utilization when expressed in *Arabidopsis*. 193, 432-444. DOI: 10.1111/j.1469-8137.2011.03929.x

Ward, B.B., Devol, A.H., Rich, J.J., Chang, B.X., Bulow, S.E., Naik, H., Pratihary, A., and Jayakumar, A. (2009). Denitrification as the dominant nitrogen loss process in the Arabian Sea. Nat. Lett. 461, 78-81. DOI: 10.1038/nature08276

Wei, X., Lyu, S., Yu, Y., Wang, Z., Liu, H., Pan, D., and Chen, J. (2017). Phylloremediation of air pollutants: Exploiting the potential of plant leaves and leaf-associated microbes. Front. Plant Sci. 8, 1318. DOI: 10.3389/fpls.2017.01318

Weger, H.G., and Turpin, D.H. (1989). Mitochondrial respiration can support NO_3^- and NO_2^- reduction during photosynthesis interactions between photosynthesis, respiration, and N-assimilation in the N-limited green alga *Selenastrum minutum*. Plant Physiol. 89, 409-415. DOI: 10.1104/pp.89.2.409

Weinstein, J.A. (2019). DNA Microscopy: Optics-free spatio-genetic imaging by a stand-alone chemical reaction. Cell 178, 229-241. DOI: 10.1016/j.cell.2019.05.019

Weiske, A., Benckiser, G., Herbert, T., and Ottow, J.C.G. (2001). Influence of nitrification inhibitor 3,4-dimethylpyrazole phosphate (DMPP) in comparison to dicyandiamide (DCD) on nitrous oxide emission and methane oxidation during 3 years repeated application in field experiments. Biol. Fertil. Soils 34, 109-117. DOI: 10.1007/s003740100386

Wen, Z., and Kaiser, B. (2018). Unraveling the functional role of NPF6 transporters. Front. Plant Sci. 9, 973. DOI: 10.3389/fpls.2018.00973

Weng, J.-K. (2014). The evolutionary paths towards complexity: a metabolic perspective. New Phytol. 201, 1141-1149. DOI: 10.1111/nph.12416

Wertz, S., Dandie, C.E., Goyer, C., Trevors, J.T., and Patten, C.L. (2009). Diversity of *nirK* denitrifying genes and transcripts in an agricultural soil. Appl. Environ. Microbiol. 75, 7365-7377.

White, P.J., and Brown, P.H. (2010). Plant nutrition for sustainable development and global health. Ann. Bot. 105, 1073-1080. DOI: 10.1093/aob/mcq085

Wiesler, F. (1998). Comparative assessment of the efficacy of various nitrogen fertilizers. J. Crop Prod. 1, 81-114. DOI: 10.1300/J144v01n02_04

Wiesler, F., Behrens, T., and Horst, W.J. (2001). The role of nitrogen-efficient cultivars in sustainable agriculture. In Optimizing Nitrogen Management in Food and Energy Production and Environmental Protection. Scientific World 1, 61-69. DOI: 10.1100/tsw.2001.264

Wrage-Mönnig, N., Velthof, G.L., van Beusichem, M.L., and Oenema, O. (2001). Role of nitrifier denitrification in the production of nitrous oxide. Soil Biol. Biochem. 33, 1723-1732. DOI: 10.1016/S0038-0717(01)00096-7

Wright, R.C., and Nemhauser, J. (2019). Plant synthetic biology: quantifying the "known unknowns" and discovering the "unknown unknowns". Plant Phys. 179, 885-893. DOI: 10.1104/pp.18.01222

Wu, D., Senbayram, M., Well, R., Brüggemann, N., Pfeiffer, B., Loick, N., Stempfhuber, B., Dittert, K., and Bol, R. (2017). Nitrification inhibitors mitigate N_2O emissions more effectively under straw-induced conditions favoring denitrification. Soil Biol. Biochem. 104, 197-207. DOI: 10.1016/j.soilbio.2016.10.022

Wu, L., Ning, D., Zhang, B., Li, Y., Zhang, P., Shan, X., Zhang, Q., Brown, M., Li, Z., Van Nostrand, J.D., Ling, F., Xiao, N., Zhang, Y., Vierheilig, J., Wells, G.F., Yang, Y., Deng, Y., Tu, Q., Wang, A., Global Water Microbiome Consortium, Zhang, T., He, Z., Keller, J., Nielsen, P.H., Alvarez, P.J.J., Criddle, C.S., Wagner, M., Tiedje, J.M., He, Q., Curtis, T.P., Stahl, D.A., Alvarez-Cohen, L., Rittmann, B.E., Wen, X., and Zhou, J. (2019). Global diversity and biogeography of bacterial communities in wastewater treatment plants. Nat. Microbiol. 4, 1183-1195. DOI: 10.1038/s41564-019-0426-5

Wüst, P.K., Horn, M.A., and Drake, H.L. (2009). In situ hydrogen and nitrous oxide as indicators of concomitant fermentation and denitrification in the alimentary canal of the earthworm *Lumbricus terrestris*. Appl. Environ. Microbiol. 75, 1852-1859. DOI: 10.1128/AEM.02745-08

Wüst, P.K., Horn, M.A., and Drake, H.L. (2011). *Clostridiaceae* and *Enterobacteriaceae* as active fermenters in earthworm gut content. ISME J. 5, 92-106. DOI: 10.1038/ismej.2010.99

Yan, Y., Yang, J., Dou, Y., Chen, M., Ping, S., Peng, J., Lu, W., Zhang, W., Yao, Z., Li, H., Liu, W., He, S., Geng, L., Zhang, X., Yang, F., Yu, H., Zhan, Y., Li, D., Lin, Z., Wang, Y., Elmerich, C., Lin, M., and Jin, Q. (2008). Nitrogen fixation island and rhizosphere competence traits in the genome of root-associated *Pseudomonas stutzeri* A1501. Proc. Natl. Acad. Sci. U.S.A. 105, 7564-7569.

Yin, Z., Bi, X., and Xu, C. (2018). Ammonia-oxidizing archaea (AOA) play with ammonia-oxidizing bacteria (AOB) in nitrogen removal from wastewater. Archaea, 8429145. DOI: 10.1155/2018/8429145

Yu, R., Perez-Garcia, O., Lu, H., and Chandran, K. (2018). *Nitrosomonas europaea* adaptation to anoxic-oxic cycling: Insights from transcription analysis, proteomics and metabolic network modelling. Sci. Total Environ. 615, 1566-1573. DOI: 10.1016/j.scitotenv.2017.09.142

Zerulla, W., Barth, T., Dressel, J., Barth, T., Dressel, J., Erhardt, K., Horchler von Locquenghien, K., Pasda, G., Rädle, M.,and Wissemeier, A. (2001). DMPP - a new nitrification inhibitor for agriculture and horticulture: an introduction. Biol. Fertil. Soils 34, 79-84. DOI: 10.1007/s003740100380

Zhang, H., Rong, H., and Pilbeam, D. (2007). Signalling mechanisms underlying the morphological responses of the root system to nitrogen in *Arabidopsis thaliana*. J. Exp. Bot. 58, 2329-2338. DOI: 10.1093/jxb/erm114

Zhang, Y., Liu, S., Cheng, Y., Cai, Z., Müller, C., and Zhang, J. (2019). Composition of soil recalcitrant C regulates nitrification rates in acidic soils. Geoderma 337, 965-972. DOI: 10.1016/j.geoderma.2018.11.014

Zhang L, Wüst A, Prasser B, Müller C, Einsle O (2019)Functional assembly of nitrous oxide reductase provides insights into copper site maturation PNAS 116, 12822-12827 https://doi.org/10.1073/pnas.1903819116

Zhao, K., Tung, C.W., Eizenga, G.C., Wright, M.H., Ali, M.L., Price, A.H., Norton, G.J., Islam, M.R., Reynolds, A., Mezey, J., McClung, A.M., Bustamante, C.D., and

McCouch, S.R. (2011). Genome-wide association mapping reveals a rich genetic architecture of complex traits in *Oryza sativa*. Nat. Commun. 2, 467. DOI: 10.1038/ncomms1467

Zheng, M., He, D., Ma, T., Chen, Q., Liu, S., Ahmad, M., Gui, M., and Ni, J. (2014). Reducing NO and N_2O emission during aerobic denitrification by newly isolated *Pseudomonas stutzeri* PCN-1. Bioresour. Technol. 162, 80-88. DOI: 10.1016/j.biortech.2014.03.125

Zhu, T.-B., Meng, T., Zhang, J., Zhong, W., Müller, C., and Cai, Z. (2014). Fungi-dominant heterotrophic nitrification in a subtropical forest soil of China. J. Soils Sedim. 15, 705-709. DOI: 10.1007/s11368-014-1048-4

Zistl-Schlingmann, M., Feng, J., Kiese, R., Stephan, R., Zuazo, P., Willibald, G., Wang, C., Butterbach-Bahl, K., and Dannenmann, M. (2019). Dinitrogen emissions: an overlooked key component of the N balance of montane grasslands. Biogeochemistry 143, 15-30. DOI: 10.1007/s10533-019-00547-8

Zumft, W.G. (1997). Cell biology and molecular basis of denitrification. Microbiol. Rev. 61, 533-616.

Chapter 12

Changes in Precipitation Patterns: Responses and Strategies from Streambed Sediment and Soil Microbes

Giulia Gionchetta[1,*], Aline Frossard[2], Luis Bañeras[3] and Anna Maria Romaní[1]

[1] GRECO, Institute of Aquatic Ecology, University of Girona, 17003 Girona, Spain; [2] Swiss Federal Research Institute WSL, Department of Forest Soils and Biogeochemistry, Zürcherstrasse 111, 8903 Birmensdorf, Switzerland; [3] Group of Molecular Microbial Ecology, Institute of Aquatic Ecology, University of Girona, 17003 Girona, Spain; [*] corresponding author

Email: giulia.gionchetta@eawag.ch, aline.frossard@wsl.ch, lluis.banyeras@udg.edu, anna.romani@udg.edu

DOI: https://doi.org/10.21775/9781913652579.12

Abstract

Sediments in intermittent watersheds as well as soil systems, especially in arid and semi-arid regions, suffer an increasing pressure of drought events and water scarcity, affecting the ecosystems and the microorganisms inhabiting them. Such microbial communities contribute greatly to global biogeochemical cycles and thus it is crucial to understand their response mechanisms to increasing dryness. Microorganisms show responses to drought at different organizational levels, from the cell (e.g. spores formation, osmolytes production, cell wall thickening) to the whole community (e.g. production of extracellular polymeric substances, community composition changes). At the same time, dryness induces functional modifications such as changes in microbial respiration and organic matter degradation capabilities in both streambed sediments and soils. Water content is a major contributor to the observed responses and is highly associated to the soil/sediment water holding capacity as well as to the water solutes concentration. Knowledge from studies on microbes inhabiting extreme habitats, like deserts and hypersaline arid zones, further stress the importance of minimal water inputs (e.g. fog, dew or light rains) to support the microbial functions.

Conservation of habitat heterogeneity, with diverse water holding capacities, might support the microbial resistance and resilience against the intensification of dry-wet extreme episodes, as suggested by recent studies in the field. Moreover, space heterogeneity, including arid and semiarid systems and tight bacterial networks at highest level will significantly contribute to ecosystem fitness in about-to-come climate scenarios.

Changing patterns: droughts and rainstorms intensification on soils and sediments

Over the past thirty years, the frequency and intensity of droughts and rewetting episodes, such as short and intense rainfalls, has increased dramatically across Europe (Collins et al., 2009). Drought is one of the major if not the main climate change disturbance affecting most ecosystems (Dai, 2011; Sherwood and Fu, 2014). In most biomes, the increase of drought is associated with an increase of temperature. At the same time, across the majority of warm and humid climates, precipitation periods are expected to become less frequent but more intense (Trenberth et al., 2014; Prein et al., 2017). Drought occurrence is directly linked to the amount of precipitation, the water infiltration/evaporation into and from soil/sediment, and plant evaporation-transpiration rates. However, it depends on the amount of water available in soil, sediment or hydrological systems.

Freshwater ecosystems, already damaged by the overuse of water by growing human populations, stand to be further affected by widespread shifts in rainfall patterns and intensifying droughts in some regions. Temperate and Mediterranean intermittent rivers are among those systems suffering greatest natural and human-based effects of climate change and the reality of the water scarcity is endangering their status moving closer to a terrestrial system (Datry et al., 2017).

Unlike the sediment in flowing watercourses, intermittent dry streambeds are transitional habitats acquiring similar features to nearby soils moving from wet to extremely dry conditions (Thorp et al., 2006; Elosegi et al., 2010; Arce et al., 2019). The degree of similarity of intermittent streambeds to soils depends on the duration of the dry phase, the rewetting episodes frequency and a corollary of natural environmental factors (e.g. solar radiation, water sediment content, shadow cover) which may help on the preservation of the aquatic features (Harms and Grimm, 2012; Mori et al., 2017). Nowadays, the duration and the frequency of dry-wet alternation is sharpened and it could endanger the functioning of the entire ecosystem mostly mediated by the microbial communities associated to the sediment (Findlay, 2010).

In terrestrial environments, anthropogenic factors causing global warming seemed to be not (yet) important in determining the location and timing of drought events (Trenberth et al., 2014), and recorded past droughts were all considered of natural causes (Dai, 2011). However, the extra heat caused by global warming could increase the rate of drying which will generate more drought events of higher intensity (Trenberth et al., 2014). Increasing drought periods frequency and intensity would enhance the expansion of drylands or semi-arid and arid zones by the intensification of desertification (Huang et al., 2016). For instance, dry periods lasting from weeks to months are common in various regions, but the magnitude and extent of drought events vary from weeks to decades in the most arid areas (Schimel, 2018).

At the large scale, patterns of precipitation are usually driven by evaporation over land, and soil moisture variations do not always follow precipitation changes (Dai, 2011). The evaporation rates tend to be homogeneously distributed over large scales while precipitations are more heterogeneously happening across the landscape. At the regional scale, variation in precipitations between wet and dry regions, or between wet and dry seasons will probably increase (Trenberth et al., 2014). Intense and abrupt precipitations after prolonged dry periods will eventually cause dramatic changes in soil conditions that will invariably affect soil physiological properties.

As main consequence, the episodic rewetting events enhance soil carbon and nitrogen release, and produce a pulsed soil respiration that leads to a loss of carbon from the system (Navarro-Garcia et al., 2012). Indeed multiple and intensified dry-wet cycles can mobilize substantial amounts of carbon (Schimel et al., 2012). Different mechanisms could explain the enhanced respiration observed after rewetting events. Among them i) a metabolic process related to the intracellular osmolytes accumulated during drought (Halverson et al., 2000), or ii) the physical hypothesis explaining that drying/rewetting fluctuation promotes physical breaking of dry soil/sediment microstructures (e.g. aggregate slaking) which allow microorganisms to access previously inaccessible substrates and thus enhance respiration (Denef et al., 2001). Therefore, the intensification and the increased strength of rainstorm episodes occurring after long periods of dryness have effects for soil-sediment nutrients cycle reducing the nutrients storage. Among the most vulnerable ecosystems, semi-arid and Mediterranean regions will likely undergo substantial disturbance caused by such changes in the rainfall pattern.

The microbial compartment: streambed biofilms and soil microbial aggregates

Microbes colonizing streambed sediments are organized in consortia assembled in biofilms within the sediment particles (Pusch et al., 1998). Microbial biofilms contribute to ecosystem processes participating to different in-stream processes related to biogeochemical fluxes, uptake or retention of inorganic and organic nutrients, water quality, gases emission, and preservation of microbial genetic diversity supporting those functions (Romaní et al., 2004; Teissier et al., 2007; Febria et al., 2012; Zeglin et al., 2015; Battin et al., 2016). In streams, microorganisms colonizing various compartments (e.g. superficial benthic, hyporheic, aquifer sediments, leaf packs material) are suitable indicators of ecosystem dynamics under a specific or multiple pressures (Romaní et al., 2016), given their short generation time and fast functional and compositional responses (Sabater et al., 2007).

Heterotrophic microorganisms constitute the majority in streambed sediment biofilms. Bacterial and fungal communities colonize both the surface and deep streambed sediments, and they are able to cope with different conditions, such as low oxygenation, flow intermittency, variable solar radiation and changes in organic matter compounds availability (Brablcova et al., 2013; Timoner et al., 2014). Contrarily, the microbial photoautotrophic compartment (algae and cyanobacteria) commonly less developed on the sediment than on the streambed rocks and cobbles due to less favorable conditions of light and stability (Romaní and Sabater 2001).

Up to now, flow intermittency is known to influence the microbial groups inhabiting the large variety of habitats within the streambed sediment (such as lentic pools, leaves packs, moist and dry spots of sediment), affecting both their functions and composition (Lake et al., 2003; Larned et al., 2010; Datry et al., 2016). However, other results report the limited capacity of the microorganisms inhabiting the intermittent streambed to adapt their functions and composition under the wet-to-dry alternations (Barthes et al., 2015; Febria et al., 2015). Nevertheless, pressing climate patterns could reduce these adaptations, when several co-occurring factors such as UV radiation, conductivity, temperature, oxygen, act on the microbial streambed communities together with the altered hydrological conditions.

The characteristics and texture of soils and sediments is usually distinct. Sediment texture, resulting from transport and deposition processes, tend to include coarser particles (more sand and less clay) compared to soils, with general lower organic matter content and patch spots of nutrients than that of soils (Frossard et al., 2015; Stacy et al., 2015). In general, the difference in texture between sediments and soils results in distinct loss of water from evaporation and in higher temperature oscillations and water infiltration in sediments (Zribi et al., 2015; Blume et al., 2016). In dry terrestrial habitats, fungi and bacteria colonizing larger and smaller pores within soil particles are able to form aggregates, which enhance water-holding capacity and reduce moisture stress for the embedded microorganisms, drying more slowly (Denef et al., 2001). The formation of aggregates is a key property of soils, and is usually applied as soil quality reference, gaining particular significance under dry conditions (Mardhiah et al., 2014; Totsche et al., 2017). Specifically, the micro- and macro-aggregates are groups of soil particles that adhere to one another more strongly than to surrounding soil particles (Arce et al., 2019). Depending on their size and porosity, aggregates influence the movement and storage of water in soils, as well as the diffusion of solutes, redox gradients, microbial community structure and vegetation development (Mora-Gómez et al., 2015; Arce et al., 2019).

Consequently, aggregation is considered as a keystone of soil architecture, because of their influence to soil physical properties. Aggregates formation depends on several factors as water availability, temperature, the presence of plants and plant debris, mycorrhizal hyphae, invertebrates and microorganisms (Miltner et al., 2012) and plays a central role in stabilizing organic nutrients (Bronick and Lal, 2005).

The relevance of water holding capacity and soil/sediment moisture

Water infiltration in soil is a function of soil texture that determines the soil water holding capacity. Physical properties of the soil, such as the silt to clay ratio and the soil organic carbon (SOC) content may exert a large effect on the water holding capacity. In general, SOC is positively correlated to the soil water holding capacity and has a direct impact on plant growth (crops) and microbial activities. Under a warmer climate scenario, increases of soil temperature may enhance soil respiration rates (He et al., 2012) affecting SOC turnover and lowering soil carbon sequestration capacity, especially in high latitude environments (Karhu et al., 2014; Van Gestel et al., 2018). In this sense, the observable effects of prolonged drought events may differ not only according to the soil moisture, but also to the organic matter content. Furthermore, if adequate moisture is available, the increase of surface heating will potentially result in an increase in evaporation or evapotranspiration of plants

(Trenberth et al., 2014). Although larger amount of precipitation is expected in regions with a warmer climate, it was estimated that the evaporative demand of the atmosphere would be so high that precipitation will not be sufficient to compensate its increase (Sherwood and Fu, 2014).

Moisture regulates several biogeochemical reactions, such as soil or dry streambed sediment respiration. Upon sudden rewetting, a pulse of CO_2 flux greater than the basal respiration is systematically observed in different soils and sediment types. This increased of CO_2 production rate is known as the "Birch effect" (Birch 1958). Although this effect has been observed a long time ago, the mechanisms behind it remain unclear (Schimel, 2018). The increased respiration rates might be of different source, such as: stable soil carbon mobilized by rewetting and respired by surviving microbes, dead microbial organic matter respired by the surviving ones or osmolytes respired because they become available with the sudden increased soil moisture. Experimental evidences of a direct relationship of "Birch" CO_2 pulses and a rapid mineralization of either dead microbial biomass or osmoregulatory substances have been reported for a Mediterranean woodland soil after measurements of the isotopic $\delta^{13}C$ signatures (Unger et al., 2010). Nevertheless, the dominance of a mechanism over another has a potential important outcome in terms of carbon cycle and climate feedback. Dry-wet cycles might reduce long-term carbon soil storage if the carbon fuelling microbial communities' results from the mobilization of stabilized soil pool. In contrast, if the "Birch effect" results from carbon released from dead microbial material, dry-wet cycles would tend to reduce old carbon loss from soil.

Streambed sediment moisture also depends on several factors, such as the physical properties of sediment particles, clay content, solar irradiance and vegetation development among others. Altogether, sediment moisture, the amount of organic matter stored and the sediment texture could influence the microbial resistance to drought. Therefore, the microbial responses might vary when considering different types of sediments, or streams surrounded by different landscapes (Berggren, Laudon, and Jansson, 2007; Berggren and del Giorgio, 2015). The increase of streambed aridity in intermittent streams and the consecutive reduction of sediment water content due to the extended air-exposition may affect the hosted microbiota and compromise the microbial metabolism (Rees et al., 2006; Marxsen et al., 2010; Febria et al., 2015; Gionchetta et al., 2019).

Consequently, the amount of soil/sediment moisture as well as their water holding capacity could result in different microbial functional (and structural) responses under hydrological stressed conditions. The comparison and assemblage of knowledge available from freshwater streambeds and from soils systems could meliorate the prediction of the consequences resulting from changes in climate. Indeed, most studies conducted on dried and rewetted freshwater sediments focused the attention on the microbial functional and diversity responses whereas literature from the dried and rewetted soil usually added important information about the physical consequences of these effects on the habitat properties (e.g. release of osmolytes and breakdown of soil aggregates). The similarities reported from dried and (re-)wetted sediment and soil studies increased the necessity to improve and expand defined ecosystem boundaries to integrate knowledge for specific

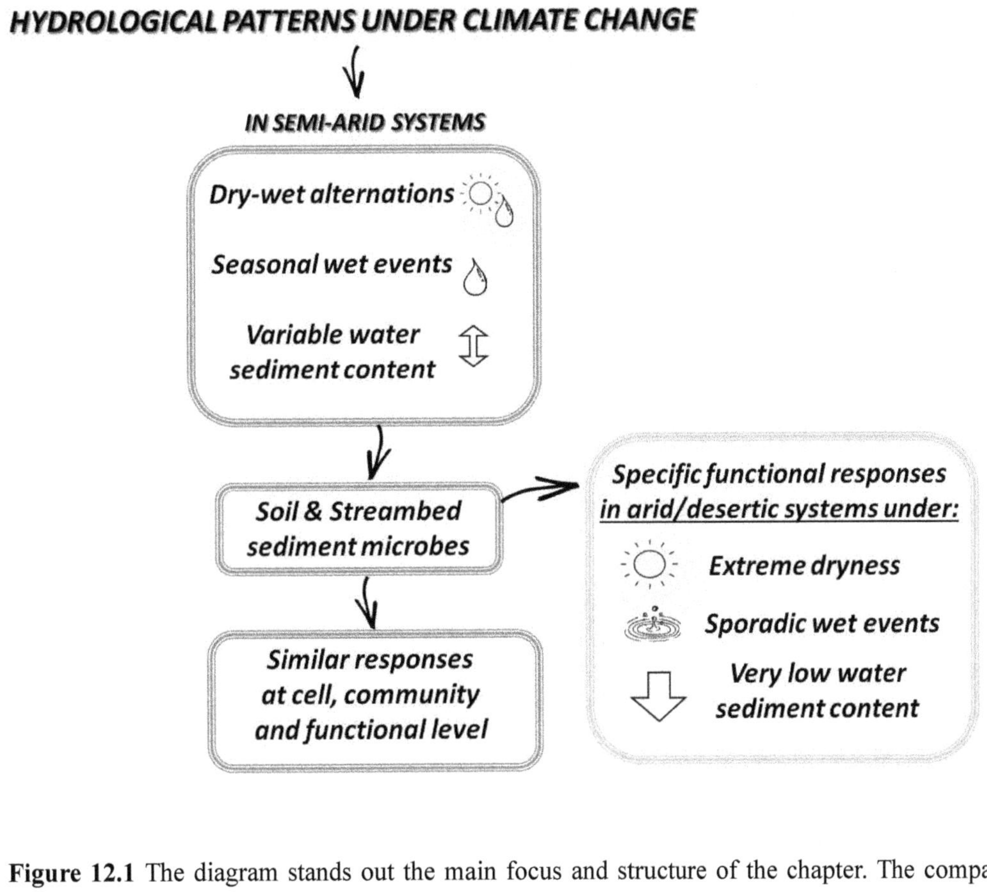

Figure 12.1 The diagram stands out the main focus and structure of the chapter. The comparison of soil and streambed sediment microbial responses is profiled to reveal the consequences of the pressing global changes in hydrological patterns.

management required for the freshwater and terrestrial ecosystems under pressing water deficit and altered precipitation patterns.

Sediments and soils may share key controlling factors of biogeochemical processes and their rates under dry-wet conditions (Arce et al., 2019). Integrate knowledge from soil and sediment streambed studies would definitely gain a complete understanding of ecosystems where connections between terrestrial and aquatic elements are narrow (Grimm et al., 2003). In the following sections, the expected strategies described by the streambed sediment and soil microbes to survive hydrological changes are presented.

Prokaryote and eukaryote responses to changes in water availability: cell level

Soil and sediment microorganisms have developed several strategies to cope with xeric stress and desiccation, such as osmoregulation, dormancy or reactivation and extracellular enzyme synthesis (Barnard et al., 2013). The microbial response to desiccation stress is, however, complex and very costly for the microbial cells. For instance, the water/moisture availability in dry soils and sediments could help in the movement of solutes and cells, which is strictly water-dependent and essential for all the intra- and extracellular reactions that support microbial life (Schimel et al., 2007; Frossard et al., 2012; Moyano et al., 2013). Humidity may also allow fungi to move vertically through the sediment column, favouring the increase in fungal biomass in the hyporheic zone (Cornut et al., 2010). Most studies emphasized that sediment moisture is essential for microorganisms, and may control microbial activity (Marxsen et al., 2010) as similarly found in arid soil ecosystems, where microbial growth is limited to specific wet periods (Kieft et al., 1987).

Specific microbial strategies at the cell level to survive desiccation include the accumulation of compatible solutes for osmotic balance (Potts, 1994) or production of dormant life forms, such as spores (Yuste et al., 2011). Production of osmolytes (solute organic molecules) is particularly efficient to reduce the internal solute potential in order to retain water inside the cell as soil dries but is energetically demanding for the cells (Wood, 2015). Therefore, cells synthesize such molecules when there is no other way to take up or synthesize other osmolytes. Differential physiological characteristic of fungi versus bacteria drive their survival rate to drought. In general, fungi were observed to survive drought more successfully than bacteria in grassland soil (Evans and Wallenstein, 2012). Survival differences between bacteria and fungi might rely also on their ability to synthesize osmolytes:

fungi depend mostly on polyols (organic compound containing multiple hydroxyl groups) or simple carbohydrates, such as glycerol and mannitol, to reduce its internal solute potential, whereas bacteria usually use nitrogenous osmolytes such as proline and glycine (Wood, 2015). In addition, fungal hyphae may expand over air-filled sediment pores to find nutrients and water in a desiccated environment and promote connectivity between individuals in a complex network, thus increasing their resistance to drought in comparison to bacteria (Barnard et al., 2013; de Vries et al., 2018). However, fungi are less resilient to desiccation in most soil environments, probably due to their slower growth rate compared to bacteria.

Dormancy is another strategy that microbial cells adopt when conditions in soil become too harsh (Jones and Lennon, 2010). However, dormancy is also energetically costly, as cells must invest resources to create structures such as spores to survive the physical stress. Spore formation has been studied in bacterial model species, such as *Bacillus cereus*, and involves a tightly modulated series of external and internal signals to trigger sporulation. In most bacterial species, sporulation is a population-based process involving many cells in the microcolony, most of them not evolving in a spore but providing enough nutrients, directed by a quorum sensing cell lysis process, to sporulating cells (Higgins and Dworkin, 2012). In contrast to filamentous bacteria (e.g. *Streptomyces*) and fungi, spore formation in *Bacillus* and *Clostridium*, require a minimum number of cells for sporulation to take place.

The increase in extracellular polymeric substances (EPS) content in streambed biofilms could be also a strategy applied to cope with harsh conditions, mainly related to low water availability (Roberson and Firestone, 1992; Gionchetta et al., 2019). For microorganisms in streambed sediment, the production of EPS such as exopolysaccharides allowed to keep the water solution at a low matric potential (Waldrop and Firestone, 2006). Through the production of biopolymers of different microbial origin, microorganisms inhabiting the stream sediment achieve larger resistance and stability, creating their own biofilm structure (Flemming et al., 2016). The composition of the EPS matrix varies widely, mainly depending on its provenance and local environment, and it performs important functions for the secreting microorganisms, including physical resistance against abrupt changes in the sediment stability (e.g., under flood and desiccation events), the augmentation of nutrients sequestration and the increase of resistance to toxins (Gerbersdorf et al., 2015). Under desiccation scenario, the increased EPS production could preserve the

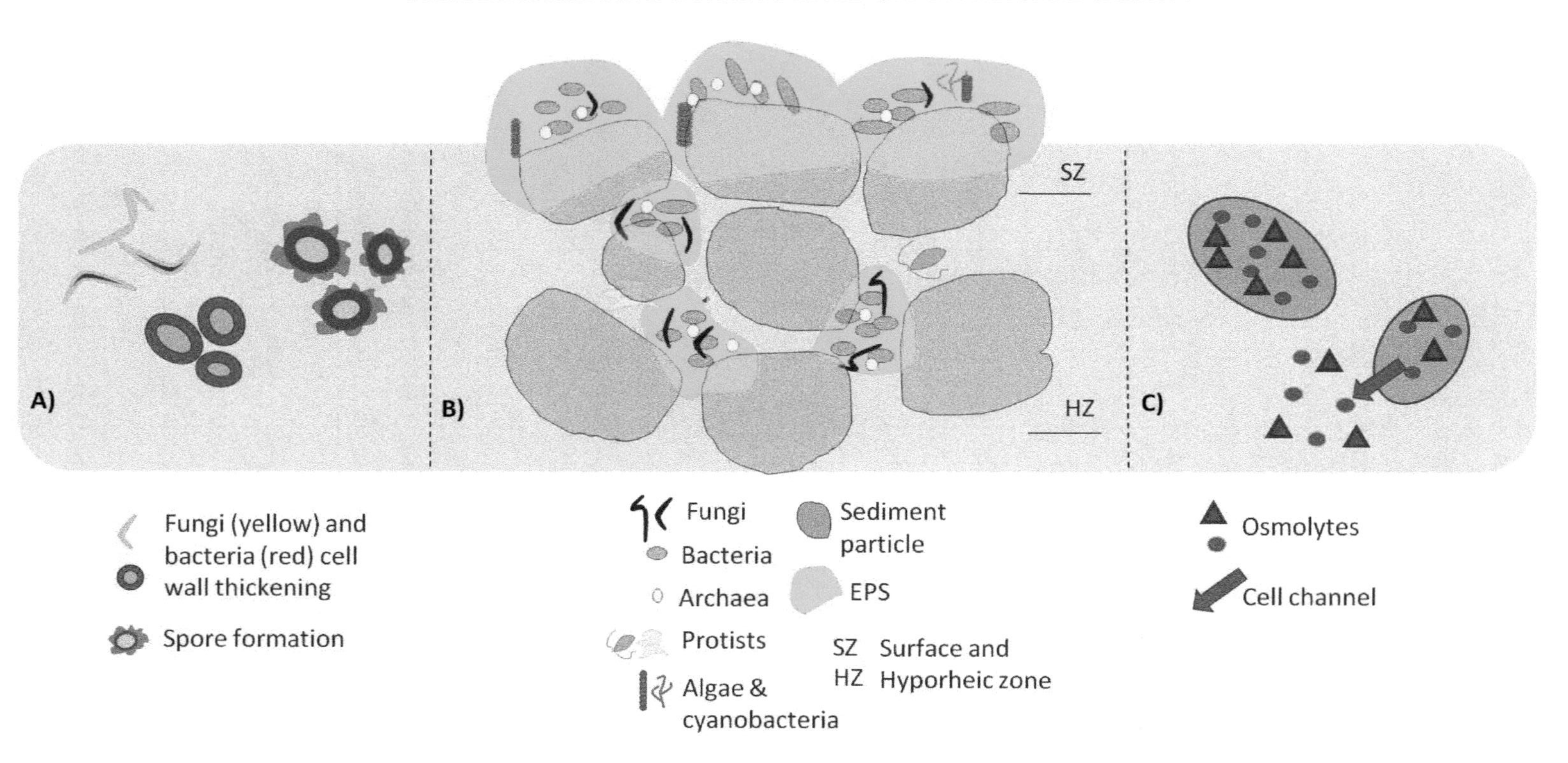

Figure 12.2 Increasing dryness duration and frequency may induce the activation of strategies and adaptations, at cell and community level, in soil and streambed microbes to survive desiccation. Some of these strategies are represented in this figure: A) cell wall thickening and spore formation; B) enhanced production of extracellular polymeric substances (EPS matrix) to increase water holding capacity; C) accumulation and release of microbial osmolytes as osmotic regulation mechanism during drought and rewetting episodes; cells accumulate osmolytes to counteract water deficit and osmolytes are released when rewetting to avoid water influx.

microorganisms, ensuring enough humidity in the sediment and performing as storage for organic matter materials (Rosenzweig et al., 2012; Sabater et al., 2016). The enhanced EPS production response observed in both, stream sediments and soils, gives reasons to correctly identify the EPS as strategic microbial response against desiccation (Roberson and Firestone, 1992; Or et al., 2007; Manzoni et al., 2012).

Prokaryote and eukaryote responses to changes in water availability: community level

In intermittent streams, the hydrological fluctuations from basal to extremely dry condition usually causes the fragmentation of the stream path and the quality and quantity of organic matter vary, leading to changes in the streambed microbial diversity and composition (Marxsen et al., 2010). Generally, the hydrology exerts a strong influence over the microbial community composition that includes aquatic, semi-aquatic, desiccation-resistant or terrestrial taxa. As the microbial diversity tends to decrease when the dry period persists, the community composition usually shifts and these changes mainly consist in variations in the relative abundance of specific taxa. Even short drought events provoke distinct changes in bacterial community composition in temperate stream sediments submitted to unusual flow reductions, increasing proportions of Actinobacteria and α-Proteobacteria, and decreasing Bacteroidetes and β-Proteobacteria (Pohlon et al., 2013b).

Soil bacterial community composition and structure could be disturbed by drying and rewetting events in different ecosystems (Evans and Wallenstein, 2012; Barnard et al., 2013; Schimel, 2018), although the assessment of changes in the microbial community during these events are challenged by the techniques used. For example, soil diversity analyses based on DNA may capture old DNA from dead microorganisms, and not only the living ones (Carini et al., 2017). Microbial diversity based on RNA, accounting for only living organisms may give a better picture of the active microbiome, providing it is based on messenger RNA (mRNA) and not on ribosomal RNA (rRNA), since dormant cells can still contain high number of ribosomes (Blazewicz et al., 2013; Gionchetta et al., 2020).

As previously observed from different studies on soil and streambed sediment, hydrological changes could influence the microbial taxa selection (e.g. relative abundance of specific classes) able to cope with osmotic stress (Pohlon et al., 2013a). Some classes reduced their abundances when increasing drought duration (e.g. β- and δ-Proteobacteria and Bacteroiidia), whereas other classes increased their abundances under the prolonged dry period, such as Actinobacteria, Bacilli and Thermoleophilia (Schimel et al., 2007, Manzoni el al., 2012, Barnard et al., 2013, Meisner et al., 2018; Gionchetta et al., 2019, 2020). The general patterns mainly consist in increasing the abundance of gram-positive bacterial and fungal cells with thicker walls, enhancing their capacity to cope with the osmotic stress (Schimel et al., 2007; Yuste 2011; Zeglin et al., 2015).

In soil ecology, resilience thresholds are defined as limits in environmental stressors that will cause a change in soil functionality and ecosystem services (Desjardins et al., 2015). Due to the complexity of microbial communities and the heterogeneity of soil, it is not surprising that changes in either composition or activities can be observed in response to a disturbance, and most studies show that conditions rarely return to pre-disturbance conditions (Shade et al., 2012). Basically, soils and sediments harbouring a highly complex microbial community (i.e. high diversity and richness indices) tend to be more resilient to environmental changes compared to low-diversity systems, especially when the system, soil or sediment, activity is considered. Reasons for this are complex but basically related to the high functional redundancy of highly-diverse microbial communities, meaning that broad system functions, such as respiration or organic matter decomposition, are not largely affected despite significant changes in the microbial community structure are recorded (de Vries et al., 2018).

Sediments and soil microbial functional responses to changes in water content

Microbial growth rates and biomass could be reduced under water stress in drying soils and sediments, as microbial cells would spend most of resources and energy to apply survival strategies to the detriment of reproduction (Schimel et al., 2007). Recent studies observed that desiccation decreases microbial organic matter decomposition activity in streambed sediments. This phenomenon was related to a decrease in the number of living bacteria, a greater investment of energy on resistance strategies (such as extracellular polymeric substances production), a decrease in autochthonous organic matter availability, such as fewer algal exudates, or a reduction of enzyme and substrate diffusion (Zoppini and Marxsen 2011; Gionchetta et al., 2019). The reduction of soil moisture can limit enzyme and substrate diffusion (Liu et al., 2000) affecting the enzyme efficiencies, might because processes such as enzyme immobilization increased while diffusion rates are reduced (Henry, 2013).

Effects of drought on soils and sediments microbial biogeochemical processes are highly linked to water content, but also, as suggested by recent studies, to the alternation of dry and wet episodes (Schimel, 2018). These alternations as well as the amount of water sediment content and organic matter available can influence the microbial extracellular enzyme activities and the biogeochemical transformations of the organic matter in stream sediment (Timoner et al., 2014). The stream flow fragmentation in intermittent streambeds triggers the development of a mosaic of habitats, which provide different environmental conditions and variable organic matter quality and quantity. Isolated pools formation and carbon limitation conditions can benefit autotrophic production, especially in open stream reaches, being hot spots of microbial activity associated to algal growth, and acting as microbial refuges (Larned et al., 2010). In forested streams, the accumulation of recalcitrant material due to extra-seasonal leaf fall (occurring in summer due to water stress to vegetation), can stimulate specific microbial activities related to plant decomposition (Acuña et al., 2007). For instance, Ylla et al. (2010) measured a reduced capacity to decompose peptides but an increased capacity to decompose polysaccharides (such as cellulose) in a drying intermittent streambed. Conversely, in a field experiment, where stream sites of a wide range of intermittency were compared, the sites suffering the longest dry periods showed a reduced β-xylosidase activity, but an increase in phenol oxidase activity was also measured (Gionchetta et al., 2020). These findings have been related to the potential change of the quality of the organic matter, stored in the superficial sediment during the desiccation phase (Ylla et al., 2010; Romaní et al., 2013), as shown by the greater use of lignin-like materials (phenol oxidase activity) in the most dried sediments (Gionchetta 2019, 2020). The greater accumulation of allochthonous plant material (Romaní et al., 2006; Sinsabaugh, 2012; Burns et al., 2013) as well as the initial shift of the microbial community towards a "terrestrial-like" composition might be the causes of these observations.

Under the intermittent flow condition, the organic matter on the dried streambed increase (e.g., from animal or vegetal debris or leaves fall) compared to the amount present during the flowing period or to that of the adjacent soils, especially in arid, semiarid and Mediterranean regions where the vegetation is limited to the riparian zones (Fossati et al., 1999; Steward, 2012). In addition, solar radiation can cause chemical oxidation reactions in the organic matter accumulated in the streambed, changing its quality and thus, affecting the microbial colonization and

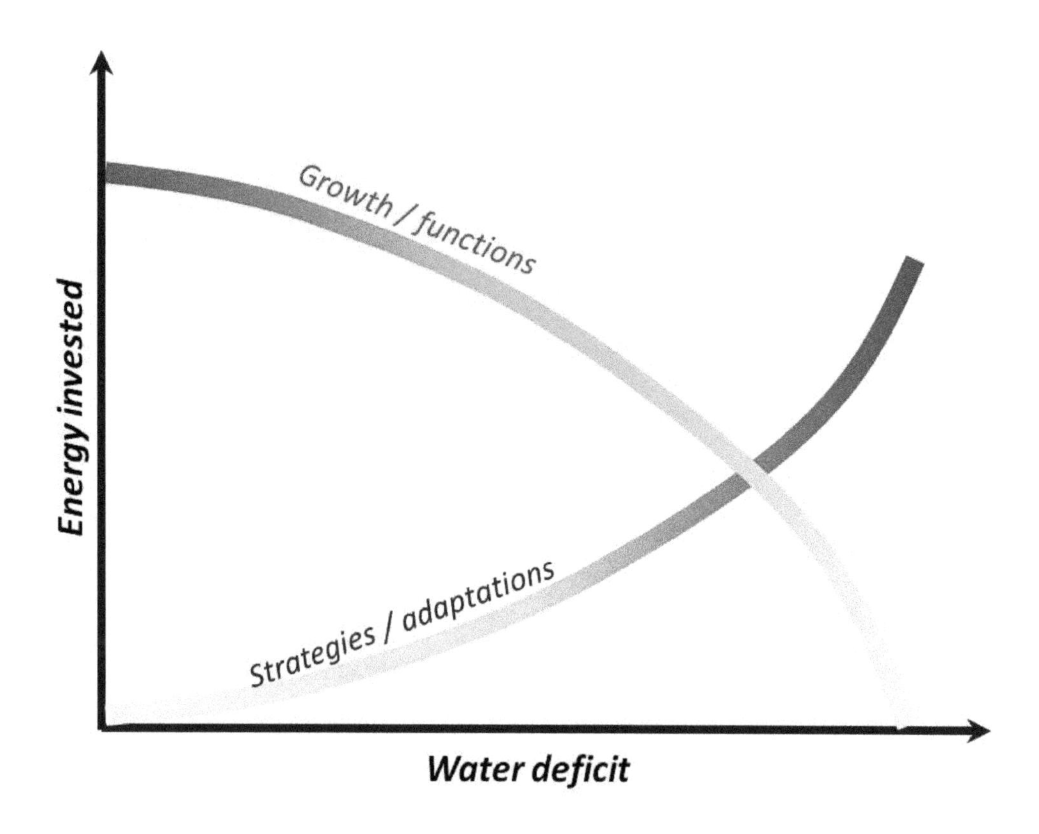

Figure 12.3 Conceptual framework indicating how microorganisms may invest their energy depending on the water available in soil/streambed sediment. The conceptual model shows that: when the conditions are favourable (high water availability) microorganisms invest their energy to increase their biomass (growth) or to develop functions, such as organic matter degradation (orange line); contrarily, under intensifying water deficit, large part of cell energy is devoted to the development of strategies and adaptations to survive harsh conditions. Note that the conceptual framework operates under the spatial concurrence of dried soils and streambed sediment involving similar climate, topography and organisms.

decomposition when rewetting (Mora-Gómez et al., 2019). Generally, the rewetting period enhances the microbial activity reactivating key microbial processes. The enzyme activity of the streambed microbial crust of a Mediterranean stream instantaneously recovered once rewetted, reaching values similar to those of the flowing period, whereas the photosynthetic activity lasted a longer time to be recovered (Romaní and Sabater 1997). However, depending on the duration of the previous desiccation and on the occurrence or not of wet events interrupting the long dry period, the resilience of the enzyme activities could be largely variable (Pohlon et al., 2013; Gionchetta et al., 2019). In addition, the quality of materials transported by the first rains, the type of sediment, and the drought historical can influence the microbial responses largely (Evans et al., 2012; Shumilova et al., 2019).

Knowledge from extreme environments: deserts and hypersaline systems in arid zones

Arid and semi-arid environments specially suffer from water scarcity and this stress condition is exacerbated in areas with high salinity conditions that may limit water availability for microbes to survive. Known microbial responses and adaptations in these extreme systems shed light on expected responses in intermittent streambeds when extreme dry and warming episodes occur together with soil salinization.

In arid systems, the availability of resources for microbes is driven by water (Noy-Meir, 1973). In these systems, microbial activities are expected to vary according to daily, seasonal and yearly variation of soil moisture and precipitation (Schwinning et al., 2004; Belnap et al., 2005). In a 3-year observational study in the Chichuahuan desert, Bell et al. (2008) reported that patterns of microbial biomass peaked with soil moisture levels. Soil microbes showed a dramatic change when soil moisture became available after 3 years of drought conditions. This study revealed that microbial biomass in desert soil responds to soil moisture primarily, regardless of previous temperature levels or temperature stress conditions of the soil. This study also indicated that microbial communities inhabiting hot desert environment are resilient enough to remain viable during long periods of drought and respond quickly as soon as water became available.

Nevertheless, the viability of the microbial cells is sometimes more limited by the harsh conditions of hot desert than their functions. For example, the activity of extracellular enzymes, particularly oxidative enzymes, is less restricted by moisture than the organisms themselves. Enzyme activity is associated with viable cells, but once in the edaphic environment, it is stabilized by soil colloids. Thus, functional diversity is likely to remain more stable than microbial diversity during periods of drought (Collins et al., 2008; Marxsen et al., 2010). In a study following the activity of enzyme activity seasonally, it was concluded that substrate availability was the most important factor affecting soil microbial diversity and activity. Soil moisture only affected microbial enzyme activity under particular stress conditions at a specific time of the year (Liu et al., 2000).

In arid environments, water can be present in two forms: either rainfall or fog with resulting dew formation. Rain events are usually relatively short lived with intermittent periods of drought. Fog or dew can be a substantial source of water in certain deserts (e.g. Namib, Atacama, and Sonoran Deserts), triggering and prolonging biological activity during the long dry periods. However, the contribution of dew and fog to the soil moisture that activates microbial activity is not clear. In the Negev desert (Israel), the effect of dew on cyanobacterial crusts was shown to be negligible (Kidron et al., 2002). Only about 10% of dew or fog events in the Negev desert contained sufficient water to trigger photosynthetic activity of cyanobacterial crusts (Kidron et al., 2002). Similarly, a study conducted in the Hopq Desert (China), showed a clearly significant correlation between the amount of dew and biological soil crust moisture content, but not between dew amount and cyanobacterial biomass (Rao et al., 2009), suggesting the limited role of the dew on crust biomass growth. Dew penetration in the desert edaphic system depends on the soil or biological crust texture (Chamizo et al., 2012). In particular, extracellular polymeric substances (EPS), produced in part by cyanobacteria, are known to improve moisture retention capacity within the biological crust (Mager and Thomas, 2011). Consequently, microbial communities present in the biological crust could promote the increase of moisture absorbance and the decrease of dew evaporation rate.

Soil moisture and temperature in hot desert environments vary following a diurnal variation pattern (Agam and Berliner, 2004). A study conducted in the Namib Desert, reported significant shifts in active bacterial community composition within 24-hours diel cycles over several days of monitoring (Gunnigle et al., 2017). For instance, taxonomic groups, such as Actinobacteria, decreased under the night conditions, as the temperature dropped and the humidity increased, whereas during daytime other groups, such as the Proteobacteria, were dominant. Interestingly, cyanobacteria have a critical role of during the diel cycle, reporting a strong adaptation in terms of growth and metabolism to diel variation of desert temperature (Wang et al., 2013; Gunnigle et al., 2017).

To comprehend the dynamics of microbial activities in arid systems, we need to understand how microbial community structure and function respond to water pulses during dry-wet cycles. Drying-rewetting cycles (i.e. drought period followed by wet event) may influence the composition of specific microbial communities and subsequently affect ecosystem functions (Clark et al., 2009). Thus, long term changes in microbial communities due to alteration of water regime may have large implications for the future trajectory of arid ecosystem. The moisture legacy is also very important to determine the microbial response to dry-wet cycles in soil and sediment. This was observed in the Namib Desert where long-term environmental adaptations of both bacterial and fungal communities to soil origins with different soil water regime history (i.e. desert gravel plain soil vs. dry streambed) were observed to attenuate short-term response towards the wetting events (Frossard et al., 2015). Furthermore, when microbial communities where affected by the dry-wet cycles, it was by the intensity rather than the frequency of the wetting events.

Long periods of drought can result in high microbes' mortality, due to desiccation and radiation damages, leaving fewer microbes to respond to the pulse event. However, atmospheric nutrient deposits which accumulate during dry periods may mitigate resource limitation to those drought-resistant microbes when precipitation occurs. Consequently, when water becomes available, microbial activity associated with rainfall may shift from water to resource limitation (Belnap et al., 2005). In arid systems, soil microbes are typically the first responders to moisture pulses and the microbial functional capabilities in these systems are directly linked to precipitation and resources availability (Austin et al., 2004; Schwinning et al., 2004). Just small amounts of precipitation can activate soil surface microbial processes whereas large rainfalls are required by higher organisms to be activated, such as vascular plants. The magnitude of a water pulse event is related to the ecological response it triggers (Schwinning and Sala, 2004), and it is associated with nutrient dynamics in arid systems (Austin et al., 2004). Then, as small rains may simply redistribute materials and reallocate resources (Fisher et al., 1998), more intense precipitations, such as floods, can end up gaining or losing nutrients from the system (Belnap et al., 2005). As a result, scattered rainfall intensity, which does not cause runoff or nutrient and organic matter removal, boost the soil microbial activities.

In arid and semi-arid saline systems, salinity prompts feedbacks on other critical factors such as reducing oxygen solubility and water availability, the organisms in these systems being defined as polyextremophiles as they are facing by a triad of stress factors, i.e. salinity, water deficit, and high temperatures (Ma et al., 2010). In salty arid soils, higher salinity decreased soil respiration and microbial biomass, resulting in a smaller and stressed microbial community with reduced metabolic efficiency (Yuan et al., 2007). Similar responses and protection mechanisms are those in saline arid ones, such as the production of extracellular polymeric substances (Nicolaus et al., 2010). At the same time, the extreme conditions also trigger the production of specific enzymes from Bacteria and Archaea able to support high salinity and temperature conditions and thus being of interest for biotechnological purposes (Ma et al., 2010).

Conclusions and future perspectives

Our knowledge suggests that microbial responses to dryness in streambed sediments and soils are sensitive to low range of moisture. As a result, little variations in water availability may influence certain microbial functions and thus, change the related ecosystem functional services. From this perspective, two research directions could be pointed out: i) the relevance of the microhabitats that might enhance moisture preservation and water holding capacity through sediment/soil texture heterogeneity, and ii) the significance of short watering episodes in leading organic matter mobilization and augmenting the water availability.

Several examples of microhabitats exist from intermittent watersheds. Frequently, a fluvial network at the final stage of the drying process consists in a fragmented landscape of isolated water pools where sediments and organic detritus accumulate without being exported (Lake, 2003). The resulting spotty distribution of habitats (e.g., lentic pools, leaves packs, moist and dry patches of sediment) may act as natural filters, selecting for the most adapted species, and as refuge for the most sensitive ones (Larned et al., 2010; Datry et al., 2016). Generally, the hyporheic zone and the leaf packs buried within the streambed often act as refuges for prokaryotes and eukaryotes under persistent drying condition, retaining larger amount of water, organic matter and nutrients compared to the surface sediment (Ylla et al., 2010; Febria et al. 2012). As consequence, the microbial communities inhabiting those sub-surface streambed habitats may manifest improved resistance and resilience to prolonged dryness. Similarly, in desert soils, specific microhabitats able to maintain some humidity, such as the below-rocks, can be crucial for microbial development (Azúa-Bustos et al., 2010).

In relation to the relevance of wet episodes, these may prompt the recovery of microbes, making suddenly water, as well as relevant nutrient resources available (Schimel 2018). From this standpoint, it is essential to improve the understanding of the different microbial responses through the comparison between soil and streambed sediment microbiota (including Bacteria, fungi and Archaea) that may determine distinct or similar functional changes to the altered hydrology (Barnard et al., 2013).

Future research direction should focus on the conservation of intermittent watersheds and soils of arid and semiarid regions, involving the management of the flow regime and land use modifications (Bunn and Arthington, 2002). Research at the micro-scale including ranges of water availability as well as considering different soil/sediment properties (such as organic matter, particle size, nutrients, salinity) needed to define thresholds or tipping points for the hosted microbial functioning. Thereby, new investigations involving different expertise could promote the knowledge of processes and mechanisms at the very fine scale, in the context of the boundary between aquatic and terrestrial systems.

References

Acuña, V., Giorgi, A., Muñoz, I., Sabater, F., and Sabater, S. (2007). Meteorological and riparian influences on organic matter dynamics in a forested Mediterranean stream. J. N. Am. Benthol. Soc. 26, 54–69. DOI: 10.1899/0887-3593.26 [54: marioo]2.0.co;2

Agam, N., and Berliner, P.R. (2004). Diurnal water content changes in the bare soil of a coastal desert. J. Hydrometeorol. 5, 922-933.

Arce, M.I., Mendoza-Lera, C., Almagro, M., Catalan, N., Romani, A.M., Martí, E., et al. (2019). A conceptual framework for understanding the biogeochemistry of dry riverbeds through the lens of soil science. Earth-Sci. Rev. 188, 441-453. DOI: 10.1016/j.earscirev.2018.12.001

Austin, A., Yahdjian, L., Stark, J., Belnap, J., Porporato, A., Norton, U. et al. (2004). Water pulses and biogeochemical cycles in arid and semiarid ecosystems. Oecologia 141, 221-235.

Azúa-Bustos, A., González-Silva, C., Mancilla, R.A. et al. (2011). Hypolithic cyanobacteria supported mainly by fog in the coastal range of the Atacama Desert. Microb. Ecol. 61, 568–581. DOI: 10.1007/s00248-010-9784-5

Barnard, R., Osborne, C., and Firestone, M. (2013). Responses of soil bacterial and fungal communities to extreme desiccation and rewetting. ISME J. 7, 2229–2241. DOI: 10.1038/ismej.2013.104

Barthès, A., Ten-Hage, L., Lamy, A., Rols, J.L., and Leflaive, J. (2015). Resilience of aggregated microbial communities subjected to drought—small-scale studies. Microb. Ecol. 70, 9–20. DOI: 10.1007/s00248-014-0532-0

Battin, T.J., Besemer, K., Bengtsson, M.M., Romani, A.M., and Packmann, A.I. (2016). The ecology and biogeochemistry of stream biofilms. Nat. Rev. Microbiol. 14, 251-263.

Bell, C., McIntyre, N., Cox, S., Tissue, D., and Zak, J. (2008). Soil microbial responses to temporal variations of moisture and temperature in a Chihuahuan Desert Grassland. Microb. Ecol. 56, 153-167.

Belnap, J., Welter, J.R., Grimm, N.B., Barger, N., and Ludwig, J.A. (2005). Linkages between microbial and hydrologic processes in arid and semiarid watersheds. Ecology 86, 298-307.

Berggren, M., Laudon, H., and Jansson, M. (2007). Landscape regulation of bacterial growth efficiency in boreal freshwaters. Global Biogeochem. Cycles 21, 1–9. DOI: 10.1029/2006GB002844

Berggren, M., and Giorgio, P. del (2015). Distinct patterns of microbial metabolism associated to riverine dissolved organic carbon of different source and quality. J. Geophys. Res. Biogeosciences 120, 989–999. DOI: 10.1002/2015JG002963

Birch, H.F. (1958). The effect of soil drying on humus decomposition and nitrogen availability. Plant Soil X, 9–10.

Blazewicz, S.J., Barnard, R.L., Daly, R.A., and Firestone, M.K. (2013). Evaluating rRNA as an indicator of microbial activity in environmental communities: limitations and uses. The ISME Journal 7, 2061.

Blume, H.P., Brümmer, G.W., Fleige, H., Horn, R., Kandeler, E., Kögel-Knabner, I., Kretzschmar, R., Stahr, K., and Wilke, B.M., 2016. Scheffer/Schachtschabel Soil Science (Berlin, Heidelberg: Springer-Verlag).

Bronick, C.J., Lal, R. (2005). Soil structure and management: a review. Geoderma 124, 3–22.

Bunn, S., and Arthington, A. (2002). Basic principles and ecological consequences of altered flow regimes for aquatic deflation basin lakes. Environ. Manage. 30, 492-507. 10.1007/s00267-002-2737-0

Burns, R.G., DeForest, J.L., Marxsen, J., Sinsabaugh, R.L., Stromberger, M.E., Wallenstein, M.D., Weintraub, M.N., and Zoppini, A. (2013). Soil enzymes in a changing environment: Current knowledge and future directions. Soil Biol. Biochem. 58, 216–234. DOI: 10.1016/j.soilbio.2012.11.009

Brablcová, L., Buriánková, I. Badurová, P., and Rulík, M. (2013). The phylogenetic structure of microbial biofilms and free-living bacteria in a small stream. Folia Microbiol. (Praha) 58, 235–243.

Carini, P., Marsden, P.J., Leff, J.W., Morgan, E.E., Strickland, M.S., and Fierer, N. (2017). Relic DNA is abundant in soil and obscures estimates of soil microbial diversity. Nat. Microbiol. 2, 16242.

Chamizo, S., Canton, Y., Miralles, I., and Domingo, F. (2012). Biological soil crust development affects physicochemical characteristics of soil surface in semiarid ecosystems. Soil Biol. Biochem. 49, 96-105.

Clark, J.S., Campbell, J.H., Grizzle, H., Acosta-Martinez, V., and Zak, J.C. (2009). Soil microbial community response to drought and precipitation variability in the Chihuahuan Desert. Microb. Ecol. 57, 248-260.

Collins, S.L., Sinsabaugh, R.L., Crenshaw, C., Green, L., Porras-Alfaro, A., Stursova, M., and Zeglin, L.H. (2008). Pulse dynamics and microbial processes in aridland ecosystems. J. Ecol. 96, 413-420.

Cornut, J., Elger, A., Lambrigot, D., Marmonier, P., and Chauvet, E. (2010). Early stages of leaf decomposition are mediated by aquatic fungi in the hyporheic zone of woodland streams. Freshw. Biol. 55, 2541–2556. DOI: 10.1111/j.1365-2427.2010.02483.x

Dai, A. (2011). Drought under global warming: a review. Clim. Change 2, 45-65.

Datry, T., Fritz, K., and Leigh, C. (2016). Challenges, developments and perspectives in intermittent river ecology. Freshw. Biol. 61, 1171–1180. DOI: 10.1111/fwb.12789

Datry, T., Bonada, N., and Boulton, A.J. (2017). Intermittent Rivers and Ephemeral Streams - Ecology and Management (Waltham, MA: Elsevier).

Denef, K., Six, J., Bossuyt, H., Frey, S.D., Elliott, E.T., Merckx, R., and Paustian, K. (2001). Influence of dry - wet cycles on the interrelationship between aggregate, particulate organic matter, and microbial community dynamics. Soil Tillage Res. 33, 1599–1611. DOI: 10.1016/S0038-0717(01)00076-1

de Vries, F.T., Griffiths, R.I., Bailey, M., Craig, H., Girlanda, M., Gweon, H.S. et al. (2018). Soil bacterial networks are less stable under drought than fungal networks. Nat. commun. 9, 3033.

Desjardins, E., Barker, G., Lindo, Z., Dieleman, C., and Dussault, A.C. (2015) Promoting resilience. Q. Rev. Biol. 90, 147-165.

Elosegi, A., Díez, J., and Mutz, M. (2010). Effects of hydromorphological integrity on biodiversity and functioning of river ecosystems. Hydrobiologia 657, 199–215. DOI: 10.1007/s10750-009-0083-4

Evans, S.E., and Wallenstein, M.D. (2012). Soil microbial community response to drying and rewetting stress: does historical precipitation regime matter? Biogeochemistry 109, 101-116.

Febria, C.M., Beddoes, P., Fulthorpe, R.R., and Williams, D.D., (2012). Bacterial community dynamics in the hyporheic zone of an intermittent stream. ISME J. 6, 1078-1088.

Findlay, S. (2010). Stream microbial ecology. J. N. Am. Benthol. Soc. 29, 170–181. DOI: 10.1899/09-023.1

Fisher, S.G., Grimm, N.B., Marti, E., and Gomez, R. (1998). Hierarchy, spatial configuration, and nutrient cycling in a desert stream. Aust. J. Ecol. 23, 41-52.

Flemming, H.-C., Wingender, J., Szewzyk, U., Steinberg, P., Rice, S.A., and Kjelleberg, S. (2016). Biofilms: an emergent form of bacterial life. Nat. Rev. Microbiol. 14, 563–575. DOI: nrmicro.2016.94

Fossati, J., Pautou, G., and Peltier, J.P. (1999). Water as resource and disturbance for wadi vegetation in a hyperarid area (Wadi Sannur, Eastern Desert, Egypt). J. Arid Environ. 43, 63–77. 10.1006/jare.1999.0526

Frossard, A., Gerull, L., Mutz, M., and Gessner, M.O. (2012). Disconnect of microbial structure and function: enzyme activities and bacterial communities in nascent stream corridors. ISME J. 6, 680–691. DOI: 10.1038/ismej.2011.134

Frossard, A., Ramond, J. B., Seely, M., and Cowan, D. A. (2015). Water regime history drives responses of soil Namib Desert microbial communities to wetting events. Sci. Rep. 5, 12263.

Gerbersdorf, S.U., and Wieprecht, S. (2015). Biostabilization of cohesive sediments: revisiting the role of abiotic conditions, physiology and diversity of microbes, polymeric secretion, and biofilm architecture. Geobiology 13, 68e97

Gionchetta, G., Oliva, F., Menéndez, M., Lopez Laseras, P., and Romaní, A.M. (2019). Key role of streambed moisture and flash storms for microbial resistance and resilience to long-term drought. Freshw. Biol. 64, 306–322. DOI: 10.1111/fwb.13218

Gionchetta, G., Romaní, A.M., Oliva, F., and Artigas, J. (2019). Distinct responses from bacterial, archaeal and fungal streambed communities to severe hydrological disturbances. Sci, Rep. 13506, 9. DOI 10.1038/s41598-019-49832-4

Gionchetta, G., Oliva, F., Romaní, A.M., and Bañeras, L. (2020). Hydrological variations shape diversity and functional responses of streambed microbes. Sci. Total Environ 714, 136838. DOI: 10.1016/j.scitotenv.2020.136838

Gionchetta, G., Artigas, J., Arias-Real, R., Oliva, F., and Romaní, A.M. (2020). Multi-model assessment of hydrological history impact on microbial structural and functional responses in Mediterranean catchments. Environ. Microbiol. 22, 2213–2229. DOI: 10.1111/1462-2920.14990

Gunnigle, E., Frossard, A., Ramond, J.-B., Guerrero, L., Seely, M., and Cowan, D.A. (2017). Diel-scale temporal dynamics recorded for bacterial groups in Namib Desert soil. Sci. Rep. 7, 40189.

Halverson, L.J., Jones, T.M., and Firestone, M.K. (2000). The release of intracellular solutes by four soil bacteria exposed to dilution stress. Soil Sci. Soc. Am. J. 64, 1630–1637.

Harms, T.K., and Grimm, N.B. (2012). Responses of trace gases to hydrologic pulses in desert floodplains. J. Geophys. Res. Biogeosci. 117, 1–14. DOI: 10.1029/2011JG001775

He, N., Chen, Q., Han, X., Yu, G., and Li, L. (2012). Warming and increased precipitation individually influence soil carbon sequestration of Inner Mongolian grasslands, China. Agric. Ecosyst. Environ. 158, 184-191.

Henry, H.A.L. (2013). Soil extracellular enzyme dynamics in a changing climate. Soil Biol. Biochem. 56, 53–59. DOI: 10.1016/j.soilbio.2012.10.022

Higgins, D., and Dworkin, J. (2012). Recent progress in *Bacillus subtilis* sporulation. FEMS Microbiol. Rev. 36, 131-148.

Huang, J., Yu, H., Guan, X., Wang, G., and Guo, R. (2016). Accelerated dryland expansion under climate change. Nat. Clim. Change 6, 166.

Jones, S.E., and Lennon, J.T. (2010). Dormancy contributes to the maintenance of microbial diversity. Proc. Natl. Acad. Sci. U.S.A. 107, 5881-5886.

Karhu, K., Auffret, M.D., Dungait, J.A., Hopkins, D.W., Prosser, J.I., Singh, B.K. et al. (2014). Temperature sensitivity of soil respiration rates enhanced by microbial community response. Nature 513, 81.

Kidron, G.J., Herrnstadt, I., and Barzilay, E. (2002). The role of dew as a moisture source for sand microbiotic crusts in the Negev Desert, Israel. J. Arid Environ. 52, 517-533.

Kieft, T.L., Soroker, E., and Firestone, M.K. (1987). Microbial biomass response to a rapid increase in water potential when dry soil is wetted. Soil Biol. Biochem. 19, 119–126. DOI: 10.1016/0038-0717(87)90070-8

Lake, P.S. (2003). Ecological effects of perturbation by drought in flowing waters. Freshw. Biol. 48, 1161-1172. DOI: 10.1046/j.1365-2427.2003.01086.x

Larned, S.T., Datry, T., Arscott, D.B., and Tockner, K. (2010). Emerging concepts in temporary-river ecology. Freshw. Biol. 55, 717-738. DOI: 10.1111/j.1365-2427.2009.02322.x

Liu, X.Y., Lindemann, W.C., Whitford, W.G., and Steiner, R.L. (2000). Microbial diversity and activity of disturbed soil in the northern Chihuahuan Desert. Biol. Fertil. Soils 32, 243-249.

Ma, Y., Galinski, E. A., Grant, W. D., Oren, A., and Ventosa, A. (2010). Halophiles 2010: Life in saline environments. Appl. Environ. Microbiol. 76, 6971-6981.

Mager, D.M., and Thomas, A.D. (2011). Extracellular polysaccharides from cyanobacterial soil crusts A review of their role in dryland soil processes. J. Arid Environ. 75, 91-97.

Manzoni, S., Schimel, J.P., and Porporato, A. (2012). Responses of soil microbial communities to water stress : results from a meta-analysis. Ecology 93, 930–938. DOI: 10.1890/11-0026.1

Mardhiah, U., Caruso, T., Gurnell, A., and Rillig, M.C., (2014). Just a matter of time: Fungi and roots significantly and rapidly aggregate soil over four decades along the Tagliamento River, NE Italy. Soil Biol. Biochem. 75 (Supplement C), 133–142.

Marxsen, J., Zoppini, A., and Wilczek, S. (2010). Microbial communities in streambed sediments recovering from desiccation. FEMS Microbiol. Ecol. 71, 374-386. DOI: 10.1111/j.1574-6941.2009.00819.x

Meisner, A., Jacquiod, S., Snoek, B.L., Ten Hooven, F.C., and van der Putten, W.H. (2018). Drought legacy effects on the composition of soil fungal and prokaryote communities. Front. Microbiol. 9, 1–12. DOI: 10.3389/fmicb.2018.00294

Miltner, A., Bombach, P., Schmidt-Brücken, B., and Kästner, M., (2012). SOM genesis: microbial biomass as a significant source. Biogeochemistry 111, 41–55.

Moyano, F.E., Manzoni, S., and Chenu, C. (2013). Responses of soil heterotrophic respiration to moisture availability: An exploration of processes and models. Soil Biol. Biochem. 59, 72–85.

Mora-Gómez, J., Elosegi, A., Mas-Martí, E., and Romaní, A.M. (2015). Factors controlling seasonality in leaf-litter breakdown in a Mediterranean stream. Freshw. Sci. 34, 1245–1258. DOI: 10.1086/683120

Mora-Gómez, J., Boix, D., Duarte, S., Cássio, F., Pascoal, C., Elosegi, A., and Romaní, A.M. (2019). Legacy of summer drought on autumnal leaf litter processing in a temporary Mediterranean stream. Ecosystems 23, 989–1003. DOI: 10.1007/s10021-019-00451-0

Mori, N., Simčič, T., Brancelj, A., Robinson, C.T., and Doering, M. (2017). Spatiotemporal heterogeneity of actual and potential respiration in two contrasting floodplains. Hydrol. Process. 31, 2622–2636. DOI: 10.1002/hyp.11211

Navarro-Garcia, F., Casermeiro, M.A. and Schimel, J.P. (2012). When structure means conservation: Effect of aggregate structure in controlling microbial responses to rewetting events. Soil Biol. Biochem. 44, 1–8.

Nicolaus, B., Kambourova, M., and Oner, E.T. (2010). Exopolysaccharides from extremophiles: from fundamentals to biotechnology. Environ. Technol. 31, 1145–1158. DOI: 10.1080/09593330903552094

Noy-Meir, I. (1973) Desert ecosystems: environment and producers. Annu. Rev. Ecol. Evol. Syst. 4, 25-51.

Or, D., Smets, B.F., Wraith, J.M., Dechesne, A., and Friedman, S.P. (2007). Physical constraints affecting bacterial habitats and activity in unsaturated porous media - a review. Adv. Water Resour. 30, 1505–1527. DOI: 10.1016/j.advwatres.2006.05.025

Pohlon, E., Mätzig, C., and Marxsen, J. (2013a). Desiccation affects bacterial community structure and function in temperate stream sediments. Fundam. Appl. Limnol. / Arch. Hydrobiol. 182, 123–134. DOI: 10.1127/1863-9135/2013/0465

Pohlon, E., Ochoa Fandino, A., and Marxsen, J. (2013b). Bacterial community composition and extracellular enzyme activity in temperate streambed sediment during drying and rewetting. PLoS One 8, e83365. DOI: 10.1371/journal.pone.0083365

Potts, M. (1994). Desiccation tolerance of prokaryotes. Microbiol. Rev. 58, 755–805. DOI: 10.1093/icb/45.5.800

Prein, A.F., Rasmussen, R.M., Ikeda, K., Liu, C., Clark, M.P., and Holland, G.J. (2017). The future intensification of hourly precipitation extremes. Nat. Clim. Change 7: 48-52.

Pusch, M., Fiebig, D., Brettar, I., Eisenmann, H., Ellis, B.K., Kaplan, L.A., et al., (1998). The role of micro-organisms in the ecological connectivity of running waters. Freshw. Biol. 40, 453-495.

Rao, B., Liu, Y., Wang, W., Hu, C., Dunhai, L., and Lan, S. (2009). Influence of dew on biomass and photosystem II activity of cyanobacterial crusts in the Hopq Desert, northwest China. Soil Biol. Biochem. 41, 2387-2393.

Rees, G.N., Watson, G.O., Baldwin, D.S., and Mitchell, A.M. (2006). Variability in sediment microbial communities in a semipermanent stream: impact of drought. J. North Am. Benthol. Soc. 25, 370–378. DOI: 10.1899/0887-3593(2006)25[370:vismci]2.0.co;2

Roberson, E.B., and Firestone, M.K. (1992). Relationship between desiccation and exopolysaccharide production in a soil *Pseudomonas* sp. Appl. Environ. Microbiol. 58, 1284–1291.

Romaní, A.M., and Sabater, S., (1997). Metabolism recovery of a stromatolitic biofilm after drought in a Mediterranean stream. Arch. Hydrobiol. 140, 261-271.

Romaní, A. M., and Sabater, S. (2001). Structure and activity of rock and sand biofilms in a Mediterranean stream. Ecology 82, 3232-3245.

Romaní, A.M., Giorgi, A., Acuña, V., and Sabater, S. (2004). The influence of substratum type and nutrient supply on biofilm organic matter utilization in streams. Limnol. Oceanogr. 49, 1713-1721.

Romaní, A.M., Vázquez, E., and Butturini, A., (2006). Microbial availability and size fractionation of dissolved organic carbon after drought in an intermittent stream: biogeochemical link across the stream-riparian interface. Microb. Ecol. 52, 501-512.

Romaní, A.M., Amalfitano, S., Artigas, J., Fazi, S., Sabater, S., Timoner, X., Ylla, I., and Zoppini, A. (2013). Microbial biofilm structure and organic matter use in Mediterranean streams. Hydrobiologia 719, 43-58.

Rosenzweig, R., Shavit, U., and Furman, A., (2012). Water retention curves of biofilm-affected soils using xanthan as an analogue. Soil Sci. Soc. Am. J. 76, 61–69.

Sabater, S., Guasch, H., Ricart, M., Romani, A., Vidal, G., Klunder, C., and Schmitt-Jansen, M., (2007). Monitoring the effect of chemicals on biological communities. The biofilm as an interface. Anal. Bioanal. Chem. 387, 1425–1434.

Sabater, S., Timoner, X., Borrego, C., and Acuña, V. (2016). Stream biofilm responses to flow intermittency: from cells to ecosystems. Front. Environ. Sci. 4, 1–10. DOI: 10.3389/fenvs.2016.00014

Scheff, J., and Frierson, D.M. (2014). Scaling potential evapotranspiration with greenhouse warming. J. Clim. 27, 1539-1558.

Schimel, J., Balser, T.C., and Wallenstein, M. (2007). Microbial stress-response physiology and its implications for ecosystem function. Ecology 88, 1386-1394.

Schimel, J.P., and Schaeffer, S.M. (2012). Microbial control over carbon cycling in soil. Front. Microbiol. 3, 1–11. DOI: 10.3389/fmicb.2012.00348

Schimel, J.P. (2018). Life in dry soils: effects of drought on soil microbial communities and processes. Annu. Rev. Ecol. Evol. Syst. 49, 409-432.

Schwinning, S., and Sala, O.E. (2004). Hierarchy of responses to resource pulses in arid and semi-arid ecosystems. Oecologia 141, 211-220. DOI: 10.1007/s00442-004-1520-8

Schwinning, S., Sala, O.E., Loik, M.E., and Ehleringer, J.R. (2004). Thresholds, memory, and seasonality: understanding pulse dynamics in arid/semi-arid ecosystems. Oecologia 141, 191-193. DOI: 10.1007/s00442-004-1683-3

Shade, A., Peter, H., Allison, S., Baho, D., and Berga, M. (2012). Fundamentals of microbial community resistance and resilience. Front. Microbiol. 3, 417.

Sherwood, S., and Fu, Q. (2014). A drier future? Science 343, 737-739.

Shumilova, O., Zak, D., Datry, T., von Schiller, D., Corti, R., Foulquier, A., Obrador, B., Tockner, K., Allan, D.C., et al. (2019). Simulating rewetting events in intermittent rivers and ephemeral streams: A global analysis of leached nutrients and organic matter. Glob. Chang. Biol. 25, 1591–1611. DOI: 10.1111/gcb.14537

Sinsabaugh, R.L. (2010). Phenol oxidase, peroxidase and organic matter dynamics of soil. Soil Biol. Biochem. 42, 391–404. DOI: 10.1016/j.soilbio.2009.10.014

Stacy, E.M., Hart, S.C., Hunsaker, C.T., Johnson, D.W., and Berhe, A.A. (2015). Soil carbon and nitrogen erosion in forested catchments: implications for erosion-induced terrestrial carbon sequestration. Biogeosciences 12, 4861–4874.

Steward, A.L. (2012). When the river runs dry: the ecology of dry river beds. Thesis (Queensland, Australia: Griffith University).

Thorp, J.H., Thoms, M.C., and Delong, M.D. (2006). The riverine ecosystem synthesis: Biocomplexity in river networks across space and time. River Res. Appl. 22, 123–147. DOI: 10.1002/rra.901

Timoner, X., Borrego, C.M., Acuña, V., and Sabater, S. (2014). The dynamics of biofilm bacterial communities is driven by flow wax and wane in a temporary stream. Limnol. Oceanogr. 59, 2057-2067.

Todman, L., Fraser, F., Corstanje, R., Deeks, L., Harris, J.A., Pawlett, M. et al. (2016). Defining and quantifying the resilience of responses to disturbance: a conceptual and modelling approach from soil science. Sci. Rep. 6, 28426.

Totsche, K.U., Amelung, W., Gerzabek, M.H., Guggenberger, G., Klumpp, E., Knief, C., Lehndorff, E., Mikutta, R., Peth, S., Prechtel, A., Ray, N., and Kögel-Knabner, I., (2017). Microaggregates in soils. J. Plant Nutr. Soil Sci. 181, 104–136.

Trenberth, K.E., Dai, A., van der Schrier, G., Jones, P.D., Barichivich, J., Briffa, K.R., and Sheffield, J. (2014). Global warming and changes in drought. Nat. Clim. Change 4, 17-22.

Unger, S., Máguas, C., Pereira, J.S., David, T.S., and Werner, C. (2010). The influence of precipitation pulses on soil respiration–Assessing the "Birch effect" by stable carbon isotopes. Soil Biol. Biochem. 42, 1800-1810.

Van Gestel, N., Shi, Z., Van Groenigen, K.J., Osenberg, C.W., Andresen, L.C., Dukes, J.S. et al. (2018). Predicting soil carbon loss with warming. Nature 554, E4.

Waldrop, M., and Firestone, M. (2006). Response of microbial community composition and function to soil climate change. Microb. Ecol. 52, 716-724.

Wang, W., Wang, Y., Shu, X., and Zhang, Q. (2013). Physiological responses of soil crust-forming cyanobacteria to diurnal temperature variation. J. Basic Microbiol. 53, 72-80.

Wood, J.M. (2015). Bacterial responses to osmotic challenges. J. Gen. Physiol. 145, 381-388.

Ylla, I., Sanpera-Calbet, Vázquez, E., Romaní, A.M., Muñoz, I., Butturini, A., et al., (2010). Organic matter availability during pre- and post-drought periods in a Mediterranean stream. Hydrobiologia 657, 217-232.

Yuan, B.C., Li, Z.Z., Liu, H., Gao, M., and Zhan, Y.Y. (2007). Microbial biomass and activity in salt affected soils under arid conditions. Appl. Soil Ecol. 35, 319–328.

Yuste, J.C., Peñuelas, J., Estiarte, M., Garcia-Mas, J., Mattana, S., Ogaya, R., Pujol, M., and Sardans, J. (2011). Drought-resistant fungi control soil organic matter decomposition and its response to temperature. Glob. Chang. Biol. 17, 1475–1486. DOI: 10.1111/j.1365-2486.2010.02300.x

Zeglin, L.H. (2015). Stream microbial diversity in response to environmental changes: review and synthesis of existing research. Front. Microbiol. 6, 454. DOI: 10.3389/fmicb.2015.00454

Zoppini, A., and Marxsen, J. (2011). Importance of extracellular enzymes for biogeochemical processes in temporary river sediments during fluctuating dry–wet conditions. In Soil Enzymology, G. Shukla and A. Varma, eds. (Heidelberg, Germany: Springer). pp. 103–117. DOI: 10.1007/978-3-642-14225-3_6

Zribi, W., Aragüés, R., Medina, E., and Faci, J.M. (2015). Efficiency of inorganic and organic mulching materials for soil evaporation control. Soil Tillage Res. 148 (Supplement C), 40–45.

Chapter 13

Groundwater Microbial Communities in Times of Climate Change

Alice Retter, Clemens Karwautz and Christian Griebler*

University of Vienna, Department of Functional & Evolutionary Ecology, Althanstrasse 14, 1090 Vienna, Austria; * corresponding author

Email: alice.retter@univie.ac.at, clemens.karwautz@univie.ac.at, christian.griebler@univie.ac.at

DOI: https://doi.org/10.21775/9781913652579.13

Abstract

Climate change has a massive impact on the global water cycle. Subsurface ecosystems, the earth largest reservoir of liquid freshwater, currently experience a significant increase in temperature and serious consequences from extreme hydrological events. Extended droughts as well as heavy rains and floods have measurable impacts on groundwater quality and availability. In addition, the growing water demand puts increasing pressure on the already vulnerable groundwater ecosystems. Global change induces undesired dynamics in the typically nutrient and energy poor aquifers that are home to a diverse and specialized microbiome and fauna. Current and future changes in subsurface environmental conditions, without doubt, alter the composition of communities, as well as important ecosystem functions, for instance the cycling of elements such as carbon and nitrogen. A key role is played by the microbes. Understanding the interplay of biotic and abiotic drivers in subterranean ecosystems is required to anticipate future effects of climate change on groundwater resources and habitats. This chapter summarizes potential threats to groundwater ecosystems with emphasis on climate change and the microbial world down below our feet in the water saturated subsurface.

Introduction

Groundwater ecosystems contain 97 % of the non-frozen freshwater resources and as such provide an important water supply for irrigation of agricultural land, industrial use (e.g. cooling agent), as well as for production of potable water. Worth mentioning, in Europe, around 50 – 70 % of all drinking water stems from groundwater (Zektser and Everett, 2004). Groundwater constitutes a major component of the hydrological cycle and sustains streams, lakes, and wetlands, many of which would not be perennial without the direct connection to an aquifer. Groundwater ecosystems are typically covered by vegetated soil and vadose sediment layers of varying thickness. Where the protective layers are thin and perforated (e.g. in mountainous areas and karstic rock) or even absent, and where groundwater reaches land surface (e.g. wetlands and springs), the down below systems are highly vulnerable to disturbance from above. Moreover, due to the comparably long residence time of groundwater in the subsurface (Danielopol et al., 2003), its response time to external impacts can be delayed and sometimes masked by complex hydrologically patterns (Alley et al., 2002; Kløve et al., 2014). The subsurface is a naturally light deprived and nutrient limited environment, hence the energy required for sustaining groundwater ecosystems is largely derived from the surface. In fact, the groundwater ecosystems balance is susceptible to several external influences. It is on one hand largely dependent on energy import from the surface, but on the other hand responding sensitively to increased input of organics and nutrients as well as various types of contaminants including heavy metals and heat. While in the past, groundwater pollution mainly resulted from source contaminations with deposited or spilled petroleum hydrocarbons and halogenated solvents as well as leaking landfills, modern impacts to groundwater ecosystems are numerous. A major threat still (and ongoing) comes from intensive land use in terms of agriculture and urbanization (Saccò et al., 2019a). Application of fertilizers and pesticides to agricultural land is a major issue besides surface sealing, down below infrastructure and geothermal energy use in big cities. Accelerated import of organic matter, nutrients and chemicals cause eutrophication as well as acute and chronic toxicological effects. In urban areas, a groundwater warming constitutes a typical phenomenon.

Related to wastewater discharge, a long list of emerging pollutants including pharmaceuticals, anticorrosion agents, artificial sweeteners, drugs, or sanitary products comprise a future challenge not only for groundwater. Finally, the application of manure to arable land and the infiltration of wastewater influenced surface water occasionally lead to groundwater contamination with biological agents such as human pathogenic germs and viruses (Krauss and Griebler, 2011). Not

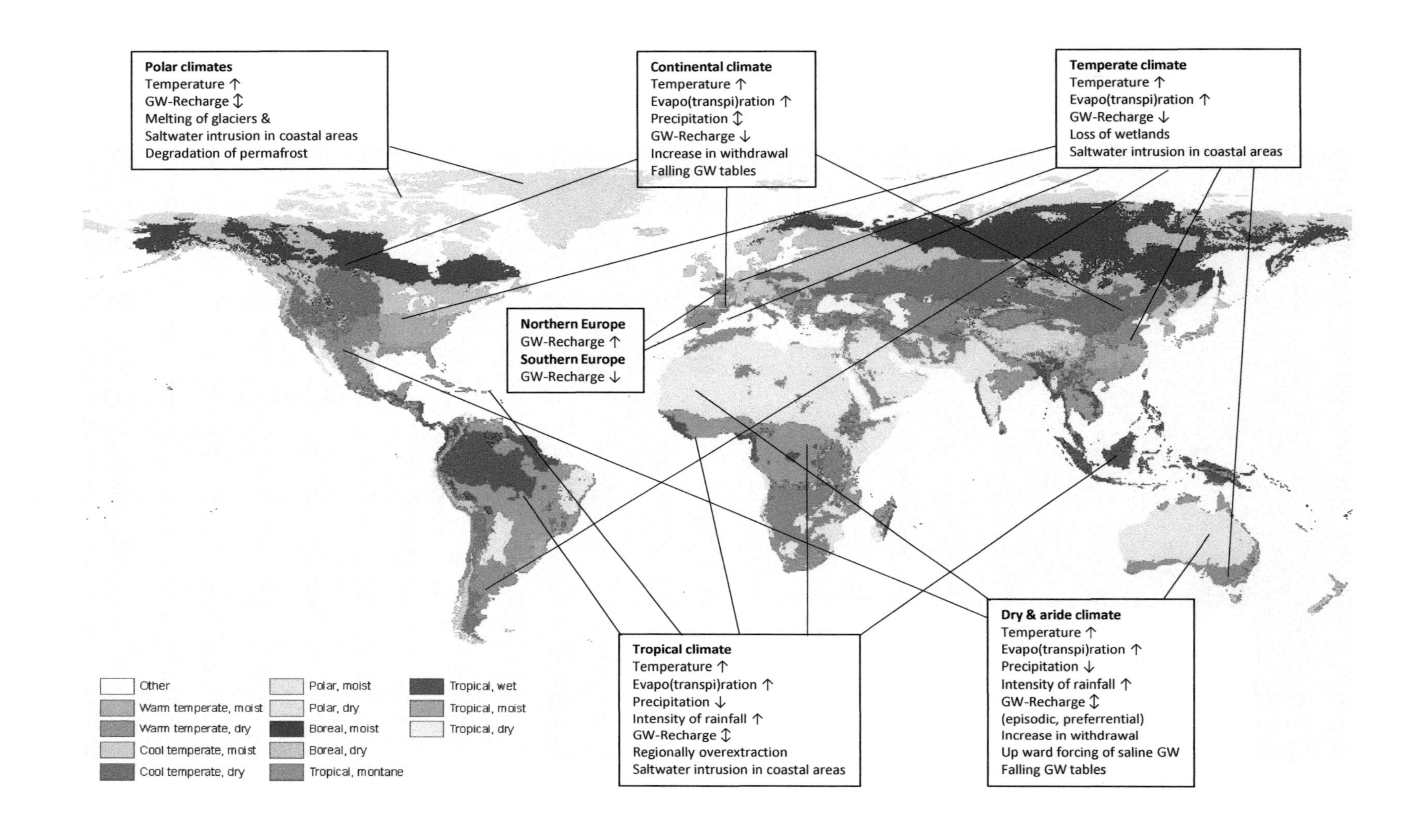

Figure 13.1 Major hydrological and meteorological effects with relation to groundwater quantity and quality caused by climate change. Information compiled mainly from Treidel et al. (2011). Map of world climate zones from https://ec.europa.eu/jrc/en/.

surprisingly, also excessive water withdrawal constitutes a major threat to groundwater ecosystems. In this context, it is timely to evaluate the possible effects of climate change to the groundwater microbiome.

Climate change impacts on groundwater ecosystems

Abiotic effects of global warming

Climate change and its consequences for global warming present one of the biggest environmental and societal challenges of our time. The pace at which it is advancing varies across ecosystems in different parts of the world and is determined to a large extent by geographic location and atmospheric chemistry e.g. greenhouse gas concentrations (Yvon-Durocher et al., 2010). The changes in seawater levels, climate, and environmental conditions alter the recharge, flow direction, storage, and discharge capacities of groundwater (Edmunds and Milne, 2001). Current climate change models predict more extreme weather events in the future that will directly affect the hydrological dynamics of groundwater (Alley at el., 2002). Fig. 13.1 compiles available information on predicted changes in precipitation and groundwater recharge for the world's major climate zones.

Some consequences of climate change already appear globally (e.g. increase in mean air temperature) while others show up geographically distinct or act only on a local scale. Prominent effects include increasing frequency of extreme hydrological events, i.e. droughts and floods, and heavy rainfall, increasing evapo(transpi)ration rates, vertical expanding of the vadose zone, horizontal expanding of dry areas including modified soil properties, decline of groundwater tables, rising sea levels and salt water intrusion, loss of glaciers as water stores, among others (Fig. 13.2) (Bates et al., 2008; Treidel et al., 2011; Richts and Vrba, 2016; Bastin et al., 2017; Betts et al., 2018). And without doubt, global warming also affects groundwater temperatures on a regional to global scale (Fig. 13.3).

Currently, it is difficult to provide quantitative predictions to which extent climate change is modifying the frequency and intensity of these events and how this directly and indirectly feeds back on groundwater quantity and quality, and in consequence on microbial processes and the sequestration and cycling of matter in the subsurface. This leaves huge uncertainties to the full impact of climate change on groundwater ecosystems (Boulton, 2005). In fact, most recent studies indicate an overall deteriorating trend with respect to groundwater quantity and quality, with only few exceptions; i.e. increased groundwater recharge in northern regions of Europe

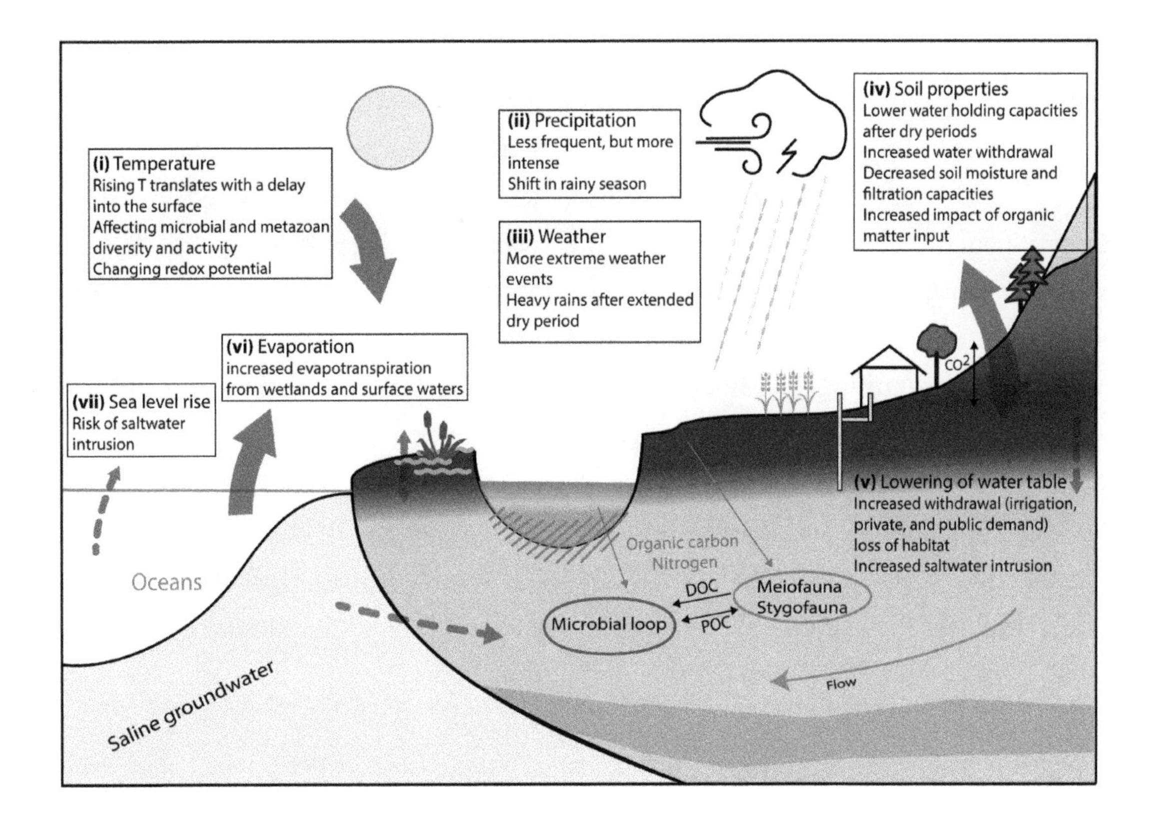

Figure 13.2 Cross-section scheme highlighting impacts of climate change on the hydrological cycle with regards to the groundwater ecosystems, water tables, physicochemical states, as well as groundwater communities and food webs.

and partly in regions where ice stocks and permafrost is melting (e.g. polar regions) or some dry and arid regions that may receive intensified rain falls (Fig. 13.1; Treidel et al., 2011).

Groundwater aquifers, as well as connected surface ecosystems (i.e. groundwater associated aquatic ecosystems [GWAAES] and groundwater dependent terrestrial ecosystems [GWDTES]) are believed to be highly vulnerable to consequences of climate change and related pressures (Nathan and Evans, 2011; Barthel and Banzhaf, 2016; Jasperson et al., 2018). With the GWAAES and the GWDTES it is mainly quantitative issues, i.e. the temporal disconnection to groundwater that matters besides consequences from altered groundwater quality that may affect the connected surface ecosystems. The vulnerability of groundwater ecosystems is related to various basic conditions. First, their low productivity goes hand in hand with a low resilience to disturbance. Second, any negative impact that traced into an aquifer stays there for long, because water residence times in the subsurface may be orders of magnitude higher than in surface waters (Danielopol et al. 2003). Third, groundwater ecosystems are typically characterized by stable environmental conditions (e.g. temperature, water chemistry, oligotrophy) all year round. In such systems, minor disturbances can have major effects on the well-adapted communities therein and the processes they mediate.

Global warming is influencing the various subterranean habitats at different spatial and temporal scales. Depending on groundwater depth, changes in the annual mean air temperature will translate into the subsurface with a time lag of years to decades compared to surface habitats, and daily or yearly fluctuations are usually attenuated (Fig. 13.3) (Menberg et al., 2014; Mammola, 2019 and citations therein). Accordingly, shallow groundwater aquifers are most exposed to external impacts and

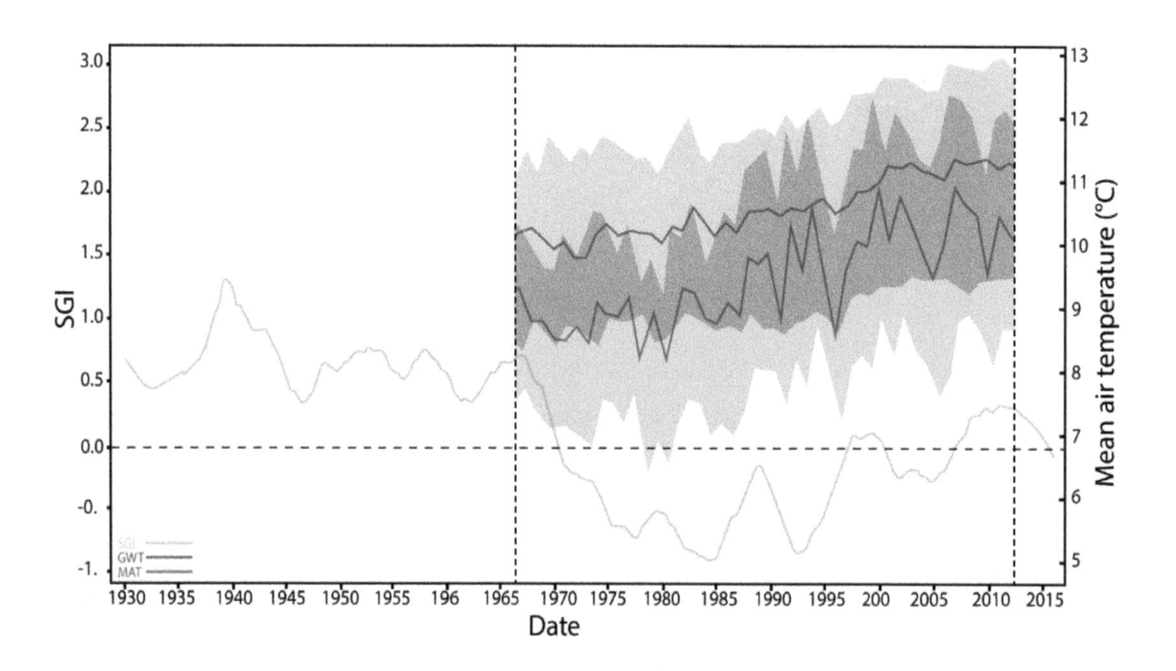

Figure 13.3 An exemplary time series from Austrian depicting trends in mean surface air temperature (MAT) and mean groundwater temperatures (GWT) (modified from Benz et al., 2018), as well as the trend of the standardized groundwater index (SGI) for Austrian groundwater (modified from Haas and Birk, 2019), showing the deviation from long-term average water levels, where negative values (<0.0) indicate declining groundwater levels; so called groundwater droughts (Uddameri et al., 2019). The general deviation between MAT and GWT relates to temperature measurements mainly in very shallow groundwater (see Benz et al., 2018).

are expected to respond more rapidly to changing temperatures than deeper aquifers. Especially in highly urbanized and agricultural areas and areas of low vegetation cover, shallow groundwater temperatures can be expected to rise even within a relatively short period of time (Henriksen and Kirkhusmo, 2000; Menberg et al., 2014; Kurylyk et al., 2015). Elevated subsurface temperatures directly alter the physical- and chemical properties of the groundwater, by lowering the solubility of gases (e.g. dissolved oxygen), as well as leading to a higher variation in dissolved organic carbon (DOC) concentrations and increased mineralization rates (Bonte et al., 2013). Additionally, a change in temperature can alter the organic matter (OM) composition and quantity of above-ground OM sources (e.g., surface flow-paths, the riparian zone, soils) that subsequently enter the groundwater via seepage water (Griebler et al., 2016; Stegen et al., 2016; Moldovan et al., 2018).

Indeed, climate change may affect the composition and quantity of OM ultimately reaching the groundwater table. As depicted in Figs. 13.1 and 13.2, changes in spatiotemporal precipitation dynamics, soil properties and groundwater recharge initiate shifts in the vadose zone thickness, which in turn influences the passage of OM, nutrients, but also contaminants into the aquifer (Ward et al., 2017). Thus, the OM from the surface dynamically entering during recharge events potentially alters the composition and activity of groundwater microbial communities on different temporal scales (Hofmann and Griebler, 2018; Benk et al., 2019). The same applies to allochthonous microorganisms introduced into the subsurface during recharge, possibly invading and re-assembling indigenous microbial communities, or even acting as a source for the groundwater microbiome (Fillinger et al., 2019; Yan et al., 2020).

The soil layer acts as a natural filter reducing or even impeding the seepage of contaminants from the surface into shallow groundwater. Extreme rainfall events have been linked to impaired groundwater quality and waterborne diseases (Ebi et al., 2017). Environmental factors such as the geographical region, land use or season do modify these effects (Guzman Herrador et al., 2015; Sonthiphand et al., 2019). Rain events can mobilize antibiotic resistant microbes from soils increasing their levels in riverine microbial communities and potentially groundwater (Di Cesare et al., 2017).

The groundwater is intricately linked to other components of the hydrological cycle through refined flow- and recharge dynamics (Havril et al., 2018). Climate change related modifications in hydrological settings thus will in some areas lead to less, but more intense carbon, nutrient, as well as contaminant inputs into the groundwater (Kaushal et al., 2014). At the same time, an associated change in groundwater table depth cause the deterioration of groundwater and linked freshwater ecosystems. This is already leading to altered carbon source, sink, and carbon cycle feedback behavior in some parts of the world (Mander et al., 2015; Dhillon et al., 2019). Besides the fast disappearance of wetlands that has been going on for centuries, hydrological dynamics of GWDTEs are dramatically changing, affecting their size and permanence. For instance, there is an ongoing trend of peatland desiccation (Swindles et al., 2019), as well as an increase of surface water areas elsewhere, for example high altitude lakes (Moore and Knowles, 1989; Zhang et al., 2017; Davidson et al., 2019), partly a leftover from disappearing glaciers.

Drought periods are expected to increase. Extended dry periods lead to desiccated surface soils which entail two main effects in terms of groundwater recharge. On one hand, dry soils have a decreased capacity to take up and hold water. In case of heavy precipitation, a large proportion of rainwater is thus directly lost from recharge by surface run-off. Simultaneously, dry cracks in desiccated soil allow some of the freshly precipitated water together with material collected at the surface to infiltrate fast via preferential flow paths, missing the natural attenuation during slow sediment passage. Thus, the decreased water retention capacity of soils and by that associated formation of cracks could lead to stronger seepage water pulses which were shown to be an important driver in cell transport and microbial community assembly in groundwater (Yan et al., 2020).

A temperature related rise in sea-level, while mostly unrecognized, will lead to increasing groundwater levels in coastal regions with devastating effects (Colombani et al., 2016). The impact of groundwater salinization can be two-fold. First, with a rise in sea-level, saline groundwater will ingress into subterranean freshwater aquifers, and secondly due to an extended demand and a resulting increase in groundwater withdrawal, a decreasing groundwater table is pulling saline groundwater inland (Treidel et al., 2011). This is of particular concern when it comes to small islands, where fresh groundwater habitats are especially vulnerable due to their confinement by the sea. The groundwater ecosystems therein especially in proximity to shorelines could thus be threatened, and anchialine habitats will probably shift inland or disappear completely (Rotzoll and Fletcher, 2012; Moritsch et al., 2014; Rogers et al., 2019).

Warming of permafrost and associated ground ice melting, as well as englacial temperatures have increased within the last decades (Ding et al., 2019). The melting of arctic permafrost leads to elevated emissions of carbon dioxide and other greenhouse gases that until now have been stored within the arctic soil. Conversely, a study of Voigt et al. (2019) showed that dry permafrost peatlands store methane rather than emitting it under warming scenarios. Without doubt, discontinuous permafrost and melting of glaciers alter the hydrology, chemistry and biology of the underlying groundwater bodies, which may experience a change from confined to unconfined aquifers. The latter will introduce the risk of groundwater pollution in case of existing above ground contaminant sources. Groundwater below glaciers and surface water ecosystems fed by them are influenced by several factors, such as thermal regimes (Ravier and Buoncristiani, 2017), the production of meltwater, and groundwater recharge, which changes when glacier size decreases (Haldorsen et al., 2010). The overall extent of climate change on aquatic habitats in polar regions can so far only be roughly estimated. They seem to harbour a distinct microbiome (Wurzbacher et al., 2017), but further research on biodiversity as well as climate change induced dynamics of carbon- and nutrient in these remote regions is urgently needed.

The groundwater food web

Groundwater ecosystems provide habitats with constant darkness and stable temperatures, narrow space, as well as low carbon, nutrient, and energy levels. In fact, the subterranean biosphere is inhabited by a diverse selection of well adapted organisms, including microbes and metazoan. Because groundwater ecosystems are among the least-studied habitats on earth (Devitt et al., 2019), the majority of the members in subterranean communities, estimated at 50,000 to 100,000 obligate subterranean metazoan species and millions of microbial taxa, still await formal description (Culver and Holsinger, 1992; Griebler and Lueders, 2009). In light of the ongoing trend of global species extinction, there is a great risk of losing species even before their discovery.

Most dominant in groundwater food webs are microbes such as bacteria, archaea, fungi and protozoa - most of them sharing a heterotrophic lifestyle. However, as already stated, groundwater also harbours micro-, meio-, and macrofauna, including a multitude of crustaceans (i.e. copepods, isopods, amphipods, ostracods), worms (oligochaetes, polychaetes, nematodes, platyhelminths), gastropods and mites. In cave waters even salamander or fish are home. Metazoans depend on microbes as food source, although opportunistic predation within their own trophic levels occurs as well (Saccò et al., 2019b). Groundwater fauna often displays highest diversity and abundance at the upper boundaries of aquatic subterranean habitats where they take advantage of increased food availability (Fišer et al., 2014). Groundwater fauna in general is highly adapted to a narrow temperature range. There are exceptions with some taxa that have a relatively broad temperature tolerance, however, always with a much better survival under temperature conditions resembling those of their natural habitats (Eme et al., 2014).

With the absence of major primary producers (i.e. photoautotrophs), groundwater food webs are typically truncated (Gibert and Deharveng, 2002). Trophic interactions appear to be bottom-up regulated, driven by the import of dissolved organic carbon, being important electron donor for microbial redox processes in groundwater (Hofbauer and Griebler, 2018; Saccò et al., 2019b). In situ microbial OM production in the light-deprived subsurface by chemolithoautotrophs is currently not considered a major contribution to the global carbon cycle. However, chemolithoautotrophic bacteria have been shown to account for a large proportion of expressed functional genes in shallow alluvial groundwater and may share functionally redundant metabolic pathways (Jewell et al., 2016). Several studies (Sarbu et al., 1996; Hutchins et al., 2016; Lau et al., 2016; Galassi et al., 2017) underline that locally, i.e. in caves and deep groundwater systems, chemolithoautotrophic production is playing a substantial role in driving trophic complexity and species abundances over long timescales. In any case, due to its huge dimensions, groundwater ecosystems represent a significant storage of global organic carbon, for example, in form of microbial biomass (Whitman et al., 1998; Magnabosco et al., 2018).

Climate change effects to the groundwater microbiome

Much has been said about the 'abiotic' effects of climate change and global warming to groundwater in terms of quantity and physical-chemical quality. But what do these expected and already ongoing changes in salinity, temperature, as well as carbon import to the subsurface mean to groundwater microbial communities? For some of the effects that work on groundwater ecosystems or that will gain importance in the future we have strong evidence, while in other cases we currently can only speculate.

It is well documented that a change in temperature alters microbial metabolic rates and affects microbial community composition. Elevated water temperatures, especially in temperate- and subpolar zones can be expected to influence the generation times and abundance of microorganisms in subsurface- and groundwater ecosystems (Rodó et al., 2013; Karthe, 2015; Shocket et al., 2018). Higher temperatures will speed up microbial mediated processes including the turnover of organic matter and respiration (redox processes). However, key to the expected effects is the magnitude of temperature change and the time scale. A temperature increase by only 1-2°C may not exhibit short-term (acute) changes in metabolic rates at the significance level taking the natural heterogeneity in groundwater systems into account. On a long-term perspective even slightly elevated rates of carbon turnover may exhaust carbon and nutrient pools available to the microbial communities if not replenished in time. There are complex scenarios to be considered, including as examples (1) elevated microbial activities along with decreasing pools of energy or (2) slightly increasing activities along with higher loads of carbon (including contaminants) arriving in pulses. In a field study targeting the influence of groundwater warming on various microbial pattern, it was concluded that in very clean groundwater, i.e. groundwater that lacks labile organic carbon as well as toxic compounds, a rise in temperature does not significantly affect the abundance and growth of prokaryotic cells (Brielmann et al., 2009). When overriding the energy limitation, as has been evaluated in subsequent lab experiments, even moderate temperature changes lead to shifts in microbial growth, biomass and respiration (Brielmann et al., 2011; Griebler et al., 2016). The limitation by an essential nutrient (e.g. phosphorus) may have similar effects. Bacterial cells that have enough bioavailable organic carbon but lack an element to build new biomass continue respiring OM without growth (Hofmann and Griebler, 2018). In consequence, OM is turned over without the production of prokaryotic biomass needed by other members of the (microbial) food web such as protozoa or meiofauna. Indeed, predator - prey interactions have shown to be temperature dependent, and the degree of dependency changes according to taxonomic affiliation, as well as species- and food web traits (Hutchins et al., 2016; Archer et al., 2019). Not to forget, the solubility of gases is also temperature dependent. Even small temperature changes may turn groundwater with already low concentrations of dissolved oxygen into hypoxic or even anoxic habitats. There is ample evidence from lab experiments and small scale field studies that the switch from oxic to anoxic conditions in groundwater triggers a cascade of effects including (1) the onset of anaerobic processes, (2) the accumulation of unwanted solutes in the water (ferrous iron, hydrogen sulfide, methane), and (3) a dramatic shift in community composition. In summary, already small changes in groundwater thermal regimes may lead to a decoupling of balanced food web interactions with consequences to the cycling of elements and the composition and structure of communities. In addition to the direct effects of a changing climate, human activities intended to counteract climate change, such as the sequestration of CO_2 in deep geological layers and the use of geothermal energy as sustainable alternative, pose threats to groundwater communities and ecosystems (Bonte et al., 2013; Griebler et al., 2016; Lawter et al., 2017; Westphal et al., 2017; Oelkers et al., 2019).

As already mentioned briefly, changes in the availability of organic matter will modulate microbial activities and processes. Groundwater microorganisms drive carbon- and nutrient cycling, catalyse the breakdown of contaminants, and modulate soil fertility and consequently global food webs (Cavicchioli et al., 2019). According to the scenario of irregular and more frequent extreme hydrological events, we may assume organic matter import into the subsurface on one hand to pause during extended periods of droughts and, on the other hand, to rapidly increase during times of heavy rain and floods. In the latter case, DOM may enter the subsurface in short-term pulses that cannot be processes spontaneously but are subject to a delayed microbial degradation (Foulquier et al., 2011; Reiss and Schmid-Araya, 2011; Saccò et al., 2019b). New degradation products will cause a reorganization of microbial communities in a compositional and functional manner. For instance, microbial communities in alkaline groundwaters were found to strongly relate to environmental factors such as pH, carbon monoxide, and methane concentrations (Twing et al., 2017). This will also impact the groundwater carbon cycle feedback to global warming (Allison and Martiny, 2009; He et al., 2010). Still, it is largely unclear to what extent groundwater habitats will experience a shift in the carbon sink-to-source ratio, and where to position groundwater within the global carbon cycle (Chapelle et al., 2016). Without doubt, changes in subsurface organic matter turnover will translate into changes in greenhouse gas emissions from the aquifers, a subject that so far received hardly any attention.

Changes in environmental conditions, i.e. temperature, salinity, carbon availability, all have in common that they directly or indirectly influence the composition of microbial communities. Microbial community composition in response to altered environmental conditions is linked to changes in ecosystem functioning. *Vice versa*, increased, or decreased process rates affect microbial diversity (Jesußek et al., 2013, Stegen et al., 2016). It is a paradigm that ecosystems rich in microbial species diversity exhibit a higher stability and resilience when facing disturbances due to a high functional redundancy. Species richness and biodiversity, on the other hand, are somehow connected to the energy available. There is conclusive evidence from several studies conducted in different aquatic environments that changes in organic matter supply in terms of quantity and quality steers shifts in microbial community composition (Shi et al., 1999; Baker et al., 2000; Findlay et al., 2003; Carlson et al., 2004; Judd et al., 2006; Kritzberg et al., 2006; Li et al., 2012; Wu et al., 2018). In particular, if the altered DOM supply lasts for longer than just a couple of hours or days (Herzyk et al., 2014; 2017; Grösbacher et al., 2016). However, we are only beginning to understand whether and how energy–diversity relationships known from macroecology apply to complex natural microbial communities. Diversity–productivity patterns from dozens of natural systems were found either negative (35%), positive (28%) or humped (23%) (Smith, 2007). From experimental studies that supplied sediment microbial communities with DOM, a change in community composition was reported, i.e. an increase in relative abundance of Betaproteobacteria with higher biodegradable dissolved organic carbon (BDOC) concentrations (Li et al., 2012). A lower BDOC was accompanied by a higher relative abundance of Firmicutes, Planctomycetes, and Actinobacteria, a pattern that mirrored our own findings. Overall, a decrease of microbial diversity in terms of richness and Shannon-Wiener diversity with higher feed concentrations was observed (Li et al., 2012; Li et al., 2013). In contrary, a positive relationship between bacterial richness and bioavailable organic matter was observed in Arctic deep-sea sediments, hinting at a positive energy–diversity relationship in oligotrophic environments (Bienhold et al., 2012). In fact, we can be sure that altered dynamics of organic matter input to groundwater systems will affect microbial community composition and mediated processes. The same holds true for nutrients (e.g. from fertilization) and in particular for contaminants. Undisturbed groundwater ecosystems are typically energy poor; they have a comparably low microbial biomass, activity and diversity (Griebler and Lueders, 2009) and, in consequence, are assumed to be less resilient to impacts. In other words, groundwater ecosystems and their communities are highly vulnerable.

Climate change will, without doubt, force us to adapt land use and agricultural practices. Several studies indicate that vegetation, land use and irrigation are important driver of the community composition and activity in shallow groundwater (Stein et al., 2010; Korbel et al., 2013; Iepure et al., 2016). The causality behind these effects is often not fully clear but the stressor 'land use' may be dissected into (1) fertilization and the application of pesticides that lead to an altered import of carbon, nutrients and contaminants, (2) deforestation and clearance of vegetation that alter conditions for groundwater recharge in terms of quantity and quality, (3) irrigation that changes the structure, moisture and water holding capacity of top soil layers and may lead to the accumulation of salts in case groundwater from deep aquifers is used as water source. Moreover, the application of manure to arable land constitutes a source of pathogenic microbes and viruses which with seepage water may enter shallow aquifers. With the serious changes in hydrological conditions and a higher frequency of extreme weather events such as heavy rainfall, there is an increasing risk of a hygienic contamination (Krauss and Griebler, 2011).

There are significant differences in the composition of microbial communities in saltwater and freshwater habitats. While there are individual studies targeting wetland biodiversity in regard to sea-level rise, little research has been directed to groundwater systems so far. In general, freshwater microbial communities and communities in low-salinity habitats are prone to a loss in diversity in case of saltwater intrusion. As shown for wetlands, microbial community shift went along with changes in the carbon biogeochemistry (Franklin et al., 2017; Dang et al., 2019). In tidal wetlands, for example, the altered carbon cycling directly affected nitrogen turnover. A loss in the ability to remove excess nitrogen leads to an excess of nutrients and increased greenhouse gas emissions in these habitats. Even under reversed climate scenarios, it may take years to decades for these ecosystems to rebalance (Franklin et al., 2017; Dang et al., 2019). A shift in soil and sediment salt content was shown to lead to an altered microbial community functioning, which in turn is depending on the duration and frequency of the salinity exposure (Liu et al., 2017; Rath et al., 2019). A study of Edmonds et al. (2009) underline the different time scales associated with changes in microbial community pattern. While a fast change in the functional component of the community with increased salinity was observed, taxonomic composition and metabolic potential on DNA level remain unchanged in response to salt stress. In summary, saltwater intrusion to fresh groundwater reservoirs poses complex changes that can be long lasting in zones with increasing salinity. Although only little is known for groundwater environments, similar effects as observed in other aquatic habitats can be expected to the groundwater biome.

Conclusions and Outlook

Climate change affects groundwater in many ways. On the one hand, it seriously alters the interplay of all important components of the hydrological cycle. Groundwater, at least the shallow portion, is changing in quantity and quality. On the other hand, climate change triggers a more intensive use of groundwater for the production of potable water, irrigation and cooling purposes. Both aspects have direct and indirect impacts to the groundwater communities in terms of composition and functions. This will have several consequences. Ecosystem functions and services mediated by microbes will change. In particular, this will apply to the subterranean carbon cycle. We may further expect that increasing temperatures impair groundwater food webs and effects to microbes will also translate into higher trophic levels.

While there is a multitude of models that try to predict climate change scenarios and related changes in carbon cycling or the loss of local to global biodiversity, groundwater ecosystems have so far been hardly considered. Although a huge amount of carbon is stored in the terrestrial subsurface, groundwater ecosystems are not yet implemented in global scale carbon models. As a first step we suggest to analyse climate variability and change in relation to hydrological, meteorological and ecological patterns on the long term (50-100 years or more) including groundwater systems to identify drivers within the oceanic-terrestrial-atmospheric coupling to understand and to be able to predict spatiotemporal changes in precipitation, evapotranspiration, recharge, discharge and groundwater storage, biogeochemical processes, and the fate of contaminants, be it abiotic or biotic (Treidel et al., 2011). For numerous sites, long-term data series are available that await targeted analysis. We also need to have an eye on the climate change induced dynamics of the vadose zone that highly determines what ends up in the aquifer in terms of energy and matter. We further need to quantify groundwater recharge and discharge and the individual pools of carbon and other elements on different spatial scales. Moreover, information on carbon turnover rates including the production and emission of greenhouse gases from groundwater systems is urgently needed. Later, these details can be implemented into scale dependent groundwater models and coupled to already existing models on carbon cycling and climate change.

Currently, our knowledge about the effects of climate change on groundwater microbial communities and interlinked food webs and cycling of matter, as well as their response to global warming is still very limited. As mentioned above, many groundwater taxa, including microorganisms such as bacteria, archaea, fungi, and protozoa are still lacking a detailed taxonomic description (Nawaz et al., 2018; Mulec and Engel, 2019; Savio et al., 2019). Only a minute portion of microbes could yet be isolated and physiologically studied. Almost all information on the microbes' physiology and biochemistry comes from genome, transcriptome, proteome, and metabolome data. This situation can be illustrated with some examples. A study by Hug and co-workers (2016) discovered novel, unrelated lineages and phyla of bacteria and archaea active in carbon fixation, as well as ammonia- and sulphur oxidation in sediment cores from the deep subsurface. Another study reported a novel lineage of nitrogen fixing bacteria, sister to Cyanobacteria that is living in anoxic zones in the subsurface (Di Rienzi et al., 2013). Also, a study by Anantharaman et al. (2016) found a large number of undescribed bacterial phyla in groundwater which were affiliated to Proteobacteria and Candidate Phyla Radiation (CPR), taking on many already known biogeochemical processes closely linked to carbon, hydrogen, and sulfur cycling, but also partly unknown metabolic pathways. Compared to other aboveground freshwater ecosystems, the knowledge about fungi in groundwater habitats, even though we have evidence that the groundwater in fact hosts numerous fungal taxa, is limited regarding species interactions, taxonomic composition and abundance, and community ecology (Grossart et al., 2019). Consequently, many known functions of the biogeochemical cycles in groundwater cannot be unambiguously linked to a certain species or group of organisms and process rates are entirely missing. As such, we are far from integrating taxonomic and functional knowledge of the groundwater microbiome into climate change models. However, this is essential to estimate the contribution of microorganisms to carbon and nutrient cycling in time and space (Antwis et al., 2017; Amend et al., 2019; Cavicchioli et al., 2019).

In the future, new bioinformatics approaches will help to refine ecological traits of microorganism and their interplay with climate change (Simonsen et al., 2019). A comprehensive and standardized study and sampling design followed by the evaluation of genomic and metadata is asked for to allow meta-analysis (Field et al., 2008). These methods need to deal with the environmental complexity and interspecies interactions that often interfere with climate-associated patterns which become apparent on the large scale (Simonsen et al., 2019).

References

Alley, W.M., Healy, R.W., LaBaugh, J.W., and Reilly, T.E. (2002). Flow and storage in groundwater systems. Science, *296*, 1985–1990. DOI: 10.1126/science.1067123

Allison, S.D., and Martiny, J.B.H. (2009). Resistance, resilience, and redundancy in microbial communities. Light Evol. *2*, 149–166. DOI: 10.17226/12501

Amend, A., Burgaud, G., Cunliffe, M., Edgcomb, V.P., Ettinger, C.L., Gutiérrez, M.H., Heitman, J., Hom, E.F.Y., Ianiri, G., Jones, A.C., et al. (2019). Fungi in the marine environment: Open questions and unsolved problems. MBio *10*, 1–15. DOI: 10.1128/mBio.01189-18

Anantharaman, K., Brown, C.T., Hug, L.A., Sharon, I., Castelle, C.J., Probst, A.J., Thomas, B.C., Singh, A., Wilkins, M.J., Karaoz, U., Brodie, E.L., Williams, K.H., Hubbard, S.S., and Banfield, J.F. (2016). Thousands of microbial genomes shed light on interconnected biogeochemical processes in an aquifer system. Nat. Commun. *7*, 1–11. DOI: 10.1038/ncomms13219

Antwis, R.E., Griffiths, S.M., Harrison, X.A., Aranega-Bou, P., Arce, A., Bettridge, A.S., Brailsford, F.L., de Menezes, A., Devaynes, A., Forbes, K.M., et al. (2017). Fifty important research questions in microbial ecology. FEMS Microbiol. Ecol. *93*, 1–10. DOI: 10.1093/femsec/fix044

Archer, L.C., Sohlström, E.H., Gallo, B., Jochum, M., Woodward, G., Kordas, R.L., Rall, B.C., and O'Gorman, E.J. (2019). Consistent temperature dependence of functional response parameters and their use in predicting population abundance. J. Anim. Ecol. *88*, 1670–1683. DOI: 10.1111/1365-2656.13060

Baker, M.A., Valett, H.M., and Dahm, C.N. (2000). Organic carbon supply and metabolism in a shallow groundwater ecosystem. Ecology *81*, 3133–3148. DOI: 10.1890/0012-9658(2000)081[3133:OCSAMI]2.0.CO;2

Barthel, R., and Banzhaf, S. (2016). Groundwater and Surface Water Interaction at the Regional-scale – A Review with Focus on Regional Integrated Models. Water Resour. Manag. *30*, 1–32. DOI: 10.1007/s11269-015-1163-z

Bastin, J.-F., Berrahmouni, N., Grainger, A., Maniatis, D., Mollicone, D., Moore, R., Patriarca, C., Picard, N., Sparrow, B., Abraham, E.M., et al. (2017). The extent of forest in dryland biomes. Science, *356*, 635–638. DOI: 10.1126/science.aam6527

Bates, B., Kundzewicz, Z.W., Wu, S. and Palutikof, J.P. (2008). Climate Change and Water: Technical Paper of the Intergovernmental Panel on Climate Change. Geological Society special publication No. 189, (Geneva: IPCC Secretariat).

Benk, S.A., Yan, L., Lehmann, R., Roth, V.N., Schwab, V.F., Totsche, K.U., Küsel, K., and Gleixner, G. (2019). Fueling Diversity in the Subsurface: Composition and Age of Dissolved Organic Matter in the Critical Zone. Front. Earth Sci. *7*, 1–12. DOI: 10.3389/feart.2019.00296

Benz, S.A., Bayer, P., Winkler, G., and Blum, P. (2018). Recent trends of groundwater temperatures in Austria. Hydrol. Earth Syst. Sci. *22*, 3143–3154. DOI: 10.5194/hess-22-3143-2018

Betts, R.A., Alfieri, L., Bradshaw, C., Caesar, J., Feyen, L., Friedlingstein, P., Gohar, L., Koutroulis, A., Lewis, K., Morfopoulos, C., Papadimitriou, L., Richardson, K.J., Tsanis, I., and Wyser, K. (2018). Changes in climate extremes, fresh water availability and vulnerability to food insecurity projected at 1.5°C and 2°C global warming with a higher-resolution global climate model. Philos. Trans. R. Soc. A Math. Phys. Eng. Sci. *376*. DOI: 10.1098/rsta.2016.0452

Bienhold, C., Boetius, A., and Ramette, A. (2012). The energy-diversity relationship of complex bacterial communities in Arctic deep-sea sediments. ISME J. *6*, 724–732. DOI: 10.1038/ismej.2011.140

Bonte, M., Röling, W.F.M., Zaura, E., Van Der Wielen, P.W.J.J., Stuyfzand, P.J., and Van Breukelen, B.M. (2013). Impacts of shallow geothermal energy production on redox processes and microbial communities. Environ. Sci. Technol. *47*, 14476–14484. DOI: 10.1021/es4030244

Boulton, A.J. (2005). Chances and challenges in the conservation of groundwaters and their dependent ecosystems. Aquat. Conserv. Mar. Freshw. Ecosyst. *15*, 319–323. DOI: 10.1002/aqc.712

Brielmann, H., Griebler, C., Schmidt, S.I., Michel, R., and Lueders, T. (2009). Effects of thermal energy discharge on shallow groundwater ecosystems. FEMS Microbiol. Ecol. *68*, 273–286. DOI: 10.1111/j.1574-6941.2009.00674.x

Brielmann, H., Lueders, T., Schreglmann, K., Ferraro, F., Avramov, M., Hammerl, V., Blum, P., Bayer, P., and Griebler, C. (2011). Oberflächennahe Geothermie und ihre potenziellen Auswirkungen auf Grundwasserökosysteme. Grundwasser, *16*, 77–91. DOI: 10.1007/s00767-011-0166-9

Carlson, C.A., Giovannoni, S.J., Hansell, D.A., Goldberg, S.J., Parsons, R., and Vergin, K. (2004). Interactions among dissolved organic carbon, microbial processes, and community structure in the mesopelagic zone of the northwestern Sargasso Sea. Limnol. Oceanogr. *49*, 1073–1083. DOI: 10.4319/lo.2004.49.4.1073

Cavicchioli, R., Ripple, W.J., Timmis, K.N., Azam, F., Bakken, L.R., Baylis, M., Behrenfeld, M.J., Boetius, A., Boyd, P.W., Classen, A.T., et al. (2019). Scientists' warning to humanity: microorganisms and climate change. Nat. Rev. Microbiol. *17*, 569–586. DOI: 10.1038/s41579-019-0222-5

Chapelle, F.H., Shen, Y., Strom, E.W., and Benner, R. (2016). The removal kinetics of dissolved organic matter and the optical clarity of groundwater. Hydrogeol. J. *24*, 1413–1422. DOI: 10.1007/s10040-016-1406-y

Colombani, N., Osti, A., Volta, G., and Mastrocicco, M. (2016). Impact of Climate Change on Salinization of Coastal Water Resources. Water Resour. Manag. *30*, 2483–2496. DOI: 10.1007/s11269-016-1292-z

Culver, D.C., and Holsinger, J.R. (1992). How many species of troglobites are there. Natl Speleol. Soc. Bull. *54*, 79-80.

Dang, C., Morrissey, E.M., Neubauer, S.C., and Franklin, R.B. (2019). Novel microbial community composition and carbon biogeochemistry emerge over time following saltwater intrusion in wetlands. Glob. Chang. Biol. *25*, 549–561. DOI: 10.1111/gcb.14486

Danielopol, D.L., Griebler, C., Gunatilaka, A., and Notenboom, J. (2003). Present state and future prospects for groundwater ecosystems. Environ. Conserv. *30*, 104–130. DOI: 10.1017/S0376892903000109

Davidson, S.J., Strack, M., Bourbonniere, R.A., and Waddington, J.M. (2019). Controls on soil carbon dioxide and methane fluxes from a peat swamp vary by hydrogeomorphic setting. Ecohydrology *12*, 1–8. DOI: 10.1002/eco.2162

Devitt, T.J., Wright, A.M., Cannatella, D.C., and Hillis, D.M. (2019). Species delimitation in endangered groundwater salamanders: Implications for aquifer management and biodiversity conservation. Proc. Natl. Acad. Sci. U. S. A. *116*, 2624–2633. DOI: 10.1073/pnas.1815014116

Dhillon, M.S., Kaur, S., and Aggarwal, R. (2019). Delineation of critical regions for mitigation of carbon emissions due to groundwater pumping in central Punjab. Groundw. Sustain. Dev. *8*, 302–308. DOI: 10.1016/j.gsd.2018.11.010

Di Cesare, A., Eckert, E.M., Rogora, M., and Corno, G. (2017). Rainfall increases the abundance of antibiotic resistance genes within a riverine microbial community. Environ. Pollut. *226*, 473–478. DOI: 10.1016/j.envpol.2017.04.036

Di Rienzi, S.C., Sharon, I., Wrighton, K.C., Koren, O., Hug, L.A., Thomas, B.C., Goodrich, J.K., Bell, J.T., Spector, T.D., Banfield, J.F., et al. (2013). The human gut and groundwater harbor non-photosynthetic bacteria belonging to a new candidate phylum sibling to Cyanobacteria. Elife *2*, 1–26. DOI: 10.7554/elife.01102

Ebi, K.L., Ogden, N.H., Semenza, J.C., and Woodward, A. (2017). Detecting and attributing health burdens to climate change. Environ. Health Perspect. *125*, 1–8. DOI: 10.1289/EHP1509

Edmonds, J.W., Weston, N.B., Joye, S.B., Mou, X., and Moran, M.A. (2009). Microbial community response to seawater amendment in low-salinity tidal sediments. Microb. Ecol. *58*, 558–568. DOI: 10.1007/s00248-009-9556-2

Edmunds, W.M., and Milne, C.J. (2001). Palaeowaters in Coastal Europe: Evolution of Groundwater since the Late Pleistocene (London: Geological Society of London). DOI: 10.1144/GSL.SP.2001.189

Eme, D., Malard, F., Colson-Proch, C., Jean, P., Calvignac, S., Konecny-Dupré, L., Hervant, F., and Douady, C.J. (2014). Integrating phylogeography, physiology and habitat modelling to explore species range determinants. J. Biogeogr. *41*, 687–699. DOI: 10.1111/jbi.12237

Field, D., Garrity, G., Gray, T., Morrison, N., Selengut, J., Sterk, P., Tatusova, T., Thomson, N., Allen, M.J., Angiuoli, S.V, et al. (2008). The minimum information about a genome. Nat. Biotechnol. *26*, 541–547. DOI: 10.1038/1360

Fillinger, L., Hug, K., and Griebler, C. (2019). Selection imposed by local environmental conditions drives differences in microbial community composition across geographically distinct groundwater aquifers. FEMS Microbiol. Ecol. *95*, 1–12. DOI: 10.1093/femsec/fiz160

Findlay, S.E.G., Sinsabaugh, R.L., Sobczak, W. V, and Hoostal, M. (2003). Metabolic and structural response of hyporheic microbial communities to variations in supply of dissolved organic matter. Limnol. Oceanogr. *48*, 1608–1617. DOI: 10.4319/lo. 2003.48.4.1608

Fišer, C., Pipan, T., and Culver, D.C. (2014). The vertical extent of groundwater metazoans: An ecological and evolutionary perspective. Bioscience *64*, 971–979. DOI: 10.1093/biosci/biu148

Foulquier, A., Malard, F., Mermillod-Blondin, F., Montuelle, B., Dolédec, S., Volat, B., and Gibert, J. (2011). Surface Water Linkages Regulate Trophic Interactions in a Groundwater Food Web. Ecosystems *14*, 1339–1353. DOI: 10.1007/s10021-011-9484-0

Franklin, R.B., Morrissey, E.M., and Morina, J.C. (2017). Changes in abundance and community structure of nitrate-reducing bacteria along a salinity gradient in tidal wetlands. Pedobiologia (Jena). *60*, 21–26. DOI: 10.1016/j.pedobi.2016.12.002

Griebler, C., and Lueders, T. (2009). Microbial biodiversity in groundwater ecosystems. Freshw. Biol. *54*, 649–677. DOI: 10.1111/j.1365-2427.2008.02013.x

Griebler, C., Brielmann, H., Haberer, C.M., Kaschuba, S., Kellermann, C., Stumpp, C., Hegler, F., Kuntz, D., Walker-Hertkorn, S., and Lueders, T. (2016). Potential impacts of geothermal energy use and storage of heat on groundwater quality, biodiversity, and ecosystem processes. Environ. Earth Sci. *75*, 1–18. DOI: 10.1007/s12665-016-6207-z

Grossart, H.P., Van den Wyngaert, S., Kagami, M., Wurzbacher, C., Cunliffe, M., and Rojas-Jimenez, K. (2019). Fungi in aquatic ecosystems. Nat. Rev. Microbiol. *17*, 339–354. DOI: 10.1038/s41579-019-0175-8

Grösbacher, M., Spicher, C., Bayer, A., Obst, M., Karwautz, C., Pilloni, G., Wachsmann, M., Scherb, H., and Griebler, C. (2015). Organic contamination versus mineral properties: Competing selective forces shaping bacterial community assembly in aquifer sediments. Aquat. Microb. Ecol. *76*, 243–255. DOI: 10.3354/ame01781

Guzman Herrador, B.R., De Blasio, B.F., MacDonald, E., Nichols, G., Sudre, B., Vold, L., Semenza, J.C., and Nygård, K. (2015). Analytical studies assessing the association between extreme precipitation or temperature and drinking water-related waterborne infections: A review. Environ. Heal. A Glob. Access Sci. Source *14*. DOI: 10.1186/s12940-015-0014-y

Haas, J.C., and Birk, S. (2019). Trends in Austrian groundwater – Climate or human impact? J. Hydrol. Reg. Stud. *22*. DOI: 10.1016/j.ejrh.2019.100597

Haldorsen, S., Heim, M., Dale, B., Landvik, J.Y., van der Ploeg, M., Leijnse, A., Salvigsen, O., Hagen, J.O., and Banks, D. (2010). Sensitivity to long-term climate change of subpermafrost groundwater systems in Svalbard. Quat. Res. *73*, 393–402. DOI: 10.1016/j.yqres.2009.11.002

He, Z., Xu, M., Deng, Y., Kang, S., Kellogg, L., Wu, L., Van Nostrand, J.D., Hobbie, S.E., Reich, P.B., and Zhou, J. (2010). Metagenomic analysis reveals a marked divergence in the structure of belowground microbial communities at elevated CO2. Ecol. Lett. *13*, 564–575. DOI: 10.1111/j.1461-0248.2010.01453.x

Henriksen, A., and Kirkhusmo, L.A. (2000). Effect of clear-cutting of forest on the chemistry of a shallow groundwater aquifer in southern Norway.

Herzyk, A., Maloszewski, P., Qiu, S., Elsner, M., and Griebler, C. (2014). Intrinsic potential for immediate biodegradation of toluene in a pristine, energy-limited aquifer. Biodegradation *25*, 325–336. DOI: 10.1007/s10532-013-9663-0

Hofmann, R., and Griebler, C. (2018). DOM and bacterial growth efficiency in oligotrophic groundwater: Absence of priming and co-limitation by organic carbon and phosphorus. Mol. Plant-Microbe Interact. *31*, 311–322. DOI: 10.3354/ame01862

Hug, L.A., Thomas, B.C., Sharon, I., Brown, C.T., Sharma, R., Hettich, R.L., Wilkins, M.J., Williams, K.H., Singh, A., and Banfield, J.F. (2016). Critical biogeochemical functions in the subsurface are associated with bacteria from new phyla and little studied lineages. Environ. Microbiol. *18*, 159–173. DOI: 10.1111/1462-2920.12930

Hutchins, B.T., Engel, A.S., Nowlin, W.H., Schwartz, B.F., Hutchins, B.T., Engel, A.S., Nowlin, W.H., and Schwartz, B.F. (2016). Chemolithoautotrophy supports macroinvertebrate food webs and affects diversity and stability in groundwater communities. Ecology *97*, 1530–1542. DOI: 10.1111/1462-2920.12930

Iepure, S., Feurdean, A., Bădăluţă, C., Nagavciuc, V., and Perşoiu, A. (2016). Pattern of richness and distribution of groundwater Copepoda (Cyclopoida: Harpacticoida) and Ostracoda in Romania: an evolutionary perspective. Biol. J. Linn. Soc. *119*, 593–608. DOI: 10.1111/bij.12686

Jasperson, J.L., Gran, K.B., and Magner, J.A. (2018). Seasonal and Flood-Induced Variations in Groundwater–Surface Water Exchange in a Northern Coldwater Fishery. J. Am. Water Resour. Assoc. *54*, 1109–1126. DOI: 10.1111/1752-1688.12674

Jesußek, A., Grandel, S., and Dahmke, A. (2013). Impacts of subsurface heat storage on aquifer hydrogeochemistry. Environ. Earth Sci. *69*, 1999–2012. DOI: 10.1007/s12665-012-2037-

Jewell, T.N.M., Karaoz, U., Brodie, E.L., Williams, K.H., and Beller, H.R. (2016). Metatranscriptomic evidence of pervasive and diverse chemolithoautotrophy relevant to C, S, N and Fe cycling in a shallow alluvial aquifer. ISME J. *10*, 2106–2117. DOI: 10.1038/ismej.2016.25

Judd, K.E., Crump, B.C., and Kling, G.W. (2006). Variation in Dissolved Organic Matter Controls Bacterial Production and Community Composition. Ecol. Soc. Am. *87*, 2068–2079. DOI: 10.1890/0012-9658(2006)87[2068:VIDOMC]2.0.CO2

Karthe, D. (2015). Bedeutung hydrometeorologischer Extremereignisse im Kontext des Klimawandels für die Trinkwasserhygiene in Deutschland und Mitteleuropa. Hydrol. Wasserbewirtsch., *59*, 264-270. DOI: 10.5675/HyWa-2015,5-7

Kaushal, S.S., Mayer, P.M., Vidon, P.G., Smith, R.M., Pennino, M.J., Newcomer, T.A., Duan, S., Welty, C., and Belt, K.T. (2014). Land use and climate variability amplify carbon, nutrient, and contaminant pulses: A review with management implications. J. Am. Water Resour. Assoc. *50*, 585–614. DOI: 10.1111/jawr.12204

Kløve, B., Ala-Aho, P., Bertrand, G., Gurdak, J.J., Kupfersberger, H., Kværner, J., Muotka, T., Mykrä, H., Preda, E., Rossi, P., Uvo, C.B., Velasco, E., and Pulido-Velazquez, M. (2014). Climate change impacts on groundwater and dependent ecosystems. J. Hydrol. *518*, 250–266. DOI: 10.1016/j.jhydrol.2013.06.037

Korbel, K.L., Hancock, P.J., Serov, P., Lim, R.P., and Hose, G.C. (2013). Groundwater Ecosystems Vary with Land Use across a Mixed Agricultural Landscape. J. Environ. Qual. *42*, 380–390. DOI: 10.2134/jeq2012.0018

Krauss, S., and Griebler, C. (2011). Pathogenic Microorganisms and Viruses in Groundwater. Acatech. https://www.acatech.de/publikation/pathogenic-microorganisms-and-viruses-in-groundwater/

Kritzberg, E.S., Langenheder, S., and Lindström, E.S. (2006). Influence of dissolved organic matter source on lake bacterioplankton structure and function - Implications for seasonal dynamics of community composition. FEMS Microbiol. Ecol. *56*, 406–417. DOI: 10.1111/j.1574-6941.2006.00084.x

Kurylyk, B.L., MacQuarrie, K.T.B., Caissie, D., and McKenzie, J.M. (2015). Shallow groundwater thermal sensitivity to climate change and land cover disturbances: Derivation of analytical expressions and implications for stream temperature modeling. Hydrol. Earth Syst. Sci. *19*, 2469–2489. DOI: 10.5194/hess-19-2469-2015

Lau, M.C.Y., Kieft, T.L., Kuloyo, O., Linage-Alvarez, B., Van Heerden, E., Lindsay, M.R., Magnabosco, C., Wang, W., Wiggins, J.B., Guo, L., et al. (2016). An oligotrophic deep-subsurface community dependent on syntrophy is dominated by sulfur-driven autotrophic denitrifiers. Proc. Natl. Acad. Sci. U. S. A. *113*, E7927–E7936. DOI: 10.1073/pnas.1612244113

Lawter, A.R., Qafoku, N.P., Asmussen, R.M., Bacon, D.H., Zheng, L., and Brown, C.F. (2017). Risk of Geologic Sequestration of CO2 to Groundwater Aquifers: Current Knowledge and Remaining Questions. Energy Procedia *114*, 3052–3059. DOI: 10.1016/j.egypro.2017.03.1433

Li, D., Sharp, J.O., Saikaly, P.E., Ali, S., Alidina, M., Alarawi, M.S., Keller, S., Hoppe-Jones, C., and Drewes, J.E. (2012). Dissolved organic carbon influences microbial community composition and diversity in managed aquifer recharge systems. Appl. Environ. Microbiol. *78*, 6819–6828. DOI: 10.1128/AEM.01223-12

Li, D., Alidina, M., Ouf, M., Sharp, J.O., Saikaly, P., and Drewes, J.E. (2013). Microbial community evolution during simulated managed aquifer recharge in response to different biodegradable dissolved organic carbon (BDOC) concentrations. Water Res. *47*, 2421–2430. DOI: 10.1016/j.watres.2013.02.012

Liu, X., Ruecker, A., Song, B., Xing, J., Conner, W.H., and Chow, A.T. (2017). Effects of salinity and wet–dry treatments on C and N dynamics in coastal-forested wetland soils: Implications of sea level rise. Soil Biol. Biochem. *112*, 56–67. DOI: 10.1016/j.soilbio.2017.04.002

Magnabosco, C., Lin, L.H., Dong, H., Bomberg, M., Ghiorse, W., Stan-Lotter, H., Pedersen, K., Kieft, T.L., van Heerden, E., and Onstott, T.C. (2018). The biomass and biodiversity of the continental subsurface. Nat. Geosci. *11*, 707–717. DOI: 10.1038/s41561-018-0221-6

Mammola, S. (2019). Finding answers in the dark: caves as models in ecology fifty years after Poulson and White. Ecography (Cop.). *42*, 1331–1351. DOI: 10.1111/ecog.03905

Mander, Ü., Maddison, M., Soosaar, K., Teemusk, A., Kanal, A., Uri, V., and Truu, J. (2015). The impact of a pulsing groundwater table on greenhouse gas emissions in riparian grey alder stands. Environ. Sci. Pollut. Res. *22*, 2360–2371. DOI: 10.1007/s11356-014-3427-1

Menberg, K., Blum, P., Kurylyk, B.L., and Bayer, P. (2014). Observed groundwater temperature response to recent climate change. Hydrol. Earth Syst. Sci. *18*, 4453–4466. DOI: 10.5194/hess-18-4453-2014

Moldovan, O.T., Kováč, Ľ., and Halse, S. (Eds.). (2018). Cave ecology (Basel, Switzerland: Springer International Publishing).

Moritsch, M.M., Pakes, M.J., and Lindberg, D.R. (2014). How might sea level change affect arthropod biodiversity in anchialine caves: A comparison of Remipedia and Atyidae taxa (Arthropoda: Altocrustacea). Org. Divers. Evol. *14*, 225–235. DOI: 10.1007/s13127-014-0167-5

Mulec, J., and Summers Engel, A. (2019). Karst spring microbial diversity differs across an oxygen-sulphide ecocline and reveals potential for novel taxa discovery. Acta Carsologica *48*, 129–143. DOI: 10.3986/ac.v48i1.4949

Nathan, R., and Evans, R. (2011). Groundwater and surface water connectivity. In Water Resources Planning and Management, R. Grafton, and K. Hussey, eds. (Cambridge: Cambridge University Press), pp. 46–67. DOI: 10.1017/CBO9780511974304.006

Nawaz, A., Purahong, W., Lehmann, R., Herrmann, M., Totsche, K.U., Küsel, K., Wubet, T., and Buscot, F. (2018). First insights into the living groundwater mycobiome of the terrestrial biogeosphere. Water Res. *145*, 50–61. DOI: 10.1016/j.watres.2018.07.067

Oelkers, E.H., Butcher, R., Pogge von Strandmann, P.A.E., Schuessler, J.A., von Blanckenburg, F., Snæbjörnsdóttir, S., Mesfin, K., Aradóttir, E.S., Gunnarsson, I., Sigfússon, B., et al. (2019). Using stable Mg isotope signatures to assess the fate of magnesium during the in situ mineralisation of CO2 and H2S at the CarbFix site in SW-Iceland. Geochim. Cosmochim. Acta *245*, 542–555. DOI: 10.1016/j.gca.2018.11.011

Rath, K.M., Fierer, N., Murphy, D. V., and Rousk, J. (2019). Linking bacterial community composition to soil salinity along environmental gradients. ISME J. *13*, 836–846. DOI: 10.1038/s41396-018-0313-8

Ravier, E., and Buoncristiani, J.F. (2018). Glaciohydrogeology. In Past Glacial Environments, J. Menzies, and J.J.M. van der Meer, eds. (Elsevier Ltd), pp. 431–466. DOI: 10.1016/B978-0-08-100524-8.00013-0

Reiss, J., and Schmid-Araya, J.M. (2011). Feeding response of a benthic copepod to ciliate prey type, prey concentration and habitat complexity. Freshw. Biol. *56*, 1519–1530. DOI: 10.1111/j.1365-2427.2011.02590.x

Richts, A., and Vrba, J. (2016). Groundwater resources and hydroclimatic extremes: mapping global groundwater vulnerability to floods and droughts. Environ. Earth Sci. *75*, 1–15. DOI: 10.1007/s12665-016-5632-3

Rodó, X., Pascual, M., Doblas-Reyes, F.J., Gershunov, A., Stone, D.A., Giorgi, F., Hudson, P.J., Kinter, J., Rodríguez-Arias, M.À., Stenseth, N.C., et al. (2013). Climate change and infectious diseases: Can we meet the needs for better prediction? Clim. Change *118*, 625–640. DOI: 10.1007/s10584-013-0744-1

Rogers, K., Kelleway, J.J., Saintilan, N., Megonigal, J.P., Adams, J.B., Holmquist, J.R., Lu, M., Schile-Beers, L., Zawadzki, A., Mazumder, D., et al. (2019). Wetland carbon storage controlled by millennial-scale variation in relative sea-level rise. Nature *567*, 91–95. DOI: 10.1038/s41586-019-0951-7

Rotzoll, K., and Fletcher, C.H. (2013). Assessment of groundwater inundation as a consequence of sea-level rise. Nat. Clim. Chang. *3*, 477–481. DOI: 10.1038/nclimate1725

Saccò, M., Blyth, A., Bateman, P.W., Hua, Q., Mazumder, D., White, N., Humphreys, W.F., Laini, A., Griebler, C., and Grice, K. (2019a). New light in the dark - a proposed multidisciplinary framework for studying functional ecology of groundwater fauna. Sci. Total Environ. *662*, 963–977. DOI: 10.1016/j.scitotenv.2019.01.296

Saccò, M., Blyth, A.J., Humphreys, W.F., Kuhl, A., Mazumder, D., Smith, C., and Grice, K. (2019b). Elucidating stygofaunal trophic web interactions via isotopic ecology. PLoS One *14*, 1–25. DOI: 10.1371/journal.pone.0223982

Sarbu, S.M., Kane, T.C., and Kinkle, B.K. (1996). A Chemoautotrophically Based Cave Ecosystem. Science, *272*, 1953–1955. DOI: 10.1126/science.272.5270.1953

Savio, D., Stadler, P., Reischer, G.H., Demeter, K., Linke, R.B., Blaschke, A.P., Mach, R.L., Kirschner, A.K.T., Stadler, H., and Farnleitner, A.H. (2019). Spring water of an alpine karst aquifer is dominated by a taxonomically stable but discharge-responsive bacterial community. Front. Microbiol. *10*. DOI: 10.3389/fmicb.2019.0002

Shi, Y., Zwolinski, M.D., Schreiber, M.E., and Hickey, W.J. (1999). Molecular analysis of microbiol community structures in pristine and contaminated aquifers:

Field and laboratory microcosm experiments. Environ. Pollut. *108*, 2143–2150. DOI: 10.1128/AEM.65.5.2143-2150.1999

Shocket, M.S., Strauss, A.T., Hite, J.L., Šljivar, M., Civitello, D.J., Duffy, M.A., Cáceres, C.E., and Hall, S.R. (2018). Temperature drives epidemics in a zooplankton-fungus disease system: A trait-driven approach points to transmission via host foraging. Am. Nat. *191*, 435–451. DOI: 10.1086/696096

Simonsen, A.K., Barrett, L.G., Thrall, P.H., and Prober, S.M. (2019). Novel model-based clustering reveals ecologically differentiated bacterial genomes across a large climate gradient. Ecol. Lett. *22*, 2077–2086. DOI: 10.1111/ele.13389

Smith, V.H. (2007). Microbial diversity-productivity relationships in aquatic ecosystems. FEMS Microbiol. Ecol. *62*, 181–186. DOI: 10.1111/j.1574-6941.2007.00381.x

Sonthiphand, P., Ruangroengkulrith, S., Mhuantong, W., Charoensawan, V., Chotpantarat, S., and Boonkaewwan, S. (2019). Metagenomic insights into microbial diversity in a groundwater basin impacted by a variety of anthropogenic activities. Environ. Sci. Pollut. Res. *26*, 26765–26781. DOI: 10.1007/s11356-019-05905-5

Stegen, J.C., Fredrickson, J.K., Wilkins, M.J., Konopka, A.E., Nelson, W.C., Arntzen, E.V., Chrisler, W.B., Chu, R.K., Danczak, R.E., Fansler, S.J., Kennedy, D.W., Resch, C.T., and Tfaily, M. (2016). Groundwater-surface water mixing shifts ecological assembly processes and stimulates organic carbon turnover. Nat. Commun. *7*, 1–12. DOI: 10.1038/ncomms11237

Stein, H., Kellermann, C., Schmidt, S.I., Brielmann, H., Steube, C., Berkhoff, S.E., Fuchs, A., Hahn, H.J., Thulin, B., and Griebler, C. (2010). The potential use of fauna and bacteria as ecological indicators for the assessment of groundwater quality. J. Environ. Monit. *12*, 242–254. DOI: 10.1039/b913484k

Swindles, G.T., Morris, P.J., Mullan, D.J., Payne, R.J., Roland, T.P., Amesbury, M.J., Lamentowicz, M., Turner, T.E., Gallego-Sala, A., Sim, T., et al. (2019). Widespread drying of European peatlands in recent centuries. Nat. Geosci. *12*, 922–928. DOI: 10.1038/s41561-019-0462-z

Treidel, H., Martin-Bordes, J.L., and Gurdak, J.J. (2011). Climate change effects on groundwater resources: A global synthesis of findings and recommendations. In International Contributions to Hydrogeology (Boca Raton, Fla: CRC Press).

Twing, K.I., Brazelton, W.J., Kubo, M.D.Y., Hyer, A.J., Cardace, D., Hoehler, T.M., McCollom, T.M., and Schrenk, M.O. (2017). Serpentinization-influenced groundwater harbors extremely low diversity microbial communities adapted to high pH. Front. Microbiol. *8*. DOI: 10.3389/fmicb.2017.00308

Uddameri, V., Singaraju, S., and Hernandez, E.A. (2019). Is Standardized Precipitation Index (SPI) a Useful Indicator to Forecast Groundwater Droughts? — Insights from a Karst Aquifer. J. Am. Water Resour. Assoc. *55*, 70–88. DOI: 10.1111/1752-1688.12698

Voigt, C., Marushchak, M.E., Mastepanov, M., Lamprecht, R.E., Christensen, T.R., Dorodnikov, M., Jackowicz-Korczyński, M., Lindgren, A., Lohila, A., Nykänen, H., Oinonen, M., Oksanen, T., Palonen, V., Treat, C.C., Martikainen, P.J., and Biasi, C. (2019). Ecosystem carbon response of an Arctic peatland to simulated permafrost thaw. Glob. Chang. Biol. *25*, 1746–1764. DOI: 10.1111/gcb.14574

Ward, N.D., Bianchi, T.S., Medeiros, P.M., Seidel, M., Richey, J.E., Keil, R.G., and Sawakuchi, H.O. (2017). Where Carbon Goes When Water Flows: Carbon Cycling across the Aquatic Continuum. Front. Mar. Sci. *4*, 1–27. DOI: 10.3389/fmars.2017.00007

Westphal, A., Kleyböcker, A., Jesußek, A., Lienen, T., Köber, R., and Würdemann, H. (2017). Aquifer heat storage: abundance and diversity of the microbial community with acetate at increased temperatures. Environ. Earth Sci. *76*. DOI: 10.1007/s12665-016-6356-0

Whitman, W.B., Coleman, D.C., and Wiebe, W.J. (1998). Prokaryotes: The unseen majority. Proc. Natl. Acad. Sci. U. S. A. *95*, 6578–6583. DOI: 10.1073/pnas.95.12.6578

Wu, X., Wu, L., Liu, Y., Zhang, P., Li, Q., Zhou, J., Hess, N.J., Hazen, T.C., Yang, W., and Chakraborty, R. (2018). Microbial interactions with dissolved organic matter drive carbon dynamics and community succession. Front. Microbiol. *9*, 1–12. DOI: 10.3389/fmicb.2018.0123

Wurzbacher, C., Nilsson, R.H., Rautio, M., and Peura, S. (2017). Poorly known microbial taxa dominate the microbiome of permafrost thaw ponds. ISME J. *11*, 1938–1941. DOI: 10.1038/ismej.2017.54

Yan, L., Herrmann, M., Kampe, B., Lehmann, R., Totsche, K.U., and Küsel, K. (2020). Environmental selection shapes the formation of near-surface groundwater microbiomes. Water Res. *170*, 115341. DOI: 10.1016/j.watres.2019.115341

Yvon-Durocher, G., Allen, A.P., Montoya, J.M., Trimmer, M., and Woodward, G. (2010). The temperature dependence of the carbon cycle in aquatic ecosystems. Adv. Ecol. Res. *43*, 267–313. DOI: 10.1016/B978-0-12-385005-8.00007-1

Zektser, I.S., and Everett, L.G. (2004). Groundwater resources of the world and their use. IHP-VI, Series on Groundwater No. 6, (Paris: United Nations Educational, Scientific and Cultural Organization).

Zhang, G., Yao, T., Shum, C.K., Yi, S., Yang, K., Xie, H., Feng, W., Bolch, T., Wang, L., Behrangi, A., et al. (2017). Lake volume and groundwater storage variations in Tibetan Plateau's endorheic basin. Geophys. Res. Lett. *44*, 5550–5560. DOI: 10.1002/2017GL073773

Chapter 14

Ecosystem Metabolism in River Networks and Climate Change

Vicenç Acuña[1,2,*], Anna Freixa[1,2], Rafael Marcé[1,2] and Xisca Timoner[1,2]

[1] Catalan Institute for Water Research, Carrer Emili Grahit 101, Girona 17003, Spain; [2] University of Girona, Plaça de Sant Domènec, 17004 Girona, Spain;
[*] corresponding author

Email: vicenc.acuna@icra.cat, afreixa@icra.cat, rmarce@icra.cat, xisca.timoner@gmail.com

DOI: https://doi.org/10.21775/9781913652579.14

Abstract

Primary production and ecosystem respiration are key processes for turnover of organic carbon, inorganic substances, and energy in river networks. Ecosystem respiration is commonly the dominant process because of the fueling by organic carbon from terrestrial origin. In fact, the mineralization of organic carbon within river networks shapes to a large extent the regional and global carbon balances, and will be highly sensitive to change owing to major increases in the extent of the non-flowing periods as well as in flood frequency and magnitude. Existing evidence points out that these alterations in the flow regime might increase organic carbon export rates, whereas temperature alterations will increase mineralization of organic carbon. The specific roles of lotic and lentic water bodies within river networks might also change, as the lakes and reservoirs might increase their roles in the carbon balance and partly counteract the effects of flow extremes on organic carbon export rates.

Introduction

Primary production (P) and ecosystem respiration (R) are key processes for turnover of organic carbon, inorganic substances, and energy in river networks. Primary production is usually the dominant process converting abiotic energy to chemical binding energy of organic carbon and thus supporting growth and maintenance of ecosystems. Ecosystem respiration is the sum of the dissipation of this energy by all organisms of the ecosystem. Lotic ecosystems rely on two sources of organic carbon: autochthonous inputs by aquatic primary producers and allochthonous inputs of dead organic carbon from terrestrial ecosystems. This chapter aims to summarize current knowledge on the carbon metabolism in river networks and is structured as follows: 1) we review the role of biofilms in carbon metabolism; 2) we then introduce the basics of P and R, their variation along the river continuum, the dominance of R in most systems, and the differences between aerobic and anaerobic metabolism; 3) after that, we review existing information on the role of freshwater ecosystems in the global carbon balance, and then discuss the specific role of the carbon metabolism; 4) last, we develop two separate sections on the specific roles of the temperature and flow regime on the carbon metabolism, the current trends in the temperature and flow regimes, and the expected changes owing to the current trends. In the section of the effects of flow regime on the carbon metabolism, we devote a separate sub-section for the carbon metabolism in temporary streams. Conclusions as well as future trends are drawn in the final section.

Terrestrial inputs of dead organic matter in form of coarse, fine, and dissolved organic carbon are important in many lotic settings, especially in small streams with a riparian forest, where R is fostered and P tends to be light-limited (Allan and Castillo, 2007). P is however expected to be of greater importance at roughly fourth- through sixth-order streams, and then less so in larger, deeper, and more turbid lowland rivers. The relationships between P and R along the river continuum were initially described by Vannote et al. (1980) and Minshall et al. (1983), and apply considerably well for pristine temperate systems, but several studies have reported different patterns for river networks with different landscape or climate settings (Wiley et al., 1990; Young and Huryn, 1997; Hotchkiss et al., 2015). Overall, R is in most cases the dominant process in river networks (Webster and Meyer, 1997; Mulholland et al., 2001; Hoellein et al., 2013), which in fact rely to a large extent on the allochthonous organic carbon inputs from terrestrial ecosystems to support heterotrophic R.

Heterotrophic organisms from freshwater ecosystems - microorganisms, meiofauna, and macrofauna - decompose and consume allochthonous supplies of organic carbon, ultimately mineralizing some fraction of the total as carbon dioxide (CO_2) or releasing methane (CH_4), but also exporting substantial quantities to downstream ecosystems and, ultimately, to the oceans. Thus, mineralization and export are the two principal fates of organic carbon in lotic ecosystems. In regards to storage, it is considered to be negligible on timescales longer than months to years in lotic ecosystems of low stream order, whereas this is not the case for lotic ecosystems of higher stream order and even less for lentic ecosystems, where storage might be as important as mineralization and export (Tranvik et al., 2009). Overall, a high rate of mineralization relative to transport and storage means that organic carbon is contributing to ecosystem metabolism and that ecosystems are efficiently mineralizing organic carbon inputs from terrestrial ecosystems.

The mineralization of organic carbon in lotic ecosystems is usually through aerobic R, which means that oxygen (O_2) is used as electron acceptor and CO_2 and water (H_2O) are the products of this process. In anaerobic conditions, R also occurs using other electron acceptors such as metals (i.e. iron and manganese), nitrate (NO_3^-), sulphate (SO_4^{2-}), or CO_2. These anaerobic R processes play major roles in the biogeochemical cycles of iron, manganese, nitrogen, sulphur, and carbon through the reduction of the oxyanions of metals, nitrogen, sulphur, and carbon to more-reduced compounds (Stets et al., 2017), and their contribution to total R is relevant from a global warming perspective, as the resulting gases such as methane (CH_4) and nitrous oxide (N_2O) have a major global warming potential than CO_2 (Forster et al., 2007). Anaerobic R might occur in certain compartments or microenvironments within lotic ecosystems such as the saturated zone below the riparian forest or backwaters where organic sediments accumulate. In contrast, anaerobic R is more common in sediments accumulated in lentic ecosystems, where it might be the dominant type of R. In consequence, river networks are not only a significant source of CO_2 (Battin et al., 2008; Tranvik et al., 2009), but of CH_4 (Bastviken et al., 2004; Comer-Warner et al., 2018) and N_2O (Beaulieu et al., 2011).

Biofilms and carbon metabolism

Biofilms are complex assemblages of microorganisms, such as algae, bacteria, fungi and protozoa, embedded in a matrix of polysaccharides, exudates and detritus (Wetzel, 1983; Lock et al., 1984). They are abundant in a multitude of natural environments (Wimpenny et al., 2000) conditioning the bare surfaces for life. In lotic ecosystems, biofilms colonize rocks, cobbles, sand, aquatic plants, submerged leaves, and wood in a variety of structural and compositional configurations (Lock, 1993). Under light-limited conditions, biofilms are dominated by bacteria and fungi (heterotrophs), whereas autotrophs (algae and cyanobacteria) can prevail when light

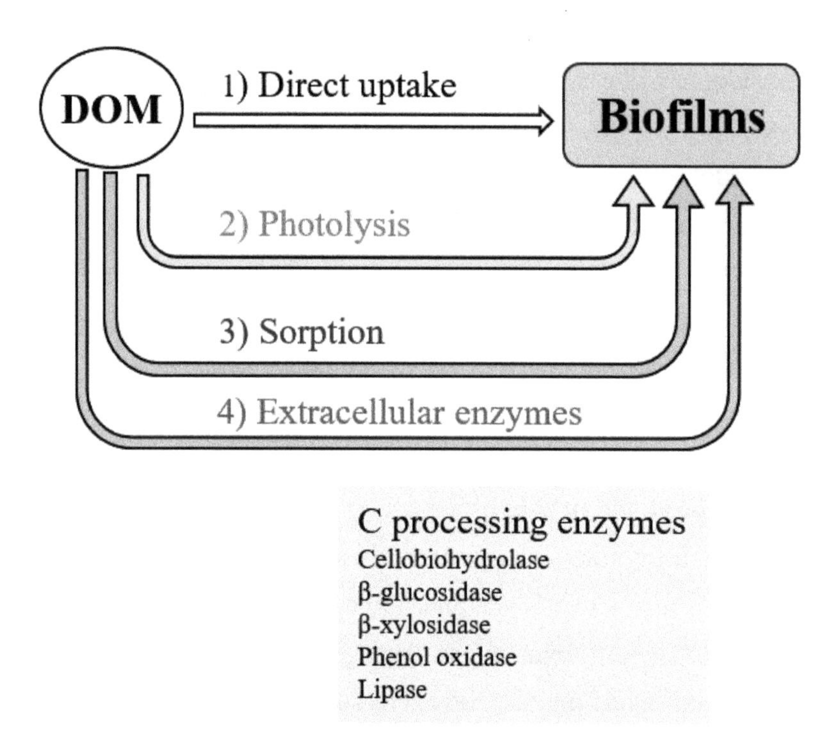

Figure 14.1 Pathways of DOC incorporation by biofilms including the list of extracellular enzymes-mediated DOC uptake.

is available (Sigee, 2005). Other physical factors (e.g. temperature and flow velocity) as well as the type of colonized substrata (e.g., organic or inorganic) further influence the relative importance of heterotrophs or autotrophs in stream biofilms (Peterson, 1996). The environmental gradients created inside biofilms are also relevant for defining microenvironments where anaerobic R processes can progress.

Several studies have pointed out the important role for the microbial community in the carbon metabolism in river networks, since they are involved in the degradation of organic carbon and other compounds (Bretschko, 1995; Romaní and Sabater, 2001; Battin et al., 2008; Findlay, 2010). Their role in the organic carbon mineralization is mainly through dissolved organic carbon (DOC) acquisition (i.e. microbial biomass) and mineralization, which can occur by four different pathways (Findlay and Sinsabaugh, 1999) (Fig. 14.1): 1) the direct uptake consists in the assimilation of organic low molecular weight molecules without mediation by an external process, however these molecules are present in low concentrations in stream water; 2) photolysis-mediated uptake is the generation of microbial assimilable molecules from the photo oxidation of DOC; 3) sorption-mediated uptake consists in the concentration of carbon and other nutrients by the sorption of dissolved molecules onto the biofilm polysaccharide matrix and cell fragments; 4) extracellular enzymes-mediated uptake is the most important pathway of the DOC acquisition, consists in the transformation of polymeric DOC (high molecular weight molecules, macromolecules), which are not directly assimilable for microorganism metabolism (Hoppe, 1993), into assimilable molecules (low molecular weight) by the action of extracellular enzymes (Chróst, 1990). Extracellular enzymes reaction products may include sources of carbon, nitrogen, and phosphorous (Fig. 14.1).

Most extracellular enzymatic activities except phosphatases are of bacterial and fungal origin, however, to a minor extent, algae and protozoa may also express enzyme activities (Vrba et al., 2004). Overall, biofilm extracellular enzyme activities mediated by biofilms play a key role inside the microbial loop, being the activity of these enzymes in many cases a limiting factor for organic carbon mineralization and microbial growth.

The role of river networks in the global carbon balance

Organic carbon inputs from terrestrial ecosystems into the river networks might be mineralized, stored or exported. Globally, organic carbon exports to the oceans account for approximately 0.9 Pg C y^{-1} (Drake et al., 2018), long-term storage accounts for between 0.15 and 0.6 Pg C y^{-1} (Drake et al., 2018; Mendonça et al., 2017), and estimates of river networks to atmosphere CO_2 flux at the global scale range between 0.65 - 4.2 Pg C y^{-1} (Cole et al., 2007; Battin et al., 2008; Tranvik et al., 2009; Aufdenkampe et al., 2011; Raymond et al., 2013; Lauerwald et al., 2015; Drake et al., 2018), which is substantial in relation to the current global carbon budget imbalance (0.4 Pg C y^{-1}) (Friedlingstein et al., 2019) (Fig. 14.2). Accordingly, if considered in the accounting of the continental greenhouse gases fluxes, river networks would partially offset the continental carbon sink (Bastviken et al., 2011). Despite the magnitude of these fluxes, it has been only recently that the role of river networks in the global carbon balance has been accounted for, and that lotic and lentic ecosystems have been recognized as hotspots of biogeochemical activity (Cole, 2013). The emerging understanding states that to properly understand the role of continental aquatic ecosystems on the global carbon balance they should be viewed as a seamless aquatic continuum from land to ocean (Regnier et al., 2013).

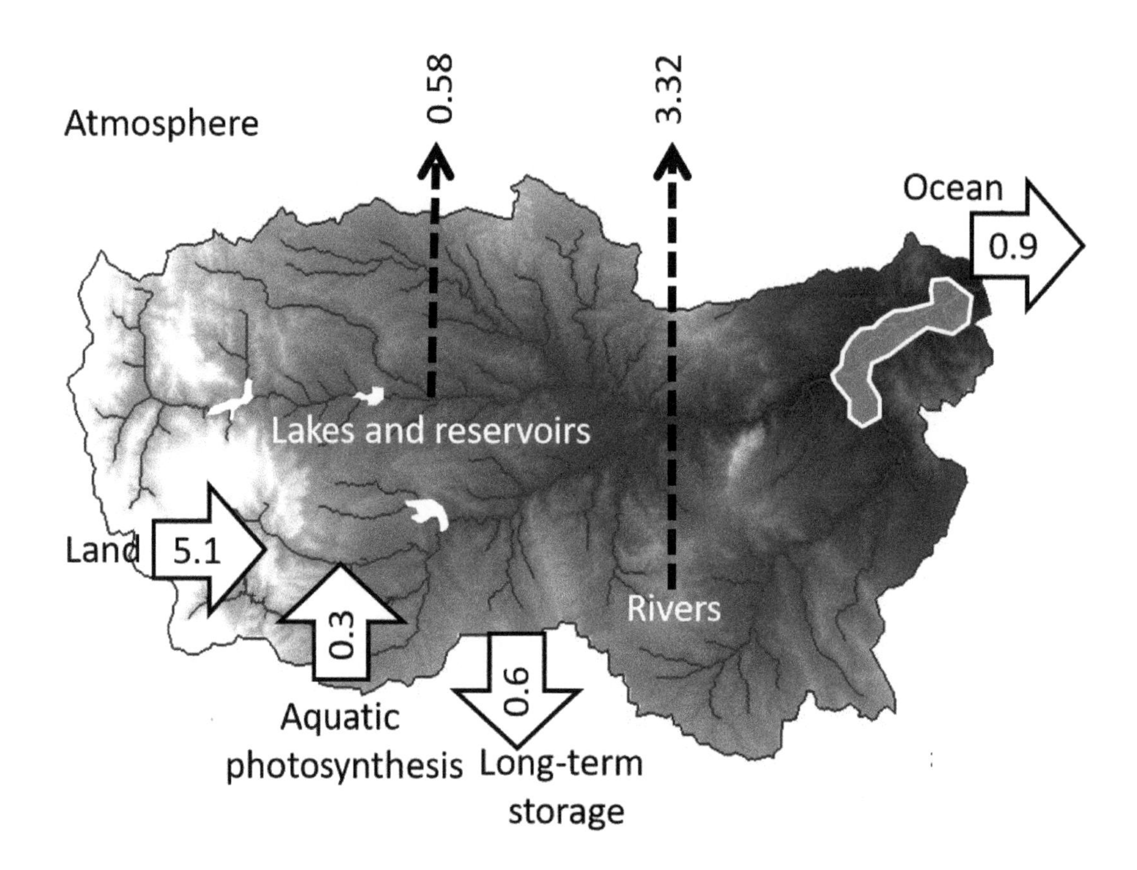

Figure 14.2 The role of river networks in the global carbon balance. Note that numbers next to the dotted arrows indicate CO_2 emissions to the atmosphere that next to the upper vertical arrow indicates inputs from terrestrial ecosystems, that next to the lower vertical arrow indicates exports to the ocean, and that next to the horizontal arrow indicates long-term storage. All numbers are expressed in Pg C y^{-1}, and the figure is based on Drake et al. (2018).

Typologies of water bodies involved in the carbon metabolism

The great majority of the total length of river networks is comprised of low order or headwater streams. In fact, 50% of the total river kilometers in the contiguous United States are first order and the total for first- to third-order combined is approximately 85% (Leopold, 1964). These first order streams play a major role in the global carbon balance, as those first order streams receive most of the organic carbon input from terrestrial ecosystems, and are responsible for most of the river to atmosphere CO_2 flux (around 36% of the total flux according to Butman and Raymond, 2011). However, there are also lentic water bodies within these river networks dominated by small streams. In fact, river networks are composed of lotic water bodies of increasing stream order commonly intersected by a series of lentic water bodies such as lakes and reservoirs (Grill et al., 2019; Gardner et al., 2019). Lotic and lentic water bodies exert a very different role in the organic carbon balance because of their differences in geometry and hydraulics. Thus, transport is the dominant process in lotic water bodies, whereas storage is often the dominant process in lentic water bodies such as lakes or reservoirs. Overall, Drake et al. (2018) estimated an annual flux from streams and rivers to atmosphere CO_2 of 3.32 Pg C, and an annual flux from lakes and reservoirs of 0.58 Pg C, where stream metabolism is responsible up to 28% of the total carbon flux from streams to the atmosphere (Hotchkiss et al. 2015).

The relevance of storage respect transport lies on the balance between sedimentation and resuspension of particulate organic carbon, as sedimentation dominates in lentic water bodies (Acuña and Tockner, 2010). Once deposited through sedimentation, organic carbon can be either mineralized or buried. The organic carbon burial in their sediments represents a short- to long- term sequestration of atmospheric CO_2, while mineralization of organic carbon by aerobic or anaerobic R closes the loop by returning carbon to the atmosphere. Despite the magnitude of organic carbon buried in the sediment of lentic water bodies [up to four times as much as in the world's oceans (Downing et al., 2008), with most recent estimate at 0.15 Pg C y^{-1} (Mendonça et al., 2017)], it is the mineralization of organic carbon entering these water bodies that is the dominant process affecting the carbon dynamics. Algesten et al. (2004) reported that in boreal lakes, 30 - 80 % of the organic carbon input from terrestrial ecosystems was mineralized in lakes, and the variation among different boreal lakes was strongly influenced by the water residence time within each basin. Overall, the importance of storage and mineralization in lentic water bodies implies that the bulk of organic carbon might be mineralized in lentic water bodies, and that those might profoundly alter the organic carbon dynamics in downstream units (Acuña and Tockner, 2010), creating therefore discontinuities in the organic carbon dynamics along the river networks.

The configuration of river networks in terms of spatial arrangement of lentic water bodies within the river network, as well as the dimensions and abundance of lentic water bodies, influence the overall metabolism (Proia et al., 2016), the rate of mineralization relative to transport and storage (Crawford et al., 2016) as well as the chemical form in which organic carbon is released to the atmosphere (CO_2:CH_4) (Gómez-Gener et al., 2018). For example, lentic water bodies with long water residence time located near the headwaters might trap most coarse particulate organic carbon and mineralize it through anaerobic R, therefore showing high rates of mineralization relative to transport and storage and releasing significant amounts of the organic carbon as CH_4 (Maexk et al., 2013). In contrast, lentic water bodies with shorter water residence time located near the river mouth might be unable to trap and mineralize most of the predominant forms of organic carbon in lowland large rivers (i.e., fine particulate organic carbon and dissolved organic carbon). Finally, very large dams located in lowland sections of large rivers may effectively trap huge amounts of fine and dissolved organic carbon, fostering anaerobic R in the sediments (Maavara et al., 2020). Considering that the global warming potential of CH_4 is 86 times greater than that of CO_2 in a 20 years horizon, this implies that lentic systems efficient in trapping organic carbon may behave as net greenhouse gas emitters to the atmosphere (Tranvik et al., 2009) irrespective of the high P frequently found in those systems, a situation already observed in productive wetlands that also accumulate sediments (Whiting and Chanton, 2001).

Overall, the role of river networks in the global carbon balance may be changed by human activities (Maavara et al., 2017), including construction of impoundments such as large hydroelectric reservoirs, small weirs, and innumerable farm ponds, which tend to accumulate large amounts of carbon in sediments regardless of their position in the river network and emit large amounts of carbon to the atmosphere as CH_4 (Maavara et al., 2020). Nonetheless, lateral exchanges between floodplain wetlands and running waters in large rivers constitute another relevant interaction between lotic and lentic systems at the river basin scale. Lateral exchanges between the main channel of rivers and their associated wetlands are crucial in determining the metabolic balance in systems as big as the Amazon (Abril et al., 2013). Besides their role as sedimentary systems, wetlands tend to be highly productive, pumping atmospheric CO_2 to the main channel of the river in the form of both CO_2 and organic carbon, which will eventually be mineralized supporting most of the R and carbon evasion to the atmosphere in the river network downstream (Borges et al., 2019).

Global warming and carbon metabolism

Temperature dependence of metabolism

Metabolic processes such as P and R obey the physical and chemical principles that govern transformations of energy and materials, the most important being the laws of energy and mass balance and thermodynamics (Gillooly et al., 2001; Brown et al., 2004). Boltzmann's factor ($e^{-E/kT}$) describes the temperature dependence of metabolic rate, where E is the activation energy (eV), k is the Boltzmann's constant (8.616 x 10^{-5} eV°K^{-1}) and T is absolute temperature (°K) (Boltzmann, 1872; Arrhenius, 1915). This exponential relationship is valid within a limited range of biologically relevant temperatures (in general 0 - 40 °C), and describes the temperature dependence of the whole-body metabolism of virtually all organisms (Gillooly et al., 2001). This relationship also applies to ecosystem processes, such as P and R (Silver and Miya, 2001; Enquist et al., 2003; López-Urrutia et al., 2006; Acuña et al., 2008). P and R react differently to changes in temperature because of their different E values. Thus, with a temperature increase from 0 to 30 °C, P increases fourfold (as predicted from the E of Rubisco carboxylation, E_p = 0.33 eV), while R increases 16-fold (as predicted from the E of the respiratory complex, E_r = 0.62 eV) (Enquist et al., 2003). These predictions are solely based on the theoretical E values for P and R, but there is also empirical data supporting the differential response of P and R to temperature (Demars et al., 2011). Regardless of the balance between P and R, the temperature sensitivity of methanogenesis is much higher than that of aerobic respiration, as the average apparent E values for methanogenesis in the literature are close to 0.9 eV (Daniels et al., 1977; Westermann, 1993), implying that methanogenesis increases 44-fold from 0 to 30 °C. This suggests that warming may promote a substantial increase of the rate of CH_4:CO_2 emissions from R processes in hot-spots of anaerobic respiration. However, the response of CO2 and CH4 emissions from freshwater ecosystems seems to be more complex according to a recent study, which stressed non-linear responses of emissions to warming, owing mainly to the influence of other variables such as streambed sediment size, organic matter availability and sediments geological origin (Comer-Warner et al., 2018). Moreover, recent studies suggest that the increase of CO2 emissions by enhanced respiration due to global warming might be countered by increasing primary production (Demars et al., 2016).

Temperature does not only influence the role of river networks in the organic carbon balance through its effects on metabolism, but also on its effects on other key processes such as the reaeration coefficients (Raymond et al., 2013), as well as the inorganic carbon chemical equilibrium. Thus, river to atmosphere CO_2 fluxes depend to a large extent on the reaeration flux, which depends on the gradient between CO_2 concentrations in the river and the atmosphere, as well as on the reaeration coefficient. The temperature dependence of the reaeration coefficient has been commonly described with the simplified Arrhenius equation, which provides an increase of the reaeration coefficient at a geometric rate of 2.41% per °C (Elmore and West, 1961), although recent studies pointed out that the rate might range between 0.5 and 4.1% per °C within a temperature range of 0 - 35 °C (Demars and Manson, 2013). Temperature also influences the chemical equilibrium between the different forms of inorganic carbon in water. The proportion of CO_2 relative to the other forms of inorganic carbon (carbonate and bicarbonate) rises at higher temperatures, promoting CO_2 supersaturation. However, although this process was described long ago for rivers (Park et al., 1969), we still lack evidence of a global impact of this process on freshwater CO2 concentration and emissions.

Current global trends in temperature

Global climate change has altered, and will further alter, the temperature of freshwater ecosystems. Surface air temperature has already increased by 0.6 °C in the past century, and is predicted to further increase by approximately 3 °C (range: 2 - 4.5 °C) by 2100 (Stocker et al., 2013) or higher according to recent studies (Forster et al., 2020). These changes in air temperature determine in turn the water temperature, as water temperature tracks air temperature (shown for European streams by Webb and Nobilis, 1994; Langan et al., 2001; Mouthon and Daufresne, 2006). In this direction, a recent analysis of river temperature trends in northern Germany reported a mean warming trend of 0.03 °C $year^{-1}$ over the period 1985-2010 (Arora et al., 2016), that is, 1.05 °C increase over the considered period. However, meteorological time series reveal not only a trend of increasing air temperature values but also an increase in the frequency and magnitude of extreme events (i.e., heat waves; Benestad, 2004).

Expected changes in the carbon metabolism by temperature trends

The response of R to future changes in temperature will be critical in determining the role of river networks in the global carbon cycle. Experimental studies predict that a 2 - 3 °C temperature increase will lead to a 20 - 40 % increase in heterotrophic R (Sand-Jensen and Pedersen, 2005; Acuña et al., 2008; Song et al., 2018). In fact, these temperature increases in temperature will also lead to increases in P, but these increases may be of lesser extent because P is less sensitive to temperature changes than R (Enquist et al., 2003; Allen et al., 2005; Demars et al., 2011) becoming streams more heterotrophic (i.e. lower P/R) as climate continues to warm. Differences in the temperature dependences of R and P at the organismal level have been pointed out as the underlying reason for likely increases of CO_2 emissions from the epipelagic oceans by López-Urrutia et al. (2006), after they scaled the metabolic balance of oceans according to the metabolic theory of ecology (Gillooly et al., 2001; Brown et al., 2004). In this direction, Demars et al. (2011) compared P and R in several Icelandic streams only differing in their mean temperature, and reported a negative relationship between the net ecosystem metabolism (balance between P and R) and warming, meaning that the higher the temperature, the higher the amount of carbon released to the atmosphere from river networks. Specifically, Demars et al. (2011) predicted that the amount of carbon released to the atmosphere might double with a mean temperature increase of 5 °C. Similarly, a recent study across six stream biomes predicted a decrease of 23.6% of overall net ecosystem production (i.e. balance between P and R) due to an increase of 1 °C of stream temperature (Song et al., 2018). At the global scale, this would imply an increase of 0.0194 Pg C y^{-1} emitted from river networks (Song et al., 2018). To our knowledge, we lack comprehensive assessments of the potential impact of warming on different R pathways across river networks, but considering the high sensitivity of methanogenesis to temperature, this is a topic that deserves much more attention in the future.

Regardless of their different sensitivity to temperature increases, likely changes in P and R may be distorted by resource limitation (i.e., light and nutrients for P and organic carbon for R). Thus, several studies of terrestrial and marine ecosystems stressed that predicted changes in R may be overestimated due to a lack of consideration of resource availability (e.g., Gifford, 1995; Janssens et al., 2001; López-Urrutia and Morán, 2007; Demars et al., 2016). However, lotic and lentic ecosystems are not commonly limited by the supply of organic carbon, because inputs from terrestrial systems commonly exceed R. Recent studies indicate that it is not the total availability of organic carbon which might limit R, but the quality of the available carbon (Jane and Rose, 2018; Romejin et al., 2019). In this direction, temperature increases might alter the balance between storage and mineralization of organic carbon in lakes (Gudasz et al., 2010), which in fact are not limited by the organic carbon input from terrestrial ecosystems. Specifically, these authors estimated that temperature increases following the latest scenarios presented by the IPCC could result in a 4 - 27% decrease in annual organic carbon storage in boreal lakes because of the temperature-led increases in heterotrophic R (Gudasz et al., 2010).

Most experimental and modelling studies of freshwater ecosystem responses to climate change have focused on responses to a trend in climate, but have mostly ignored the complex responses to alterations of the regime type (but see Dang et al., 2009; Miao et al., 2009; Kirschbaum, 2010). Furthermore, these studies have mainly ignored likely adaptation mechanisms to higher water temperatures, which might offset predictions of the expected changes in the organic carbon metabolism in river networks by temperature trends. In regard to the alteration of other components of the temperature regime different than the mean, an experimental study by Dang et al. (2009) testing the effects of a 5 ºC temperature increase, with and without temperature oscillations, on litter decomposition by fungal communities in artificial streams showed the relevance of the alterations of the temperature regime at the daily scale. The acceleration of decomposition was larger with stronger temperature oscillation, showing that daily temperature patterns around the average value should be also considered when assessing the consequences of global warming on ecosystem processes. In this regard, Freixa et al. (2017) observed how the temperature daily patterns influenced biofilm functioning where the DOC content increased during light-time hours and it was microbially consumed during night-time. A longer temporal scale, Kirschbaum (2010) stressed that for global estimates of the likely global warming led increase of heterotrophic R, it is critically important to explicitly consider seasonal temperature variations.

Adaptation mechanisms to recent temperature conditions might be caused by biofilm community structural changes, or by thermal physiological adaptation to current or antecedent water temperature. Adaptation by microbial communities to the antecedent temperature regime has been reported by Fierer et al. (2006). These authors characterized the temperature dependence of microbial R from 77 soils collected from a wide array of ecosystem types, and types and reported that 17 % of the Q_{10} variability could be explained by the mean monthly temperature at the time of sampling. In addition, several studies stressed that the temperature sensitivity of soil R, in general, is negatively related to temperature (e.g. Kirschbaum, 1995; Atkin et al., 2000; Qi et al., 2002). In aquatic ecosystems, Acuña et al. (2008) reported a significant relationship between the mean temperature over the last 2 weeks before measurement and the estimated E_r values (range: 0.3 - 0.8 eV), meaning that there was a short-term thermal adaptation of the sensitivity of R to recent temperature conditions. The reported differences in terms of mean water temperatures within the studied river network involved that equivalent changes of 2.5 °C in water temperature in the colder headwaters caused a R increase of 23%, whereas the increase in warmer lowland reaches was of only 18 %. However, there are also empirical evidences of consistent temperature dependence of R across aquatic ecosystems contrasting in thermal history (Perkins et al., 2012).

Overall, the rate of mineralization relative to transport and storage may increase following temperature increases, and therefore likely increase the river to atmosphere CO_2 flux at the global scale. This may be further enhanced by the fact that mineralization in sediments stored in lakes and reservoirs is strongly dependent on temperature (Gudasz et al., 2010), favoring the export of CO_2 to downstream rivers and eventually to the atmosphere. If less terrestrial organic carbon is exported through river networks, there are probably to be effects on estuarine and coastal marine productivity (e.g. Findlay et al., 1998; Sigleo and Macko, 2002). In addition, if less organic carbon is trapped and preserved in coastal sediments (Hedges and Oades, 1997), which act as a sink compartment in the global carbon cycle, an increase in CO_2 release from river networks is probably to be a positive feedback for global warming. Biotic feedbacks from a warming of the Earth have already been stressed for terrestrial and marine ecosystems (Woodwell et al., 1998; López-Urrutia et al., 2006).

Flow regime and carbon metabolism

The role of extreme flow events on the carbon metabolism

The flow regime controls the hydraulic retention of particulate organic carbon, and therefore determines the substrate availability of particulate organic carbon for heterotrophic R. Flow is the main driver determining the balance between sedimentation and resuspension of particulate organic carbon, so that higher flows decrease sedimentation and increase resuspension, therefore reducing the availability of particulate organic carbon. Accordingly, flow variation along the year influences the substrate availability for heterotrophic R in lotic ecosystems. Flow is also the main driver of the dissolved organic carbon availability in lotic systems, as dissolved organic carbon supply from terrestrial ecosystems increases with rainfall, stimulating aquatic R, and CO_2 emissions (Demars 2019). However, extreme flow events play a major role in the relationship between flow and carbon metabolism, as non-flow events cause the cessation of transport and major accumulations of particulate organic carbon in streambeds, whereas floods involve catastrophic removals of commonly all stored particulate organic carbon (Acuña et al., 2004; Manzoni and Porporato, 2011), and decreases in the metabolic rates R and P (Reisinger et al. 2017).

Non-flow events involve an abrupt change in the balance between mineralization, storage and export, given the lack of transport and the accumulation of particulate organic carbon. Particulate organic carbon only accumulates if there is a roughly continuous supply of detritus from surrounding terrestrial ecosystems (i.e. litterfall from the riparian forest), and if the rate of input is higher than the rate of mineralization. Studies performed on low order streams draining forested areas reported accumulations on the dry streambeds proportional to the length of the non-flow period (Acuña et al., 2007). Despite the multiple studies analyzing the organic carbon mineralization in temporary lotic ecosystems, their specific role in the carbon metabolism in river networks remains unclear and we devote a specific section to it. In contrast with that, there is a general agreement on the effect of floods on the carbon metabolism. Indeed, floods events commonly involve a major catastrophic removal of accumulated particulate organic carbon, and their effect is commonly separately modelled when aiming to characterize the relationship between flow regime and carbon metabolism (Acuña and Tockner, 2010).

Overall, it is crucial to account not only for the changes in the mean values of flow, but changes on the flow regime, paying special attention to the role of extreme flow events given their relevance for the carbon metabolism. Predictions on how changes in flow regime might alter carbon metabolism should therefore consider changes in the frequency and magnitude of extreme flow events, the timing of these events, and the role of lentic systems as potential sinks for organic carbon. In this sense, an experimental study by Miao et al. (2009) analyzing the vegetation responses to extreme hydrological events, stressed that the impacts of multiple extreme events cannot be modelled by simply summing the projected effects of individual extreme events, but rather that models should consider the event sequences.

Current global trends in flow regime

Several studies reported an increasing frequency and intensity of extreme hydrological events (New et al., 2001; Huntington, 2006; Hirabayashi et al., 2008; Gudmundsson et al., 2019), and other events such as El Niño Southern Oscillation over the past decades (Huntington, 2006). Accordingly, more extreme and unpredictable hydrological events, such as flooding and drought, are anticipated, thereby creating novel environmental conditions in lotic and lentic ecosystems.

Temporary lotic ecosystems and carbon metabolism

Temporary lotic ecosystems are those defined as waterways that cease to flow at some points in space and time along their course (Acuña et al., 2014), that include those referred to as "intermittent" and "ephemeral". Temporary lotic ecosystems experience at certain locations or periods of time, the previously described changes in the balance between mineralization, storage, and export. Accordingly, the dominant processes shift from flowing to non-flowing phases and determine major differences between perennial and temporary streams (Acuña and Tockner, 2010; Corti and Datry, 2012; Dieter et al., 2013).

Temporary lotic ecosystems occur in every terrestrial biome on earth, and recent work indicates that 69% of first order streams below 60° latitude flow only intermittently as do even a significant fraction (34%) of fifth order streams (Raymond et al., 2013). In regards to their role in the carbon metabolism, temporary waterways during the non-flowing phase are considered to merely accumulate particulate organic matter because of low mineralization rates (Langhans and Tockner, 2006; Acuña et al., 2007; Bruder et al., 2011). Accordingly, the role of temporary streams during the non-flowing phase has been considered negligible in recent attempts to estimate the role of river networks on the global carbon balance (Raymond et al., 2013). In contrast, other studies reported low but continuous mineralization of organic carbon during the non-flowing phase (Timoner et al., 2012; Gionchetta et al., 2019a). Specifically, extracellular enzyme activity β-glucosidase, responsible for the use of polysaccharides, was detected at relatively high levels during the non-flowing phase in a Mediterranean temporary stream (Fig. 14.3). Extracellular enzyme ratios revealed that there was a preferential carbon demand during non-flow phase, while a preferential nitrogen demand occurred during the last weeks of the flow phase (Fig. 14.4). These results indicate the importance of extracellular enzymes in the maintenance of biofilms during the non-flowing phase (Zoppini and Marxsen, 2011; Burns et al., 2013) showing their potential roles as a survival strategy. The maintenance of heterotrophic processes during the non-flow phase enabled high quality allochthonous organic matter available during the rewetting phase, enhancing the overall stream heterotrophy (Romaní et al., 2006, 2013). A recent study analyzing the effects of rewetting on 200 temporary lotic ecosystems reported that the rewetting involves an R increase between 32 and 66 times respect the non-flow phase (von Schiller et al., 2019).

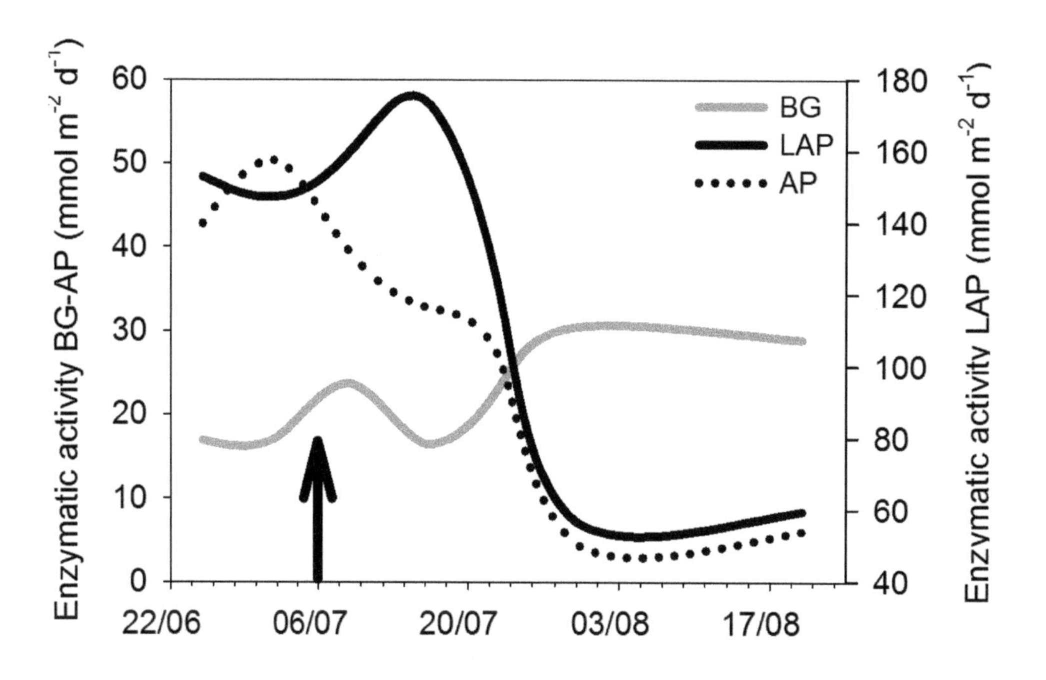

Figure 14.3 Temporal changes in BG (β-glucosidase activity), AP (phosphatase activity), and LAP (leucine-aminopeptidase activity) of the stream biofilm in the Fuirosos stream during the summer of 2009. Note that the vertical arrow indicates flow cessation, and therefore differentiates the flow (left) and the non-flow (right) phases (based on Timoner et al., 2012).

Timoner et al. (2012) found that during the non-flowing phase, between 30 - 40 % of the total bacterial cells remained alive in the streambed sediments (Fig. 14.4), probably contributing to the maintenance of the heterotrophic activity during that period, suggesting that the bacterial community was able to adapt to the new environmental conditions.

Due to the irregular flows of temporary streams, microbial communities might be composed by a set of species capable to resist and persist under these extreme conditions (Febria et al., 2011; Zeglin et al., 2011). However, depending on the duration, intensity and frequency of the non-flow periods, some studies reported changes in microbial community composition (i.e. bacteria and archaea) during the non-flowing phase (Marxsen et al., 2010; Timoner et al., 2014; Gionchetta et al. 2019b). In fact, bacterial communities in the biofilms growing on the coarse sediments (i.e. rocks) experienced dramatic changes and became dominated by

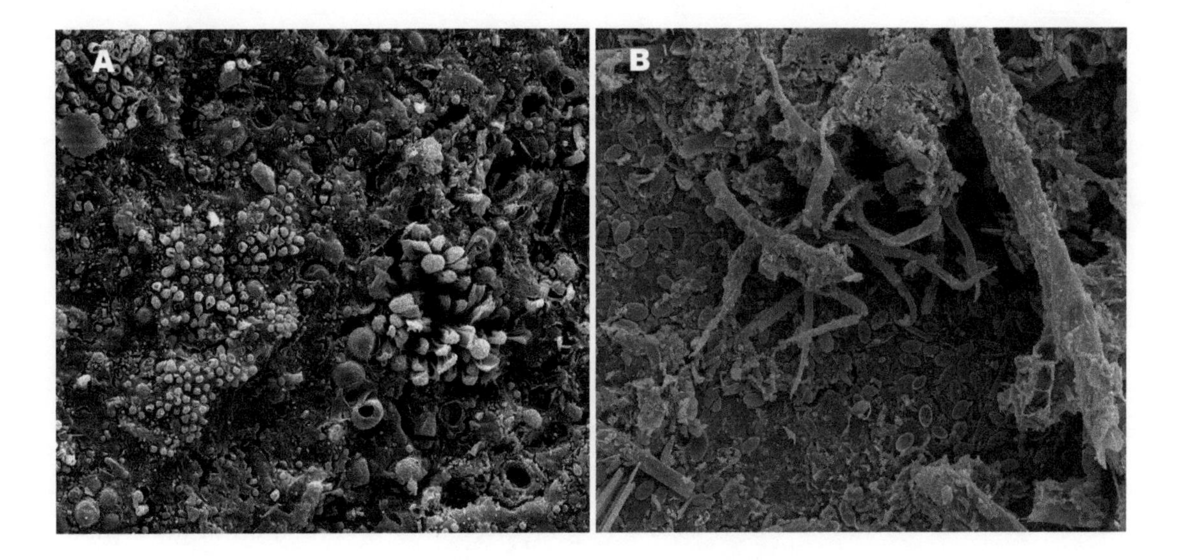

Figure 14.4 Biofilm appearances under the scanning microscope during the flowing (A) and non-flowing (B) phases in the Fuirosos stream. Note the structures such as matrix of polysaccharides developed in the biofilm during the flowing phase resist the non-flowing phase.

terrestrial Firmicutes during the non-flowing phase; whereas biofilms growing on finer sediments (i.e. sands) were richer compared to those growing on rocks and experienced fewer changes during the non-flowing phase. Overall, either through resistant species or through community changes, the bacterial community remained alive and active during the non-flowing phase, threatening the estimates of the role of the river networks on the global carbon balance assuming a negligible role of temporary waterways during the non-flowing phase.

Expected changes in the carbon metabolism by flow trends

The climate change driven likely increase in the frequency and magnitude of extreme flow events might enhance exports respect mineralization, and alter the relative role of lotic and lentic water bodies (Acuña and Tockner, 2010). Thus, less organic carbon might be mineralized in lotic water bodies, but more might be mineralized in lentic water bodies, partly counteracting the effects of flow regime alteration on the carbon dynamics at the river network scale. In other words, reservoirs and lakes might partly counteract the effects of flow extremes on carbon export rates, and their role in the carbon dynamics increases with increasing flow variability (Tranvik et al., 2009; Acuña and Tockner, 2010). Another relevant impact of flow trends on the interplay between lotic and lentic systems is the loss of dilution capacity of point sources of labile carbon in human-impacted watersheds. This is especially relevant for river networks interrupted by dams, where longer residence times facilitate the mineralization of labile carbon loads (Marcé et al., 2008). Actually, O_2 consumption and extent of anoxia in human-impacted reservoirs have increased in response of recent decreasing flow trends, a consequence of an enhanced R activity mostly fueled by human-derived labile organic carbon (Marcé et al., 2010). Whether mineralization of labile organic carbon from human activities is enhanced by decreasing flow trends also in rivers is a question that is yet to be answered.

Both floods and non-flow events might depress mineralization of organic carbon temporarily, and non-flow events might lead to an accumulation of particulate organic carbon in the temporary sections of river networks because of slower rates of mineralization respect the supply from terrestrial ecosystems. Accordingly, we expect a reduction in CO_2 emissions from river networks (Marcé et al., 2019) and an increase in organic carbon export. Furthermore, because of the likely increase of the role of lentic water bodies in the overall mineralization of organic carbon within river networks, we also expect a relative increase of the CH_4 emissions respect those of CO_2. The alteration of the organic carbon dynamics in terms of spatial and temporal trends might certainly alter the compartments and periods when organic carbon is stored or mineralized. In other words, the changes in the residence time of organic carbon in either lotic or lentic water bodies, and the residence time of organic carbon during downstream transport will influence the efficiency with which river networks mineralize organic carbon (Battin et al., 2008).

The alteration of the flow regime might also alter the relationship between river networks and neighboring ecosystems such as riparian forests and peatlands (Tranvik et al., 2009). In fact, changes in the rainfall regime might profoundly alter the timing of the major organic carbon fluxes, as well as the relative relevance of the surface versus subsurface connections, or the relative abundance of particulate versus dissolved organic carbon inputs (Likens and Bormann, 1974; Lohse et al., 2009; Manzoni and Porporato, 2011). Furthermore, changes in rainfall regime might also alter the functioning of these neighboring ecosystems, altering for example the magnitude and timing of litterfall from riparian forests (Acuña et al., 2007), or the magnitude, timing and forms of organic carbon entering the river network from peatlands (Fenner and Freeman, 2011).

Conclusions and future trends

The carbon metabolism in river networks shapes to a large extent the regional and global carbon balances, and will be highly sensitive to Global climate change owing to major increases in the extent of the non-flow periods, as well as in flood frequency and magnitude. Existing evidence points out that these alterations in the flow regime might increase organic carbon export rates, whereas temperature alterations will increase mineralization of organic carbon. The specific roles of lotic and lentic water bodies within river networks might also change, as the lakes and reservoirs might increase their roles in the carbon balance and partly counteract the effects of flow extremes on organic carbon export rates. Furthermore, construction of new impoundments might also increase their relative role in the global carbon balance, altering not only the fraction of the mineralized organic carbon processes in lentic water bodies, but the type of emissions given the predominance of CH_4 emissions in those systems. Because of the relevance of the river networks in the global carbon balance, there are many issues that need to be addressed in future work if we aim to better comprehend their current and future roles.

Given that the CO_2 emissions predicted by Raymond et al. (2013) are largest from tropical rivers and lakes in Southeast Asia and Amazonia, and that freshwater ecosystems from tropical regions are seriously under-represented in global data sets, additional studies of carbon dynamics in the predicted hotspot areas in the tropics are urgently needed (Wehrli, 2013). Another weakness of the estimates reported by Raymond et al. (2013) is the specific role of temporary waterways in the global carbon balance, and more specifically, the role of the non-flow phase of these systems. We believe that it needs to be quantified and thus included in the estimates of the role of river networks in the global carbon balance.

Regardless of region or flow regime, organic carbon mineralization is influenced by temperature, but we need to clarify the potential influence of acclimation mechanisms of heterotrophic R to warmer temperatures. In this respect, the differential sensitivity of aerobic and anaerobic respiration to temperature deserves more attention in the future, because the huge warming potential of CH_4 combined to a higher sensitivity to warming may substantially amend our conclusions about the relevance of flow regime versus temperature changes in carbon emissions. This is certainly crucial to predict metabolism changes owing to global warming, and more empirical data from different temporal and spatial scales is needed to clarify the current divergence between different types of ecosystems, as well as within aquatic ecosystems.

Changes in the flow regime seem to be far more relevant than those in the temperature regime, and these changes in the flow regime will not only influence the metabolism within river networks, but the interactions with neighboring ecosystems, which need to be better understood. In fact, most studies focus on single ecosystems, whereas the interactions among ecosystems are crucial in determining the overall carbon balance, and the particulate carbon fractions that are processed or transported. Mechanistic models should therefore account for fluxes of organic carbon in its different forms from terrestrial to aquatic ecosystems, considering the specific role of lotic and lentic water bodies, as well as the organic mineralization in aerobic and anaerobic conditions. In this direction, and given the higher global warming potential respect CO_2, CH_4 emissions need to be better quantified and included in the regional and global carbon balances (Abril et al., 2005; Wehrli, 2013).

References

Abril, G., Guérin, F., Richard, S., Delmas, R., Galy-Lacaux, C., Gosse P., et al. (2005). Carbon dioxide and methane emissions and the carbon budget of a 10-year old tropical reservoir (Petit Saut, French Guiana). Glob. Biogeochem. Cycles 13, 19:GB4007.

Abril, G., Martinez, J.-M., Artigas, L.F., Moreira-Turcq, P., Benedetti, M.F., Vidal, L., et al. (2013). Amazon River carbon dioxide outgassing fueled by wetlands. Nature 505, 395-398.

Acuña, V., Giorgi, A., Muñoz, I., Sabater, F., and Sabater, S. (2007). Meteorological and riparian influences on organic matter dynamics in a forested Mediterranean stream. J. N. Am. Benthol. Soc. 26, 54–69.

Acuña, V., Giorgi, A., Muñoz, I., Uehlinger, U., and Sabater, S. (2004). Flow extremes and benthic organic matter shape the metabolism of a headwater Mediterranean stream. Freshw. Biol. 49, 960–971.

Acuña, V., and Tockner, K. (2010). The effects of alterations in temperature and flow regime on organic carbon dynamics in Mediterranean river networks. Glob. Change Biol. 16, 2638–2650.

Acuña, V., Wolf, A., Uehlinger, U., and Tockner, K. (2008). Temperature dependence of stream benthic respiration in an Alpine river network under global warming. Freshw. Biol. 53, 2076–2088.

Algesten, G., Sobek, S., Bergström, A., Agren, A., Tranvik, L.J., and Jansson, M. (2004). Role of lakes for organic carbon cycling in the boreal zone. Glob. Change Biol. 10, 141–147.

Allan, J.D., and Castillo, M.M. (2007). Stream Ecology, 2nd ed. (Dordrecht, The Netherlands: Springer).

Allen, A.P, Gillooly, J.F., and Brown, J.H. (2005). Linking the global carbon cycle to individual metabolism. Funct. Ecol. 19, 202–213.

Arora, R., Tockner, K., and Venohr, M. (2016). Changing river temperatures in northern Germany: trends and drivers of change. Hydrol. Process. 30, 17.

Arrhenius, S. (1915). Quantitative Laws in Biological Chemistry (London, UK: Bell).

Atkin, O.K., Edwards, E.J., and Loveys, B.R. (2000). Response of root respiration to changes in temperature and its relevance to global warming. New Phytol. 147, 141–154.

Aufdenkampe, A.K., Mayorga, E., Raymond, P.A., Melack, J.M., Doney, S.C., Alin, S.R., et al. (2011). Rivering coupling of biogeochemical cycles between land, oceans and atmosphere. Front. Ecol. Environ. 9, 53–60.

Bastviken, D., Cole, J., Pace, M., and Tranvik, L. (2004). Methane emissions from lakes: Dependence of lake characteristics, two regional assessments, and a global estimate. Glob. Biogeochem. Cycles 20, GB4009.

Bastviken, D., Tranvik, L.J., Downing, J.A., Crill, P.M., and Enrich-Prast, A. (2011). The Continental Carbon Sink. Science 331, 50.

Battin, T.J., Kaplan, L.A., Findlay, S., Hopkinson, C.S., Marti, E., Packman, A.I., et al. (2008). Biophysical controls on organic carbon fluxes in fluvial networks. Nat. Geosci. 1, 95–100.

Beaulieu, J.J., Tank, J.L., Hamilton, S.K., Wollheim, W.M., Hall, R.O., Mulholland, P.J., et al. (2011). Nitrous oxide emission from denitrification in stream and river networks. Proc. Natl. Acad. Sci. U.S.A. 108, 214–219.

Benestad, R.E. (2004). Record-values, nonstationarity tests and extreme value distributions. Glob. Planet Change 44, 11–26.

Boltzmann, L. (1872). Weitere Studien über das Wärmegleichgewicht unter Gasmolekülen. Wiener Berichte 66, 275–370.

Boulêtreau, S., Salvo, E., Lyautey, E., Mastrorillo, S., and Garabetian, F. (2012). Temperature dependence of denitrification in phototrophic river biofilms. Sci. Total Environ. 416, 323-328.

Bretschko, G. (1995). Running water ecosystems-a bare field for modelling? Ecol. Modell. 78, 77–81.

Brown, J.H., Gillooly, J.F., Allen, A.P., Savage, V.M., and West, G.B. (2004). Towards a metabolic theory of ecology. Ecology 85, 1771–1789.

Bruder, A., Chauvet, E., and Gessner, M.O. (2011). Litter diversity, fungal decomposers and litter decomposition under simulated stream intermittency. Funct. Ecol. 25, 1269-1277.

Burns, R.G., DeForest, J.L., Marxsen, J., Sinsabaugh, R.L., Stromberger, M.E., Wallenstein, M.D., Weintraub, M.H., and Zoppini, A. (2013). Soil enzymes in a changing environment: current knowledge and future directions. Soil Biol. Biochem. 58, 216–234. DOI: 10.1016/j.soilbio.2012.11.009

Butman, D., and Raymond, P.A. (2011). Significant efflux of carbon dioxide from streams and rivers in the United States. Nat. Geosci. 4, 839–842.

Chróst, R.J. (1990). Microbial ectoenzymes in aquatic environments. In Aquatic Microbial Ecology: Biochemical and Molecular Approaches, J. Overbeck and R.J. Chróst, eds. (New York, NY: Springer-Verlag) pp. 47–78.

Cole, J.J. (2013). Freshwater in flux. Nat. Geosci. 6, 13–14.

Cole, J.J., Prairie, Y.T., Caraco, N.F., McDowell, W.H., Tranvik, LJ., Striegl, R.G., et al. (2007). Plumbing the Global Carbon Cycle: Integrating Inland Waters into the Terrestrial Carbon Budget. Ecosystems 10, 172–185.

Comer-Warner, S.A., Romejin, P., Goody, D.C., Ullah, S., Kettridge, N., Marchant, B., Hannah, D.M., and Krause, S. (2018). Thermal sensitivity of CO_2 and CH_4 emissions varies with streambed sediment properties. Nature Commun. 9, 2803.

Corti, R., and Datry, T. (2012). Invertebrates and sestonic matter in an advancing wetted front travelling down a dry riverbed (Albarine, France). Freshw. Sci. 31, 1187–201.

Crawford, J.T., Loken, L.C., Stanley, E.H., Stets, E.G., Dornblaser, M.M., and Striegl, R.G. (2016). Basin scale controls on CO_2 and CH_4 emissions from the Upper Mississippi River. Geophys. Res. Lett. 43, 1973–1979.

Dang, C.K., Schindler, M., Chauvet, E., and Gessner, M.O. (2009). Temperature oscillation coupled with fungal community shifts can modulate warming effects on litter decomposition. Ecology 90, 122–131.

Daniels, L., Fuchs, G., Thauer, R.K., and Zeikus, J.G. (1977). Carbon monoxide oxidation by methanogenic bacteria. J. Bacteriol. 132, 118–126.

Demars, B.O.L., and Manson, J.R. (2013). Temperature dependence of stream aeration coefficients and the effect of water turbulence: a critical review. Water Res. 47, 1–15.

Demars, B.O.L., Russell-Manson, J., Ólafsson, J.S., Gíslason, G.M., Gudmundsdóttir, R., Woodward, G., et al. (2011). Temperature and the metabolic balance of streams. Freshw. Biol. 56, 1106–1121.

Demars, B.O.L., Gíslason, G.M., Ólafsson, J.S., Manson, J.R., Friberg, N., Hood, J.M., Thompson, J.J.D., and Freitag, T.E. (2016). Impact of warming on CO_2 emissions from streams countered by aquatic photosynthesis. Nature Geosci. 9, 758-761.

Demars, B.O.L. (2019). Hydrological pulses and burning of dissolved organic carbon by stream respiration. Limnol. Oceanogr. 64, 406-421.

Denman, K.L., and Brasseur, G. (2007). Couplings Between Changes in the Climate System and Biogeochemistry. In Climate Change 2007. The Physical Science Basis. Contribution of Working Group I to the Fourth Assessment Report of the Intergovernmental Panel on Climate Change, S. Solomon, D. Qin, M. Manning, Z. Chen, M. Marquis, K.B. Averyt et al., eds. (Cambridge, UK: Cambridge University Press).

Dieter, D., Frindte, K., Krüger, A., Wurzbacher, C. (2013). Preconditioning of leaves by solar radiation and anoxia affects microbial colonisation and rate of leaf mass loss in an intermittent stream. Freshw. Biol. 58, 1918–1931.

Downing, J.A., Cole, J.J., Middelburg, J.J., Striegl, R.G., Duarte, C.M., Kortelainen, P., et al. (2008). Sediment organic carbon burial in agriculturally eutrophic impoundments over the last century. Global Biogeochem. Cycles 22, GB1018.

Drake, T.W., Raymond, P.A., and Spencer, R.G.M. (2018). Terrestrial carbon inputs to inland waters: A current synthesis of estimates and uncertainty. Limnol. Oceanogr. 3, 132-142. DOI: 10.1002/lol2.10055.

Elmore, H.L., and West, W.F. (1961). Effect of water temperature on stream reaeration. J. Sanit. Eng. Div. ASCE 87, 59–71.

Enquist, B.J., Economo, E.P., Huxman, T.E., Allen, A.P., Ignace, D.D., and Gillooly, J.F. (2003). Scaling metabolism from organisms to ecosystems. Nature 423, 639–642.

Febria, C.M., Beddoes, P., Fulthorpe, R.R., and Williams, D.D. (2011). Bacterial community dynamics in the hyporheic zone of an intermittent stream. ISME J. 6, 1078–1088.

Fenner, N., and Freeman, C. (2011). Drought-induced carbon loss in peatlands. Nat. Geosci. 4, 895–900.

Fierer, N., Colman, B.P., Schimel, J.P., and Jackson, R.B. (2006). Predicting the temperature dependence of microbial respiration in soil: A continental-scale analysis. Global Biogeochem. Cycles. 20, GB3026.

Findlay, S.E.G. (2010). Stream microbial ecology. J. N. Am. Benthol. Soc. 29, 170–181.

Findlay, S.E.G., and Sinsabaugh, R.L. (1999). Unravelling the sources and bioavailability of dissolved organic matter in lotic aquatic ecosystems. Mar. Freshw. Res. 50, 781-790.

Findlay, S.E.G., Sinsabaugh, R.L., Fisher, D.T., and Francini, P. (1998). Sources of dissolved organic carbon supporting planktonic bacterial production in the tidal freshwater Hudson River. Ecosystems 1, 227–239.

Forster, P., Ramaswamy, V., Artaxo, P., Berntsen, T., Betts, R., Fahey, D.W., et al. (2007). Changes in Atmospheric Constituents and in Radiative Forcing. In Climate Change 2007. The Physical Science Basis. Contribution of Working Group I to the Fourth Assessment Report of the Intergovernmental Panel on Climate Change, S. Solomon, D. Qin, M. Manning, Z. Chen, M. Marquis, K.B. Averyt et al., eds. (Cambridge, UK: Cambridge University Press).

Forster, P.M., Maycock, A.C., McKenna, C.M., and Smith, C.J. (2010). Latest climate models confirm need for urgent mitigation. Nat. Clim. Change 10, 7-10.

Freixa, A., Acuña, V., Casellas, M., Pecheva, S., and Romaní, A.M. (2017). Warmer night-time temperature promotes microbial heterotrophic activity and modifies stream sediment community. Glob. Change Bio. 23, 3825-3837.

Friedlingstein, P., Jones, M.W., O'Sullivan, M., Andrew, R.M., Hauck, J., Peters, G.P., Peters, W., Pongratz, J., Sitch, S., Le Quéré, C., Bakker, D.C.E., Canadell, J.G., Ciais, P., Jackson, R.B., Anthoni, P., Barbero, L., Bastos, A., Bastrikov, V., Becker, M., Bopp, L., Buitenhuis, E., Chandra, N., Chevallier, F., Chini, L.P., Currie, K.I., Feely, R.A., Gehlen, M., Gilfillan, D., Gkritzalis, T., Goll, D.S., Gruber, N., Gutekunst, S., Harris, I., Haverd, V., Houghton, R.A., Hurtt, G., Ilyina, T., Jain, A.K., Joetzjer, E., Kaplan, J.O., Kato, E., Klein Goldewijk, K., Korsbakken, J.I.,

Landschützer, P., Lauvset, S.K., Lefèvre, N., Lenton, A., Lienert, S., Lombardozzi, D., Marland, G., McGuire, P.C., Melton, J.R., Metzl, N., Munro, D.R., Nabel, J.E.M.S., Nakaoka, S.I., Neill, C., Omar, A.M., Ono, T., Peregon, A., Pierrot, D., Poulter, B., Rehder, G., Resplandy, L., Robertson, E., Rödenbeck, C., Séférian, R., Schwinger, J., Smith, N., Tans, P.P., Tian, H., Tilbrook, B., Tubiello, F.N., van der Werf, G.R., Wiltshire, A.J., and Zaehle, S. (2019). Global Carbon Budget 2019, Earth Syst. Sci. Data 11, 1783–1838. DOI: 10.5194/essd-11-1783-2019

Gardner, J.R., Pavelsky, T.M., and Doyle, M.W. (2019). The abundance, size, and spacing of lakes and reservoirs connected to river networks. Geophysical Res. Lett. 46, 2592–2601.

Gifford, R.M. (1995). Whole plant respiration and photosynthesis of wheat under increased CO_2 concentration and temperature: long-term vs. short-term distinctions for modelling. Glob. Change Biol. 1, 385–396.

Gillooly, J.F., Brown, J.H., West, G.B., Savage, V.M., and Charnov, E.L. (2001). Effects of size and temperature on metabolic rate. Science. 293, 2248–2251.

Gionchetta, G., Oliva, F., Menéndez, M., Lopez Laseras, P., and Romaní, A.M. (2019). Key role of streambed moisture and flash storms for microbial resistance and resilience to long-term drought. Freshw. Biol. 64, 306-322.

Gionchetta, G., Romaní, A.M., Oliva, F., and Artigas, J. (2019). Distinct responses from bacterial, archaeal and fungal streambed communities to severe hydrological disturbances. Sci. Rep. 9, 1-13.

Grill, G., Lehner, B., Thieme, M. et al. (2019). Mapping the world's free-flowing rivers. Nature 569, 215–221. DOI: 10.1038/s41586-019-1111-9

Guarch-Ribot, A., and Butturini, A. (2016). Hydrological conditions regulate dissolved organic matter quality in an intermittent headwater stream. From drought to storm analysis. Sci. Total Environ. 571, 1358-1369.

Gudasz, C., Bastviken, D., Steger, K., Premke, K., Sobek, S., and Tranvik, L.J. (2010). Temperature-controlled organic carbon mineralization in lake sediments. Nature 466, 478–481.

Gudmundsson, L., Leonard, M., Do, H.X., Westra, S., and Seneviratne, S.I. (2019). Observed trends in global indicators of mean and extreme streamflow. Geophys. Res. Lett. 46, 756–766.

Hedges, J.I., and Oades, J.M. (1997). Comparative organic geochemistries of soils and marine sediments. Org. Geochem. 27, 319–361.

Hirabayashi, Y., Kanae, S., and Emori, S. (2008). Global projections of changing risks of floods and droughts in a changing climate. Hydrol. Sci. – J. Sci. Hydrol. 53, 754–772.

Hoellein, T., Bruesewitz, J., Denise, A., and Richardson, D.C., (2013). Revisiting Odum (1956): A synthesis of aquatic ecosystem metabolism. Limnol. Oceanogr. 58, 2089-2100. DOI: 10.4319/lo.2013.58.6.2089

Hotchkiss, E.R., Hall, J.R.O., Sponseller, R.A., Butman, D., Klaminder, J., Laudon, H., and Karlsson, J. (2015). Sources of and processes controlling CO_2 emissions change with the size of streams and rivers. Nat. Geosci. 8, 696-699.

Hoppe, H.-G. (1993). Use of Fluorogenic Model Substrates for Extracellular Enzyme Activity (EEA) Measurement of Bacteria. In Handbook of Methods in Aquatic Microbial Ecology, P.F. Kemp, B.F. Sherr, E.B. Sherr and J.J. Cole, eds. (Boca Raton, FL: Lewis Publishers), pp. 423–431.

Huntington, T.G. (2006). Evidence for intensification of the global water cycle: Review and synthesis. J. Hydrol. 319, 83–95.

Jane, S.F., and Rose, K.C. (2018). Carbon quality regulates the temperature dependence of aquatic ecosystem respiration. Freshw. Biol. 63, 1407-1419.

Janssens, I.A., Lankreijer, H., Matteucci, G., Kowalski, A.S., Buchmann, N., Epron, D., et al. (2001). Productivity overshadows temperature in determining soil and ecosystem respiration across European forests. Glob. Change Biol. 7, 269–278.

Kirschbaum, M.U.F. (1995). The temperature dependence of soil organic matter decomposition, and the effect of global warming on soil organic carbon storage. Soil Biol. Biochem. 27, 753–760.

Kirschbaum, M.U.F. (2010). The temperature dependence of organic matter decomposition: seasonal temperature variations turn a sharp short-term temperature response into a more moderate annually averaged response. Glob. Change Biol. 16, 2117–2129.

Langan, S.J., Johnston, L., Donaghy, M.J., Youngson, A.F., Hay, D.W., and Soulsby, C. (2001). Variation in river water temperatures in an upland stream over a 30-year period. Sci. Total Environ. 265, 195–207.

Langhans, S.D., and Tockner, K. (2006). The role of timing, duration, and frequency of inundation in controlling leaf litter decomposition in a river-floodplain ecosystem (Tagliamento, northeastern Italy). Oecologia 147, 501–509.

Lauerwald, R., Laruelle, G.G., Hartmann, J., Ciais, P., and Regnier, P.A. (2015). Spatial patterns in CO_2 evasion from the global river network. Global Biogeochem. Cycles 29, 534–554. DOI: 10.1002/2014GB004941

Leopold, L.B. (1964). Fluvial Processes in Geomorphology (San Francisco, CA: W.H. Freeman).

Likens, G.E., and Bormann, F.H. (1974). Linkages between terrestrial and aquatic ecosystems. Bioscience 24, 447–456.

Lock, M.A. (1993). Attached Microbial Communities in Rivers. In Aquatic Microbiology, An Ecological Approach, T.E. Ford, ed. (Oxford, UK: Blackwell), pp. 113–138.

Lock, M.A., Wallace, R.R., Costerton, J.W., Ventullo, R.M., and Charlton, S.E. (1984). River epilithon: toward a structural-functional model. Oikos 42, 10–22.

Lohse, K.A., Brooks, P.D., McIntosh, J.C., Meixner, T., and Huxman, T.E. (2009). Interactions between biogeochemistry and hydrologic systems. Annu. Rev. Environ. Resour. 34, 65–96.

López-Urrutia, A., and Morán, X.A. (2007). Resource limitation of bacterial production distorts the temperature dependence of oceanic carbon cycling. Ecology 88, 817–822.

López-Urrutia, A., San Martin, E., Harris, R.P., and Irigoien, X. (2006). Scaling the metabolic balance of the oceans. Proc. Natl. Acad. Sci. U.S.A. 103, 8739–8744.

Maavara, T., Lauerwald, R., Regnier, P. and Van Cappellen, P. (2017). Global perturbation of organic carbon cycling by river damming. Nat. Commun. 8, 15347.

Maavara, T., Chen, Q., Van Meter, K. et al. (2020). River dam impacts on biogeochemical cycling. Nat. Rev. Earth Environ. 1, 103–116. DOI: 10.1038/s43017-019-0019-0

Maeck, A., Delsontro, T., McGinnis, D.F., Fischer, H., Flury, S., Schmidt, M., Fietzek, P., and Lorke, A. (2013). Sediment trapping by dams creates methane emission hot spots. Environ. Sci. Technol. 47, 8130–8137.

Manzoni, S., and Porporato, A. (2011). Common hydrologic and biogeochemical controls along the soil-stream continuum. Hydrol. Process. 25, 1355–1360.

Marcé, R., Moreno-Ostos, E., López, P., and Armengol, J. (2008). The role of allochthonous inputs of dissolved organic carbon on the hypolimnetic oxygen content of reservoirs. Ecosystems 11, 1035–1053.

Marcé, R., Rodríguez-Arias, M.A., García, J.C., and Armengol, J. (2010). El Niño Southern oscillation and climate trends impact reservoir water quality. Glob. Change Biol. 16, 2857–2865.

Marcé, R., Obrador, B., Gómez-Gener, L., Catalán, N., Koschorreck, M., Arce, M. I., et al. (2019). Emissions from dry inland waters are a blind spot in the global carbon cycle. Earth Sci. Rev. 188, 240–248. DOI: 10.1016/j.earscirev.2018.11.012

Marxsen, J., Zoppini, A., and Wilczek, S. (2010). Microbial communities in streambed sediments recovering from desiccation. FEMS Microbiol. Ecol. 71, 374-386. DOI: 10.1111/j.1574-6941.2009.00819.x

Mendonça, R., Müller, R.A., Clow, D. et al. (2017). Organic carbon burial in global lakes and reservoirs. Nat. Commun. 8, 1694. 10.1038/s41467-017-01789-6

Miao, S., Zou, C.B., and Breshears, D.D. (2009). Vegetation responses to extreme hydrological events: sequence matters. Am. Nat. 173, 113–118.

Minshall, G.W., Petersen, R.C., Cummins, K.W., Bott, T.L., Sedell, J.R., Cushing, C.E., et al. (1983). Interbiome comparison of stream ecosystem dynamics. Ecol. Monogr. 53, 1–25.

Mouthon, J., and Daufresne, M. (2006). Effects of the 2003 heatwave and climatic warming on mollusc communities of the Saone: a large lowland river and of its two main tributaries (France). Glob. Change Biol. 12, 441–449.

Mulholland, P.J., Fellows, C.S., Tank, J.L., Grimm, N.B., Webster, J.R., Hamilton, S.K., et al. (2001). Inter-biome comparison of factors controlling stream metabolism. Freshw. Biol. 46, 1503–1517.

New, M., Todd, M., Hulme, M., and Jones, P. (2001). Precipitation measurements and trends in the twentieth century. Int. J. Climatol. 21, 1899–1922.

Park, P.K., Gordon, L.I., Hager, S.W., and Cissell, M.C. (1969). Carbon dioxide partial pressure in the Columbia River. Science 166, 867-868.

Perkins, D.M., Yvon-Durocher, G., Demars, B.O.L., Reiss, J., Pichler, D.E., Friberg, N., et al. (2012). Consistent temperature dependence of respiration across ecosystems contrasting in thermal history. Glob. Change Biol. 18, 1300–1311.

Peterson, C.G. (1996). Response of Benthic Algal Communities to Natural Disturbance. In Algal Ecology, R.J. Stevenson, M.L. Bothwell, and R.L. Lowe, R.L., eds. (San Diego, USA: Academic Press Inc), pp. 375–402.

Proia, L., von Schiller, D., Gutierrez, C., Casas-Ruiz, J.P., Gómez-Gener, L., Marcé, R., Obrador, B., Acuña, V., and Sabater, S. (2016). Microbial carbon processing along a river discontinuum. Freshw. Sci. 35, 1133–1147.

Qi, Y., Xu, M., and Wu, J. (2002). Temperature sensitivity of soil respiration and its effects on ecosystem carbon budget: nonlinearity begets surprises. Ecol. Modell. 153, 131–142.

Raymond, P.A., Hartmann, J., Lauerwald, R., Sobek, S., McDonald, C., Hoover, M., et al. (2013). Global carbon dioxide emissions from inland waters. Nature 503, 355–359.

Reisinger, A.J., Rosi, E.J., Bechtold, H.A., Doody, T.R., Kaushal, S.S., and Groffman, P.M. (2017). Recovery and resilience of urban stream metabolism following Superstorm Sandy and other floods. Ecosphere 8, e01776.

Regnier, P., Friedlingstein, P., Ciais, P., Mackenzie, F.T., Gruber, N., and Janssens, I.A. (2013). Anthropogenic perturbation of the carbon fluxes from land to ocean. Nat. Geosci. 6, 597–607.

Romaní, A.M., Amalfitano, S., Artigas, J., Fazi, S., Sabater, S., Timoner, X., et al. (2013). Microbial biofilm structure and organic matter use in Mediterranean streams. Hydrobiologia 719, 43–58.

Romaní, A.M., and Sabater, S. (2001). Structure and activity of rock and sand biofilms in a Mediterranean stream. Ecology 82, 3232–3245.

Romaní, A.M., Vázquez, E., and Butturini, A. (2006). Microbial availability and size fractionation of dissolved organic carbon after drought in an intermittent stream: Biogeochemical link across the stream–riparian interface. Microb. Ecol. 52, 501–512.

Romejin, P., Comer-Warner, S.A., Ullah, S., Hannah, D.M., and Krause, S. (2019). Streambed organic matter controls on Carbon dioxide and Methane emissions from streams. Env. Sci. Technol. 53, 2364-2374.

Sand-Jensen, K., and Pedersen, N.L. (2005). Differences in temperature, organic carbon and oxygen consumption among lowland streams. Freshw. Biol. 50, 1927–1937.

Sigee, D.C. (2005). Freshwater Microbiology: Biodiversity and Dynamic Interactions of Microorganisms in the Aquatic Environment (Chichester, UK: John Wiley & Sons Ltd).

Sigleo, A.C., and Macko, S.A. (2002). Carbon and nitrogen isotopes in suspended particles and colloids, Chesapeake and San Francisco estuaries, U.S.A. Estuar. Shelf Sci. 54, 701–711.

Silver, W.L., and Miya, R.K. (2001). Global patterns in root decomposition: comparisons of climate and litter quality effects. Oecologia 129, 407–419.

Song, C., Dodds, W.K., Rüegg, J., Argerich, A., Baker, C.L., Bowden, W.B., et al. (2018). Continental-scale decrease in net primary productivity in streams due to climate warming. Nat. Geosci. 11, 415-420. DOI: 10.1038/s41561-018-0125-5

Stets, E.G., Butman, D., McDonald, C.P., Stackpoole, S.M., DeGrandpre, M.D., and Striegl, R.G. (2017). Carbonate buffering and metabolic controls on carbon dioxide in rivers. Global Biogeochem. Cycles 31, 663– 677. DOI: 10.1002/2016GB005578

Stocker, T.F., Qin, D., Plattner, G.K., et al. (2013). Technical Summary. In Climate Change 2013: The Physical Science Basis. Contribution of Working Group I to the Fifth Assessment Report of the Intergovernmental Panel on Climate Change, T.F. Stocker, D. Qin, G.K. Plattner, M. Tignor, S.K. Allen, J. Boschung, A. Nauels, Y. Xia, V. Bex, and P.M. Midgley, eds. (Cambridge, UK, and New York, NY, USA: Cambridge University Press).

Timoner, X., Acuña, V., von Schiller, D., and Sabater, S. (2012). Functional responses of stream biofilms to flow cessation, desiccation and rewetting. Freshw. Biol. 57, 1565–1578.

Timoner, X., Borrego, C.M., Acuña, V., and Sabater, S. (2014). The dynamics of biofilm bacterial communities is driven by flow wax and wane in intermittent streams. Limnol. Oceanogr. 59, 2057-2067. DOI: 10.4319/lo.2014.59.6.2057

Tranvik, L., Downing, J.A., Cotner, J.B., Loiselle, S.A., Striegl, R.G., Ballatore, T.J., et al. (2009). Lakes and reservoirs as regulators of carbon cycling and climate. Limnol. Oceanogr. 54, 2298–2314.

Vannote, R.L., Minshall, G.W., Cummins, K.W., Sedell, J.R., and Cushing, C.E. (1980). The river continuum concept. Can. J. Fish. Aquat. Sci. 37, 130–137.

Vrba, J., Callieri, C., Bittl, T., Šimek, K., Bertoni, R., Filandr, P., et al. (2004). Are bacteria the major producers of extracellular glycolytic enzymes in aquatic environments? Int. Rev. Hydrobiol. 89, 102–117.

Webb, B.W., and Nobilis, F. (1994). Water temperature behavior in the River Danube during the 20th-century. Hydrobiologia 291, 105–113.

Webster, J.R., and Meyer, J.L. (1997). Organic matter budgets for streams: a synthesis. J. N. Am. Benthol. Soc. 16, 141–161.

Wehrli, B. (2013). Conduits of the carbon cycle. Nature 503, 346–347.

Westermann, P. (1993). Temperature regulation of methanogenesis in wetlands. Chemosphere 26, 321–328.

Wetzel, R.G. (1983). Periphyton of Freshwater Ecosystems (The Hague, Boston, Lancaster: Dr. W. Junk Publishers).

Whiting, G.J., and Chanton, J.P. (2001). Greenhouse carbon balance of wetlands: methane emission versus carbon sequestration. Tellus B. 53, 521–528.

Wiley, M.J., Osborne, LL., and Larimore, R.W. (1990). Longitudinal structure of an agricultural prairie river system and its relationship to current stream ecosystem theory. Can. J. Fish. Aquat. Sci. 47, 373–384.

Wimpenny, J., Manz, W., and Szewzyk, U. (2000). Heterogeneity in biofilms. FEMS Microbiol. Rev. 24, 661–671.

Woodwell, G.M., MacKenzie, F.T., Houghton, R.A., Apps, M., Gorham, E., and Davidson, E. (1998). Biotic feedbacks in the warming of the Earth. Clim. Change 40, 495–518.

Young, R., and Huryn, A. (1997). Longitudinal patterns of organic matter transport and turnover along a New Zealand grassland river. Freshw. Biol. 38, 93–107.

Zeglin, L.H., Dahm, C.N., Barrett, J.E., Gooseff, M.N., Fitpatrick, S.K., and Takacs-Vesbach, C.D. (2011). Bacterial community structure along moisture gradients in the parafluvial sediments of two ephemeral desert streams. Microb. Ecol. 61, 543–556.

Zoppini, A., and Marxsen, J. (2011). Importance of Extracellular Enzymes for Biogeochemical Processes in Temporary River Sediments during Fluctuating Dry-wet Conditions. In Soil Enzymology, G. Shukla and A. Varma, eds. (Berlin, Heidelberg: Springer-Verlag), pp. 103–117. DOI: 10.1007/978-3-642-14225-3_6

Chapter 15

Microbial Communities and Processes under Climate and Land-use Change in the Tropics

Stephen A. Wood[1,2,3,*] Krista McGuire[1,4,5] and Jonathan E. Hickman[2,6]

[1] Department of Ecology, Evolution, and Environmental Biology, Columbia University, New York, NY, USA 10027; [2] Agriculture and Food Security Center, The Earth Institute, Columbia University, Palisades, NY USA 10964; [3] The Nature Conservancy, New Haven, CT USA, Yale School of the Environment, New Haven, CT USA (current address); [4] Department of Biology, Barnard College of Columbia University, New York, NY, USA 10027; [5] University of Oregon, Institute of Ecology and Evolution, Eugene, OR, USA 97403 (current address); NASA Goddard Institute for Space Studies, New York, NY-10025, USA (current address); * corresponding author

Email: stephen.wood@tnc.org, kmcguire@uoregon.edu, jonathan.e.hickman@nasa.gov

DOI: https://doi.org/10.21775/9781913652579.15

Abstract

Climate change and land-use change are two of the most important drivers of diversity loss among macrobial taxa. These pressures are especially strong in the tropics, where the effects of climate change may be severe and where economic pressures to convert land to human use are strong. The impact of these two global change drivers on microbial communities, however, is not well studied. Understanding this is important because microorganisms comprise most of the world's biological diversity and play essential roles in the biogeochemical processes that make life possible for higher orders of taxa. In this chapter we review the literature on the impact of these key global change drivers on soil microbial communities and several key microbial mediated biogeochemical processes in the tropics. We find evidence that both climate and land-use change impact the composition and functioning of tropical microbial communities. These two factors may interact, potentially amplifying the consequences of climate change. We propose research priorities for improving understanding of microbial responses to climate- and land-cover change.

Introduction

Microorganisms are the most the abundant taxonomic group on the planet, making up around half of the planet's protoplastic biomass (Roselló-Mora and Amann, 2001). Microbes are the main drivers of Earth's biogeochemical engine (Falkowski et al., 2008), regulating nutrient availability in ecosystems and directly influencing higher trophic levels (Bever et al., 2012). Like all taxa, soil microbes are impacted by changes in environmental conditions such as altered mean and variability of precipitation and temperature and changes in nutrient availability. However, less is known about the resilience and resistance of microbial communities and processes to these changes than is known for taxa from higher trophic levels (Allison and Martiny, 2008).

The extent of available information on the resilience of soil microbial communities and processes to environmental change has largely been limited to temperate zones. In this chapter we assess the impact of environmental change on microbial communities and processes in the tropics. We focus on the two global change drivers that are strongest in the tropics—climate change and land use change (Fig. 15.1).

The tropics

The tropics lie between 23.5 degrees north and south of the equator and make up 38 percent of the total land surface of the Earth. Because of Earth's tilt, the tropics are the only place on the planet where the sun passes directly overhead, resulting in lower seasonal variation in temperature than in higher latitudes. Differences among locations within the tropics are largely driven by altitude. There are four main precipitation zones in the tropics: wet (including tropical rainforests in the Congo, Malaysia, the Upper Amazon, etc); seasonal (much of the savannah zones of Africa and Brazil); dry (mainly northern Australia); and desert (Sanchez, 1977). Unlike temperature, precipitation exhibits high inter-annual variability in the tropics. The Sahel, for instance, has a characteristic drought periodicity of 10 years (Shanahan et al., 2009). The tropics also exhibit greater intra-annual variation in precipitation than temperate zones. For instance, the New England region of the United States and the Guineo-Savannah of West Africa receive similar aggregate annual rainfall, but

Figure 15.1 Diagram of two potential interacting global change drivers that may structure microbial community composition and functioning in the tropics.

rainfall in semi-arid West Africa falls over a three-to-six month rainy season, with a harsh dry season lasting the rest of the year, while in New England precipitation is distributed over most of the year. The pattern of intra-annual precipitation also differs within the tropics. The uni-modal West African savannah has different vegetation than a rainforest in western Kenya that receives a similar amount of precipitation, but distributed bi-modally, thus avoiding the stress of a harsh, long dry season. Tropical rains also vary in their intensity, which impacts the retention of water in soil. In

Surinam 80 percent of a 10 mm hr^{-1} rainfall event remains in the soil, but only 32 percent of a 50 mm hr^{-1} rainfall event (Mohr et al., 1973). Differences in soil type can lead to differences in these rates across the tropics, where different mineralogies are associated with different flow rates through soil even for a single rainfall regime. A sandy soil may be dry in a high precipitation environment if drainage is great enough; a clayey soil with low drainage may exhibit water logging during brief rainfall events in a low-precipitation environment.

Broadly, biological diversity in the tropics is higher than in temperate systems (Willig et al., 2003), though little is known about similar latitudinal gradients for soil organisms; Archaea, bacteria, and eukaryotic microbes do appear to show biogeographic patterns, though these do not seem to follow a latitudinal gradient (Green and Bohannan, 2006; Martiny et al., 2006; Hanson et al., 2012). Land surfaces in the tropics range from the oldest on Earth (ancient crystalline rocks in South America and Africa) to the most recent volcanic deposits. Older tropical soils, however, tend to be limited in phosphorous, which cannot be fixed biologically and is lost by weathering over time (Vitousek et al., 2010). Nitrogen (N) is not thought to be as limiting in the tropics, though it may be in certain areas (Hedin et al., 2009). The wide variation in climate, vegetation, age, and parent material—soil forming factors —in the tropics leads to wide variation in soil types. The only common condition to all soils in the tropics is lack of seasonal variation in temperature (Sanchez, 1977).

Projected climate changes

The tropics are expected to experience global changes differently than temperate zones. On average, global temperature is projected to increase while precipitation patterns become more variable and extreme weather events become more frequent (Stocker et al., 2013). The greatest effects of temperature increases within the tropics are likely to be in tropical Africa and central and Southeast Asia (Stocker et al., 2013). Most tropical areas are expected to have increased mean precipitation, with decreased precipitation in the sub-tropics (Stocker et al., 2013). Increasing dry season lengths and increased precipitation coefficient of variation are expected in much of the Amazon basin, southern and western Africa, and central and Western Australia, while parts of East Africa to India are likely to experience mean precipitation increases and shorter dry seasons (Hulme and Viner, 1998; Stocker et al., 2013).

Mechanisms of microbial response

There are four key mechanisms through which climate change may impact soil microorganisms. First, an increase in temperature can impact the diversity of microbes (Pold and DeAngelis, 2013) as well as their metabolic rates, influencing ecosystem process rates such as decomposition (Bradford, 2013). Second, increase in variance of precipitation—and extreme weather events—can lead to strong dry-wetting cycles, which has been shown to structure microbial community composition (Fierer et al., 2003) and functioning (Fierer and Schimel, 2002). Mean decreases in soil moisture can also modify the microbial community (Hulme and Viner, 1998). Third, the elevated carbon dioxide (CO_2) driving changes in climate also modifies microbial communities through CO_2 fertilization of the aboveground plant community. Elevated CO_2 can lead to increases primary producer biomass, changes in plant community composition, increased litter deposition, changes in litter quality, and increased root exudation (Wardle et al., 2004; Bardgett and Wardle, 2010; Norby and Zak, 2011). This change in plant composition and biomass can alter the nutrient input pathways belowground and shift communities towards dominance by either copio- or oligotrophic taxa. Fourth, use of mineral N pools to sustain CO_2 fertilization could lead to progressive N limitation for both plants and microbes, structuring the community towards more oligotrophic taxa (Rütting et al., 2010).

The other main driver of change in the tropics—and likely driver of change in microbial communities—is change in land cover associated with anthropogenic activities. In the decade leading up to the year 2000, tropical forests were lost at a rate of between 12 and 14 million hectares per year (Kates and Parris, 2003). Though in some temperate regions forest cover is increasing with abandonment of agricultural land (Jeon et al., 2014), in the tropics it is estimated that around 80% of new cropland is replacing forests (Gibbs et al., 2010); agriculture has cleared more than 50% of tropical savannahs and 30% of the world's tropical forest biome (Ramankutty et al., 2008). Tropical agriculture is responsible for 98% of CO_2 emitted from land clearing (DeFries and Rosenzweig, 2010). Agricultural intensification—growing more food on less land through the application of synthetic fertilizers and high-yielding crops, has been hypothesized to reduce pressure on land cover change (Borlaug, 2000)—but has been shown to not decrease pressure on land clearing in the tropics (Rudel et al., 2009). The effect of intensification on greenhouse gas production from agricultural landscapes is uncertain. Increased fertilization has been found to both increase

(Signor et al., 2013) and not impact (Hickman et al., in review) N-based greenhouse gas emissions; these is also evidence that in cases where intensification decreases land clearing, there is a net mitigating effect from avoided CO_2 release (Burney et al., 2010).

There are also four mechanisms through which land cover change can modify soil organisms. First, the loss of vegetation associated with land cover change can decrease organic matter, an important microbial resource, through erosion or loss of organic supply from vegetation. Second, changes in land cover—and, thus, the nutrient inputs to belowground systems—can modify pH, which is a dominant driver of microbial communities (Lauber et al., 2009). Third, changes in aboveground plant composition can lead to the direct loss of belowground taxa when microbial taxa depend on linkages with aboveground systems, such as for certain mycorrhizal relationships and symbiotic N-fixation (Wardle et al., 2004; Bardgett and Wardle, 2010; Hafich et al., 2012). Fourth, land cover changes can modify microbial communities through feedbacks with the climate system, such as increased CO_2 emissions, modified temperature due to different albedo patterns, and changing precipitation due to reduced evapotranspiration from land clearing.

Theory of microbial response

Early theories in microbial ecology assumed that microbial communities were not dispersal limited and, thus, were globally distributed and functionally redundant (Finlay, 2002). Under this set of assumptions, global changes such as climate and land-use change would not influence microbial community composition or microbial functioning since any local die-offs would immediately be recovered through dispersal. These assumptions, however, have recently been challenged, as there is increasing evidence that microbial communities are functionally distinct (Strickland et al., 2009; McGuire et al., 2010) and have clear biogeographic patterns (Green and Bohannan, 2006; Martiny et al., 2006; Hanson et al., 2012). These empirical findings suggest that the composition and functioning of microbial communities should respond to environmental disturbance in ways analogous to macrobial communities (*i.e.*, plants and animals).

How microbial communities and processes respond to global change factors depends on three key ecological factors: (1) physiological tolerance (i.e., resilience and/or resistance); (2) dispersal; and (3) functional dissimilarity. First, populations and communities that exhibit high physiological tolerance to a given set of environmental conditions should be less disturbed by changes in those environmental conditions than physiologically sensitive populations and communities. It has been shown that microbial communities that are dominated by organisms that are physiologically adapted to low-resource conditions (slow-growing, oligotrophs (K-strategists)) are more resistant, but less resilient to acute disturbances than communities dominated by species that do well in high-resource conditions (fast-growing, copiotrophs (r-strategists)) (De Vries and Shade, 2013). The broader resource environment seems to constrain this physiological response of organisms, such that under low-resource conditions shifts in (and recovery of) microbial community composition will be slower because oligotrophs have a competitive advantage, whereas in high-resource conditions communities may respond more quickly (Wallenstein and Hall, 2011). Thus, physiological tolerance and life strategy combined with background abiotic conditions play an important role in determining the response of microbial communities to perturbation (Fig. 15.2).

Second, when populations are pushed by global change factors beyond their physiological tolerances locally (*i.e.*, undergo local extinction), the new community composition will be determined in part by the recruitment of new species that have tolerance to the new environmental conditions (assuming the change is chronic, not acute) or the recruitment of locally extinct species from a regional species pool once conditions return to the original state (Lindström and Langenheder, 2012). Recruitment following local extinction may be particularly important for soil organisms given the highly complex and heterogeneous nature of the soil environment (Ritz et al., 2004). Third, communities that maintain species that differ in their response to and impact on the environment are more likely to maintain ecosystem functioning than functionally identical communities. Communities with species that differentially impact the environment are likely to contribute more to ecosystem functioning through niche complementarity (Loreau and Hector, 2001). Some redundancy is needed in responses to the environment to maintain community stability, which decreases as systems become more complex when all species are identical in environmental response (May, 1972). Thus, if a community is uniform in its response to environmental change, disturbance events could lead to shifts in community composition that is likely to modify underlying process rates (Schimel and Gulledge, 1998).

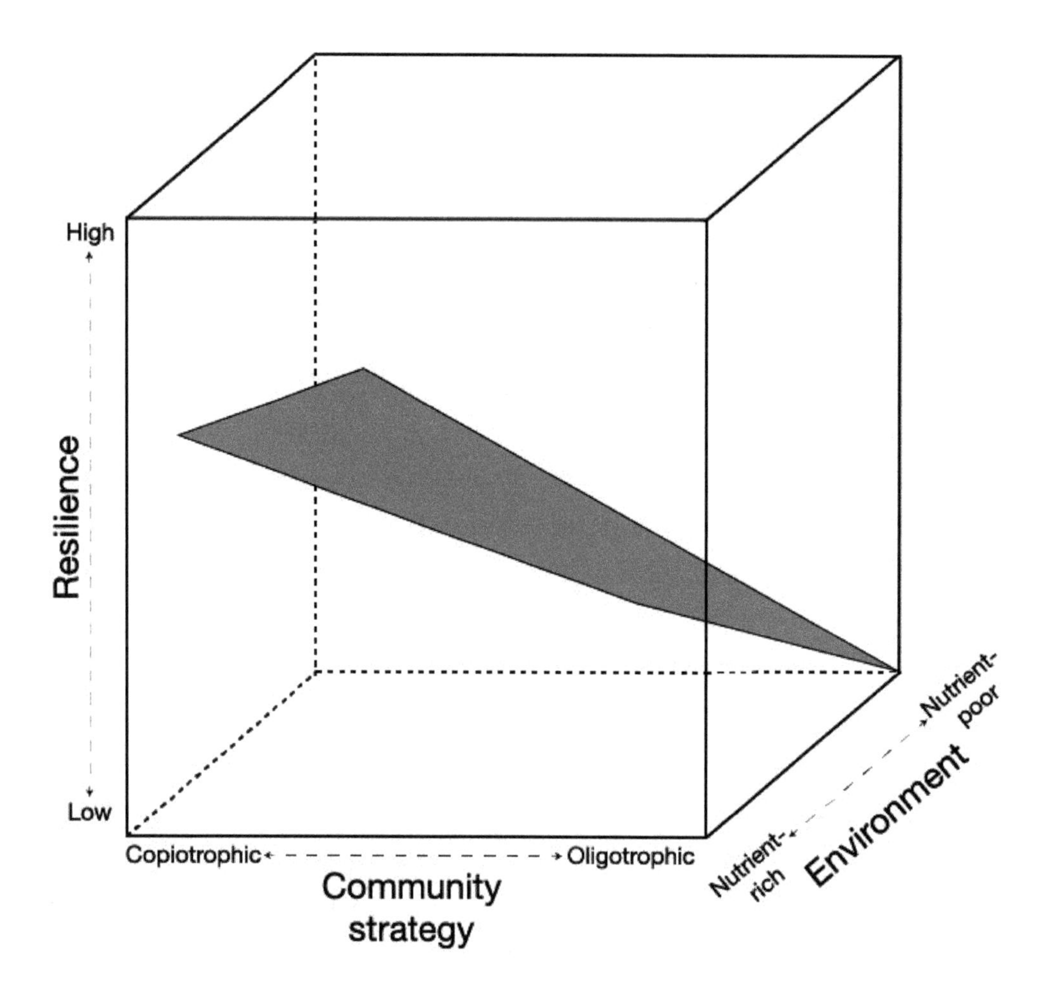

Figure 15.2 A conceptual figure of the resilience of microbial communities and process to global change drivers. The plane represents the relationship between resilience, community strategy, and environment for a particular location. The shape of this plane will likely vary by location. The resilience of microbial communities and processes, therefore, depends on the response of microbial communities to environment. Adapted from De Vries and Shade (2013).

Understanding these mechanisms of microbial responses to disturbance is further complicated by the fact that microbial communities and processes both respond to (Waldrop and Firestone, 2006) and impact (Bardgett et al., 2008) climate change, as well as aboveground communities (Wardle et al., 2004; Bardgett and Wardle, 2010). Because perturbations can interact, amplify the effects of other change drivers (Bardgett et al., 2008; Tylianakis et al., 2008; Bardgett et al., 2013), and are partially mediated through higher trophic levels (e.g., plant communities) (Wardle et al., 1998; Wardle et al., 2004; Bardgett and Wardle 2010), causality of microbial response may be difficult to disentangle. For instance, the dominant life-history strategy in a microbial community may be modified by global changes that influence resource availability, but life history strategy—represented, for instance, by fungal- vs bacterial-dominated food webs—can also influence potential global change through modifying rates of nutrient turnover and pool sizes. Rather than focus on disentangling complex mechanisms, in this review we will focus on patterns of changes in microbial composition and processes under two key global change drivers in the tropics: land cover and climate change.

Microbial community responses in the tropics

Biomass

The biomass of microorganisms as a measure of microbial productivity is often correlated with rates of microbial processes. Microbial necromass, the biomass of dead microorganisms, is also an important soil property because of its contribution to stable soil organic matter, a key ecosystem property (Schmidt et al., 2011). Changes in microbial biomass will likely have important consequences for ecosystems because of microbes' influence on nutrient cycling.

Many factors are known to generally impact soil microbial biomass, including nutrient availability, pH, texture, temperature and moisture, and plant traits (Wardle, 1992). Global change factors may affect microbial biomass both directly and indirectly, through changes in these variables. Indirect effects may occur when increased root exudation under elevated CO_2 (Bardgett et al., 2009) leads to increased anabolic activity. This process interacts with other indirect effects such as changes in litter carbon (C): nutrient ratios since microbes shift between anabolism and catabolism largely based on nutrient availability (Bradford, 2013). This process can feedback to the composition of microbial communities when increased biomass increases immobilization of key resources, modifying plant stoichiometry, and ultimately shifting microbial communities towards fungal domination (van der Heijden et al., 2008). Because fungi have a higher C use efficiency than bacteria and are more resistant to decomposition, this community change may feedback to lower respiration rates and increase C storage (Six et al., 2006).

Broadly, forest loss across biomes has been shown to lead to a 30% decrease in the biomass of both bacterial and fungal communities across multiple biomes, sites, and types of disturbance (Holden and Treseder, 2013). These changes are strongest under abiotic-driven changes to forests, such as fire, rather than biotic changes, such as pest invasion, which may become an increasingly important mechanism of landcover change due to climate shifts (Dale et al., 2001). Microbial biomass loss is thought to be mediated through losses in soil C (Holden and Treseder, 2013).

Conversion of forest to pasture or cropland has been shown to be associated with decreases in microbial biomass in the tropics (Basu and Behera, 1993). Conversion from forest to agriculture was found to decrease microbial biomass by 75% in Tahiti (Waldrop et al., 2000). However, microbial biomass was not found to decrease, and in some cases was found to be higher, with conversion of riparian forest to pasture and residential land use in Costa Rica (Groffman et al., 2001). It may be that relatively undisturbed forest provides readily available resources to the soil community, thus increasing microbial activity and biomass. This hypothesis is supported by high correlations between SOM and microbial biomass. Thus, changes that decrease SOM may increase microbial activity in the short term.

Once broad land use changes have occurred, the management of farms and pastures is important for microbial biomass. Santos et al., (2012) found that microbial biomass increased steadily among farms in Brazil that had been converted from conventional to organic farming over a variable period of time. Fertilizer application in semi-arid India was found to be associated with an increase in soil microbial biomass (Goyal et al., 1999). The conversion of managed land back to forest, not surprisingly, has positive effects on microbial biomass (Hafich et al., 2012), with longer-term regrowth having higher biomass than short-term regrowth (Maithani et al., 1996).

Compared to land use conversion, little empirical work has been done to assess the impact of climate effects on microbial biomass in the tropics. In non-tropical conditions, microbial biomass is well known to decrease upon drying and increase upon rewetting, and a significant about of C and N may be released from the microbial biomass during these cycles (Wardle, 1992). Changes in climate that increase the regularity or severity of such cycles may lead to more leaky nutrient cycles. The microbial biomass relationship with temperature appears to be more subtle, and related to the particular temperature optima of the particular taxa (Wardle, 1992).

In one study microbial biomass in moist tropical and temperate soils was measured at regular intervals after being incubated at 15 and 35°C. Microbial biomass declined more slowly in the tropical soils at both temperatures. Higher latitude areas, which are associated with greater inter-seasonal fluctuations in temperature, have been shown to be correlated with greater inter-seasonal flux of microbial biomass, though soil C, N, and pH were shown to stabilize variability in microbial biomass (Wardle, 1998). Given that microbial biomass responds less strongly to warming, that predicted warming for the tropics is lower than in higher latitudes, and temperatures lack seasonal variation, warming in the tropics appears unlikely to result in larger perturbations to microbial biomass, at least in the short-term. Changes in the magnitude and variability of precipitation and dry spells may, however, change short-term microbial biomass dynamics.

The long-term response of microbial biomass to climate will depend on physiological trade-offs. Though short-term warming decreases microbial biomass by shifting metabolic activity towards catabolism, long-term acclimation processes, such as shifts in extracellular enzyme production to enzymes with lower reactivity, and changes in underlying conditions, such as C substrate depletion, may lead to a shift back towards anabolism and investment in biomass (Bradford, 2013). Increased soil temperatures also increase enzymatic reaction rates, with the potential to accelerate nutrient cycling and availability.

Composition and Diversity

With the advent of next-generation molecular techniques, it has become possible to study microbial communities and their response to environmental changes at increasing levels of resolution. Previous methods focused on changes in short-chain phospholipid fatty acids present in soils, which are indicators of fungal to bacteria ratios and provide community fingerprint information, but unlike next-generation molecular approaches, they do not allow for deeper characterization of the taxa present in an environmental sample. As such, these approaches have limitations (Torsvik and Øvreås, 2002), but still provide an opportunity to better understand a relatively poorly understood system such as a tropical soil. The application of these techniques is quickly growing in the tropics. This section will review the recent molecular work that has been done, complemented by studies using older techniques.

Some of the largest land cover changes in the tropics are occurring as a result of tropical deforestation. Changes in plant community composition can impact microbial communities by altering the type and quality of litter and root exudate resources available in the soil and through modifications of local climate conditions through loss of forest cover. In the Amazon, much tropical forest has been converted to cattle pasture to supply an increased global demand for meat (Soares-Filho et al., 2006), and increasing amounts of the Brazilian cerrado are being converted to soybean production. Though forest conversion decreases the diversity of plant and animal communities (Soares-Filho et al., 2006), soil bacteria have been found to increase in diversity shortly after conversion (Borneman and Triplett, 1997; da C Jesus et al., 2009; Navarrete et al., 2010; Rodrigues et al., 2013). Bacterial community composition shows a trend towards greater abundance of Firmicutes in Amazonian pastureland, which has lower organic matter, and lower abundance of Acidobacteria (Rodrigues et al., 2013). Firmicutes are gram-positive bacteria that thrive in arid conditions. Acidobacteria, which are more prevalent in low-pH forest plots, are known to thrive in low-pH conditions. These groups, however, are functionally diverse and the implications of this community change for the functioning of soils is unknown. Though forest conversion increases alpha bacterial diversity in the Amazon, it is associated with a decrease in beta diversity (Rodrigues et al., 2013). Decreasing beta diversity suggests that communities in pastureland are more homogenous across space, which could make the entire community less resilient to environmental changes than forest plots.

Other studies have found that changes in tropical forest communities are associated with shifts in microbial communities. Experimental manipulation of organic C inputs to forest soils in Costa Rica showed that oligotrophic microbial taxa are significantly higher in relative abundance in low resource conditions (Nemergut et al., 2010). Resilience of these communities—measured as the community similarity between pre- and post-treatment sampling—increased with dominance by oligotrophic taxa, confirming the hypothesis that life-history strategy is a key mechanism underlying microbial resilience in the tropics (De Vries and Shade, 2013). Resilience was also found to increase with nitrate availability, supporting the hypothesis that underlying resource conditions are important for microbial resilience (De Vries and Shade, 2013). Forests in Costa Rica that varied in plant species richness also differed significantly in microbial community structure, as measured by PLFA (Carney and Matson, 2005). Microbial community structure was highly correlated with the catabolic capacity of the microbial community; plots with greater plant diversity also had higher rates of decomposition. These results suggest that changes in the composition of forests, in addition to loss of forest, may decrease the functional capacity of microbial communities. Further experimental work in tropical forests of Costa Rica have shown that differences in C availability as a result of litter removal change the diversity of microbial communities, measured by high-throughput sequencing (Leff et al., 2012). Forest plots with different levels of plant diversity also differ in their ability to use specific C substrates, though they do not differ in their use of native dissolved organic matter substrates. There is, therefore, suggestive evidence that changes in tropical forest plant communities modify soil microbial communities with cascading impacts on the potential of those communities to decompose and assimilate organic C, though whether these impacts continue with the restoration of plant diversity is unknown (Sandoval-Pérez et al., 2009). Actual rates of C use, however, may depend on the nature of the organic matter being used.

The mechanism of forest clearing can in itself modify the microbial community. Burning of forest vegetation for cropland (a practice known as slash-and-burn agriculture, which is typically part of a shifting cultivation practice in which the forest is allowed to regrow after several years of cultivation) has been shown alter belowground microbial communities (Waldrop et al., 2000; da C Jesus et al., 2009; Navarrete et al., 2010). Burning forest vegetation leads to ash deposition from the burnt vegetation and increased rates of organic matter decomposition, drawing down the soil organic matter pool (Juo and Manu, 1996). The loss of organic matter from soil can lead to lower taxonomic richness of fungal communities on soils burnt for cultivation than in natural forest soils (Navarrete et al., 2010). On farms previously cleared by slash-and-burn that have undergone regrowth, bacterial communities tend to return to resemble the original communities (da C Jesus et al., 2009).

Land use land cover change transitions *from* agricultural land have also been shown to impact microbial community composition and diversity in the tropics. Just as loss of SOM associated with slash-and-burn transitions impacted microbial community composition, an increase in SOM associated with fallow lands in Senegal was positively associated with large increases in microbial biomass, microbial enzyme activity, and small differences in community structured, measured by denaturing gradient gel electrophoresis (Sall et al., 2006).

Changes in the concentration of CO_2 has been shown to impact microbial functional composition in temperate zones (He et al., 2010), while increases in soil temperature have been shown to modify microbial diversity and composition (Frey et al., 2008) as well as increase microbial respiration rates from soil. For microbial C efflux, communities have been shown to acclimate to higher temperatures (Allison et al., 2010; Frey et al., 2013); the acclimation of community composition, however, is relatively unexplored, especially in the tropics. In Puerto Rico, microbial communities that have been exposed to drought show a 40% reduction in diversity compared to control soils. However, soil communities that have previously been exposed to drought showed no reduction in diversity when exposed to further drought (Bouskill et al., 2012). In other words, the largest impacts on microbial diversity may be expected following qualitative changes in disturbance patterns rather than changes in frequency or intensity. However, there may be thresholds in the frequency or intensity of disturbance events beyond which the community will be shifted to an alternative stable state (Jones et al., 2011). Determining the limits of these thresholds should be a research priority as few studies in tropical soils have been done evaluating the effects of repeated disturbances on soil microbial composition and function.

Microbial processes responses in the tropics

Nitrogen

Changes in precipitation in the tropics are expected to affect the activities of microbial processes carried out by obligate anaerobes, with the activities of functional groups of microbes such as denitrifiers and methanotrophs decreasing or increasing depending on the soil moisture content of the soil (Cattânio et al., 2002; Davidson et al., 2004). Changes in precipitation can also indirectly affect microbial activity through their effects on substrates used as resources by microbial communities. Precipitation-induced changes in the turnover of fine roots and leaf litter would be expected to have cascading effects on rates of decomposition, N mineralization, nitrification, and denitrification (Davidson et al., 2004).

Land-use and land-cover change in the tropics have extensive documented impacts on many of the N-cycling processes that are mediated by microbial activity. In general, deforestation tends to increase the availability of mineral N and rates of N cycling in the period immediately following clearing (e.g., Matson et al., 1990). Early evidence suggested that the greenhouse gas nitrous oxide (N_2O), which typically is primarily produced by denitrifying bacteria, was emitted at lower rates from soils in intact Amazonian forest than from soils in forest that had been cleared for pasture (Luizão et al., 1989). However, subsequent studies in Latin America almost uniformly found this increase to be short-lived and limited to roughly three years immediately following forest clearing, when high-SOM forest soils experience enhanced mineralization and heavily reduced rates of plant uptake (Veldkamp et al., 1999; Verchot et al., 1999; Melillo et al., 2001); this pattern is also observed in transitions from forest to coffee gardens (Verchot et al., 2006). In some cases (e.g., Garcia-Montiel et al., (2001, Wick et al., (2005)), N_2O fluxes did not increase in recently-cleared pastures.

The presence or absence of this initial pulse of N_2O is hypothesized to be related to differences in soil mineralogy: coarser-textured soils may be less able to protect SOM from rapid microbial mineralization. In addition, differences in the P status of soils may play a role by boosting soil fertility, and thus the production of SOM. For example, in one study coarser soils where high rates of N turnover were observed had higher levels of soil P, which appeared to be responsible for higher levels of productivity in cleared pastures, and consequently of higher SOM and N mineralization rates (Wick et al., 2005). Results from Zimbabwe offer some partial evidence that soil fertility may be the more important factor driving these patterns. In sites with clayey soils, N_2O fluxes were lower in intact woodland than in recently cleared woodland, but fluxes exhibited the opposite pattern in sites with loamy sand soils (Mapanda et al., 2012), which is the opposite of what would be expected under the hypothesis that rapid mineralization following deforestation would be higher in coarser soils. However, there was evidence that the woodland clay soils were more fertile, with higher concentrations of SOM and more productive maize crops in the clay than the loamy sand sites, though the difference was not significant. Fluxes in a fertile region of Costa Rica also exhibited elevated N_2O fluxes in pastures that had been cleared within the previous 10 years, supporting the notion that soil fertility may be an important driver of these transient enhanced fluxes (Keller et al., 1993).

Following the initial pulse of increased fluxes in recently-cleared forest, N oxide fluxes, N cycling, and mineral N concentrations tend to decline in both crop and pasture systems, often continuing their decline to levels considerably lower than emissions from the intact forest as pastures age (Keller et al., 1993; Erickson and Keller 1997; Verchot et al., 1999; Garcia-Montiel et al., 2001; Melillo et al., 2001; Neill et al., 2005; Veldkamp et al., 2008), ultimately leading to an overall reduction in

N_2O fluxes (Neill et al., 2005). However, additions of fertilizer can reverse this trend (Vitousek et al., 1989; Davidson et al., 1991; Verchot et al., 2006), as may livestock excretion and reduced burning (Pendall et al., 2010). Available N and N oxide fluxes can also increase when the change in land cover is from natural savanna to managed and sporadically fertilized pasture (Grover et al., 2012), or from subtropical evergreen to bamboo plantations (Liu et al., 2011). Deforestation is typically followed by increases in soil bulk density (Neill et al., 1999; Veldkamp et al., 2008), modifying microbial environments by changing soil structure and water infiltration.

In general, intact tropical forests appear to have a "leakier" N cycle, with mineral N pools dominated by NO_3^-, higher N oxide fluxes, and generally higher nitrification potential and rates of net nitrification, while mineral pools in older pastures tend to be dominated by NH_4^+, indicative of a more closed N cycle (Neill et al., 1995, 1999; Verchot et al., 1999; Davidson et al., 2000). N processes in secondary forest tend to be much more variable (Davidson et al., 2000). While mineral N, N cycling, and N oxide fluxes might be expected to decrease as plants utilize resources for growth (Vitousek and Reiners, 1975), this expectation is not always observed in tropical secondary forests. This variability appears to be related to the relative abundance of legumes: in their relative absence, N cycling and fluxes may be low, but in the presence of legumes, their contribution of labile, N-rich litter to the soil accelerates N cycling and N oxide fluxes (Erickson et al., 2001, 2002; Kachenchart et al., 2012).

Land-use change can also have indirect effects on the cycles of N and other processes through its effects on climate. The changes in evapotranspiration that occur with large-scale deforestation can have dramatic impacts on regional climate, reducing the frequency and magnitude of precipitation, particularly in areas where local convection is a major source of rainfall (Shukla et al., 1990). Large-scale deforestation has additional implications for global climate, as stocks of aboveground and soil C are mineralized and lost to the atmosphere at a rate of 0.5 Pg C year-1 (Ramankutty et al., 2007) in addition to the loss of tropical forests' possible role as a C sink (Stephens et al., 2007). In turn, the Amazon is at risk of additional forest loss due to climate change, further accelerating these interactive effects of land use and climate change on microbial processes (Malhi et al., 2008).

Carbon

Enrichment of atmospheric CO_2 is known to increase the release of labile sugars, organic acids, and amino acids from plant roots, providing a nutrient source for microbes (Bardgett et al., 2009), which are frequently C limited. This increase in C availability belowground can stimulate microbes to invest in growth by assimilating extra C into body mass or to increase respiration when mineral nutrients are not available in sufficient quantity to assimilate available C (Prescott, 2010). Whether extra C stimulates soil respiration depends on nutrient availability, where microbes respire more in nutrient-poor conditions because of waste metabolism (Bradford, 2013). Soil warming can increase respiration rates by stimulating metabolic activity, which is a thermodynamically regulated process. This process, however, is constrained by soil moisture, which is projected to change with precipitation patterns, thereby making the net effect of climate on soil respiration unclear. The balance of these effects is likely to vary by region, with increased drought conditions potentially playing a greater role than increased temperature in tropical environments, as drought may cause large reductions in microbial biomass or shifts in community structure (De Deyn et al., 2008).

The short- and long-term persistence of soil organic matter in the soil was long-thought to be determined primarily by the molecular structure of C. Evidence has mounted in recent years that the long-term persistence of organic matter is more a property of the ecosystem, depending on biotic inputs and the spatial arrangement of organic matter molecules, mineral particles, and microorganisms in the soil (Schmidt et al., 2011).

The literature on changes in C dynamics as a result of land use change and climate is beyond the scope of this review. It is well known that conversion of forest to agricultural land leads to decreases of soil C and other nutrients in the tropics (Awiti et al., 2008), though this loss does not necessarily come from conversion of forest to pasture (Murty et al., 2002). We focus here on studies that include specific study of some component of the microbial mechanisms of C dynamics.

Microbial C was shown to decline with land conversion in dry forest soils in India (Srivastava and Singh, 1991). Luizao et al (1992) found that soil biomass C fell by 64% after slash-and-burn in the Brazilian Amazon, though it increased slightly after re-establishment to pasture. Microbial biomass C increased in semi-arid India under the use of inorganic fertilizers paired with organic amendments, compared to no organic amendments (Goyal et al., 1999). In Bangladesh, cultivated land had significantly less active and total microbial biomass C, as well as lower amounts of labile organic C, than forested plots, reforested plots, and grasslands (Islam and Weil, 2000). Proportion of C in microbial biomass was found to be lowest on a Eucalyptus plantation site in India, as opposed to under natural forest regeneration (Behera and Sahani, 2003). In moist tropical forest soils in India, land use conversion from forest to agriculture was shown to have a significant negative impact on microbial biomass C, suggesting that deforestation and cultivation due to decreased C turnover and nutrient availability (Dinesh et al., 2004).

Climate can impact turnover of C in soil through mediating rates of decomposition, which is a thermodynamically constrained process (Bradford et al., 2008). Rates of decomposition in tropical soils are expected to be especially sensitive to changes in temperature (Silver, 1998). In fact, tropical soils have been shown to have elevated rates of decomposition that is likely due to greater metabolic activity under higher temperature and unique physicochemical soil properties (Rasche and Cadisch, 2013). The mechanism by which temperature impacts decomposition is through changes in extracellular enzyme activity (Allison et al., 2010).

Recent theoretical work has explored this relationship through variation in microbial C use efficiency by temperature (Allison et al., 2010) and by nutrient availability (which is impacted by land use change) (Sinsabaugh et al., 2013), but there is little support—empirical or theoretical—for whether this microbial parameter is in fact temperature- and resource- sensitive in ways that differ by taxa (Bradford and Crowther, 2013; Bradford 2013). Ultimately, global change-mediated changes in extracellular enzyme activity could lead to impacts on rates of decomposition and SOM formation, which also depends on enzyme production (Schmidt et al., 2011), though little is known about the relative balance of decomposition versus SOM formation or the resulting patterns in long-term SOM. Metagenomic and functional gene approaches have been used to explore the mechanisms of soil C turnover under temperature, finding that soil respiration acclimatizes under warming as a result of community composition shift to less temperature-sensitive taxa, activating genes associated with labile C use thus leaving recalcitrant C in the soil (Zhou et al., 2011).

There seem to be broad differences in the relative abundances of certain extra-cellular enzymes in the tropics vs. temperate areas that reflect lower P availability in tropics (Waring et al., 2014). There is also some evidence for changes in extracellular enzyme potential in tropical soils under global change drivers. For instance, decreased microbial biomass from land use change in Tahiti decreased the potential activity of several extracellular enzymes as well as percent C and percent N (Waldrop et al., 2000). Conversion to agriculture from moist tropical forest in India was also associated with significant decreases in extracellular enzyme activity in soils (Dinesh et al., 2004). In Mexico, P-related enzymes were consistently affected by conversion from forest to pasture (Sandoval-Pérez et al., 2009). Enzyme activities also correlate with climate variables, such as mean annual temperature and precipitation, suggesting that climate acts as an important control of enzyme activity in the tropics (Waring et al., 2014). This poses the potential for climate change to modify enzymatic activity and, thus, decomposition; though there is little data from tropical systems describing the response of decomposition to climate change in the tropics (Wood et al., 2012).

Microbially mediated C dynamics may also be modified by global change-induced modifications to the aboveground subsystem. There is evidence that tropical forests respond to increased stress conditions, such as changing climate conditions, by increasing litter fall (Wood et al., 2012). An increase of litter inputs to the soil has been shown to increase soil respiration rates through priming the microbial community (Kuzyakov, 2010). It remains unknown how microbial thermal acclimation in the tropics will interact with priming that is driven by litter fall under climate stress conditions. In addition, increased litter fall that is induced by physiological stress may modify overall primary production, which could also impact belowground C dynamics.

Priorities for future research

The tropics are a frontier of land-cover change, largely driven by conversion of land to agriculture (DeFries and Rosenzweig, 2010). Many parts of the tropics are also predicted to experience higher temperatures and increased variation in precipitation—with some locations increasing and others decreasing on average. In this chapter we document extant theoretical and empirical evidence for changes in microbial communities and processes as a result of these two global change drivers. There are several key remaining research priorities.

It remains to be established whether changes in the composition of microbial communities resulting from global changes will further contribute to differences in ecosystem process rates that are also modified by land-cover and climate change. For instance, decomposition is known to be driven by land-cover and climate change (see

Carbon), but may (or may not) also be impacted by the composition of microbial communities (Schimel and Schaeffer, 2012), which changes under the two global change drivers. Thus, there may be effects of changes in community composition that are independent of the ultimate control on process rates by land-cover and climate change. Future research will therefore need to identify not only whether microbial communities and processes change in the tropics under global change, but whether changes in microbial communities contribute to changes in processes.

Spatial dynamics of soil are also likely to be important in determining rates of ecosystem processes (Schimel and Schaeffer, 2012). Soils with highly complex spatial arrangements of organo-mineral aggregates may protect organic matter molecules and thus serve as an ultimate control over process rates under global change. The role of spatial dynamics in soil likely differs by taxa and may be more important for passively dispersed bacteria and fungi than for actively dispersed microbes, which are able to access occluded and sorbed organic matter. Thus, understanding the particularities of soil conditions in various parts of the tropics—such as mineralogy, texture, and relative abundance of bacteria and fungi—is likely to be crucial in developing a predictive understanding of how particular locations respond to global changes.

Global change drivers, such as climate and land-cover change, do not occur separately; they interact in presenting simultaneous and potentially non-additive pressure on biological communities and ecosystem processes. Multi-factor experiments have played an important role in understanding how multiple change drivers interact to drive patterns in plant communities (Reich et al., 2006; Bradford et al., 2012; Kardol et al., 2012). The multi-factor framework has yet to be largely employed in understanding how global change will impact microbial communities in tropical systems. One example of such a study from a temperate system found that in years of normal precipitation, warming decreased bacterial diversity; in drought years, warmed plots had higher microbial diversity, suggesting that under warmed conditions it is water budgets that determined microbial diversity (Sheik et al., 2011). These sorts of studies will need to be conducted in the tropics across multiple change factors—not just climate-related ones. Much multi-factor work focuses on two-levels —ambient and elevated—of either CO_2 or N enrichment; multi-level multi-factor work has been used to a lesser degree in exploring plant response (Bradford et al.,

2012) though this approach is potentially more informative in understanding the range of conditions over which communities and processes change.

Microbial systems have long been an interesting test system for evolutionary questions because, due to short generation lengths, they are highly suitable model organisms. Microbial evolution has largely been omitted from studies of microbial responses to environmental change and the resulting impacts on ecosystem functioning. Initial work has begun in the context of thermal adaptation (Frey et al., 2013), but none of this research has been conducted in the tropics. Microbial evolution, and particularly the evolution of cooperative behavior related to exo-enzyme production and decomposition (Bachmann et al., 2011; Allen and Nowak 2013), could play an important role in how the functioning of soil communities responds to global change.

Finally, the importance of microbial natural history in tropical (and temperate) zones should not be overlooked. The predictive, quantitative science into which ecology has evolved has a deep culture of natural history and observation. Understanding which microbial species are present and what biotic and abiotic factors determine their biogeographical distributions and demographic rates is a crucial part of ecological research—and a significant challenge in microbial ecology. Without knowing what microbes are where today, it is difficult to have an appreciation for what climate change and land use change mean for what microbes will be where tomorrow. And yet this project remains only its nascent stages in ecology. Advanced molecular techniques will need to be coupled with rigorous sampling strategies to identify the distribution of microbial taxa throughout the tropics.

Conclusions

Understanding how global change in the tropics impacts microbial communities and the processes they mediate can be important for understanding consequences to systems that are needed for human wellbeing. For instance, smallholder farmers in the tropics typically depend largely on internal nutrient cycling processes and SOM to sustain productivity and meet food needs over time. Changes in microbial processes that control the availability of these organic and mineral nutrients could drastically impact food production systems. This is issue is particularly pressing because smallholder farmers tend to be concentrated in tropical regions (Mendelsohn et al., 2006) and capital-constrained and, thus, less likely to be able to afford adaptation strategies that requires capital inputs (Wood et al., 2014).

In this review, we show overall evidence suggesting that patterns of the changes in microbial communities and microbially mediated processes in the tropics reflect theory and previously established patterns from temperate systems. Climate and land use change will both impact microbial communities and processes, through both direct and indirect mechanisms. Understanding and predicting the degree of these responses will likely depend on abiotic conditions, such as soil mineralogy and texture. It may also depend on properties of the community of microorganisms present in a particular biogeographic zone—such as their relative growth rate and resource requirements. Microbes regulate ecosystem processes in a coupled plant-soil system. Most of our understanding of the response of this system to global change is based on single drivers. More work needs to be done to understand the impacts of multi-factor global change drivers that, through de-coupling aboveground-belowground linkages, could drastically impact microbes and their processes.

References

Allen, B., and Nowak, M.A. (2013). Cooperation and the Fate of Microbial Societies. PLoS Biol. *11*, e1001549.

Allison, S.D., and Martiny, J.B.H. (2008). Resistance, resilience, and redundancy in microbial communities. Proc. Natl. Acad. Sci. USA *105*, 11512-11519.

Allison, S.D., Wallenstein, M.D., and Bradford, M.A. (2010). Soil-carbon response to warming dependent on microbial physiology. Nat. Geosci. *3*, 336-340.

Awiti, A.O., Walsh, M.G., and Kinyamario, J. (2008). Dynamics of topsoil carbon and nitrogen along a tropical forest-cropland chronosequence: Evidence from stable isotope analysis and spectroscopy. Agricult. Ecosyst. Environ. *127*, 265-272.

Bachmann, H., Molenaar, D., Kleerebezem, M., and van Hylckama Vlieg, J.E.T. (2011). High local substrate availability stabilizes a cooperative trait. ISME J. *5*, 929-932.

Bardgett, R.D., De Deyn, G.B., and Ostle, N.J. (2009). Plant-soil interactions and the carbon cycle. J. Ecol. *97*, 838-839.

Bardgett, R.D., Freeman, C., and Ostle, N.J. (2008). Microbial contributions to climate change through carbon cycle feedbacks. ISME J. *2*, 805-814.

Bardgett, R.D., Manning, P., Morrien, E., and De Vries, F.T. (2013). Hierarchical responses of plant-soil interactions to climate change: consequences for the global carbon cycle. J. Ecol. *101*, 334-343.

Bardgett, R.D., and Wardle, D.A. (2010). Aboveground-belowground linkages: biotic interactions, ecosystem processes, and global change. (Oxford, UK: Oxford University Press).

Basu, S., and Behera, N. (1993). The effect of tropical forest conversion on soil microbial biomass. Biol. Fertil. Soil *16*, 302-304.

Behera, N., and Sahani, U. (2003). Soil microbial biomass and activity in response to Eucalyptus plantation and natural regeneration on tropical soil. For. Ecol. Manag. *174*, 1-11.

Bever, J. D., Platt, T.G., and Morton, E.R. (2012). Microbial Population and Community Dynamics on Plant Roots and Their Feedbacks on Plant Communities. Annu. Rev. Microbiol. *66*, 265-283.

Borlaug, N.E. (2000). The Green Revolution Revisited and the Road Ahead. Special 30th Anniversary Lecture. (Oslo, Norway: Norwegian Nobel Institute).

Borneman, J., and Triplett, E.W. (1997). Molecular microbial diversity in soils from eastern Amazonia: evidence for unusual microorganisms and microbial population shifts associated with deforestation. Appl. Environ. Microbiol. *63*, 2647-2653.

Bouskill, N. J., Lim, H.C., Borglin, S., Salve, R., Wood, T.E., Silver, W.L., and Brodie, E.L. (2012). Pre-exposure to drought increases the resistance of tropical forest soil bacterial communities to extended drought. ISME J. *7*, 384-394.

Bradford, M.A., and Crowther, T. (2013). Carbon use efficiency and storage in terrestrial ecosystems. New Phytol. *199*, 7-9.

Bradford, M.A. (2013). Thermal adaptation of decomposer communities in warming soils. Front. Microbiol. *4*.

Bradford, M.A., Davies, C.A., Frey, S.D., Maddox, T.R., Melillo, J.M., Mohan, J.E., Reynolds, J.F., Treseder, K.K., and Wallenstein, M.D. (2008). Thermal adaptation of soil microbial respiration to elevated temperature. Ecol. Lett. *11*, 1316-1327.

Bradford, M.A., Wood, S.A., Maestre, F.T., Reynolds, J.F., and Warren, R.J. (2012). Contingency in ecosystem but not plant community response to multiple global change factors. New Phytol. *196*, 462-471.

Burney, J.A., Davis, S.J., and Lobell, D.B. (2010). Greenhouse gas mitigation by agricultural intensification. Proc. Natl. Acad. Sci. USA. *107*, 12052-12057.

Carney, K.M., and Matson, P.A. (2005). Plant communities, soil microorganisms, and soil carbon cycling: Does altering the world belowground matter to ecosystem functioning? Ecosystems *8*, 928-940.

Cattânio, J.H., Davidson, E.A., Nepstad, D.C., Verchot, L.V., and Ackerman, I.L. (2002). Unexpected results of a pilot throughfall exclusion experiment on soil emissions of CO2, CH4, N2O, and NO in eastern Amazonia. Biol. Fertil. Soils *36*, 102–108.

da C Jesus, E., Marsh, T.L., Tiedje, J.M., and de S Moreira, F.M. (2009). Changes in land use alter the structure of bacterial communities in Western Amazon soils. ISME J. *3*, 1004-1011.

Dale, V.H., Joyce, L.A., McNulty, S., Neilson, R.P., Ayres, M.P., Flannigan, M.D., Hanson, P.J., Irland, L.C., Lugo, A.E., and Peterson, C.J. (2001). Climate Change and Forest Disturbances: Climate change can affect forests by altering the frequency, intensity, duration, and timing of fire, drought, introduced species, insect and pathogen outbreaks, hurricanes, windstorms, ice storms, or landslides. BioScience *51*, 723-734.

Davidson, E.A., Ishida, F.Y., and Nepstad, D.C. (2004). Effects of an experimental drought on soil emissions of carbon dioxide, methane, nitrous oxide, and nitric oxide in a moist tropical forest. Glob. Change Biol. *10*, 718–730.

Davidson, E.A., Keller, M., Erickson, H.E., Verchot, L.V., and Veldkamp, E. (2000). Testing a conceptual model of soil emissions of nitrous and nitric oxides. BioScience *50*, 667–680.

Davidson, E.A., Vitousek, P.M., Matson, P.A., Riley, R., García-Méndez, G., and Maass, J.M. (1991). Soil emissions of nitric oxide in a seasonally dry tropical forest of México. J. Geophys. Res. *96*, 15439–15445.

De Deyn, G.B., Cornelissen, J.H.C., and Bardgett, R.D. (2008). Plant functional traits and soil carbon sequestration in contrasting biomes. Ecol. Lett. *11*, 516-531.

De Vries, F.T., and Shade, A. (2013). Controls on soil microbial community stability under climate change. Front. Microbiol. *4*.

DeFries, R., and Rosenzweig, C. (2010). Toward a whole-landscape approach for sustainable land use in the tropics. Proc. Natl. Acad. Sci. USA *107*, 19627-19632.

Dinesh, R., Ghoshal Chaudhuri, S., and Sheeja, T. (2004). Soil biochemical and microbial indices in wet tropical forests: effects of deforestation and cultivation. J. Plant Nut. Soil Sci. *167*, 24-32.

Erickson, H.E., and Keller, M. (1997). Tropical land use change and soil emissions of nitrogen oxides. Soil Use Manage. *13*, 278–287.

Erickson, H.E., Davidson, E.A., and Keller, M. (2002). Former land-use and tree species affect nitrogen oxide emissions from a tropical dry forest. Oecologia *130*, 297–308.

Erickson, H.E., Keller, M., and Davidson, E.A. (2001). Nitrogen Oxide Fluxes and Nitrogen Cycling during Postagricultural Succession and Forest Fertilization in the Humid Tropics. Ecosystems *4*, 67–84.

Falkowski, P.G., Fenchel, T., and Delong, E.F. (2008). The microbial engines that drive Earth's biogeochemical cycles. Science *320*, 1034-1039.

Fierer, N., Schimel, J., and Holden, P. (2003). Influence of drying-rewetting frequency on soil bacterial community structure. Microb. Ecol. *45*, 63-71.

Fierer, N., and Schimel, J.P. (2002). Effects of drying-rewetting frequency on soil carbon and nitrogen transformations. Soil Biol. Biochem. *34*, 777-787.

Finlay, B.J. (2002). Global dispersal of free-living microbial eukaryote species. Science *296*, 1061-1063.

Frey, S.D., Drijber, R., Smith, H., and Melillo, J. (2008). Microbial biomass, functional capacity, and community structure after 12 years of soil warming. Soil Biol. Biochem. *40*, 2904-2907.

Frey, S.D., Lee, J., Melillo, J.M., and Six, J. (2013). The temperature response of soil microbial efficiency and its feedback to climate. Nat. Clim. Change *3*, 395-398.

Garcia-Montiel, D.C., Steudler, P.A., Piccolo, M.C., Melillo, J.M., Neill, C., and Cerri, C.C. (2001). Controls on soil nitrogen oxide emissions from forest and pastures in the Brazilian Amazon. Glob. Biogeochem. Cy. *15*, 1021–1030.

Gibbs, H., Ruesch, A., Achard, F., Clayton, M., Holmgren, P., Ramankutty, N., and Foley, J. (2010). Tropical forests were the primary sources of new agricultural land in the 1980s and 1990s. Proc. Natl. Acad. Sci. USA *107*, 16732-16737.

Goyal, S., Chander, K., Mundra, M., and Kapoor, K. (1999). Influence of inorganic fertilizers and organic amendments on soil organic matter and soil microbial properties under tropical conditions. Biol. Fertil. Soils *29*, 196-200.

Green, J.L., and Bohannan, B.J.M. (2006). Spatial scaling of microbial biodiversity. Trends Ecol. Evol. *21*, 501-507.

Groffman, P.M., McDowell, W.H., Myers, J.C., and Merriam, J.L. (2001). Soil microbial biomass and activity in tropical riparian forests. Soil Biol. Biochem. *33*, 1339-1348.

Grover, S.P.P., Livesley, S.J., Hutley, L.B., Jamali, H., Fest, B., Beringer, J., Butterbach-Bahl, K., and Arndt, S.K. (2012). Land use change and the impact on greenhouse gas exchange in north Australian savanna soils. Biogeosciences *9*, 423–437.

Hafich, K., Perkins, E., Hauge, J., Barry, D., and Eaton, W.D. (2012). Implications of land management on soil microbial communities and nutrient cycle dynamics in the lowland tropical forest of northern Costa Rica. Trop. Ecol. *53*, 215-224.

Hanson, C.A., Fuhrman, J.A., Horner-Devine, M.C., and Martiny, J.B.H. (2012). Beyond biogeographic patterns: processes shaping the microbial landscape. Nat. Rev. Microbiol. *10*, 497-506.

He, Z., Xu, M., Deng, Y., Kang, S., Kellogg, L., Wu, L., Van Nostrand, J.D., Hobbie, S.E., Reich, P.B., and Zhou, J. (2010). Metagenomic analysis reveals a marked divergence in the structure of belowground microbial communities at elevated CO_2. Ecol. Lett. *13*, 564-575.

Hedin, L.O., Brookshire, E.N.J., Menge, D.N.L., and Barron, A.R. (2009). The Nitrogen Paradox in Tropical Forest Ecosystems. Annu. Rev. Ecol. Evol. Syst. *40*, 613-635.

Hickman, J.E., Palm, C.A., Tully, K.L., Diru, W., and Groffman, P.M. (in review). A nitrogen tipping point in tropical agriculture: avoiding rapid increases in nitrous oxide fluxes in an African Green Revolution. Ecol. Appl.

Holden, S.R., and Treseder, K.K. (2013). A meta-analysis of soil microbial biomass responses to forest disturbances. Front. Microbiol. *4*.

Hulme, M., and Viner, D. (1998). A climate change scenario for the tropics. Clim. Change *39*, 145-176.

Islam, K., and Weil, R. (2000). Land use effects on soil quality in a tropical forest ecosystem of Bangladesh. Agric. Ecosyst. Environ. *79*, 9-16.

Jeon, S.B., Olofsson, R., and Woodcock, C.E. (2014). Land use change in New England: a reversal of the forest transition. J. Land Use Sci. *9*, 1-26.

Juo, A.S., and Manu, A. (1996). Chemical dynamics in slash-and-burn agriculture. Agric. Ecosyst. Environ. *58*, 49-60.

Kachenchart, B., Jones, D.L., Gajaseni, N., Edwards-Jones, G., and Limsakul, A. (2012). Seasonal nitrous oxide emissions from different land uses and their controlling factors in a tropical riparian ecosystem. Agric. Ecosyst. Environ. *158*, 15–30.

Kardol, P., De Long, J.R., and Sundqvist, M.K. (2012). Crossing the threshold: the power of multi-level experiments in identifying global change responses. New Phytol. *196*, 323-326.

Kates, R.W., and Parris, T.M. (2003). Long-term trends and a sustainability transition. Proc. Natl. Acad. Sci. USA *100*, 8062-8067.

Keller, M., Veldkamp, E., Weitz, A.M., and Reiners, W.A. (1993). Effect of pasture age on soil trace-gas emissions from a deforested area of Costa Rica. Nature *365*, 244-246.

Kuzyakov, Y. (2010). Priming effects: interactions between living and dead organic matter. Soil Biol. Biochem. *42*, 1363-1371.

Lauber, C.L., Hamady, M., Knight, R., and Fierer, N. (2009). Pyrosequencing-Based Assessment of Soil pH as a Predictor of Soil Bacterial Community Structure at the Continental Scale. Appl. Environ. Microbiol. *75*, 5111-5120.

Leff, J., Nemergut, D., Grandy, S.A., O'Neill, S., Wickings, K., Townsend, A., and Cleveland, C. (2012). The Effects of Soil Bacterial Community Structure on Decomposition in a Tropical Rain Forest. Ecosystems *15*, 284-298.

Lindström, E.S., and Langenheder, S. (2012). Local and regional factors influencing bacterial community assembly. Environ. Microbiol. Rep. *4*, 1-9.

Liu, J., Jiang, P., Li, Y., Zhou, G., Wu, J., and Yang, F. (2011). Responses of N_2O Flux from Forest Soils to Land Use Change in Subtropical China. Bot. Rev. *77*, 320–325.

Loreau, M., and Hector, A. (2001). Partitioning selection and complementarity in biodiversity experiments. Nature *412*, 72-76.

Luizão, F., Matson, P., Livingston, G., Luizão, R.C., and Vitousek, P.M. (1989). Nitrous oxide flux following tropical land clearing. Glob. Biogeochem. Cy. *3*, 281–285.

Luizão, R.C., Bonde, T.A., and Rosswall, T. (1992). Seasonal variation of soil microbial biomass,Äîthe effects of clearfelling a tropical rainforest and establishment of pasture in the Central Amazon. Soil Biol. Biochem. *24*, 805-813.

Maithani, K., Tripathi, R., Arunachalam, A., and Pandey, H. (1996). Seasonal dynamics of microbial biomass C, N and P during regrowth of a disturbed subtropical humid forest in north-east India. Appl. Soil Ecol. *4*, 31-37.

Malhi, Y., Roberts, J.T., Betts, R.A., Killeen, T.J., Li, W., and Nobre C.A. (2008). Climate Change, Deforestation, and the Fate of the Amazon. Science *319*, 169–172.

Mapanda, F., Wuta, M., Nyamangara, J., Rees, R.M., and Kitzler, B. (2012). Greenhouse gas emissions from Savanna (Miombo) woodlands: responses to clearing and cropping. Afr. Crop Sci. J. *20*, 385–400.

Martiny, J.B.H., Bohannan, B.J.M., Brown, J.H., Colwell, R.K., Fuhrman, J.A., Green, J.L., Horner-Devine, M.C., Kane, M., Krumins, J.A., Kuske, C.R., Morin, P.J., Naeem, S., Ovreås, L., Reysenbach, A.-L., Smith, V.H., and Staley, J.T. (2006). Microbial biogeography: putting microorganisms on the map. Nat. Rev. Microbiol. *4*, 102-112.

Matson, P.A., Vitousek, P.M., Livingston, G.P., and Swanberg, N.A. (1990). Sources of variation in nitrous oxide flux from Amazonian ecosystems. J. Geophys. Res. *95*, 16789–16798.

May, R. (1972). Will a large complex system be stable? Nature *238*, 413-414.

McGuire, K.L., Bent, E., Borneman, J., Majumder, A., Allison, S.D., and Treseder, K.K. (2010). Functional diversity in resource use by fungi. Ecology *91*, 2324-2332.

Melillo, J.M., Steudler, P.A., Feigl, B.J., Neill, C., Garcia, D., Piccolo, M.C., Cerri, C.C., and Tian, H. (2001). Nitrous oxide emissions from forests and pastures of various ages in the Brazilian Amazon. J. Geophys. Res. *106*, 34179–34188.

Mendelsohn, R., Dinar, A., and Williams, L. (2006). The distributional impact of climate change on rich and poor countries. Env. Dev. Econ. *11*, 159-178.

Mohr, E.C.J., van Baren, F.A., and van Schuylenborgh, J. (1973). Tropical soils: a comprehensive study of their genesis (The Hague, The Netherlands: Mouton-Ichtiar Baru-Van Hoeve).

Murty, D., Kirschbaum, M.U., Mcmurtrie, R.E., and Mcgilvray, H. (2002). Does conversion of forest to agricultural land change soil carbon and nitrogen? A review of the literature. Glob. Change Biol. *8*, 105-123.

Navarrete, A.C.A., Cannavan, F.S., Taketani, R.G., and Tsai, S.M. (2010). A molecular survey of the diversity of microbial communities in different Amazonian agricultural model systems. Diversity *2*, 787-809.

Neill, C., Piccolo, M.C., Melillo, J.M., Steudler, P.A., and Cerri, C.C. (1999). Nitrogen dynamics in Amazon forest and pasture soils measured by 15 N pool dilution. Soil Biol. Biochem. *31*, 567–572.

Neill, C., Piccolo, M.C., Steudler, P.A., Melillo, J.M., Feigl, B.J., and Cerri C.C. (1995). Nitrogen dynamics in soils of forests and active pastures in the western Brazilian Amazon Basin. Soil Biol. Biochem. *27*, 1167–1175.

Neill, C., Steudler, P.A., Garcia-Montiel, D.C., Melillo, J.M., Feigl, B.J., Piccolo, M.C., and Cerri, C.C. (2005). Rates and controls of nitrous oxide and nitric oxide emissions following conversion of forest to pasture. Nutr. Cycl. Agroecosyst. *71*, 1–15.

Nemergut, D.R., Cleveland, C.C., Wieder, W.R., Washenberger, C.L., and Townsend, A.R. (2010). Plot-scale manipulations of organic matter inputs to soils correlate with shifts in microbial community composition in a lowland tropical rain forest. Soil Biol. Biochem. *42*, 2153-2160.

Norby, R.J. and Zak, D.R. (2011). Ecological lessons from free-air CO_2 enrichment (FACE) experiments. Annu. Rev. Ecol. Evol. Syst. *42*, 181.

Pendall, E., Schwendenmann, L., Rahn. T., Miller, J.B., Tans, P.P., and White, J.W.C. (2010). Land use and season affect fluxes of CO_2, CH_4, CO, N_2O, H_2 and isotopic source signatures in Panama: evidence from nocturnal boundary layer profiles. Glob. Change Biol. *16*, 2721–2736.

Pold, G., and DeAngelis, K.M. (2013). Up Against The Wall: The Effects of Climate Warming on Soil Microbial Diversity and The Potential for Feedbacks to The Carbon Cycle. Diversity *5*, 409-425.

Prescott, C.E. (2010). Litter decomposition: what controls it and how can we alter it to sequester more carbon in forest soils? Biogeochemistry *101*, 133-149.

Rütting, T., Clough, T.J., Müller, C., Lieffering, M., and Newton, P.C. (2010). Ten years of elevated atmospheric carbon dioxide alters soil nitrogen transformations in a sheep-grazed pasture. Glob. Change Biol. *16*, 2530-2542.

Ramankutty, N., Gibbs, H.K., Achard, F., DeFries, R., Foley, J.A., and Houghton, R.A. (2007). Challenges to estimating carbon emissions from tropical deforestation. Glob. Change Biol. *13*, 51–66.

Ramankutty, N., Evan, A.T., Monfreda, C., and Foley, J.A. (2008). Farming the planet: 1. Geographic distribution of global agricultural lands in the year 2000. Glob. Biogeochem. Cy. *22*.

Rasche, F., and Cadisch, G. (2013). The molecular microbial perspective of organic matter turnover and nutrient cycling in tropical agroecosystems-What do we know? Biol. Fertil. Soils *49*, 251-262.

Reich, P.B., Hungate, B.A., and Luo, Y. (2006). Carbon-Nitrogen Interactions in Terrestrial Ecosystems in Response to Rising Atmospheric Carbon Dioxide. Annu. Rev. Ecol. Evol. Syst. *37*, 611-636.

Ritz, K., McNicol, J.W., Nunan, N., Grayston, S., Millard, P., Atkinson, D., Gollotte, A., Habeshaw, D., Boag, B., Clegg, C.D., Griffiths, B.S., Wheatley, R.E., Glover, L.A., McCaig, A.E., and Prosser, J.I. (2004). Spatial structure in soil chemical and microbiological properties in an upland grassland. FEMS Microbiol. Ecol. *49*, 191-205.

Rodrigues, J.L., Pellizari, V.H., Mueller, R., Baek, K., da C Jesus, E., Paula, F.S., Mirza, B., Hamaoui, G.S., Tsai, S.M., Feigl, B., Tiedje, J.M., Bohannan, B.J.M., and Nüsslein, K. (2013). Conversion of the Amazon rainforest to agriculture results in biotic homogenization of soil bacterial communities. Proc. Natl. Acad. Sci. USA *110*, 988-993.

Roselló-Mora, R., and Amann, R. (2001). The species concept for prokaryotes. FEMS Microbiol. Rev. *25*, 39-67.

Rudel, T.K., Schneider, L., Uriarte, M., Turner, B.L., DeFries, R., Lawrence, D., Geoghegan, J., Hecht, S., Ickowitz, A., Lambin, E.F., Birkenholtz, T., Baptista, S., and Grau, R. (2009). Agricultural intensification and changes in cultivated areas, 1970-2005. Proc. Natl. Acad. Sci. USA *106*, 20675-20680.

Sall, S.N., Masse, D., Ndour, N.Y.B., and Chotte, J.-L. (2006). Does cropping modify the decomposition function and the diversity of the soil microbial community of tropical fallow soil? Appl. Soil Ecol. *31*, 211-219.

Sanchez, P.A. (1977). Properties and Management of Soils in the Tropics (Wiley).

Sandoval-Pérez, A., Gavito, M., García-Oliva, F., and Jaramillo, V. (2009). Carbon, nitrogen, phosphorus and enzymatic activity under different land uses in a tropical, dry ecosystem. Soil Use Manage. *25*, 419-426.

Santos, V.B., Araújo, A.S., Leite, L.F., Nunes, L.S.A., and Melo, W.J. (2012). Soil microbial biomass and organic matter fractions during transition from conventional to organic farming systems. Geoderma *170*, 227-231.

Schimel, J.P., and Schaeffer, S. (2012). Microbial control over carbon cycling in soil. Front. Microbiol. *3*.

Schimel, J.P., and Gulledge, J. (1998). Microbial community structure and global trace gases. Glob. Change Biol. *4*, 745-758.

Schmidt, M.W.I., Torn, M.S., Abiven, S., Dittmar, T., Guggenberger, G., Janssens, I.A., Kleber, M., Kogel-Knabner, I., Lehmann, J., Manning, D.A.C., Nannipieri, P., Rasse, D.P., Weiner, S., and Trumbore, S.E. (2011). Persistence of soil organic matter as an ecosystem property. Nature *478*, 49-56.

Shanahan, T. M., Overpeck, J.T., Anchukaitis, K., Beck, J.W., Cole, J.E., Dettman, D.L., Peck, J.A., Scholz, C.A., and King, J.W. (2009). Atlantic forcing of persistent drought in West Africa. Science *324*, 377-380.

Sheik, C. S., Beasley, W.H., Elshahed, M.S., Zhou, X., Luo, Y., and Krumholz, L.R. (2011). Effect of warming and drought on grassland microbial communities. ISME J *5*, 1692-1700.

Shukla, J., Nobre, C., and Sellers, P. (1990). Amazon deforestation and climate change. Science. *247*, 1322–1325.

Signor, D., Cerri, C., and Conant, R. (2013). N2O emissions due to nitrogen fertilizer applications in two regions of sugarcane cultivation in Brazil. Environ. Res. Lett. *8*, 015013.

Silver, W.L. (1998). The potential effects of elevated CO_2 and climate change on tropical forest soils and biogeochemical cycling. Clim. Change *39*, 337-361.

Sinsabaugh, R. L., Manzoni, S., Moorhead, D.L., and Richter, A. (2013). Carbon use efficiency of microbial communities: stoichiometry, methodology and modelling. Ecol. Lett. *16*, 930-939.

Six, J., Frey, S., Thiet, R., and Batten, K. (2006). Bacterial and fungal contributions to carbon sequestration in agroecosystems. Soil Sci. Soc. Amer. J. *70*, 555-569.

Soares-Filho, B.S., Nepstad, D.C., Curran, L.M., Cerqueira, G.C., Garcia, R.A., Ramos, C.A., Voll, E., McDonald, A., Lefebvre, P., and Schlesinger, P. (2006). Modelling conservation in the Amazon basin. Nature. *440*, 520-523.

Srivastava, S., and Singh, J. (1991). Microbial C, N and P in dry tropical forest soils: Effects of alternate land-uses and nutrient flux. Soil Biol. Biochem. *23*, 117-124.

Stephens, B.B., Gurney, K.R., Tans, P.P., Sweeney, C., Peters, W., Bruhwiler, L., Ciais, P., Ramonet, M., Bousquet, P., Nakazawa, T., Aoki, S., Machida, T., Inoue, G., Vinnichenko, N., Lloyd, J., Jordan, A., Heimann, M., Shibistova, O., Langenfelds, R.L., Steele, L.P., Francey, R.J., and Denning, A.S. (2007). Weak Northern and Strong Tropical Land Carbon Uptake from Vertical Profiles of Atmospheric CO_2. Science. *316*, 1732–1735.

Stocker, T., Qin, D., and Platner, G. (2013). Climate Change 2013: The Physical Science Basis. Working Group I Contribution to the Fifth Assessment Report of the Intergovernmental Panel on Climate Change. Summary for Policymakers (IPCC).

Strickland, M.S., Lauber, C., Fierer, N., and Bradford, M.A. (2009). Testing the functional significance of microbial community composition. Ecology. *90*, 441-451.

Torsvik, V., and Øvreås, L. (2002). Microbial diversity and function in soil: from genes to ecosystems. Curr. Op. Microbiol. *5*, 240-245.

Tylianakis, J.M., Didham, R.K., Bascompte, J., and Wardle, D.A. (2008). Global change and species interactions in terrestrial ecosystems. Ecol. Lett. *11*, 1351-1363.

van der Heijden, M.G.A., Bardgett, R.D., and van Straalen, N.M. (2008). The unseen majority: soil microbes as drivers of plant diversity and productivity in terrestrial ecosystems. Ecol. Lett. *11*, 296-310.

Veldkamp, E., Davidson, E.A., Erickson, H., Keller, M., and Weitz, A. (1999). Soil nitrogen cycling and nitrogen oxide emissions along a pasture chronosequence in the humid tropics of Costa Rica. Soil Biol. Biochem. *31*, 387–394.

Veldkamp, E., Purbopuspito, J., Corre, M.D., Brumme, R., and Murdiyarso, D. (2008). Land use change effects on trace gas fluxes in the forest margins of Central Sulawesi, Indonesia. J. Geophys. Res. *113*, G02003.

Verchot, L.V., Davidson, E.A., Cattânio, H., Ackerman, I.L., Erickson, H.E., and Keller, M. (1999). Land use change and biogeochemical controls of nitrogen oxide emissions from soils in eastern Amazonia. Glob. Biogeochem. Cy. *13*, 31–46.

Verchot, L.V., Hutabarat, L., Hairiah, K., and van Noordwijk, M. (2006). Nitrogen availability and soil N2O emissions following conversion of forests to coffee in southern Sumatra. Global Biogeochem Cycl. *20*.

Vitousek, P. M., and Reiners, W.A. (1975). Ecosystem succession and nutrient retention: a hypothesis. BioScience *25*, 376–381.

Vitousek, P.M., Matson, P., Volkmann, C., Maass, J.M., and Garcia, G. (1989). Nitrous oxide flux from dry tropical forests. Glob. Biogeochem. Cy. *3*, 375–382.

Vitousek, P.M., Porder, S., Houlton, B.Z., and Chadwick, O.A. (2010). Terrestrial phosphorus limitation: mechanisms, implications, and nitrogen-phosphorus interactions. Ecol. Appl. *20*, 5-15.

Waldrop, M., Balser, T., and Firestone, M. (2000). Linking microbial community composition to function in a tropical soil. Soil Biol. Biochem. *32*, 1837-1846.

Waldrop, M., and Firestone, M. (2006). Response of microbial community composition and function to soil climate change. Microb. Ecol. *52*, 716-724.

Wallenstein, M.D., and Hall, E.K. (2011). A trait-based framework for predicting when and where microbial adaptation to climate change will affect ecosystem functioning. Biogeochemistry. *109*, 35-47.

Wardle, D.A. (1992). A comparative assessment of factors which influence microbial biomass carbon and nitrogen levels in soil. Biol. Rev. *67*, 321-358.

Wardle, D.A. (1998). Controls of temporal variability of the soil microbial biomass: a global-scale synthesis. Soil Biol. Biochem. *30*, 1627-1637.

Wardle, D.A., Bardgett, R.D., Klironomos, J.N., Setälä, H., Van Der Putten, W.H., and Wall, D.H. (2004). Ecological linkages between aboveground and belowground biota. Science. *304*, 1629-1633.

Wardle, D.A., Verhoef, H.A., and Clarholm, M. (1998). Trophic relationships in the soil microbial food-web: Predicting the responses to a changing global environment. Glob. Chang. Biol. *4*, 713-727.

Waring, B.G., Weintraub, S.R., and Sinsabaugh, R.L. (2014). Ecoenzymatic stoichiometry of microbial nutrient acquisition in tropical soils. Biogeochemistry *117*, 101-113.

Wick, B., Veldkamp, E., De Mello, W., Keller, M., and Crill, P. (2005). Nitrous oxide fluxes and nitrogen cycling along a pasture chronosequence in Central Amazonia, Brazil. Biogeosci. *2*, 499–535.

Willig, M.R., Kaufman, D.M., and Stevens, R.D. (2003). Latitudinal gradients of biodiversity: pattern, process, scale, and synthesis. Annu. Rev. Ecol. Evol. Syst. *34*, 273-309.

Wood, S.A., Jina, A.S., Jain, M., Kristjanson, P., and DeFries, R.S. (2014). Smallholder farmer cropping decisions related to climate variability across multiple scales. Glob. Environ. Chang. *25*, 163-172.

Wood, T.E., Cavaleri, M.A., and Reed, S.C. (2012). Tropical forest carbon balance in a warmer world: a critical review spanning microbial-to ecosystem-scale processes. Biol. Rev. *87*, 912-927.

Zhou, J., Xue, K,. Xie, J., Deng, Y., Wu, L., Cheng, X., Fei, S., Deng, S., He, Z., Van Nostrand, J.D., and Luo, Y. (2011). Microbial mediation of carbon-cycle feedbacks to climate warming. Nat. Clim. Chang. *2*, 106-110.

Chapter 16

Geoengineering the Climate via Microorganisms: a Peatland Case Study

Christian Dunn*, Nathalie Fenner, Anil Shirsat and Chris Freeman

Wolfson Carbon Capture Laboratory, School of Natural Sciences, Bangor University, Bangor LL57 2UW, UK; *author for correspondence

Email: c.dunn@bangor.ac.uk, n.fenner@bangor.ac.uk, a.h.shirsat@bangor.ac.uk, c.freeman@bangor.ac.uk

DOI: https://doi.org/10.21775/9781913652579.16

Abstract

Peatlands contain more than double the amount of carbon than is found in the biomass of the world's forests. Such stores are due to the build-up of dead plant material, resulting from restraints on microbial decomposition in the peat-substrate: in particular the inhibitory effects of phenolic compounds create an 'enzymic latch' on the breakdown of organic matter. We propose that this mechanism could be harnessed for a number of peatland-based geoengineering schemes. Such strategies would involve using molecular, agronomical and biogeochemical approaches to manipulate microbial activities in peatlands – maximising their abilities to store and capture carbon. Although like all geoengineering proposals, peatland geoengineering does not offer a 'magic bullet' in reversing the effects of climate change, it potentially has numerous advantages over other suggested schemes. Moreover, recent research indicates that this stored carbon can be made far more resilient to future global warming than had previously been appreciated. Most of the technologies and knowledge are already established, the projects are reversible, and they do not compete with other land uses such as food production. It can therefore be argued that peatland geoengineering offers a realistic 'Plan B' to save the planet from the effects of anthropogenic climate change.

Introduction

The United Nations' (UN) climate panel, the Intergovernmental Panel on Climate Change (IPCC), states that evidence for the warming of average global temperatures is 'unequivocal' (IPCC, 2013). To blame for this is the ever increasing concentrations of greenhouses gases (GHGs) in the Earth's atmosphere. In a groundbreaking report in 2013, the IPCC stated that atmospheric concentrations of the key GHG, carbon dioxide (CO_2), had increased to its highest level in 800,000 years. This, the IPCC says, is due primarily to the increased burning of fossil fuels seen since 1750, and humans are therefore the 'dominant cause' of global warming. Indeed, despite the growing understanding of climate change over the past decades annual CO_2 have continued to rise 9.5 Gt C/year contributing to the GHG's record concentration of 391 ppm in the atmosphere. If this concentration continues to rise, as is expected, then by 2100 the surface temperature of the Earth is likely to be around 2°C warmer - sufficient to cause significant problems ranging from rising sea levels, drought and loss of habitat, to severe storms and increased ocean acidity (IPCC, 2013).

Crucially, the IPCC predicts that many aspects of climate change will persist for many centuries even if all emissions of CO_2 are immediately halted. A conclusion also arrived at by the UK's Royal Society - which, founded in 1660 is one of the oldest scientific learned society in the world and advisors the British Government on a range of issues. They highlighted that global emission reductions will not be sufficient to avoid the dangers of global warming and suggested that geoengineering could help reduce the future extent of climate change or provide more time to address the challenges posed by mitigation, such as phasing out CO_2-emitting fossil fuel energy technologies and developing/deploying energy sources that are carbon neutral (Royal Society, 2009).

Several geoengineering techniques have been put forward to achieve this, all with a distinct set of advantages and disadvantages. Few methods though have looked at exploiting the functions of microorganisms to sequester increased levels of carbon. Considering around half of the carbon flux through terrestrial ecosystems is mediated by heterotrophic metabolism of microorganisms this may seem surprising. However, relatively little is known about microbial mechanisms regulating carbon exchange compared to our knowledge of the primary production and its response to climate change (Bahn et al., 2008). Indeed, since global climate change includes changes in the key controlling variables (e.g. temperature rise, changes in precipitation, elevated

CO_2, and sea level rise) for microbial composition and process rates, it is unclear whether climate-microbial feedbacks could amplify or dampen future climate change by increasing or decreasing GHG releases through microbial processes (Heimann and Reichstein, 2008).

However, work by Freeman et al. (2012), Fenner and Freeman (2020) has suggested that microbial process in peatlands could be modified in order to suppress decomposition of organic matter; thereby storing climate-changing amounts of carbon and offering an effective, safe and novel geoengineering technique.

Geoengineering

Geoengineering is "the deliberate large-scale intervention in the Earth's climate system, in order to moderate global warming" (Royal Society, 2009). Geoengineering methods can be divided into two basic classes:

1. **Carbon dioxide removal** (CDR) techniques which remove CO_2 from the atmosphere.
2. **Solar radiation management** (SRM) techniques that reflect a small percentage of the sun's light and heat back into the space.

Suggested SRM techniques include increasing the reflectivity of the planet (known as albedo) by brightening human structures (e.g. painting them white), planting crops with high reflectivity (Singarayer and Davies-Barnard, 2012), covering desserts with reflective materials, enhancing marine cloud reflectivity (Latham et al., 2012), injecting sulphate aerosols into the lower stratosphere to mimic the effects of volcanic eruptions (Davidson et al., 2012) and even placing deflectors in space to reduce the amount of solar energy reaching the Earth (Goldblatt and Watson, 2012). It has been suggested that such techniques would take only a few years to have an effect on the climate once deployed but as they do not treat the root cause of climate change (increased levels of GHG in the atmosphere) they would create an artificial balance between increased GHG concentrations and reduced solar radiation. This would therefore need constant maintenance - perhaps for centuries (Shepherd, 2012).

The use of geoengineering techniques remain highly controversial due to the fact their effects, both intended and unintended, are not yet fully understood as no large scale projects have been undertaken (Bertram, 2010). However, it has been argued that fighting the effects of climate change should involve a comprehensive three-point approach embracing mitigation, adaptation and geoengineering (referred to as the MAG approach). Indeed in their comprehensive report on the subject the

Royal Society stated that while the overall reduction of GHG emissions is the safest and most predictable method of moderating climate change (and should therefore be policy-makers' priority) geoengineering is a potentially viable method for supporting these aims. They concluded that further research should be conducted, as a matter of some urgency, to investigate whether proposed techniques could be made available if it becomes necessary to reduce the rate of warming this century (Royal Society, 2009).

Manipulating the biogeochemistry of peatlands, would be classified as CDR geoengineering technique as it would involve the increased sequestration of carbon. Other land use management projects have been proposed including afforestation and reforestation (Bonan, 2008); however, unlike these methods peatlands are not vulnerable to rapid re-release of sequestered carbon when trees die and decompose (Korner, 2003).

Decomposition in peatlands

Peatland is generally considered to be a generic term for any freshwater wetland that accumulates partially decayed plant matter peat (usually to a depth of greater than 40cm), referred to as peat (Mitsch and Gosselink, 2000a). They are classified as organic wetlands and are estimated to be the most widespread group of wetlands (Fig. 16.1) differing from mineral wetlands by having a living plant layer plus thick accumulations of preserved plant detritus. Recent research suggests peatlands are underappreciated in climate mitigation strategies (Leifeld & Menichetti 2018) and are hotspots for drinking water supply (Xu et al 2018). Still, more importantly, peatlands accrete globally significant amounts of atmospheric CO_2, lowering global temperatures by up to 3°C (Holden 2005). They do so, as they have higher carbon storage densities per unit area of ecosystem than either the oceans or truly terrestrial systems (Cole et al., 2007; Dean and Gorham, 1998) and despite only covering 3% of the Earth's land they store between 390 and 528 Pg of carbon (Bragazza et al., 2006; Freeman et al., 2004b; Gorham, 1991; Immirzi and Maltby, 1992; Page et al., 2011). This is equivalent to around 60 to 78 percent of the entire atmospheric carbon pool and is twice the amount of carbon found in the entire world's forest biomass (IPCC, 2007; Oechel et al., 1993; Parish et al., 2008). Despite many areas of peatlands suffering from anthropogenic damage, such as drainage and burning, they have been considered a net sink of atmospheric carbon throughout the Holocene period to the

Figure 16.1 World peatland distribution (Pravettoni, 2009).

present day (Kayranli et al., 2010; Moore, 2002; Smith et al., 2004; Waddington et al., 2010).

This carbon storage capacity occurs because rates of primary production exceed particularly slow decomposition rates, resulting in the long-term accumulation of the partially decomposed peat, or organic matter known (Gorham, 1991; Limpens et al., 2008). **T**raditionally this impaired decay has been attributed to anoxia (Gorham, 1991; McLatchey and Reddy, 1998), low nutrients, low temperatures and low pH (Gorham, 1991; Laiho, 2006). But it was recently appreciated that oxygen constraints on a single class of enzymes exert a particularly potent control over the vast carbon stock held in peatlands (Fenner and Freeman, 2011; Freeman et al., 2001b). Phenol oxidases are among the few enzymes able to fully degrade inhibitory phenolic compounds (derived from plant breakdown products) but require oxygen to operate efficiently (McLatchey and Reddy, 1998). The predominantly anoxic conditions in peat allow an accumulation of phenolic compounds, which, in turn, prevent the major agents of nutrient and carbon cycling, namely hydrolase enzymes, from carrying out their normal processes of decay (Freeman et al, 1990; Appel, 1993; Wetzel, 1992) suppressing microbial enzymic decomposition of senescent vegetation and thus promoting sequestration of huge stores of carbon. While this highlights the potential for carbon loss as a result of increased drought frequency (brought about through climate change), the recognition of this 'enzymic latch' mechanism (Freeman et al., 2001b), also for the proposal of valuable new geoengineering techniques.

The enzymic latch mechanism

A key rate limiting step in the decomposition of organic matter in peat-soil is the activity of extracellular hydrolase enzymes (Burns, 1982; Sinsabaugh, 1994). These are produced by heterotrophic soil organisms in order to cleave large, complex, high molecular weight, organic matter macromolecules into their respective monomers, which can then cross cell membranes and be assimilated by the microbes (Rogers, 1961; Swift et al., 1979). This is important as microbial uptake of organic matter is governed by molecular size and composition (Kaplan and Bott, 1983), with microorganisms generally being unable to assimilate organic molecules with a molecular weight over 1,000 Da (Daltons; Confer and Logan, 1998) and preferentially utilising low molecular weight molecules (Meyer et al., 1987). The activities of extracellular hydrolysing enzymes can therefore prevent the excessive accumulation of detrital OM in the environment, whilst also supplying photosynthetic organisms with nutrients and give an indication of growth and development of microbial communities in the peat matrix (Gajewski and Chrost, 1995). However, the production and activity of these essential extracellular hydrolase enzymes in soil can be inhibited by various biotic and abiotic factors (Burns, 1978; Ladd, 1978). In peatlands this inhibition has led to an unprecedented accumulation of high molecular weight, recalcitrant, aged organic carbon (Clymo, 1983) and, specifically, a range of compounds known as polyphenols, or phenolics (Freeman et al., 2001a; Freeman et al., 2001b).

Phenolic compounds are a diverse group of organic compounds, defined by the presence of at least one aromatic ring bearing one (phenol) or more (polyphenol) hydroxyl substituents or their derivatives e.g. esters and glycosides. Simple monomeric phenols, which include flavonoids and phenolic acids are produced universally within higher-order plants (Hattenschwiler and Vitousek, 2000), while woody plants are the main producers of polyphenols, such as lignin and tannins (Faulon and Hatcher, 1994; Haslam, 1989).

The important role phenolics have on decomposition and nutrient cycling has long been realised (Horner et al., 1988) and in particular it has been documented that polyphenols are capable of inhibiting organic decomposition in peatland-associated environments (Freeman et al., 1990; Freeman et al., 2001b; Freeman et al., 2004b). This is brought about by the phenolic compounds forming complexes with protein molecules, serving as polydentate ligands, causing inactivation of hydrolase enzymes (the main agents of decomposition) through competitive and non-competitive inhibition (Wetzel, 1992). The polyphenols also bind to the reactive sites of proteins and other organic and inorganic substrates, often rendering them inactive to further chemical activity and biological attack (Muscolo and Sidari, 2006). Soil environmental conditions may influence the potency of the inhibitory nature of polyphenols; the dissociation of polyphenol protein complexes for example, being positively correlated with pH (Basaraba and Starkey, 1966).

The extent to which such phenolic compounds accumulate within the peatland soil is in turn dependent on the activity of another set of enzymes capable of their elimination - through oxidation. Phenol oxidases are oxidative copper containing enzymes that catalyse the release of reactive oxygen-radicals that can then promote a variety of non-enzymic reactions to oxidise phenolic compounds (Claus, 2004; Thurston, 1994). Indeed phenol oxidases can instigate the oxidation of both complex polyphenols with outcomes ranging from partial oxidation and the release of oxidative intermediates, to the complete degradation of phenolic compounds to non-phenolic end products. In peatlands, phenol oxidases are produced by fungi, bacteria, Actinobacteria and, to a lesser extent, plants (Burke and Cairney, 2002; Fenner et al., 2005; Gramss et al., 1999).

The enzymic latch mechanism therefore refers to how constraints on the activity of phenol oxidases in peatlands represent a fragile 'latch' holding back the decomposition and release of a vast 455 Pg store of carbon. These constraints prevent the following sequence of events (Freeman et al., 2012), shown in Fig. 16.2, whereby: oxygen stimulates phenol oxidase [a] causing a decline in the abundance of inhibitory phenolic compounds [b] allowing stimulation of hydrolase enzymes [c] creating two responses; increased breakdown of low molecular weight labile dissolved organic carbon [d] and release of inorganic nutrients that were previously sequestered in the peat matrix [e]. Both these events provide substrates and nutrients for enhanced microbial activity [f], and hence increased production of hydrolase enzymes [g] and

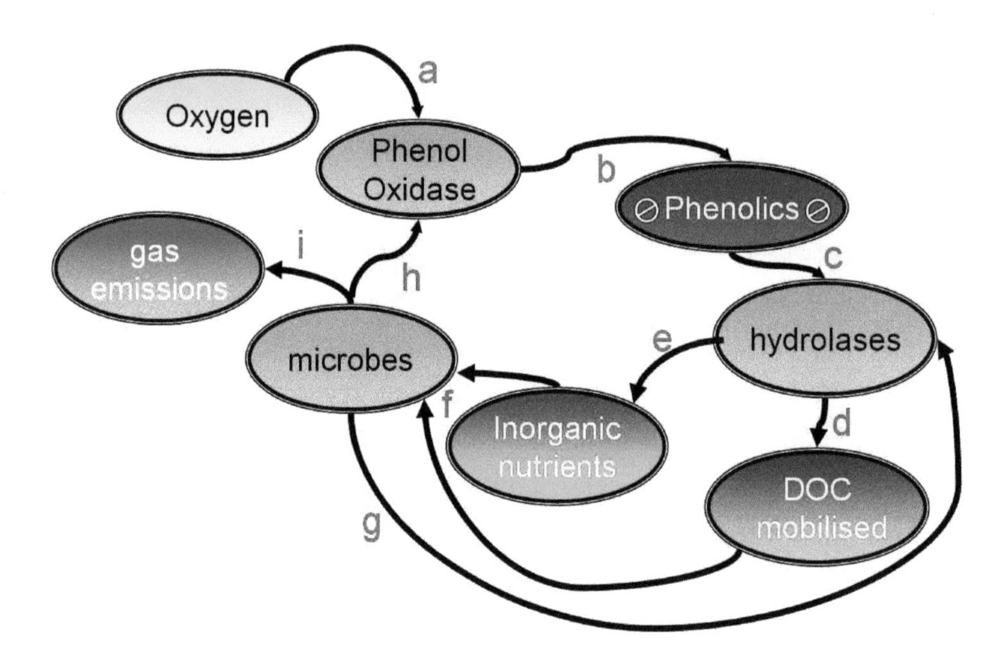

Figure 16.2 The role of phenol oxidase in regulating peatland carbon storage through the enzymic latch mechanism.

phenol oxidase [h], together with enhanced emissions of biogenic trace gases such as CO_2, CH_4 and N_2O [i]. The key regulators of this pathway are therefore the phenolics and the factors influencing the enzymes capable of depleting these inhibitory compounds.

Harnessing the enzymic latch mechanism

Northern peatlands currently capture around $1.7\text{x}10^{15}$ g carbon/year through photosynthetic litter production (Reader and Stewart, 1972). However, at present they only sequester $1\text{x}10^{14}$ g carbon/year or 9%, of that in the long-term. This inefficiency arises because substantially less carbon enters long-term storage, 34 to 52 g carbon/ m^2/year, than is initially captured from net primary production (NPP), which can reach 489g carbon/m^2/year (Reader and Stewart, 1972; Wieder, 2001). Therefore, although decomposition of organic matter is considerably suppressed in peatlands, compared to other ecosystems, it is still taking place and there is scope for amplifying the strength of the enzymic latch to further inhibit the breakdown of plant litter.

It has been proposed that the enzymic latch mechanism has the potential to be used to increase carbon sequestration in two principle ways (Fig. 16.3). The first is through strengthening the enzymic latch by increasing the abundance of phenolic inhibitors or manipulating edaphic factors that could slow down the microbial production of phenol oxidases and/or its activity (e.g. inorganic nutrient availability and pH). Secondly, by increasing the amount of carbon influenced by the enzymic latch using methods to either increase peatland plant productivity or add forms of externally captured carbon.

Strengthening the enzymic latch

Empirical testing of the enzymic latch concept across peatlands representing a nutrient gradient has identified specific points in the latch regulatory pathway that could dramatically impair decomposition in both aerobic and anaerobic pathways (Fig. 16.2); adverse pH, oxygen, labile C, and inorganic nutrient availability can all promote enzymic latch mediated suppression of decomposition, along with generation of certain particularly inhibitory phenolic compound types. In earlier studies of the enzymic latch, we have found phenol oxidase to regulate key aspects of the biogeochemical decomposition process in peatlands through the sequence of

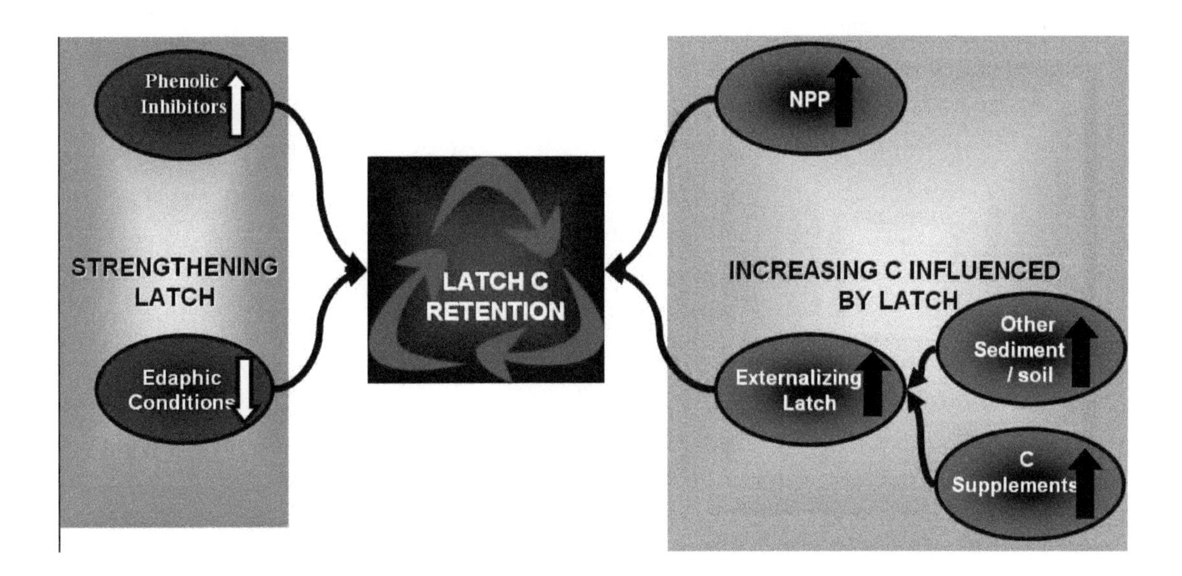

Figure 16.3 Enhancing peatland carbon sequestration using the enzymic latch.

events summarised in Fig. 16.2, key processes within which are amenable to manipulation (below) using approaches such as those illustrated in Fig. 16.4.

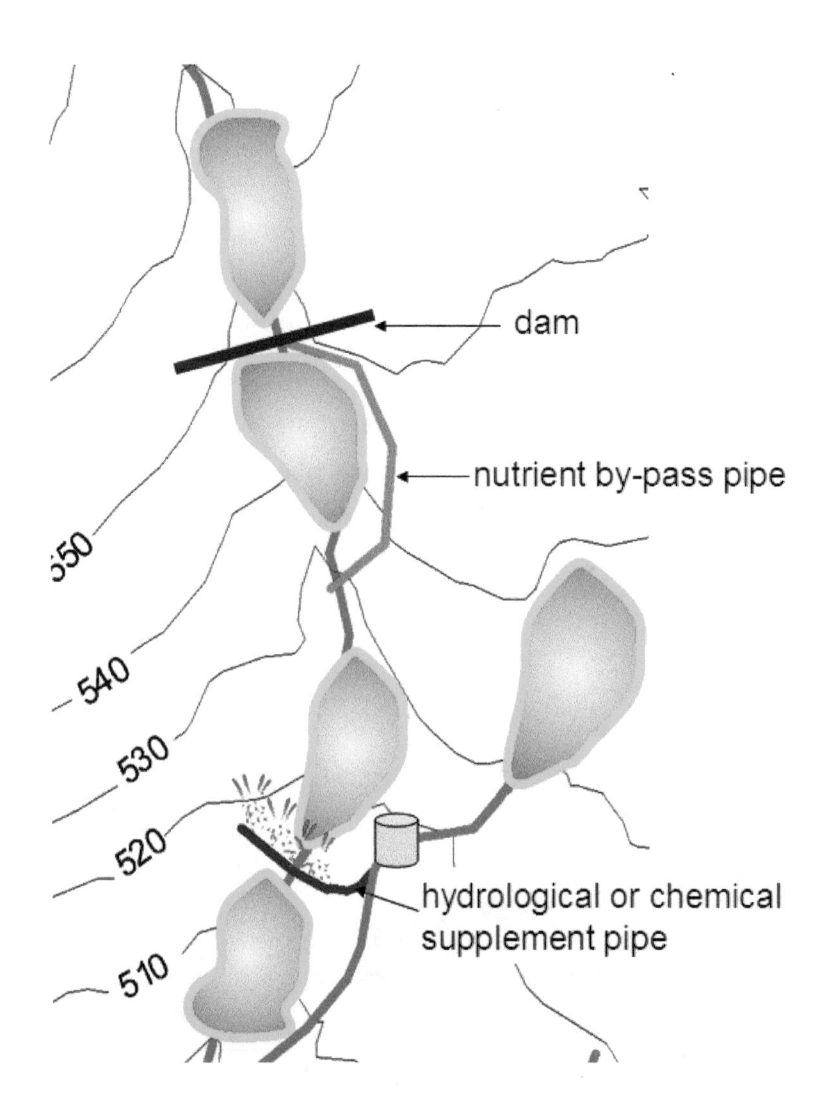

Figure 16.4 An example of an experimental peatland manipulation field site, with options for installing dams to increase waterlogging in peatlands above the dam. Pipes can take water from the dam to divert nutrients away from the wetland below to modify plant or microbial biodiversity. A supplement pipe can take waters from parallel watersheds to increase waterlogging/paludification in the test sites below. Nutrients can be added to increase *Sphagnum* productivity or other chemicals (e.g. sulphates, exogenous phenolic inhibitors) can be added to the water to strengthen the enzymic latch and suppress enzymic decomposition.

Process 1: O_2 availability can be manipulated by raising water table levels and reducing flow rates (hydrological manipulations such as the introduction of dams; Fig. 16.4) to reduce O_2 replenishment. There may also be value in adding chemical reducing agents to the soil. This will regulate activity in the form of *de novo* phenol oxidase synthesis in the microbial community and also in the form of pre-existing 'edaphic' exoenzyme activity in the peat itself.
Process 2: Acidification, by addition of sulphates (potentially from the SRM strategies described earlier), will suppress phenol enzyme activities. Phenol oxidase activity is particularly sensitive to low pH (Pind et al., 1994; Williams et al., 2000) and will have indirect effects on hydrolases enzymes by allowing accumulation of higher levels of inhibitory phenolic compounds. Indeed, the effects of ammonium sulphate, a physiological acid fertiliser, can increase wetland rice productivity while decreasing decomposition (Minamikawa et al., 2006).
Process 3: Phenolic supplements can be applied to suppress i) microbial metabolism and therefore *de novo* synthesis of enzymes (phenol oxidase and both carbon- and inorganic nutrient-cycling hydrolases), and ii) edaphic enzyme activities (phenol oxidase and hydrolases). Sources of phenolic compounds might include naturally produced anaerobic compounds in peat leachates, *trans*-sphagnum acid (see below) collected *in vitro*, or other analogue polyphenolic waste materials (e.g. coir, apple pulp from cider production and green compost) and liquids (e.g. olive waste, paper waste, green compost leachate) that release their nutrients slowly, while maintaining low soil carbon turnover rates.
Process 4: Hydrolase enzyme activities can be reduced by lowering pH, labile carbon and inorganic nutrient supply (see process 5).
Process 5: Lowering available inorganic nutrients and labile carbon, will suppress further enzyme production by the microbial community (*de novo* synthesis) by the addition of complexing agents (e.g. high molecular weight organic carbon and phenolics) and hydrological manipulation to reduce the replenishment of nutrients.

Peatland plant communities clearly have an important role in ecosystem carbon dynamics and storage, with every species possessing a unique functional phenotype (Fenner et al., 2007a; Fenner et al., 2004; Fenner et al., 2007b). *Sphagnum* spp. are particularly important in this respect, as they are characteristic of both peat bogs and nutrient poor fens (Gajewski et al., 2001), showing the greatest global ecological dominance of all bryophyte taxa (Andrus, 1986). The slow decomposition rates of

Sphagnum tissues (Kuhry and Vitt, 1996) make them the single most important C accreting species in bogs (Heijmans et al., 2002). Indeed, it has been estimated that more than half of the world's peat originated from *Sphagnum* spp., representing 10–15% of the entire terrestrial C stock (Clymo and Hayward, 1982).

The unusually low rates of decomposition of this genus are attributed to specific secondary compounds, most notably polyphenols which have a potent inhibitory effect on microbial breakdown (Freeman et al., 2001b; Verhoeven and Liefveld, 1997). The most important polyphenol in *Sphagnum* mosses is a cinnamic acid derivative called trans-sphagnum acid, which is unique to peat mosses. The quantity present varies depending on species, season and part of the plant investigated (Rudolph and Samland, 1985), but all species analysed are characterised by substantial quantities and also can secrete it into the external medium. There is considerable evidence that while *Sphagnum* lacks lignin, the phenolic *trans* sphagnum acid is synthesised *via* the phenylpropanoid pathway from phenylalanine (Rasmussen et al., 1995). By promoting production of this phenolic compound, through the genetic modification of *Sphagnum* plants responsible for phenolic synthesis, it may be possible to vastly increase the overall concentration of phenolics in the peat-matrix. The enzyme responsible for the final catalytic stage resulting in the synthesis of *trans*-sphagnum acid, *trans*-sphagnum acid synthase (tSphA) remains to be isolated. However the genes coding for the other determining enzymes of the pathway have already been isolated from a variety of plants, including the moss species *Physcomitrella patens*. The determining enzyme of the phenylpropanoid pathway - phenylalanine ammonia lyase (PAL) has already been shown to be the key step regulating flux through the pathway – an increase in PAL levels leads to an increase in the accumulation of major phenylpropanoid pathway products (Bate et al., 1994). We therefore suggest that the use of a transgenic approach, to increase levels of *trans*-sphagnum acid in *Sphagnum* by increasing the expression of *Sphagnum* PAL, could offer great benefits to geoengineering.

While at first consideration, successful introduction of a new form of *Sphagnum* into the environment may seem highly challenging, it should be noted that this would not be the first occasion on which a single *Sphagnum* plant had colonised an entire continent. A total of 100% of the gene pool of *S. subnitens* in North America was contributed by one individual plant (Karlin et al., 2011), and while this took place over 300 years as a natural process, it is likely to be achieved far more rapidly with human intervention.

Furthermore, while some of the earlier described options for peatland geoengineering of our climate represent a new concept, other forms of peatland manipulation have a long history. Drainage, for example, commenced before Roman times and was also recorded in Domesday (Darby, 1956). The peak rate of drainage in the UK has been estimated to have reached 100,000 ha yr^{-1} in 1970 (Green, 1973; Robinson and Armstrong, 1988). In Northern Ireland there are only 169 km^2 of intact peat left compared with 1190 km^2 of total peatland (Cooper et al., 1991). Such figures give a clear indication of the potential for rapid regional scale manipulation of peatlands.

Increasing the amount of carbon

The other proposed option for using microbially controlled peatland biogeochemistry as a form of geoengineering, or carbon capture, is to increase the amount of carbon entering the soil system. Strengthening the enzymic latch after increasing the amount of carbon retained, would synergistically increase the amount of carbon sequestered.

Optimising plant productivity

Unlike conventional agriculture, where productivity has more than doubled in the last 40 years (Tilman et al., 2002), no attempt has yet been made to increase the productivity (and hence carbon sequestration) rates of peatland plants. Established agronomical approaches (irrigation, warming, fertilizer addition etc.) could therefore prove viable methods to increasing peatland net primary productivity (NPP). Water is the most important factor conferring wetland properties (Mitsch and Gosselink, 2000b) and global productivity patterns indicate wetter microhabitats are the most productive in *Sphagnum* bogs (Gunnarsson, 2005). Both water-table and flow rates could then be manipulated in order to enhance NPP (Fig. 16.4). Mean annual temperature can also be manipulated as a key factor affecting global peatland productivity, with warm humid conditions being optimal for *Sphagnum* growth (Gunnarsson, 2005). This can be achieved by manipulating the insulation of peat soils (through modification of plant cover characteristics). And, while current CO_2 levels may not be limiting to bryophyte NPP due to their proximity to soil respired CO_2, other peatland plant functional groups (e.g. *grasses and sedges*) can be stimulated by elevated CO_2 (Fenner et al., 2007b). Waste CO_2 could be fixed in peatland mesocosms by plant and algal components and then used as soil amendments to 'pristine' or restored peatlands. Chapin et al (2004) found that subtle changes in N:P ratios and pH can increase above ground NPP in bogs and poor-intermediate fens during their investigations of the relative roles of nutrient limitation and pH stress in peatland plant community structure. Thus, even where the vegetation is adapted to acidic and oligotrophic conditions, the potential exists for increased carbon capture, which would then be available for long-term storage *via* the enzymic latch. Manipulation of N:P (nitrogen:phosphorous) ratios using NH_4Cl and NaH_2PO_4+Na_2HPO_4 for example could therefore be investigated. Similarly, the effect of nitrogen applied as amino acids warrants research, since plants (including bryophytes) that are able to utilise amino acids directly from the soil may have a competitive advantage in these low nitrogen ecosystems (Krab et al., 2008). Application of ammonium sulphate, a physiological acid fertiliser, has been shown to increase wetland rice productivity while decreasing decomposition (Minamikawa et al., 2006) and therefore could be investigated along with liming (addition of $CaCO_3$) to raise the pH. Finally, soil amendments could also be investigated to promote NPP. Application of organic matter in the form of *Sphagnum* litter would represent a good candidate, since this can be an important source of N for bryophytes (Gerdol et al., 2006) yet is also known to inhibit microbial decay. Similarly, anaerobically produced peat leachate (rich in phenolics) and analogue polyphenolic waste materials (e.g. coir, apple pulp from cider production and green compost) and liquids (e.g. olive waste, paper waste, green compost leachate) that release their nutrients slowly, while maintaining low soil carbon turnover rates, would require testing.

Of course, long-term storage of carbon captured by NPP is inextricably linked with peatland plant functional groups (Fenner et al., 2007a; Fenner et al., 2004; Fenner et al., 2007b) and, in addition to carbon capture and retention, the following are identified as mechanisms that could be optimised to promote latch-mediated carbon storage:

A. Transport of O_2 below ground. Minimising the abundance of plant species capable of translocating O_2 below ground (also implicated in CH_4 release), would strengthen the latch. Such plants also increase evapotranspiration compared with *Sphagnum*, the avoidance of which would reduce soil moisture loss and thus minimise aerobic carbon losses in the form of CO_2, and dissolved organic carbon (DOC).

B. Certain phenolic compounds are more potent inhibitors (Wetzel, 1992) of microbial *de novo* enzyme synthesis and/or edaphic enzyme activity. Identifying the most potent would allow us to maximise carbon storage, for example, by promoting plant functional groups rich in such compounds.
C. Root exudation: Minimising plant labile carbon can strengthen the enzymic latch (Fenner and Freeman, 2011; Shackle et al., 2000), as such exudates have been associated with priming (i.e. accelerated decomposition) in peatlands (Fenner et al., 2004; Freeman et al., 2004b) and soils (Fontaine et al., 2007; Kuzyakov et al., 2007). If priming was suppressed (e.g. *via* manipulating plant functional groups) it would represent a valuable net gain in carbon storage.
D. Promoting plant groups that compete strongly for inorganic nutrients can also impair microbial decomposition rates in peatlands (Fenner et al., 2007b) and a similar mechanism can occur in upland soils (Freeman et al., 1998).
E. Lowering pH (by encouraging paludification, manipulating plant groups and hydrological regime) can inhibit phenol oxidase activity (Pind et al., 1994; Williams et al., 2000), in turn, creating conditions favouring inhibition of hydrolases and thus, decomposition.

Supplementary carbon sources

Afforestation has long been recognised as a valuable tool for counteracting rising atmospheric CO_2 levels, but it is vulnerable to rapid re-release when trees die and decompose or are subject to pests, diseases and forest fires (Pan et al 2011). Unlike forests, peatlands are unusually effective at inhibiting decomposition (Freeman et al 2001), as illustrated by wooden artefacts and bog bodies preserved over millennia, and thus, combining forest carbon capture with peatland preservation of that externally captured carbon could represent an alternative geoengineering approach to combat climate change. But, whether such carbon would be vulnerable to rapid release due to severe drought events (IPCC 2014) remained until recently a major concern. While the concept of capturing carbon through afforestation is well established (Bonan, 2008), it is recognised as only a temporary solution - decomposition or burning re-releases that C back to the atmosphere (Marechal and Hecq, 2006). The novel approach of 'injecting' timber below the surface of peatlands, could though offer a solution to this problem *via* long-term sequestration as a beneficial consequence of enzymic-latch-mediated suppression of decomposition. CO_2 sequestered in forested areas could be "fixed" by injection of timber into deep peats where the enzymic latch would prevent re-release of that CO_2 back to the atmosphere. Oak's widespread nature, abundance and predominance as ancient archaeological remains (Guyette et al., 2008; Huisman et al., 2008) make it a candidate for research, although fast growing, non-durable, commonly used species (such as poplar) could be more useful for carbon sequestration (Yang et al., 2007). Pine has been found to possess particularly potent antimicrobial properties (Schonwalder et al., 2002) and therefore could also be useful in this role if decay was found to be particularly slow, and similarly tropical hardwoods that possess higher levels of extractives than temperate hardwood (Blanchette, 2000). Designing optimal methods of timber injection while ensuring minimal damage to the peatland ecosystem would be a prerequisite to field scale application if such a technique were found to be viable at the microcosm and mesocosm scale. Timber could for example be inserted initially to create a path to facilitate vehicular access, and then be hidden by inserting the timber below the surface as the heavy machinery withdraws from the site after inserting the timber. This approach would leave the vegetated surface in a relatively undisturbed state.

The most recent data (Fenner and Freeman 2020) show that not only is the exogenous carbon effectively preserved in peatlands, but that the extra polyphenols leached from the wood exert a "quadruple lock" on decomposition and therefore, protect the host ecosystem from severe drought effects. Furthermore, this held true across both pristine and highly degraded peatlands and a similar effect was induced by woody shrub colonisation. These findings suggest novel geo- and ecological engineering applications, harnessing natural ecosystem resilience mechanisms to help 'future proof' carbon stocks. Moreover, there is the potential for improved drinking water quality as a result of these interventions by reducing disinfection by-product precursor load, linked with cancers of the gastrointestinal tract and other undesirable effects (e.g. Goslan et al 2009), arriving at treatment works.

Synthesising from high impact work on the 'enzymic-latch' (Freeman et al 2001), drought effects (the 'biogeochemical cascade' Fenner & Freeman 2011), resilience (Wang et al 2015) and the 'iron gate' (Wang et al 2017) significantly furthers our understanding of natural carbon sequestration mechanisms in peatland systems (Fenner & Freeman 2020). And, this allows us to propose novel strategies to combat climate change that do not risk collateral damage or catastrophic reversal if stopped (Wigley 2006) and could be synergised with restoration and sustainable wet agriculture, i.e., 'carbon farming' for the future (Alshehri et al., 2020; Fenner & Freeman, 2020).

Should enzymic latch manipulation be used for geoengineering?

All currently suggested geoengineering schemes struggle to achieve the right balance between the key criterion necessary for a successful way of altering the Earth's climate system (Royal Society, 2009), which are:

Affordability: of both deployment and operation of the scheme, for a given effect. Usually calculated as cost per Gt C (gigatonne of carbon) for CDR methods and cost per W/m^2 for SRM, evaluated over century timescales.

Effectiveness: including confidence in the scientific and technological basis, technological feasibility, and the magnitude, spatial scale and uniformity of the effect available.

Safety: including the predictability and verifiability of the intended effects, the absence of predictable or unintended adverse side-effects and environmental impacts and low potential for things to go wrong on a large scale.

Timeliness: the state of readiness for implementation (e.g. have all necessary experiments been completed) and the speed with which the intended effect on the climate would occur.

Reversibility: the ability to cease a method and have its effects (including any undesired negative impacts) terminated, within a short period of time - should it be deemed necessary.

However, when compared to other popular CDR techniques (Table 16.1) suppressing microbial decomposition of organic matter in peatlands, by strengthening the enzymic latch, arguably offers the most effective method, scoring highly in all five of the principle requirements. However, it should be borne in mind that any such development must inevitably be associated with risks. Taking this to the extremes, perhaps the most disturbing of these is the prospect that we may be so successful at sequestering carbon, that we might conceivably induce runaway carbon capture. The prospect of plunging the planet into another ice age (Lacis et al., 2010) would certainly raise public concerns. But unlike many other geoengineering options, carbon sequestered in peat can always be returned to the atmosphere, should such a risk ever develop - making it a highly reversible technique. In recent times we have become extremely effective at developing approaches to destroy peatland ecosystems, through drainage and fire so re-releasing their sequestered carbon (Gorham, 1991). As highlighted in Table 16.1 there is also a threat to the stability of geoengineered peatlands in the form of droughts, which are likely to increase in many areas as the planet continues to warm (Laiho, 2006). While this must raise some concern about the fate of carbon sequestered in peatlands, there is evidence that phenolics can

Table 16.1 A comparison of different carbon dioxide removal (CDR) techniques for geoengineering the climate.

Technique	Advantages	Disadvantages
Peatland enzymic latch manipulation Suppressing microbial decomposition of organic matter in peatlands by strengthening the enzymic latch	• Low cost • Preliminary experiments successful • Low risk of unanticipated environmental effects • Reversible - CO_2 can be re-released (e.g. burning) • Several different methods of strengthening the enzymic latch which can be adapted depending on the location and type of peatland • Uses 'low value' land, so little competition with other land uses • Long-term storage of large amounts of carbon • 'Carbon stewardship' may pay for peatland conservation and management	• Restrained by area of global peatlands • May be necessary to 'damage' pristine peatlands • Droughts my negate the effect of the enzymic latch
Afforestation Establishment of a forest or stand of trees	• Low cost • Proven record of success • Begins CO_2 reductions immediately • Low risk of unanticipated environmental effects • Reversible- CO_2 can be re-released (e.g. burning)	• Restrained by availability of sequestration sites • Competition with other land uses, especially agriculture • Biodiversity implications • Limited potential for carbon removal • Sequestered carbon not secure for long periods of time
Bioenergy with carbon capture and sequestration (BECS) Biomass is harvested and used as fuel with the resulting CO_2 being captured and sequestered (e.g. in geological formations)	• Low to medium costs • Low to medium risk of unanticipated environmental effects • Uses existing knowledge and technology • Produces a useable source of energy	• Costs reliant on various factors such as fertiliser and transportation costs • More expensive than fossil fuel carbon capture and storage (CCS) as biofuels can be more expensive • Slow to reduce global temperatures • Potential land-use conflicts • Limited by plant productivity
CO_2 removal from ambient air Using an industrial process to capture CO_2 from ambient air to produce a pure CO_2 stream for use or disposal	• Very low risk of unanticipated environmental effects • No inherent limit on the size of the effect achievable • Reversible - process can be stopped and captured CO_2 re-released • Very little competition with other land uses	• Work still needed to find cost most efficient and effective method • Requires substantial infrastructure construction • Additional carbon storage locations required • Transportation costs of captured CO_2 need to be considered
Ocean fertilisation Introduction of nutrients to the upper ocean to increase marine primary productivity	• No competition with other land uses	• Not likely to be very effective • May reduce biological carbon uptake elsewhere in the oceans • Low long-term carbon storage potential • Not expected to be very cost-effective • Slow to reduce global temperatures • Substantial prior research required to investigate environmental impacts, efficacy and verifiability • High potential for unintended and undesirable ecological consequences

remain active inhibitors even under aerobic conditions (Freeman et al., 1990). Thus, provided that phenolic supplements are applied in excess, carbon loss during the more aerobic conditions associated with droughts should be minimised. A further source of potential risk lies in the fact that by expanding peatland abundance, we are encouraging an ecosystem that contributes 20% to total annual emissions of methane (CH_4), the largest natural source of a GHG with 25 times greater warming potential than CO_2 (Meehl et al., 2007). It is therefore essential to ensure that any applied enzymic latch-mediated enhancement of carbon sequestration does not add to the atmospheric burden of CH_4. Fortunately, by suppressing enzymic generation of labile low molecular weight substrates, the enzymic latch has the potential to constrain CH_4 emissions in the same way that it suppresses CO_2 emissions.

Concerns may be anticipated; however, about the application of transgenic *Sphagnum* in peatlands, although research in New Zealand suggests the approach may be accepted by the public to a greater has been the case for genetically modification foods. Genetic modification to promote a capacity for pollution control has been shown to be looked upon relatively favourably, ranking alongside genetic modification for curing disease in terms of its acceptability (Cronin and Jackson, 2004). There may also be risks to biodiversity, although whether introducing genetically modified plants such as *Sphagnum* could be considered damaging remains to be determined (Bartz et al., 2010).

Peatland geoengineering may be highly controversial in pristine ecosystems, because, for example, it may alter natural pH gradients and a large part of peatland biodiversity is dependent on pH gradients. However, the addition of natural phenolic compounds or edaphic manipulation to acidify the system by natural mechanisms could preserve pH gradients and therefore biodiversity, whilst further slowing decay. Moreover, this could be highly valuable in protecting 'pristine' peatlands from haemorrhaging carbon during and after severe drought (Freeman et al., 1993). Furthermore, one could argue that pristine systems do not exist anymore given all are now 1) bathed in high CO_2 (Schlesinger, 2004), and 2) many are either enriched with nutrients or recovering from SO_4 deposition. Application of the methods described in this paper in the burgeoning field of peatland restoration could prove much more acceptable and highly valuable both for increased carbon storage and waste nutrient sequestration. It would also allow enhanced drinking water quality in peat-dominated catchments by lowering dissolved organic carbon (DOC) exports into potable waters (Fenner et al., 2001; Freeman et al., 2004a).

The two broad approaches to strengthening the enzymic latch and increasing its influence should create a capacity for increasing carbon capture *via* NPP and the efficiency with which that carbon is retained. For example, peat litter production alone could potentially sequester around 1.7×10^{15} g/year were that primary production all sequestered effectively in the peat, even without external carbon capture. However, combining these two approaches to achieve maximum carbon accumulation may require a 'trade off' between plant growth and decomposition rates; the balance will probably differ depending on the type of peatland (bog, fen or riparian) and dominant plant functional groups (bryophytes, sedges, grasses, dwarf shrubs) involved. Direct measurement of carbon sequestration is notoriously difficult though (Niklaus and Falloon, 2006), and measuring potentially relatively small increases in carbon sequestration against high background carbon levels represented by peatlands requires careful consideration. One approach could be to combine stable isotopic labelling and soil carbon cycle modelling to partition net sequestration into changes in new, middle-aged and old carbon over the experimental period (Fenner et al., 2004; Niklaus and Falloon, 2006). This partitioning is advantageous because new and middle-aged carbon accumulates (>0), while old carbon is lost with time (<0) (Niklaus and Falloon, 2006). Using $^{13}C/^{12}C$ mass balance and inverse modelling, new and middle-aged inputs to the different carbon pools can be predicted *versus* decomposition of old carbon, allowing the estimation of carbon sequestration (Freeman et al., 2004a). However, detailed lifecycle analysis of treatments would be critical to determine the fate of carbon sequestered from the atmosphere and the wider effects on ecosystem function/services.

Conclusions

Suppressing microbial decomposition of organic matter in peatlands by strengthening the enzymic latch is not without its risks. As with all geoengineering techniques, because of the expense and any potential unforeseen consequences, the hope is it will never need to be used on a large scale due to a global consensus to reduce GHG emissions. However, such political unity is unprecedented and the evidence suggests that it may already be too late, with atmospheric GHG concentrations already having reached a tipping point. It is therefore becoming increasingly likely that some form of geoengineering will be necessary if some of the more damaging consequences of climate change are to be avoided.

Some argue that due to the natural carbon sequestering abilities of peatlands they should simply be conserved with only subtle and passive management practises being adopted to encourage carbon sequestration. However, although commendable such 'carbon stewardship' practises (which include controlled grazing and vegetation management) are unlikely to deliver the climate-altering levels of carbon capture sought. For this, large scale manipulation of microbial communities and the biogeochemical cycles they control are needed. Hopefully, using methods to strengthen the enzymic latch in peatlands will prove less controversial than other geoengineering schemes (and gain the support of those that prefer subtler methods of peatland use) because, in essence, it is simply harnessing the natural carbon-accreting and preservation properties of the ecosystem. Even injection of externally captured carbon as timber mimics the natural process of recruitment of wood into waterlogged environments (Blanchette, 2000; Guyette et al., 2008) and furthermore, many of these measures are essentially reversible.

The challenges posed by these engineering activities cannot be underestimated. Injection of timber below the peat surface will require significant investment. Large scale construction of dams is only feasible in areas that are relatively accessible by heavy machinery. With suitable incentives though, piecemeal small scale interventions can result in landscape scale changes by engaging landowners in the carbon sequestration process. Adding supplements whether in the form of acids, phenolics, nutrients or even modified *Sphagnum* plants will also require research and investment in a suitable method of deployment. Considerable infrastructure for aerial deployment of solids and liquids is already available in agricultural areas, although more would clearly be required. It may also be possible to achieve such deployment as part of proposed SRM geoengineering methods using reflective aerosols in the stratosphere. Where sulphate aerosols are used, there is an added advantage that the acid character of the material is also known to suppress methane emissions from peatlands (Freeman et al., 1994). Clearly, while some of these strategies can only be attempted at a modest scale and are limited to accessible sites, others are applicable at the largest scales (e.g. stratospheric deployment) and have the potential to impact upon peatlands in even the most inaccessible locations.

As with any geoengineering strategy, peatland geoengineering does not represent a 'magic bullet' that can reverse the effects of anthropogenic emissions (Woodward et al., 2009). That said, the technique does possess some beneficial supplementary features: it is reversible (through combustion and drainage) should a need arise to return the sequestered carbon to the atmosphere; it has great potential value as a method for producing peat carbon as a renewable energy source / biofuel, and one that does not compete with food production, due to its location within non-agricultural lands. And crucially, the technology and knowledge is established and ready go should we need formulate and act on a 'Plan B' to save the planet from the effects of anthropogenic climate change.

More work clearly needs to be done to not only look at the feasibility of manipulating microbial activities to maximise carbon sequestration in peat, but also in other ecosystems.

References

Alshehri, A., Dunn, C., Freeman, C., Hugron, S., Jones, T.G., and Rochefort, L. (2020). A potential approach for enhancing carbon sequestration during peatland restoration using low-cost, phenolic-rich biomass supplements. Front. Environ. Sci. 8, 48. DOI: 10.3389/fenvs.2020.00048

Andrus, R.E. (1986). Some aspects of *Sphagnum* ecology. Can. J. Bot. *64*, 416-426.

Appel, H.M. (1993). Phenolics in ecological interactions - the importance of oxidation. J. Chem. Ecol. *19*, 1521-1552.

Bahn, M., Rodeghiero, M., Anderson-Dunn, M., Dore, S., Gimeno, C., Drösler, M., Williams, M., Ammann, C., Berninger, F., Flechard, C., *et al.* (2008). Soil respiration in European grasslands in relation to climate and assimilate supply. Ecosystems *11*, 1352-1367.

Bartz, R., Heink, U., and Kowarik, I. (2010). Proposed Definition of Environmental Damage Illustrated by the Cases of Genetically Modified Crops and Invasive Species. Conserv. Biol. *24*, 675-681.

Basaraba, J., and Starkey, R.L. (1966). Effect of plant tannins on decomposition of organic substances. Soil Sci. *101*, 17-23.

Bate, N.J., Orr, J., Ni, W., Meromi, A., Nadler-Hassar, T., Doerner, P.W., Dixon, R.A., Lamb, C.J., and Elkind, Y. (1994). Quantitative relationship between phenylalanine ammonia-lyase levels and phenylpropanoid accumulation in transgenic tobacco identifies a rate-determining step in natural product synthesis. Proc. Natl. Acad. Sci. *91*, 7608-7612.

Bertram, C. (2010). Ocean iron fertilization in the context of the Kyoto protocol and the post-Kyoto process. Energ. Policy *38*, 1130-1139.

Blanchette, R.A. (2000). A review of microbial deterioration found in archaeological wood from different environments. Int. Biodeter. Biodegr. *46*, 189-204.

Bonan, G.B. (2008). Forests and climate change: Forcings, feedbacks, and the climate benefits of forests. Science *320*, 1444-1449.

Bragazza, L., Freeman, C., Jones, T., Rydin, H., Limpens, J., Fenner, N., Ellis, T., Gerdol, R., Hajek, M., Hajek, T., *et al.* (2006). Atmospheric nitrogen deposition promotes carbon loss from peat bogs. Proc. Natl. Acad. Sci. *103*, 19386-19389.

Burke, R.M., and Cairney, J.W.G. (2002). Laccases and other polyphenol oxidases in ecto- and ericoid mycorrhizal fungi. Mycorrhiza *12*.

Burns, R.G. (1978). Enzyme activity in soil: some of theoretical and practical considerations. In Soil enzymes, R.G. Burns, ed. (London, UK: Academic Press Inc.), pp. 295-340.

Burns, R.G. (1982). Enzyme-activity in soil: location and a possible role in microbial ecology. Soil Biol. Biochem. *14*, 423-427.

Chapin, C., Bridgham, S., and Pastor, J. (2004). pH and nutrient effects on above-ground net primary production in a Minnesota, USA, bog and fen. Wetlands *24*, 186-201.

Claus, H. (2004). Laccases: structure, reactions, distribution. Micron *35*, 93-96.

Clymo, R.S. (1983). Peat. In Mires: swamp, bog, fen and moor, A.J.P. Gore, ed. (Amsterdam, Netherlands.: Elsevier Scientific Publishing Company), pp. 159-224.

Clymo, R.S., and Hayward, P.M. (1982). The ecology of *Sphagnum*. In Bryophyte ecology, A.J.E. Smith, ed. (London, UK: Chapman and Hall), pp. 228-289.

Cole, J.J., Prairie, Y.T., Caraco, N.F., McDowell, W.H., Tranvik, L.J., Striegl, R.G., Duarte, C.M., Kortelainen, P., Downing, J.A., Middelburg, J.J., *et al.* (2007). Plumbing the global carbon cycle: Integrating inland waters into the terrestrial carbon budget. Ecosystems *10*, 171-184.

Confer, D.R., and Logan, B.E. (1998). A conceptual model describing macromolecule degradation by suspended cultures and biofilms. Water Sci. Technol. *37*, 231-234.

Cooper, A., Murray, R., and McCann, T. (1991). The environmental effects of blanket peat exploitation (UK: Ulster University).

Cronin, K., and Jackson, L. (2004). Hands Across the Water: Developing dialogue between stakeholders in the New Zealand biotechnology debate. (Wellington New Zealand: Victoria University of Wellington).

Darby, H.C. (1956). The Draining of the Fens (Cambridge, UK: Cambridge University Press).

Davidson, P., Burgoyne, C., Hunt, H., and Causier, M. (2012). Lifting options for stratospheric aerosol geoengineering: advantages of tethered balloon systems. Philos. Trans. R. Soc. A *370*, 4263-4300.

Dean, W.E., and Gorham, E. (1998). Magnitude and significance of carbon burial in lakes, reservoirs, and peatlands. Geology *26*, 535-538.

Faulon, J.L., and Hatcher, P.G. (1994). Is there any order in the structure of lignin? Energ. Fuel. *8*, 402-407.

Fenner, N., and Freeman, C. (2011). Drought-induced carbon loss in peatlands. Nat. Geosci. *4*, 895-900.

Fenner, N., and Freeman, C. (2020). Woody litter protects carbon stocks during drought. Nat. Clim. Chang. 10, 363–369. DOI: 10.1038/s41558-020-0727-y

Fenner, N., Freeman, C., Hughes, S., and Reynolds, B. (2001). Molecular weight spectra of dissolved organic carbon in a rewetted Welsh peatland and possible implications for water quality. Soil Use Manag. *17*, 106-112.

Fenner, N., Freeman, C., Lock, M.A., Harmens, H., Reynolds, B., and Sparks, T. (2007a). Interactions between elevated CO_2 and warming could amplify DOC exports from peatland catchments. Env. Sci. Technol. *41*, 3146-3152.

Fenner, N., Freeman, C., and Reynolds, B. (2005). Hydrological effects on the diversity of phenolic degrading bacteria in a peatland: implications for carbon cycling. Soil Biol. Biochem. *37*. 1277-1287. DOI: 10.1016/j.soilbio.2004.11.024

Fenner, N., Ostle, N., Freeman, C., Sleep, D., and Reynolds, B. (2004). Peatland carbon afflux partitioning reveals that *Sphagnum* photosynthate contributes to the DOC pool. Plant Soil *259*, 345-354.

Fenner, N., Ostle, N.J., McNamara, N., Sparks, T., Harmens, H., Reynolds, B., and Freeman, C. (2007b). Elevated CO_2 effects on peatland plant community carbon dynamics and DOC production. Ecosystems *10*, 635-647.

Fontaine, S., Barot, S., Barre, P., Bdioui, N., Mary, B., and Rumpel, C. (2007). Stability of organic carbon in deep soil layers controlled by fresh carbon supply. Nature *450*, 277-U210.

Freeman, C., Baxter, R., Farrar, J.F., Jones, S.E., Plum, S., Ashendon, T.W., and Stirling, C. (1998). Could competition between plants and microbes regulate plant nutrition and atmospheric CO_2 concentrations? Sci. Tot. Env. *220*, 181-184.

Freeman, C., Evans, C.D., Monteith, D.T., Reynolds, B., and Fenner, N. (2001a). Export of organic carbon from peat soils. Nature *412*, 785-785.

Freeman, C., Fenner, N., Ostle, N.J., Kang, H., Dowrick, D.J., Reynolds, B., Lock, M.A., Sleep, D., Hughes, S., and Hudson, J. (2004a). Export of dissolved organic carbon from peatlands under elevated carbon dioxide levels. Nature *430*, 195-198.

Freeman, C., Fenner, N., and Shirsat, A.H. (2012). Peatland geoengineering: an alternative approach to terrestrial carbon sequestration. Philos. Trans. R. Soc. A *370*, 4404-4421.

Freeman, C., Hudson, J., Lock, M.A., Reynolds, B., and Swanson, C. (1994). A possible role of sulfate in the supression of wetland methane fluxes following drought. Soil Biol. Biochem. *26*, 1439-1442.

Freeman, C., Lock, M.A., Marxsen, J., and Jones, S.E. (1990). Inhibitory effects of high-molecular-weight dissolved organic-matter upon metabolic processes in biofilms from contrasting rivers and streams. Freshw. Biol. *24*, 159-166.

Freeman, C., Lock, M.A., and Reynolds, B. (1993). Fluxes of carbon dioxide, methane and nitrous oxide from a Welsh peatland following simulation of water-table draw-down: potential feedback to climatic-change. Biogeochemistry *19*, 51-60.

Freeman, C., Ostle, N., and Kang, H. (2001b). An enzymic 'latch' on a global carbon store - A shortage of oxygen locks up carbon in peatlands by restraining a single enzyme. Nature *409*, 149-149.

Freeman, C., Ostle, N.J., Fenner, N., and Kang, H. (2004b). A regulatory role for phenol oxidase during decomposition in peatlands. Soil Biol. Biochem. *36*, 1663-1667.

Gajewski, A.J., and Chrost, R.J. (1995). Production and enzymatic decomposition of organic-matter by microplankton in a eutrophic lake. J. Plankton Res. *17*, 709-728.

Gajewski, K., Viau, A., Sawada, M., Atkinson, D., and Wilson, S. (2001). *Sphagnum* peatland distribution in North America and Eurasia during the past 21,000 years. Global Biogeochem. Cycles *15*, 297-310.

Gerdol, R., Bragazza, L., and Brancaleoni, L. (2006). Microbial nitrogen cycling interacts with exogenous nitrogen supply in affecting growth of *Sphagnum papillosum*. Environ. Exp. Bot. *57*, 1-8.

Goldblatt, C., and Watson, A.J. (2012). The runaway greenhouse: implications for future climate change, geoengineering and planetary atmospheres. Philos. Trans. R. Soc A *370*, 4197-4216.

Gorham, E. (1991). Northern Peatlands - Role in the carbon-cycle and probable responses to climatic warming. Ecol. Appl. *1*, 182-195.

Goslan, E.H., Krasner, S.W., Bower, M., Rocks, S., Holmes, P., Levy, L.S., and Parsons, S. A. (2009). A comparison of disinfection by-products found in chlorinated and chloraminated drinking waters in Scotland. Water Res. 43, 4698–706. DOI: 10.1016/j.watres.2009.07.029

Gramss, G., Voigt, K.D., and Kirsche, B. (1999). Oxidoreductase enzymes liberated by plant roots and their effects on soil humic material. Chemosphere *38*.

Green, F.H.W. (1973). Aspects of changing environment - some factors affecting aquatic environment in recent years. J. Environ. Manage. *1*, 377.

Gunnarsson, U. (2005). Global patterns of *Sphagnum* productivity. J. Bryol. *27*, 269-279.

Guyette, R.P., Dey, D.C., and Stambaugh, M.C. (2008). The temporal distribution and carbon storage of large oak wood in streams and floodplain deposits. Ecosystems *11*, 643-653.

Haslam, E. (1989). Plant polyphenols. Vegetable tannins revisited. (Cambridge, UK: Press Syndicate of the University of Cambridge).

Hattenschwiler, S., and Vitousek, P.M. (2000). The role of polyphenols in terrestrial ecosystem nutrient cycling. Trends Ecol. Evol. *15*, 238-243.

Heijmans, M., Klees, H., de Visser, W., and Berendse, F. (2002). Response of a *Sphagnum* bog plant community to elevated CO2 and N supply. Plant Ecol. *162*, 123-134.

Heimann, M., and Reichstein, M. (2008). Terrestrial ecosystem carbon dynamics and climate feedbacks. Nature *451*, 289-292.

Holden, J. (2005). Peatland hydrology and carbon release: why small-scale process matters. Philos. Trans. R. Soc. A 363, 2891-2913.

Horner, J.D., Gosz, J.R., and Cates, R.G. (1988). The role of carbon-based plant secondary metabolites in decomposition in terrestrial ecosystems. Am. Nat. *132*, 869-883.

Huisman, D.J., Manders, M.R., Kretschmar, E.I., Klaassen, R.K.W.M., and Lamersdorf, N. (2008). Burial conditions and wood degradation at archaeological sites in the Netherlands. Int. Biodeter. Biodegr. *61*, 33-44.

Immirzi, C.P., and Maltby, E. (1992). The Global Status of Peatlands and Their Role in Carbon Cycling, a Report for Friends of the Earth by the Wetland Ecosystems (Department of Geography, University of Exeter, Exeter, UK. Research Group).

IPCC, ed. (2007). Climate Change 2007: Impacts, Adaptation and Vulnerability (Cambridge, UK: Cambridge University Press).

IPCC (2013). Working Group Contribution to the IPCC 5th Assessment Report 'Climate Change 2013: The Physical Science Basis'. (Geneva, Switzerland.: Intergovernmental Panel on Climate Change).

IPCC (2014). AR5 Climate Change 2014: Impacts, Adaptation, and Vulnerability. The Assessment of Impacts, Adaptation, and Vulnerability in the Working Group II Contribution to the IPCC's Fifth Assessment Report (WGII AR5).

Kaplan, L.A., and Bott, T.L. (1983). Microbial heterotrophic utilisation of dissolved organic-matter in a piedmont stream. Freshwat. Biol. *13*, 363-377.

Karlin, E.F., Andrus, R.E., Boles, S.B., and Shaw, A.J. (2011). One haploid parent contributes 100% of the gene pool for a widespread species in northwest North America. Mol. Ecol. *20*, 753-767.

Kayranli, B., Scholz, M., Mustafa, A., and Hedmark, A. (2010). Carbon Storage and Fluxes within Freshwater Wetlands: a Critical Review. Wetlands *30*, 111-124.

Korner, C. (2003). Slow in, rapid out - Carbon flux studies and Kyoto targets. Science *300*, 1242-1243.

Krab, E., Cornelissen, J.C., Lang, S., and Logtestijn, R.P. (2008). Amino acid uptake among wide-ranging moss species may contribute to their strong position in higher-latitude ecosystems. Plant Soil *304*, 199-208.

Kuhry, P., and Vitt, D.H. (1996). Fossil carbon/nitrogen ratios as a measure of peat decomposition. Ecology *77*, 271-275.

Kuzyakov, Y., Hill, P., and Jones, D. (2007). Root exudate components change litter decomposition in a simulated rhizosphere depending on temperature. Plant Soil *290*, 293-305.

Lacis, A.A., Schmidt, G.A., Rind, D., and Ruedy, R.A. (2010). Atmospheric CO2: Principal Control Knob Governing Earth's Temperature. Science *330*, 356-359.

Ladd, J.N. (1978). Origin and Range of Enzymes in Soil. In Soil Enzymes, R.G. Burns, ed. (London, UK.: Academic Press Inc.), pp. 51-97.

Laiho, R. (2006). Decomposition in peatlands: Reconciling seemingly contrasting results on the impacts of lowered water levels. Soil Biol. Biochem. *38*, 2011-2024.

Latham, J., Bower, K., Choularton, T., Coe, H., Connolly, P., Cooper, G., Craft, T., Foster, J., Gadian, A., Galbraith, L., *et al.* (2012). Marine cloud brightening. Philos. Trans. R. Soc A *370*, 4217-4262.

Leifeld J. & Menichetti L. (2018). The underappreciated potential of peatlands in global climate change mitigation strategies. Nat. Commun. 9, 1071. DOI: 10.1038/s41467-018-03406-6.

Limpens, J., Berendse, F., Blodau, C., Canadell, J.G., Freeman, C., Holden, J., Roulet, N., Rydin, H., and Schaepman-Strub, G. (2008). Peatlands and the carbon cycle: from local processes to global implications - a synthesis. Biogeosciences *5*, 1475-1491.

Marechal, K., and Hecq, W. (2006). Temporary credits: A solution to the potential non-permanence of carbon sequestration in forests? Ecol. Econ. *58*, 699-716.

McLatchey, G.P., and Reddy, K.R. (1998). Regulation of organic matter decomposition and nutrient release in a wetland soil. J. Environ. Qual. *27*, 1268-1274.

Meehl, G.A., Stocker, T.F., Collins, W.D., Friedlingstein, P., Gaye, A.T., Gregory, J.M., Kitoh, A., Knutti, R., Murphy, J.M., Noda, A., *et al.* (2007). Global Climate Projections. In Climate Change 2007: The Physical Science Basis Contribution of Working Group I to the Fourth Assessment Report of the Intergovernmental Panel on Climate Change, S. Solomon, D. Qin, M. Manning, Z. Chen, M. Marquis, K.B. Averyt, M. Tignor and H.L. Miller, eds. (Cambridge: Cambridge University Press).

Meyer, J.L., Edwards, R.T., and Risley, R. (1987). Bacterial growth on dissolved organic-carbon from a blackwater river. Microb. Ecol. *13*, 13-29.

Minamikawa, K., Sakai, N., and Yagi, K. (2006). Methane Emission from Paddy Fields and its Mitigation Options on a Field Scale. Microbes Environ. *21*, 135-147.

Mitsch, W.J., and Gosselink, J.G. (2000a). Inland Wetland Ecosystems - Peatlands. In Wetlands (USA: John Wiley and Sons Inc.), pp. 419-469.

Mitsch, W.J., and Gosselink, J.G. (2000b). Wetlands - 3rd Edition (USA: John Wiley and Sons, Inc.).

Moore, P.D. (2002). The future of cool temperate bogs. Environ. Conserv. *29*, 3-20.

Muscolo, A., and Sidari, M. (2006). Seasonal fluctuations in soil phenolics of a coniferous forest: effects on seed germination of different coniferous species. Plant Soil *284*, 305-318.

Niklaus, P.A., and Falloon, P. (2006). Estimating soil carbon sequestration under elevated CO_2 by combining carbon isotope labelling with soil carbon cycle modelling. Global Change Biol. *12*, 1909-1921.

Oechel, W.C., Hastings, S.J., Vourlitis, G., Jenkins, M., Riechers, G., and Grulke, N. (1993). Recent change of Arctic tundra ecosystems from a net carbon dioxide sink to a source. Nature *361*, 520-523.

Page, S.E., Rieley, J.O., and Banks, C.J. (2011). Global and regional importance of the tropical peatland carbon pool. Global Change Biol. *17*, 798-818.

Pan Y. et al (2011) A Large and Persistent Carbon Sink in the World's Forests. Science 333, (6045), 988-993. DOI: 10.1126/science.1201609

Parish, F., Sirin, A., Charman, D., Joosten, H., Minayeva, T., Silvius, M., and Stringer, L. (2008). Assessment on Peatlands, Biodiversity and Climate Change: Main Report. In Kuala Lumpur and Wetlands International, G.E. Centre, ed. (Wageningen).

Pind, A., Freeman, C., and Lock, M.A. (1994). Enzymatic degradation of phenolic materials in peatlands - measurement of Phenol Oxidase activity. Plant Soil *159*, 227-231.

Pravettoni, R. (2009). UNEP/GRID-Arenda. Peat distribution in the World. http://www.grida.no/publications/rr/the-natural-fix.

Rasmussen, S., Peters, G., and Rudolph, H. (1995). Regulation of phenylpropanoid metabolism by exogenous precursors in axenic cultures of *Sphagnum fallax*. Physiol. Plant. *95*, 83-90.

Reader, R.J., and Stewart, J.M. (1972). Relationship between NET Primary Production and accumulation for a peatland in Southeastern Manitoba. Ecology *53*, **1024-1037**.

Robinson, M., and Armstrong, A.C. (1988). The Extent of Agricultural Field Drainage in England and Wales, 1971-80. Trans. Inst. Br. Geogr. *13*, 19-28.

Rogers, H. (1961). The dissimilation of high molecular weight organic substances. In The bacteria A treatise on structure and function, I.C. Gunsalus, and R.Y. Stainer, eds. (New York, USA,: Academic Press Inc.).

Royal Society (2009). Geoengineering the climate. Science, governance and uncertainty (London: The Royal Society).

Rudolph, H., and Samland, J. (1985). Occurrence and metabolism of sphagnum acid in the cell-walls of bryophytes. Phytochemistry *24*, 745-749.

Schlesinger, W.H. (2004). Better living through biogeochemistry. Ecology *85*, 2402-2407.

Schonwalder, A., Kehr, R., Wulf, A., and Smalla, K. (2002). Wooden boards affecting the survival of bacteria? Holz Als Roh-Und Werkstoff *60*, 249-257.

Shackle, V.J., Freeman, C., and Reynolds, B. (2000). Carbon supply and the regulation of enzyme activity in constructed wetlands. Soil Biol. Biochem. *32*, 1935-1940.

Shepherd, J.G. (2012). Geoengineering the climate: an overview and update. Philos. Trans. R. Soc A *370*, 4166-4175.

Singarayer, J.S., and Davies-Barnard, T. (2012). Regional climate change mitigation with crops: context and assessment. Philos. Trans. R. Soc A *370*, 4301-4316.

Sinsabaugh, R.L. (1994). Enzymatic analysis of microbial pattern and process. Biol. Fertil. Soil. *17*, 69-74.

Smith, L.C., MacDonald, G.M., Velichko, A.A., Beilman, D.W., Borisova, O.K., Frey, K.E., Kremenetski, K.V., and Sheng, Y. (2004). Siberian peatlands a net carbon sink and global methane source since the early Holocene. Science *303*, 353-356.

Swift, M.J., Heal, O.W., and Anderson, J.M. (1979). Decomposition in terrestrial ecosystems (Oxford, UK.: Blackwell Scientific Publications).

Thurston, C.F. (1994). The structure and function of fungal laccases. Microbiology-Sgm *140*, 19-26.

Tilman, D., Cassman, K.G., Matson, P.A., Naylor, R., and Polasky, S. (2002). Agricultural sustainability and intensive production practices. Nature *418*, 671-677.

Verhoeven, J.T.A., and Liefveld, W.M. (1997). The ecological significance of organochemical compounds in *Sphagnum*. Acta Bot. Neerl. *46*, 117-130.

Waddington, J.M., Strack, M., and Greenwood, M.J. (2010). Toward restoring the net carbon sink function of degraded peatlands: Short-term response in CO2 exchange to ecosystem-scale restoration. J. Geophys. Res. Biogeosci. *115*.

Wang, H., Richardson C. J. and Ho M. (2015). Dual controls on carbon loss during drought in peatlands. Nat. Clim. Chang. 5, 584-588.

Wang Y., Wang H., He J-S. & Feng X (2017). Iron-mediated soil carbon response to water-table decline in an alpine wetland. Nat. Commun. 8, 1-9.

Wetzel, R.G. (1992). Gradient-dominated ecosystems - sources and regulatory functions of dissolved organic-matter in fresh-water ecosystems. Hydrobiologia *229*, 181-198.

Wieder, R.K. (2001). Past, present, and future peatland carbon balance: an empirical model based on 210Pb-dated cores. Ecol. Appl. *11*, 327-342.

Wigley, T. M. L. A (2006). Combined mitigation/geoengineering approach to climate stabilization. Science 314, 452-454.

Williams, C.J., Shingara, E.A., and Yavitt, J.B. (2000). Phenol oxidase activity in peatlands in New York State: Response to summer drought and peat type. Wetlands *20*.

Woodward, F.I., Bardgett, R.D., Raven, J.A., and Hetherington, A.M. (2009). Biological Approaches to Global Environment Change Mitigation and Remediation. Curr. Biol. *19*, R615-R623.

Xu J., Morris P. J., Liu J. & Holden J. (2018). Hotspots of peatland-derived potable water use identified by global analysis. Nat. Sustain. 1, 246–253

Yang, J.S., Ni, J.R., Yuan, H.L., and Wang, E. (2007). Biodegradation of three different wood chips by Pseudomonas sp PKE117. Int. Biodeter. Biodegr. *60*, 90-95.

CPSIA information can be obtained
at www.ICGtesting.com
Printed in the USA
BVHW051612261021
619860BV00002B/10